AFFINE ALGEBRAIC GEOMETRY

Proceedings of the Conference

AFFINE ALGEBRAIC GEOMETRY

Proceedings of the Conference

Osaka, Japan 3 – 6 March 2011

Editors

Kayo Masuda
Kwansei Gakuin University, Japan

Hideo Kojima
Niigata University, Japan

Takashi Kishimoto
Saitama University, Japan

World Scientific

NEW JERSEY • LONDON • SINGAPORE • BEIJING • SHANGHAI • HONG KONG • TAIPEI • CHENNAI

Published by

World Scientific Publishing Co. Pte. Ltd.
5 Toh Tuck Link, Singapore 596224
USA office: 27 Warren Street, Suite 401-402, Hackensack, NJ 07601
UK office: 57 Shelton Street, Covent Garden, London WC2H 9HE

British Library Cataloguing-in-Publication Data
A catalogue record for this book is available from the British Library.

AFFINE ALGEBRAIC GEOMETRY
Proceedings of the Conference

ISBN 978-981-4436-69-4

Printed in Singapore by B & Jo Enterprise Pte Ltd

Conference on Affine Algebraic Geometry, March 3–6, 2011, Osaka Umeda Campus of Kwansei Gakuin University

DEDICATION

Professor Masayoshi Miyanishi was born in 1940. He entered Kyoto University in 1958 and started his career as a mathematician in the very stimulating atmosphere. Around 1960, there was a very strong school of algebraic geometry in Kyoto. Professor Nagata was one of the leaders of this school. Under the leadership of Professor Nagata, several brilliant students of Professor Miyanishi's generation worked hard to absorb new language and theories in algebraic geometry. In fact, the first book of EGA was published in 1960 and at that time the great revolution was performed in algebraic geometry. To those young mathematicians who studied with Professor Miyanishi belonged Professors T. Oda, T. Miyata, H. Sumihiro and M. Maruyama. They later became leaders of algebraic geometry in Japan.

After finishing his master thesis, Professor Miyanishi became a research assistant and then an instructor of Kyoto University. He received his Ph.D. in 1968 from Kyoto University. Afterwards he became an associate professor at Osaka University in 1973 and a professor in 1984 which he kept until he retired from Osaka University in 2003. He was a professor of mathematics in Kwansei Gakuin University from 2003 to 2009.

Professor Miyanishi visited several foreign countries and had opportunities to do fruitful collaborations with many mathematicians from various countries.

In 1965, he went to Paris and attended Chevalley seminar. Of course he met Grothendieck. Under the influence of the time, his first interest of algebraic geometry was in a general aspect of the theory. Besides research papers, he published his first book on algebraic geometry with Professors M. Nagata and M. Maruyama in 1973. This book was the first textbook on a scheme theory available in Japanese language.

Professor Miyanishi continued his collaborations with mathematicians like Professors S. S. Abhyanker, P. Russell and R. V. Gurjar. Meanwhile he gradually shifted his interest towards concrete objects during the 70's. He is a leading figure in establishing a new branch of mathematics called affine algebraic geometry. He published more than 90 research articles and his research results on affine algebraic geometry include "A characterization of the affine plane", "The cancellation problem of the affine plane" and "A theory of log Del Pezzo surfaces". His Tata lecture notes in 1978 is the first summary of his research in this area, and one of the good references of the recent development in affine algebraic geometry is his article in the book "Affine Algebraic Geometry, 2007 (T. Hibi, ed.)" published by the Osaka University Press.

Apart from the activity on research and education, he made a great contribution toward the university administration. He became a dean of the faculty of science and afterwards was a vice president of Osaka University until 2003. The time of his vice presidency corresponded to the time when Japanese universities were preparing for corporatization. He played an important role in performing this transition.

While he was a professor at Kwansei Gakuin University, he started a conference on affine algebraic geometry at the Umeda campus of Kwansei Gakuin University in March, 2008. He is still giving stimulations to younger generations.

Toru Sugie

PREFACE

A Conference on Affine Algebraic Geometry was held at Osaka Umeda Campus of Kwansei Gakuin University during the period 3–6 March, 2011 on the occasion of the seventieth birthday of Professor Masayoshi Miyanishi. Eleven researchers were invited from abroad and 19 talks were given. The present proceedings is dedicated to Professor Miyanishi. This volume contains 16 papers contributed by participants in the conference and by friends of Professor Miyanishi who were not able to participate in the conference. All papers in this volume have been refereed. Most subjects treated in the present volume are related to those Professor Miyanishi has been studying and developing for more than 40 years.

Professor Miyanishi is one of the founders of affine algebraic geometry. After the fundamental works by S. Abhyankar, M. Nagata and *et al.*, he laid the foundations in some of the theories of affine algebraic geometry. Based on his deep and wide-ranged knowledge on algebraic geometry and commutative algebra, he established the theory of open algebraic surfaces with the people including T. Fujita, S. Iitaka, Y. Kawamata and S. Tsunoda. He has authored 99 research papers so far. It is remarkable that he is still very active after the retirement of the professorships he held in Osaka University and Kwansei Gakuin University. Indeed, he authored ten papers after his retirement from Kwansei Gakuin University which he had been affiliated to untill March, 2009 after having left Osaka University in 2003. He is now a guest researcher of the Research Center for Mathematical Sciences, which he himself established at Kwansei Gakuin University. As well as his research, he has been devoted to the education. He brought up over ten students to researchers. He is generous in sharing mathematical ideas. This is testified by the fact that there are many joint papers with his colleagues and students.

The conference was held as the 7th Meeting of Affine Algebraic Geometry. The Affine Algebraic Geometry Meeting has started in March, 2008 as a small meeting organized by Professor Miyanishi and me. Since then, the meeting has been held twice a year, September and March, at Osaka Umeda

Campus of Kwansei Gakuin University, providing interested reseachers with a place of announcing new results and developments and of exchanging mathematical ideas. The meeting is open not only to affine algebraic geometers but also to researchers in algebraic geometry and commutative algebra. In September, 2012, we had the 10th Meeting. Throughout all the meetings, Professor Miyanishi has played a role of activating the meetings by giving good questions, suggestions and advice.

The conference was financially supported by Grant-in-Aid for Scientific Research (C) 22540059, JSPS and by Research Center for Mathematical Sciences, Kwansei Gakuin University. I would like to thank all the people who made the conference successful and helped me during the preparation of this volume.

Kayo Masuda
Sanda, Hyogo, Japan
24 December 2012

BIBLIOGRAPHY OF MASAYOSHI MIYANISHI

Research Articles

1. Appendix to "M. Nagata, Invariants of a group in an affine ring", J. Math. Kyoto Univ. **3** (1964), 369–377.
2. On formal rings, J. Math. Kyoto Univ. **5** (1965), 45–59.
3. On some dualities concerning abelian varieties (with H. Matsumura), Nagoya Math. J. **27** (1966), 447–462.
4. La pro-représentabilité d'un foncteur sur la catégorie des groupes formels artiniens, C. R. Acad. Sci. Paris Sér. A-B **262** (1966), A1385–A1388.
5. On the pro-representability of a functor on the category of finite group schemes, J. Math. Kyoto Univ. **6** (1966), 31–48.
6. On the extensions of linear groups by abelian varieties over a field of positive characteristic p, J. Math. Soc. Japan **19** (1967), 1–29.
7. Some remarks on a covering of an abelian variety, J. Math. Kyoto Univ. **7** (1967), 77–92.
8. On the cohomologies of commutative affine group schemes, J. Math. Kyoto Univ. **8** (1968), 1–39. (Ph.D. Thesis)
9. A remark on an iterative infinite higher derivation, J. Math. Kyoto Univ. **8** (1968), 411–415.
10. Quelques remarques sur la première cohomologie d'un préschéma affine en groupes commutatifs, Japanese J. Math. **38** (1969), 51–60.
11. Some remarks on the actions of the additive group schemes, J. Math. Kyoto Univ. **10** (1970), 189–205.
12. G_a-action of the affine plane, Nagoya Math. J. **41** (1971), 97–100.
13. On the vanishing of the Demazure cohomologies and the existence of quotient preschemes, J. Math. Kyoto Univ. **11** (1971), 399–414.
14. The theorem of the cube for principal homogeneous spaces, J. Math. Kyoto Univ. **12** (1972), 1–15.
15. On the algebraic fundamental group of an algebraic group, J. Math. Kyoto Univ. **12** (1972), 351–367.
16. Une charactérisation d'un groupe algébrique simplement connexe, Illinois J. Math. **16** (1972), 639–650.

17. Some remarks on polynomial rings, Osaka J. Math. **10** (1973), 617–624.
18. Some remarks on algebraic homogeneous vector bundles, Number theory, algebraic geometry and commutative algebra, in honor of Yasuo Akizuki, 71–93, Kinokuniya, Tokyo, 1973.
19. Some remarks on strongly invariant rings (with Y. Nakai), Osaka J. Math. **12** (1975), 1–17.
20. An algebraic characterization of the affine plane, J. Math. Kyoto Univ. **15** (1975), 169–184.
21. Unirational quasi-elliptic surfaces in characteristic 3, Osaka J. Math. **13** (1976), no. 3, 513–522.
22. Unirational quasi-elliptic surfaces, Japanese J. Math. **3** (1977), no. 2, 395–416.
23. Simple birational extensions of a polynomial ring $k[x, y]$, Osaka J. Math. **15** (1978), no. 3, 663–677.
24. On flat fibrations by the affine line (with T. Kambayashi), Illinois J. Math. **22** (1978), no. 4, 662–671.
25. Analytic irreducibility of certain curves on a nonsingular affine rational surface, Proceedings of the International Symposium on Algebraic Geometry (Kyoto Univ., Kyoto, 1977), 575–587, Kinokuniya Book Store, Tokyo, 1978.
26. On the structure of minimal surfaces of general type with $2p_g = (K^2)+2$ (with K. Nakamura), J. Math. Kyoto Univ. **18** (1978), no. 1, 137–171.
27. Regular subrings of a polynomial ring, Osaka J. Math. **17** (1980), no. 2, 329–338.
28. Generically rational polynomials (with T. Sugie), Osaka J. Math. **17** (1980), no. 2, 339–362.
29. Affine surfaces containing cylinderlike open sets (with T. Sugie), J. Math. Kyoto Univ. **20** (1980), no. 1, 11–42.
30. p-cyclic coverings of the affine space, J. Algebra **63** (1980), no. 1, 279–284.
31. On a projective plane curve whose complement has logarithmic Kodaira dimension $-\infty$ (with T. Sugie), Osaka J. Math. **18** (1981), no. 1, 1–11.
32. Singularities of normal affine surfaces containing cylinderlike open sets, J. Algebra **68** (1981), no. 2, 268–275.
33. On affine-ruled irrational surfaces, Invent. Math. **70** (1982/83), no. 1, 27–43.
34. Regular subrings of a polynomial ring. II, Osaka J. Math. **19** (1982), no. 4, 901–921.
35. The structure of open algebraic surfaces. II (with S. Tsunoda), Clas-

sification of algebraic and analytic manifolds (Katata, 1982), 499–544, Progr. Math. 39, Birkhaüser, Boston, 1983.

36. Purely inseparable coverings of exponent one of the affine plane (with P. Russell), J. Pure and Applied Algebra **28** (1983), no. 3, 279–317.
37. Algebraic methods in the theory of algebraic threefolds - surrounding the works of Iskovskikh, Mori and Sarkisov, Algebraic varieties and analytic varieties (Tokyo, 1981), 69–99, Advanced Studies in Pure Math. 1, North-Holland, Amsterdam, 1983.
38. On the affine-ruledness of algebraic varieties, Algebraic Geometry (Tokyo/Kyoto, 1982), 449–489, Lecture Notes in Math. 1016, Springer, Berlin, 1983.
39. On group actions, Recent Progress of Algebraic Geometry in Japan, 152–187, North Holland Mathematics Studies 73, North-Holland, Amsterdam, 1983.
40. An algebro-topological characterization of the affine space of dimension three, Amer. J. Math. **106** (1984), no. 6, 1469–1485.
41. Noncomplete algebraic surfaces with logarithmic Kodaira dimension $-\infty$ and with nonconnected boundaries at infinity (with S. Tsunoda), Japanese J. Math. **10** (1984), no. 2, 195–242.
42. Logarithmic del Pezzo surfaces of rank one with noncontractible boundaries (with S. Tsunoda), Japanese J. Math. **10** (1984), no. 2, 271–319.
43. Normal affine subalgebras of a polynomial ring, Algebraic and Topological Theories (Kinosaki, 1984), 37–51, Kinokuniya, Tokyo, 1986.
44. Étale endomorphisms of algebraic varieties, Osaka J. Math. **22** (1985), no. 2, 345–364.
45. Specializations of cofinite subalgebras of a polynomial ring (with O. Kawai), Osaka J. Math. **23** (1986), no. 1, 207–215.
46. Open algebraic surfaces with Kodaira dimension $-\infty$ (with S. Tsunoda), Algebraic geometry, Bowdoin, 1985 (Brunswick, Maine, 1985), 435–450, Proc. of Symposia in Pure Math. 46, Part 1, Amer. Math. Soc., Providence, RI, 1987.
47. Projective degenerations of surfaces according to S. Tsunoda, Algebraic Geometry, Sendai, 1985, 415–447, Adv. Stud. Pure Math. 10, North-Holland, Amsterdam, 1987.
48. Algebraic characterizations of the affine 3-space, Algebraic Geometry Seminar (Singapore, 1987), 53–67, World Sci. Publishing, Singapore, 1988.
49. Affine surfaces with $\bar{\kappa} \leq 1$ (with R. V. Gurjar), Algebraic Geometry and Commutative Algebra in honor of Masayoshi Nagata, 99–124, Kinoku-

niya, Tokyo, 1988.

50. Gorenstein log del Pezzo surfaces of rank one (with D. Q. Zhang), J. Algebra **118** (1988), no. 1, 63–84.
51. Cyclic étale endomorphisms of rational surfaces, Bull. Inst. Math. Acad. Sinica **16** (1988), no. 1, 57–70.
52. Finite equivalence relations on algebraic varieties, J. Pure and Applied Algebra **57** (1989), no. 1, 83–91.
53. A brief proof of Jacobian hypothesis implies flatness (with L. Robiano and S. Wang), Proc. Amer. Math. Soc. **109** (1990), no. 2, 327–330.
54. Examples of homology planes of general type (with T. Sugie), An appendix to "On T. Petrie's problem concerning homology planes by T. Sugie", J. Math. Kyoto Univ. **30** (1990), 317–342.
55. Finite group scheme actions on the affine plane (with T. Nomura), J. Pure and Applied Algebra **71** (1991), no. 2–3, 249–264.
56. Homology planes with quotient singularities (with T. Sugie), J. Math. Kyoto Univ. **31** (1991), no. 3, 755–788.
57. $\mathbf{Q}$-homology planes with $\mathbf{C}^{**}$-fibrations (with T. Sugie), Osaka J. Math. **28** (1991), no. 1, 1–26.
58. On algebras which resemble the local Weyl algebra (with C. U. Hang, K. Nishida and D. Q. Zhang), Osaka J. Math. **29** (1992), no. 2, 393–404.
59. Absence of the affine lines on the homology planes of general type (with S. Tsunoda), J. Math. Kyoto Univ. **32** (1992), no. 3, 443–450.
60. Gorenstein del Pezzo surfaces, II (with D. Q. Zhang), J. Algebra **156** (1993), no. 1, 183–193.
61. Affine lines on logarithmic $\mathbf{Q}$-homology planes (with R. V. Gurjar), Math. Ann. **294** (1992), no. 3, 463–482.
62. Vector fields on factorial schemes, J. Algebra **173** (1995), no. 1, 144–165.
63. On contractible curves in the complex affine plane (with R. V. Gurjar), Tôhoku Math. J. **48** (1996), no. 3, 459–469.
64. Minimization of the embeddings of curves into the affine plane, J. Math. Kyoto Univ. **36** (1996), no. 2, 311–329.
65. On Chern numbers of homology planes of certain types (with T. Sugie), J. Math. Kyoto Univ. **36** (1996), no. 3, 331–358.
66. On Roberts' counterexample to the fourteenth problem of Hilbert (with H. Kojima), J. Pure and Appl. Algebra **122** (1997), no. 3, 277–292.
67. Finitely generated algebras associated with rational vector fields (with H. Aoki), J. Algebra **198** (1997), no. 2, 481–498.
68. Surfaces of general type whose canonical map is composed of a pencil of genus 3 with small invariants (with Jin-Gen Yang), J. Math. Kyoto

Univ. **38** (1998), no. 1, 123–149.

69. Reduction modulo p of cyclic group actions of order p^r (with K. Kinugawa), J. Pure and Applied Algebra **141** (1999), no. 1, 37–58.
70. Invariant subvarieties of low codimension in the affine spaces (with K. Masuda), Tôhoku Math. J. **52** (2000), no. 1, 61–77.
71. Geometric construction of certain finite étale endomorphisms (with H. Aoki), Appendix to "Étale endomorphisms of smooth affine surfaces" by H. Aoki, J. Algebra **226** (2000), 15–52.
72. On the Jacobian conjecture for **Q**-homology planes (with R. V. Gurjar), J. Reine Angew. Math. **516** (1999), 115–132.
73. Étale endomorphisms of algebraic surfaces with G_m-actions (with K. Masuda), Math. Ann. **319** (2001), no. 3, 493–516.
74. Generalized Jacobian conjecture and related topics (with K. Masuda), Algebra, Arithmetic and Geometry, Part I, II (Mumbai, 2000), 427–466, Tata Inst. Fund. Res. Stud. Math. 16, Tata Inst. Fund. Res., Bombay, 2002.
75. Completely parametrized $\mathbb{A}^1_*$-fibrations on the affine plane, Computational commutative algebra and combinatorics (Osaka, 1999), 185–201, Advanced Studies in Pure Mathematics 33, Math. Soc. Japan, Tokyo, 2002.
76. Open algebraic surfaces with finite group actions (with K. Masuda), Transformation Groups **7** (2002), no. 2, 185–207.
77. The additive group actions on $\mathbb{Q}$-homology planes (with K. Masuda), Annales de l'Institut Fourier (Grenoble) **53** (2003), no. 2, 429–464.
78. Equivariant classification of Gorenstein open log del Pezzo surfaces with finite group actions (with De-Qi Zhang), J. Math. Soc. Japan **56** (2004), no. 1, 215–245.
79. Automorphisms of affine surfaces with $\mathbb{A}^1$-fibrations (with R. V. Gurjar), Michigan Math. J. **53** (2005), no. 1, 33–55.
80. Equivariant cancellation for algebraic varieties (with K. Masuda), Affine Algebraic Geometry, 183–195, Contemporary Mathematics 369, Amer. Math. Soc., Providence, RI, 2005.
81. On two recent views of the Jacobian Conjecture (with T. Kambayashi), Affine Algebraic Geometry, 113–138, Contemporary Mathematics 369, Amer. Math. Soc., Providence, RI, 2005.
82. Affine pseudo-planes and cancellation problem (with K. Masuda), Trans. Amer. Math. Soc. **357** (2005), no. 12, 4867–4883.
83. Affine pseudo-coverings of algebraic surfaces, J. Algebra **294** (2005), no. 1, 156–176.

84. Affine pseudo-planes with torus actions (with K. Masuda), Transformation Groups **11** (2006), no. 2, 249–267.
85. A geometric approach to the Jacobian conjecture in dimension two, J. Algebra **304** (2006), no. 2, 1014–1025.
86. $\mathbb{Q}$-factorial subalgebras of polynomial rings, Acta Mathematica Vietnamica **32** (2007), no. 2–3, 113–122.
87. Affine lines on affine surfaces and the Makar-Limanov invariant (with R. V. Gurjar, K. Masuda and P. Russell), Canad. J. Math. **60** (2008), no. 1, 109–139.
88. G_a-actions and completions, J. Algebra **319** (2008), no. 7, 2845–2854.
89. The Jacobian problem for singular surfaces (with R. V. Gurjar), J. Math. Kyoto Univ. **48** (2008), no. 4, 757–764.
90. Smoothness of the images of the members of a linear pencil under an endomorphism of the affine plane (with P. Cassou-Noguès), J. Pure and Appl. Algebra **213** (2009), no. 5, 711–723.
91. Additive group scheme actions on the integral schemes defined over discrete valuation rings, J. Algebra **322** (2009), no. 9, 3331–3344.
92. Lifting of the additive group scheme actions (with K. Masuda), Tohoku Math. J. **61** (2009), no. 2, 267–286.
93. Frobenius sandwiches of affine algebraic surfaces, Affine algebraic geometry, 243–260, CRM Proceedings and Lecture Notes 54, Amer. Math. Soc., Providence, RI, 2011.
94. Affine normal surfaces with simply-connected smooth locus (with R. V. Gurjar, M. Koras and P. Russell), Math. Ann. **353** (2012), no. 1, 127–144.
95. A homology plane of general type can have at most a cyclic quotient singularity (with R. V. Gurjar, M. Koras and P. Russell), to appear in J. Alg. Geom.; arXiv:1012.4120.
96. $\mathbb{A}^1$-fibrations on affine threefolds (with R. V. Gurjar and K. Masuda), J. Pure and Applied Algebra **216** (2012), no. 2, 296–313.
97. Algebraic derivations on affine domains (with K. Masuda), to appear in the proceedings of the CAAG conference in Bangalore, 2010; arXiv:1301.3259.
98. $\mathbb{A}^1_*$-fibrations on affine threefolds (with R. V. Gurjar, M. Koras, K. Masuda and P. Russell), in the present proceedings; arXiv:1211.1757.
99. Surjective derivations in small dimensions (with R. V. Gurjar and K. Masuda), to appear in a volume commemorating the eightieth birthday of C. S. Seshadri; arXiv:1211.0744.

Survey Articles

1. On homogeneous spaces and the first cohomology of group schemes (in Japanese), 数学 **22** (1970), no. 4, 252–263.
2. Polynomial rings and related topics (多項式環とその周辺) (in Japanese), 数学 **31** (1979), no. 2, 97–109.
3. Recent topics on open algebraic surfaces (開代数曲面の最近の話題) (in Japanese), 数学 **46**, no. 3 (1994), 243–257. English Translation: Recent topics on open algebraic surfaces, Selected papers on number theory and algebraic geometry, 61–76, Amer. Math. Soc. Transl. Ser. 2, 172, Amer. Math. Soc., Providence, RI, 1996.
4. New trends in the university mathematics education (大学での数学教育の新しい流れ) (in Japanese), 数学 **46** (1994), 164–170.
5. Introduction to algebraic curves (代数曲線の話) (in Japanese), 現代数学序説, 105–142, Osaka Univ. Press, 1996.
6. The Jacobian conjecture and related topics (ジャコビアン予想とその周辺) (in Japanese), 数理科学 **421** (1998), 48–54.
7. Recent developments in affine algebraic geometry, Affine Algebraic Geometry, 307–378, Osaka University Press, Osaka, 2007.
8. Discriminants and resultants (判別式と終結式) (in Japanese) (with K. Masuda), 現代数理入門, 113–146, Kwansei Gakuin Univ. Press, 2009.
9. Masayoshi Nagata (1927–2008) and his mathematics, Kyoto J. Math. **50** (2010), no. 4, 645–659.

Books and Lecture Notes

1. Introduction à la théorie des sites et son application à la construction des préschémas quotients, Séminaire de Mathématiques Supérieures no. 47 (Été 1970), Les Presse de l'Université de Montréal, Montréal, 1971, 96pp.
2. Abstract algebraic geometry (抽象代数幾何学) (in Japanese) (with M. Nagata and M. Maruyama), 共立出版, 現代の数学 10 (1972), 249pp.
3. Unipotent algebraic Groups (with T. Kambayashi and M. Takeuchi), Lecture Notes in Math. 414, Springer Verlag, Berlin-New York, 1974, v+165pp.
4. On forms of the affine line over a field (with T. Kambayashi), Lectures in Math., Kyoto University 10, Kinokuniya, Tokyo, 1977, i+80pp.
5. Curves on rational and unirational surfaces, Tata Institute of Fundamental Research, Lectures on Mathematics and Physics 60, Tata Institute of Fundamental Research, Bombay, 1978, ii+302pp.
6. Noncomplete algebraic surfaces, Lecture Notes in Math. 857, Springer-

Verlag, Berlin-New York, 1981, viii+244pp.

7. Algebraic Geometry (代数幾何学) (in Japanese), 裳華房, 数学選書 10, 1990, 332pp; English Translation: Algebraic Geometry, Translation of Mathematical Monographs 136, Americam Mathematical Society, Providence, RI, 1994, xii+246pp.
8. Introduction to contemporary mathematics I (現代数学序説 I) (in Japanese) (joint eds. with K. Kawakubo), Osaka Univ. Press, 1996.
9. Introduction to contemporary mathematics II (現代数学序説 II) (in Japanese) (joint eds. with K. Kawakubo), Osaka Univ. Press, 1998.
10. Invitation to the complex numbers (複素数への招待) (in Japanese) (with K. Masuda), 日本評論社, 1998, 181pp.
11. Scientific Technology and its relations with human beings I (科学技術と人間のかかわり I) (in Japanese) (joint eds. with K. Hatada), Osaka Univ. Press, 1998.
12. Scientific Technology and its relations with human beings II (科学技術と人間のかかわり II) (in Japanese) (joint eds. with K. Hatada), Osaka Univ. Press, 1999.
13. Open algebraic surfaces, CRM Monograph Series 12, Amer. Math. Soc., Providence, RI, 2001, viii+259pp.
14. Introduction to contemporary mathematics III (現代数学序説 III) (in Japanese) (joint eds. with N. Kawanaka), Osaka Univ. Press, 2002.
15. Lectures on Geometry and Topology of Polynomials - surrounding the Jacobian Conjecture, Lecture Note Series in Mathematics 8, Department of Mathematics, Osaka University, 2004.
16. Contemporary topics in mathematics and informatics (現代数理入門) (in Japanese) (joint eds. with T. Ibaraki), Kwansei Gakuin Univ. Press, 2009.
17. Algebra 1 - The fundamental course- (代数学 1 -基礎編-) (in Japanese), 裳華房, 2010, 278pp.
18. Algebra 2 -The advanced course- (代数学 2 -発展編-) (in Japanese), 裳華房, 2011, 323pp.

CONTENTS

Acyclic curves and group actions on affine toric surfaces

Ivan Arzhantsev[1] and Mikhail Zaidenberg[2]

[1] *Department of Algebra, Faculty of Mechanics and Mathematics, Lomonosov Moscow State University, Leninskie Gory 1, Moscow, 119991, Russia*
E-mail: arjantse@mccme.ru

[2] *Université Grenoble I, Institut Fourier, UMR 5582 CNRS-UJF, BP 74, 38402 St. Martin d'Hères cédex, France*
E-mail: Mikhail.Zaidenberg@ujf-grenoble.fr

To Masayoshi Miyanishi on occasion of his 70th birthday

We show that every irreducible, simply connected curve on a toric affine surface X over $\mathbb{C}$ is an orbit closure of a $\mathbb{G}_m$-action on X. It follows that up to the action of the automorphism group $\operatorname{Aut}(X)$ there are only finitely many non-equivalent embeddings of the affine line $\mathbb{A}^1$ in X. A similar description is given for simply connected curves in the quotients of the affine plane by small finite linear groups. We provide also an analog of the Jung-van der Kulk theorem for affine toric surfaces, and apply this to study actions of algebraic groups on such surfaces.

Keywords: Affine surface, acyclic curve, automorphism group, torus action, quotient.

Introduction

The geometry of affine toric surfaces still attracts the researches.[a] Every affine toric surface over $\mathbb{C}$ except for $\mathbb{A}^1_* \times \mathbb{A}^1_*$ and $\mathbb{A}^1 \times \mathbb{A}^1_*$, where

---

2010 *Mathematics Subject Classification*: Primary 14H45, 14M25; Secondary 14H50, 14R20

This work was done during a stay of the first author at the Institut Fourier, Grenoble and of the second author at the Max Planck Institute of Mathematics, Bonn. We thank both institutions for generous support and hospitality. The work of the first author was supported by a Simons Award.

[a] See, for instance, the recent paper [21] on the Hilbert scheme of zero-cycles on such a surface.

$\mathbb{A}^1_* = \mathbb{A}^1 \setminus \{0\}$, is the quotient of $\mathbb{A}^2$ by a small finite cyclic subgroup $G \subseteq \mathrm{GL}(2, \mathbb{C})$.[b]

Throughout the paper, by 'acyclic curve' we mean a connected and simply connected complex affine algebraic curve. A classification of acyclic curves on the affine plane, both irreducible and reducible, up to the action of the automorphism group of the plane is well known, see e.g. [1], [38], [47], [53], [55]; we recall it in subsection 1.1 below. In section 3 we classify acyclic curves on affine toric surfaces. Similarly as in [54], actions of one-parameter groups play a crucial role in this classification.

Let $\pi\colon \mathbb{A}^2 \to X = \mathbb{A}^2/G$ be the quotient morphism, and let $\mathcal{C}$ be an irreducible acyclic curve on X. Then $\pi^*(\mathcal{C})$ is an acyclic (reducible and non-reduced, in general) curve on $\mathbb{A}^2$. We show (see theorems 3.3 and 3.6) that applying an appropriate automorphism of the affine plane we can transform the curve $\pi^*(\mathcal{C})$ and the G-action on $\mathbb{A}^2$ to canonical forms simultaneously.

In subsection 2.2, given an acyclic plane curve C, we describe the stabilizer subgroup $\mathrm{Stab}(C) \subseteq \mathrm{Aut}(\mathbb{A}^2)$ of all automorphisms which preserve C. We use this description in subsections 3.1 and 3.2 in order to obtain the canonical forms of irreducible acyclic curves on an affine toric surface X. We treat separately the cases of the curves passing or do not passing through the singular point of X. This leads in subsection 3.3 to the conclusion that every irreducible acyclic curve on X is the closure of a non-closed orbit of a $\mathbb{G}_m$-action on X. Furthermore, if such a curve is contained in the smooth locus X_{reg} then it is smooth and is as well an orbit of a $\mathbb{G}_a$-action on X, hence is included in a one-parameter family of such curves. We show that any affine toric surface X possesses only finitely many equivalence classes of embedded affine lines (see theorem 3.10). This is an analog of the celebrated Abhyankar-Moh-Suzuki Embedding Theorem, which says that there is just one class of embeddings of the affine line in the affine plane (see theorem 1.1 below).

The above description enables us to classify in subsection 3.4 all reduced simply connected curves on an affine toric surface, whenever they are irreducible or not.

Section 4 is devoted to the automorphism groups of affine toric surfaces. In theorem 4.2 we obtain an analog of the classical Jung-van der Kulk theorem on a free amalgamated product structure on the group $\mathrm{Aut}(\mathbb{A}^2)$ (see theorem 1.4). Using this theorem we describe in theorems 4.15 and 4.17 (reductive) algebraic groups acting effectively on affine toric surfaces.

[b]A finite linear group is *small* if it does not contain pseudoreflections.

In the final section 5 we deal with acyclic curves on a quotient $X = \mathbb{A}^2/G$ of the affine plane $\mathbb{A}^2$ by a nonabelian small finite group G. It turns out that the only irreducible acyclic curves on X are the images of the affine lines in $\mathbb{A}^2$ passing through the origin (see theorem 5.1). In particular, every such curve is the closure of a $\mathbb{G}_m$-orbit and passes through the singular point. Since the family of these curves is preserved under automorphisms, the automorphism group $\mathrm{Aut}(X)$ is rather poor. Namely, it coincides with $N(G)/G$, where $N(G)$ stands for the normalizer of G in $\mathrm{GL}(2, \mathbb{C})$ (see theorem 5.3). Consequently, $N(G)$ coincides with the normalizer of G in the full automorphism group $\mathrm{Aut}(\mathbb{A}^2)$. As an example, we describe explicitly the affine lines on the quaternion surface $X = \mathbb{A}^2/Q_8$.

When the paper was finished, S. Kaliman kindly informed us that he also came, for different purposes, to similar conclusions, but never wrote them down. The authors thank S. Kaliman for this information and interesting discussions.

1. Preliminaries

In this section we gather some well known facts on the geometry of the affine plane $\mathbb{A}^2$ over $\mathbb{C}$ that we need in the sequel.

1.1. *Simply connected plane affine curves*

By a *curve* we mean (for short) a complex affine algebraic curve. A curve C is called *acyclic* if $\pi_0(C) = \pi_1(C) = 1$, i.e. C is connected and simply connected. Two plane curves C and C' will be called *equivalent* if $C' = \gamma(C)$ for some $\gamma \in \mathrm{Aut}(\mathbb{A}^2)$. The following theorems provide canonical forms of acyclic and, more generally, of simply connected plane curves.

Theorem 1.1. (Abhyankar-Moh [1], Suzuki [47]) *Any reduced, irreducible, smooth, acyclic plane curve is equivalent to the affine line $C_y = \{y = 0\}$. Furthermore, if C is parameterized via a map $\varphi : \mathbb{A}^1 \xrightarrow{\simeq} C$, $\varphi(t) = (p(t), q(t))$, where $p, q \in \mathbb{C}[t]$, then either $\deg p \mid \deg q$ or $\deg q \mid \deg p$.*

Using Suzuki's formula for the Euler characteristics of the fibers in a fibration on a smooth affine surface [47] (see also e.g. [20]) it is not difficult to deduce the following corollary.

Corollary 1.2. *Any disconnected, simply connected, reduced plane curve is equivalent to a union of $r \geq 2$ parallel lines.*

Theorem 1.3. (Lin-Zaidenberg [55])

(a) *Any reduced, irreducible, singular, acyclic plane curve is equivalent to one and only one of the curves* $C_{a,b} = \{y^a - x^b = 0\}$, *where* $1 < a < b$ *and* $\gcd(a, b) = 1$.

(b) *Any reduced, simply connected plane curve is equivalent to a curve given by one of the equations*

$$y^{\varepsilon_y} p(x) = 0 \tag{1}$$

or

$$x^{\varepsilon_x} y^{\varepsilon_y} \prod_{i=1}^{r} (y^a - \kappa_i x^b) = 0\,, \tag{2}$$

where $\varepsilon_x, \varepsilon_y \in \{0, 1\}$, $p \in \mathbb{C}[t]$ *is a polynomial with simple roots,* $a, b \geq 1$ *and* $\gcd(a, b) = 1$, $r > 0$, *and* $\kappa_i \in \mathbb{C}^\times$, $i = 1, \ldots, r$, *are pairwise distinct.*

1.2. *The automorphism group of the affine plane*

In this subsection the base field $\mathbb{k}$ can be arbitrary. We let $\mathbb{A}^n_{\mathbb{k}}$ denote the affine n-space over $\mathbb{k}$ and $\mathrm{Aff}(\mathbb{A}^n_{\mathbb{k}})$ the group of all affine transformations of this space. By $\mathrm{JONQ}^+(\mathbb{A}^2_{\mathbb{k}})$ ($\mathrm{JONQ}^-(\mathbb{A}^2_{\mathbb{k}})$, respectively) we denote the group of the de Jonqières transformations

$$\Phi^+ : (x, y) \longmapsto (\alpha x + f(y), \beta y + \gamma)\,, \tag{3}$$

respectively,

$$\Phi^- : (x, y) \longmapsto (\alpha x + \gamma, \beta y + f(x))\,, \tag{4}$$

where $\alpha, \beta \in \mathbb{k}^\times$, $\gamma \in \mathbb{k}$, and $f \in \mathbb{k}[t]$. The subgroup

$$\mathrm{Aff}^{\pm}(\mathbb{A}^2_{\mathbb{k}}) = \mathrm{Aff}(\mathbb{A}^2_{\mathbb{k}}) \cap \mathrm{JONQ}^{\pm}(\mathbb{A}^2_{\mathbb{k}}) \tag{5}$$

consists of all upper (lower, respectively) triangular affine transformations $\Phi^{\pm}$ with $\deg f \leq 1$. The structure of the automorphism group $\mathrm{Aut}(\mathbb{A}^2_{\mathbb{k}})$ is described by the following classical theorem.

Theorem 1.4. (Jung [26], van der Kulk [49]) *The automorphism group* $\mathrm{Aut}(\mathbb{A}^2_{\mathbb{k}})$ *is the free product of the subgroups* $\mathrm{JONQ}^+(\mathbb{A}^2_{\mathbb{k}})$ *and* $\mathrm{Aff}(\mathbb{A}^2_{\mathbb{k}})$ *amalgamated over their intersection* $\mathrm{Aff}^+(\mathbb{A}^2_{\mathbb{k}})$:

$$\mathrm{Aut}(\mathbb{A}^2_{\mathbb{k}}) = \mathrm{JONQ}^+(\mathbb{A}^2_{\mathbb{k}}) *_{\mathrm{Aff}^+(\mathbb{A}^2_{\mathbb{k}})} \mathrm{Aff}(\mathbb{A}^2_{\mathbb{k}})\,.$$

Remarks 1.5. 1. In fact Jung [26] just established, over a field $\Bbbk$ of characteristic 0, the equality

$$\mathrm{Aut}(\mathbb{A}^2_\Bbbk) = \langle U^+, \mathrm{Aff}(\mathbb{A}^2_\Bbbk)\rangle ,$$

where

$$U^+ = \{\Phi \in \mathrm{JONQ}^+(\mathbb{A}^2_\Bbbk) \mid \Phi : (x,y) \mapsto (x+f(y), y)\} . \tag{6}$$

However, $\mathrm{Aut}(\mathbb{A}^2_\Bbbk) \neq U^+ *_{U^+\cap \mathrm{Aff}(\mathbb{A}^2_\Bbbk)} \mathrm{Aff}(\mathbb{A}^2_\Bbbk)$, see Remark in [28, §2].

Over an arbitrary ground field, van der Kulk did not formulate the theorem in terms of amalgamated free products, but from his results the theorem can be deduced readily, as this is done in [40] or [28, Theorem 2]. In characteristic zero, the theorem is a consequence of the Abhyankar-Moh-Suzuki theorem 1.1. See also [2], [10], [11], [15], [16], [19], [22], [34], [36], [37], [45], [48], and [52] for different approaches.

2. Actually we have

$$\mathrm{Aut}(\mathbb{A}^2_\Bbbk) = \langle \mathrm{JONQ}^+(\mathbb{A}^2_\Bbbk), \tau\rangle ,$$

where $\tau \in \mathrm{Aff}(\mathbb{A}^2_\Bbbk)$, $\tau : (x,y) \longmapsto (y,x)$ is a twist. Notice that $\tau \in \langle \mathrm{Aff}^+(\mathbb{A}^2_\Bbbk), \mathrm{Aff}^-(\mathbb{A}^2_\Bbbk)\rangle$ and

$$\mathrm{JONQ}^-(\mathbb{A}^2_\Bbbk) = \tau\mathrm{JONQ}^+(\mathbb{A}^2_\Bbbk)\tau,\ \mathrm{Aff}^-(\mathbb{A}^2_\Bbbk) = \tau\mathrm{Aff}^+(\mathbb{A}^2_\Bbbk)\tau,\ \text{and}\ U^- = \tau U^+\tau ,$$

where the subgroup $U^- \subseteq \mathrm{JONQ}^-(\mathbb{A}^2_\Bbbk)$ is defined similarly as U^+. In particular

$$\mathrm{Aut}(\mathbb{A}^2_\Bbbk) = \langle \mathrm{JONQ}^+(\mathbb{A}^2_\Bbbk), \mathrm{JONQ}^-(\mathbb{A}^2_\Bbbk)\rangle .$$

The following theorem absorbed several previously known results. In this generality, it was first proved by Kambayashi [29] (using a result of Wright [50]) as a consequence of the Jung-van der Kulk theorem 1.4.

Theorem 1.6. (Kambayashi [29, Theorem 4.3], Wright [50], [51]) *Any algebraic subgroup of the group* $\mathrm{Aut}(\mathbb{A}^2_\Bbbk)$ *is conjugate either to a subgroup of* $\mathrm{Aff}(\mathbb{A}^2_\Bbbk)$, *or to a subgroup of* $\mathrm{JONQ}^+(\mathbb{A}^2_\Bbbk)$.

The proof exploits the following observation: every algebraic subgroup of $\mathrm{Aut}(\mathbb{A}^2_\Bbbk)$ has bounded degree, hence also a bounded length with respect to the free amalgamated product structure. However, by Serre [46] a subgroup of bounded length in an amalgamated free product $A *_C B$ is conjugate to a subgroup of one of the factors A and B.

In the next corollary we suppose that the base field $\Bbbk$ is algebraically closed of characteristic 0 (while certain assertions remain valid in the positive characteristic case).

Corollary 1.7.

(1) (Igarashi [25], Furushima [18]) *Every finite subgroup of* $\mathrm{Aut}(\mathbb{A}^2_{\Bbbk})$ *is conjugate to a subgroup of* $\mathrm{GL}(2,\Bbbk)$.

(2) (Gutwirth [23], Bialynicki-Birula [9]) *Every maximal torus in* $\mathrm{Aut}(\mathbb{A}^2_{\Bbbk})$ *has rank 2 and is conjugate to the standard maximal torus* $\mathbb{T} \subseteq \mathrm{GL}(2,\Bbbk)$.

(3) (Gutwirth [23], Bialynicki-Birula [9]) *Every one-torus in* $\mathrm{Aut}(\mathbb{A}^2_{\Bbbk})$ *is conjugate to a subtorus of* $\mathbb{T}$.

(4) (Rentschler [45]) *Every* $\mathbb{G}_a(\Bbbk)$*-action on* $\mathbb{A}^2_{\Bbbk}$ *is conjugate to an action via de Jonqières transformations*

$$t.(x,y) = (x + tf(y), y), \quad \textit{where} \quad t \in \mathbb{G}_a(\Bbbk) \quad \textit{and} \quad f \in \Bbbk[y]\,.$$

Remarks 1.8. 1. Assertions (1)-(3) follow from a more general result for reductive groups, see proposition 2.5 below.

2. Analogs of (1) and (4) fail in higher dimensions, while (3) holds in dimension 3 and is open in higher dimensions. We do not dwell on this here (see, however, [27], [43] and the survey [31]; see also [7] and [39] for the case of a positive characteristic).

2. Subgroups of de Jonquères group and stabilizers of plane curves

2.1. *Subgroups of the de Jonquères group*

In this subsection the base field $\Bbbk$ will be an algebraically closed field of characteristic zero (while some results are still valid over an arbitrary field of characteristic zero). By abuse of language, we still call *de Jonquères groups* the subgroups $\mathrm{Jonq}^{\pm}(\mathbb{A}^2_{\Bbbk}) \subseteq \mathrm{JONQ}^{\pm}(\mathbb{A}^2_{\Bbbk})$ consisting, respectively, of the transformations

$$\varphi^{+} : (x,y) \longmapsto (\alpha x + f(y), \beta y)\,, \tag{7}$$

and

$$\varphi^{-} : (x,y) \longmapsto (\alpha x, \beta y + f(x))\,, \tag{8}$$

where $\alpha, \beta \in \Bbbk^{\times}$ and $f \in \Bbbk[t]$ (so $\Phi^{\pm}$ as in (3), (4) belongs to $\mathrm{Jonq}^{\pm}(\mathbb{A}^2_{\Bbbk})$ if and only if $\gamma = 0$).

Clearly,

$$\mathrm{Jonq}^{+}(\mathbb{A}^2_{\Bbbk}) \simeq U^{+} \rtimes \mathbb{T}\,, \tag{9}$$

where $\mathbb{T} \subseteq \mathrm{GL}(2, \Bbbk)$ is the maximal torus

$$\mathbb{T} = \{\delta \in \mathrm{Jonq}^+(\mathbb{A}^2_\Bbbk) \mid \delta : (x, y) \mapsto (\alpha x, \beta y)\},$$

and $U^+ \simeq \Bbbk[t]$ as in (6) is an infinite dimensional vector group. Indeed, let $\rho : \mathrm{Jonq}^+(\mathbb{A}^2_\Bbbk) \to \mathbb{T}$ denote the canonical surjection provided by (9). Then any element $\varphi^+ \in \mathrm{Jonq}^+(\mathbb{A}^2_\Bbbk)$ as in (7) admits a decomposition $\varphi^+ = \partial \circ \delta$, where $\delta = \rho(\varphi^+) \in \mathbb{T}$ and $\partial \in U^+$, $\partial : (x, y) \mapsto (x + f(y/\beta), y)$.

In particular, $\mathrm{Jonq}^+(\mathbb{A}^2_\Bbbk)$ is a metabelian group, and U^+ can be considered as its unipotent radical in the following sense. There is a natural filtration by algebraic subgroups

$$\mathrm{Jonq}^+(\mathbb{A}^2_\Bbbk) = \bigcup_{n \in \mathbb{N}} \mathrm{Jonq}^+_n(\mathbb{A}^2_\Bbbk),$$

where

$$\mathrm{Jonq}^+_n(\mathbb{A}^2_\Bbbk) \simeq U^+_n \rtimes \mathbb{T}$$

consists of all elements $\varphi^+ \in \mathrm{Jonq}^+(\mathbb{A}^2_\Bbbk)$ with $\deg f \le n$. Then $\mathbb{T}$ is a maximal torus of $\mathrm{Jonq}^+_n(\mathbb{A}^2_\Bbbk)$, and $U^+_n = U^+ \cap \mathrm{Jonq}^+_n(\mathbb{A}^2_\Bbbk)$ is its unipotent radical.

Any algebraic subgroup G of $\mathrm{Jonq}^+(\mathbb{A}^2_\Bbbk)$ has finite degree

$$d = \min\{n \mid G \subseteq \mathrm{Jonq}^+_n(\mathbb{A}^2_\Bbbk)\}.$$

In particular, every maximal torus in $\mathrm{Jonq}^+(\mathbb{A}^2_\Bbbk)$ is conjugate with $\mathbb{T}$, and any algebraic subgroup which consists of semi-simple elements is conjugate to a subgroup of $\mathbb{T}$ (see [24, VII.19.4, VIII.21.3A]).

In the following lemma we characterize semi-simple and torsion elements in the group $\mathrm{Jonq}^+(\mathbb{A}^2_\Bbbk)$.

Lemma 2.1.

(a) *An element $\varphi^+ \in \mathrm{Jonq}^+(\mathbb{A}^2_\Bbbk)$ as in (7) is semi-simple if and only if the polynomial $f(y) = \sum_{m \ge 0} a_m y^m$ satisfies the condition*

$$a_m = 0 \quad \text{if} \quad \alpha = \beta^m, \tag{10}$$

if and only if there exists $\mu \in U^+$ such that $\mu^{-1}\varphi^+\mu = \delta \in \mathbb{T}$.[c]

(b) *An element $\varphi^+ \in \mathrm{Jonq}^+(\mathbb{A}^2_\Bbbk)$ as in (7) is of finite order if and only if it is semi-simple and $\delta = \rho(\varphi^+) \in \mathbb{T}$ is of finite order.*

[c] In fact here $\delta = \rho(\varphi^+)$.

Proof. (a) We claim that if φ^+ satisfies (10) then there exists $\mu \in U^+$, $\mu : (x, y) \mapsto (x + g(y), y)$, where $g \in \Bbbk[t]$, such that, in the notation as in lemma 2.1

$$(11)\quad \mu^{-1}\varphi^+\mu = \delta = \rho(\varphi^+) \in \mathbb{T} \quad \text{or, equivalently,} \quad \partial = [\mu, \delta] = \mu\delta\mu^{-1}\delta^{-1}.$$

Writing $g(y) = \sum_{m\geq 0} b_m y^m$, it is readily seen that μ satisfies (11) if and only if the coefficients b_m of g satisfy the conditions

$$(12)\qquad b_m = \begin{cases} \frac{a_m}{\beta^m - \alpha} & \text{if } \beta^m - \alpha \neq 0 \\ \text{arbitrary} & \text{if } \beta^m - \alpha = 0 = a_m. \end{cases}$$

Indeed, (11) can be written as

$$(x + g(y) - \alpha g(y/\beta), y) = (x + f(y/\beta), y)$$

which is equivalent to (12). This shows the existence of μ in (11) under condition (10).

On the other hand, if (10) fails i.e., there exists $m \in \mathbb{N}$ such that $\beta^m - \alpha = 0$ and $a_m \neq 0$, then μ as in (11) cannot exist. The remaining claims in (a) are easy and so we leave them to the reader.

(b) We have to show that if $(\varphi^+)^k = \text{id}$ then (10) holds and $\delta^k = \text{id}$, and vice versa. Indeed, letting $\gamma = \varphi^+$ for any $k \geq 1$ we obtain

$$\gamma^k : (x, y) \longmapsto \big(\alpha^k x + f_k(y),\ \beta^k y\big),$$

where

$$f_k(y) = \alpha^{k-1} f(y) + \alpha^{k-2} f(\beta y) + \ldots + f(\beta^{k-1} y).$$

Therefore $\gamma^k = \text{id}$ if and only if $\alpha^k = \beta^k = 1$ (i.e. $\delta^k = \text{id}$) and $f_k = 0$. However, $f_k = 0$ if and only if

$$\forall m \geq 0, \qquad \text{either} \quad a_m = 0 \quad \text{or} \quad \alpha^{k-1} + \alpha^{k-2}\beta^m + \ldots + \beta^{m(k-1)} = 0.$$

The latter equality can be written as

$$(\alpha/\beta^m)^{k-1} + (\alpha/\beta^m)^{k-2} + \ldots + 1 = 0.$$

Since $(\alpha/\beta^m)^k = 1$ this holds if and only if $\alpha \neq \beta^m$. □

Lemma 2.1(b) admits the following interpretation.

Remark 2.2. Consider a $\mathbb{Z}/d\mathbb{Z}$-grading on the polynomial ring $A = \Bbbk[t]$:

$$A = \bigoplus_{i=0}^{d-1} A_{d,i}, \quad \text{where} \quad A_{d,i} = t^i \Bbbk[t^d].$$

According to this decomposition any element $f \in A$ can be written as $f = \sum_{i=0}^{d-1} f_i$, where $f_i \in A_i$ $\forall i$. Assuming that β has finite order d we let $m(\alpha, \beta) = \min\{m \geq 0 \mid \alpha = \beta^m\}$. Then condition (10) can be expressed as $f_{m(\alpha,\beta)} = 0$, provided this quantity is well defined. (Otherwise (10) does not impose any condition.) This phenomenon can be seen on the following simple examples.

Example 2.3. Letting $d = 2$ any element $f \in \mathbb{k}[t]$ can be written as $f = f_0 + f_1$, where f_0 is even and f_1 is odd. There are the following three types of involutions $\varphi^+ : (x, y) \mapsto (\alpha x + f(y), \beta y)$ in $\mathrm{Jonq}^+(\mathbb{A}^2_{\mathbb{k}})$:

(1) $\alpha = 1,\ \beta = -1,\ f \in \mathbb{k}[t]$ is odd i.e. $f_0 = 0$;
(2) $\alpha = -1,\ \beta = 1,\ f \in \mathbb{k}[t]$ is arbitrary;
(3) $\alpha = -1,\ \beta = -1,\ f \in \mathbb{k}[t]$ is even i.e. $f_1 = 0$.

Lemma 2.4. *Consider a pair of elements* $\gamma, \tilde{\gamma} \in \mathrm{Jonq}^+(\mathbb{A}^2_{\mathbb{k}})$,

$$\gamma : (x, y) \mapsto (\alpha x + f(y), \beta y) \quad \textit{and} \quad \tilde{\gamma} : (x, y) \mapsto (\tilde{\alpha} x + \tilde{f}(y), \tilde{\beta} y),$$

where

$$f(y) = \sum_{m \geq 0} a_m y^m \quad \textit{and} \quad \tilde{f}(y) = \sum_{m \geq 0} \tilde{a}_m y^m .$$

Then γ *and* $\tilde{\gamma}$ *commute if and only if*

$$a_m(\tilde{\beta}^m - \tilde{\alpha}) = \tilde{a}_m(\beta^m - \alpha) \quad \forall m \geq 0 . \tag{13}$$

Proof. The proof is easy and is left to the reader. □

Recall that a *quasitorus* is a product of a torus and a finite abelian group. Any algebraic subgroup of a torus is a quasitorus.

Proposition 2.5. *Any reductive algebraic subgroup* G *of the group* $\mathrm{Jonq}^+(\mathbb{A}^2_{\mathbb{k}})$ *is conjugate to a subgroup of the torus* $\mathbb{T}$. *More precisely, there exists an element* $\mu \in U^+$ *such that*

$$\mu^{-1} \circ \varphi \circ \mu = \rho(\varphi) \quad \forall \varphi \in G , \tag{14}$$

where

$$\rho : \mathrm{Jonq}^+(\mathbb{A}^2_{\mathbb{k}}) \to \mathbb{T} = \mathrm{Jonq}^+(\mathbb{A}^2_{\mathbb{k}})/U^+$$

is the natural surjection. In particular $\mu^{-1} G \mu = \rho(G) \subseteq \mathbb{T}$.

Proof. Since G is reductive the unipotent radical $R_u(G)$ is trivial. Hence also the subgroup $G \cap U^+$ is trivial. Thus ρ restricts to an isomorphism $\rho|G : G \xrightarrow{\simeq} \rho(G) \subseteq \mathbb{T}$. In particular, G is abelian and consists of semi-simple elements, cf. lemma 2.1. By [24, VII.19.4, VIII.21.3A] G is contained in a maximal torus $\mathbb{T}'$ conjugate to $\mathbb{T}$. Now the first assertion follows.

Let us show the second. Since G is abelian and consists of semi-simple elements, (10) and (13) are fulfilled for any pair of elements $\gamma, \tilde{\gamma} \in G$. Thus there is $\mu \in U^+$ satisfying (12) and then also (11) for all $\varphi \in G$ simultaneously (see the proof of lemma 2.1). This μ is as desired. □

As an application we can deduce the following well known fact (see [32, Theorem 2]).

Corollary 2.6. *Every effective action of a reductive algebraic group G on the affine plane $\mathbb{A}^2_{\Bbbk}$ is linearizable. In other words, G is conjugate in the group* $\mathrm{Aut}(\mathbb{A}^2)$ *to a subgroup of* $\mathrm{GL}(2, \Bbbk)$.

Let us provide an argument following an indication in [32].

Proof. By theorem 1.6 G is conjugate in $\mathrm{Aut}(\mathbb{A}^2)$ to a subgroup of one of the groups $\mathrm{Aff}(\mathbb{A}^2_{\Bbbk})$ and $\mathrm{Jonq}^+(\mathbb{A}^2_{\Bbbk})$. In the latter case by proposition 2.5 G is conjugate to a subgroup of the torus $\mathbb{T}$ contained in $\mathrm{Aff}(\mathbb{A}^2_{\Bbbk})$. Thus we may assume that $G \subseteq \mathrm{Aff}(\mathbb{A}^2_{\Bbbk})$.

It remains to show that G has a fixed point in $\mathbb{A}^2$. Observe that $G \subseteq \mathrm{Aff}(\mathbb{A}^2_{\Bbbk})$ admits a representation in $\mathrm{GL}(3, \Bbbk)$ by matrices of the form $\begin{pmatrix} * & * & * \\ * & * & * \\ 0 & 0 & 1 \end{pmatrix}$. Since G is reductive and $\mathrm{char}\,(\Bbbk) = 0$, G is geometrically reductive. Therefore the G-invariant plane $L_0 = \{x_3 = 0\}$ in $\mathbb{A}^3_{\Bbbk}$ with coordinates (x_1, x_2, x_3) has a G-invariant complement, say, R, which meets the parallel G-invariant plane $L_1 = \{x_3 = 1\}$ in a fixed vector. This yields a fixed point of G in $\mathbb{A}^2_{\Bbbk} \simeq_G L_1$. Now the proof is completed. □

Remark 2.7. Due to corollary 2.6, when dealing with quotients of the affine plane by finite group actions it suffices to restrict to linear such actions.

2.2. *Stabilizers of acyclic plane curves*

In the sequel $\mathbb{A}^n$ stands for the affine n-space over $\mathbb{C}$. For an algebraic curve C in $\mathbb{A}^2$ we let $\mathrm{Stab}(C)$ be the stabilizing group, or *stabilizer* of C i.e. the group of all algebraic automorphisms of $\mathbb{A}^2$ that preserve C:

$$\mathrm{Stab}(C) \;=\; \{\gamma \in \mathrm{Aut}(\mathbb{A}^2) \,|\, \gamma(C) = C\}\,.$$

In this subsection we describe the stabilizers of acyclic plane curves given in one of the canonical forms (1) and (2) of theorem 1.3(b). Accordingly, we distinguish the following six types of acyclic curves C:

(I) C is irreducible and smooth, and then $C \sim C_y$;
(II) C consists of two smooth, mutually transversal components, and then $C \sim C_x \cup C_y$;
(III) C has $r \geq 2$ singular points and so is equivalent to a union of r parallel lines and a transversal line given by (1) with $\varepsilon_y = 1$;
(IV) C has an ordinary singularity of multiplicity $r \geq 3$ and so is equivalent to a union of r distinct lines through the origin;
(V) All irreducible components of C are smooth and C has a non-ordinary singular point. So C is equivalent to a curve given by equation (2) with $a = 1 < b$ and $\varepsilon_y + r \geq 2$;
(VI) C contains a singular component, and then it is equivalent to a curve given by (2) with $\min\{a, b\} > 1$ and $r \geq 1$.

We analyse each of these cases separately.

In the next proposition we study the curves of type (I). We show that the stabilizer $\mathrm{Stab}(C_y)$ of the coordinate axis $C_y = \{y = 0\}$ in $\mathbb{A}^2$ consists of the de Jonquières transformations. For the acyclic curves of types (II)-(IV) reduced to the canonical form the stabilizer is described in corollaries 2.9 and 2.10, and for those of types (V) and (VI) in propositions 2.13 and 2.14.

Proposition 2.8. *The stabilizer* $\mathrm{Stab}(C_y)$ *in* $\mathrm{Aut}(\mathbb{A}^2)$ *coincides with the subgroup* $\mathrm{Jonq}^+(\mathbb{A}^2)$, *while* $\mathrm{Stab}(C_x) = \mathrm{Jonq}^-(\mathbb{A}^2)$.

Proof. Every $\gamma \in \mathrm{Stab}(C_y)$ sends y to βy for some $\beta \in \mathbb{C}^\times$. Up to an affine transformation we may assume that $\beta = 1$ and $\gamma|_{C_y} = \mathrm{id}_{C_y}$. Suppose that γ sends x to $x + h(x, y)$. Since $h(x, 0) = 0$ we have $h(x, y) = yp(x, y)$. To show that $p(x, y)$ does not depend on x we write

$$p(x, y) \;=\; a_0(y) + a_1(y)x + \ldots + a_k(y)x^k \quad \text{with} \quad a_k(y) \neq 0\,.$$

Clearly, γ preserves every line $y = y_0$ and induces an affine automorphism of this line. Picking y_0 with $a_k(y_0) \neq 0$ we get $k \leq 1$. Letting $k = 1$ we obtain $\gamma : x \mapsto (1 + ya_1(y))x + ya_0(y)$. If y_1 is a root of the non-constant polynomial $1 + ya_1(y)$ then γ induces a constant map on the line $y = y_1$, a contradiction. Hence $k = 0$. Thus $\mathrm{Stab}(C_y) = \mathrm{Jonq}^+(\mathbb{A}^2)$. The proof of the second assertion is similar. □

In the following two corollaries we describe the stabilizers of the canonical curves of types (II)–(IV).

Corollary 2.9. *If* $C = \bigcup_{i=1}^r L^i$ *is a union of* $r \geq 2$ *affine lines in* $\mathbb{A}^2$ *through the origin then* $\mathrm{Stab}(C) \subseteq \mathrm{GL}(2,\mathbb{C})$.

Proof. We may suppose that $L_1 = C_x$ and $L_2 = C_y$. For any $g \in \mathrm{Stab}(C)$ we can find $h \in \mathrm{GL}(2,\mathbb{C})$ such that $g(L_i) = h(L_i)$, $i = 1, 2$. It follows by proposition 2.8 that $\gamma = h^{-1}g \in \mathrm{GL}(2,\mathbb{C})$. Hence also $g = h\gamma \in \mathrm{GL}(2,\mathbb{C})$. □

The following corollary is immediate.

Corollary 2.10.

(a) Let $C = \{f(y) = 0\}$, *where* $f \in \mathbb{C}[y]$ *is a polynomial of degree* ≥ 2 *with simple roots. If* $K \subseteq \mathbb{C}$ *denotes the set of these roots, then*

$$\mathrm{Stab}(C) = \mathbb{T}_{1,0} \cdot U^+ \cdot \mathrm{Stab}(K)\,,$$

where $U^+ \subseteq \mathrm{Aut}(\mathbb{A}^2)$ *is as in (6),*

$$\mathbb{T}_{1,0} = \{\lambda \in \mathbb{T} \mid \lambda : (x,y) \mapsto (\alpha x, y),\, \alpha \in \mathbb{C}^\times\}\,,$$

and the stabilizer $\mathrm{Stab}(K) \subseteq \mathrm{Aut}(\mathbb{A}^1) \hookrightarrow \mathrm{Aut}(\mathbb{A}^2)$ *acts naturally on the symbol* y.

(b) If C *of type (II) is the coordinate cross* $\{xy = 0\}$ *in* $\mathbb{A}^2$ *then* $\mathrm{Stab}(C) = N(\mathbb{T})$ *is the normalizer of the maximal torus* $\mathbb{T}$ *in the group* $\mathrm{GL}(2,\mathbb{C})$.

(c) If C *of type (IV) is a union of* r *affine lines through the origin, where* $r \geq 3$, *then* $\mathrm{Stab}(C) \subseteq \mathrm{GL}(2,\mathbb{C})$ *is a finite extension of the group* $\mathbb{T}_{1,1} = \mathbb{C}^\times \cdot \mathrm{id}$ *of scalar matrices.*

(d) If C *of type (III) is given by equation* $xyf(x) = 0$, *where* $f \in \mathbb{C}[x]$ *is a polynomial of degree* ≥ 1 *with simple roots such that* $f(0) \neq 0$, *then* $\mathrm{Stab}(C) \subseteq \mathbb{T}$ *is a finite extension of the one-torus* $\mathbb{T}_{0,1} \subseteq \mathbb{T}$.

Thus in (b)-(d) the stabilizer $\mathrm{Stab}(C)$ is a linear group, while the group in (a) is infinite dimensional. The group $\mathrm{Stab}(C)$ in (b) is nonabelian. It can occur to be nonabelian also in (c), as in the following simple example.

Example 2.11. Given a nonabelian finite subgroup $G \subsetneq \mathrm{GL}(2, \mathbb{C})$ consider the curve $C = \bigcup_{g \in G} g(C_y)$. Since $G \subseteq \mathrm{Stab}(C)$ the latter group is nonabelian.

Consider further an irreducible acyclic curve $C_{a,\,b}$ given in $\mathbb{A}^2$ by equation $y^a - x^b = 0$, where $a, b \geq 1$ and $\gcd(a, b) = 1$. In the following proposition we describe the stabilizer $\mathrm{Stab}(C_{a,\,b})$ for a singular such curve. Consider a one-parameter subgroup $\mathbb{T}_{a,b}$ of the torus $\mathbb{T}$,

$$\mathbb{T}_{a,b} = \{\gamma_{a,b}(t) \,|\, t \in \mathbb{G}_m\} \subseteq \mathrm{Stab}(C_{a,\,b}), \qquad (15)$$
$$\text{where} \quad \gamma_{a,b}(t) : (x, y) \longmapsto (t^a x,\, t^b y)\,.$$

Proposition 2.12. *If* $\min\{a, b\} > 1$ *then* $\mathrm{Stab}(C_{a,\,b}) = \mathbb{T}_{a,b}$.

Proof. Letting $C = C_{a,\,b}$ and $\Gamma = \mathrm{Stab}(C)$, we consider the pointwise stabilizer

$$\Gamma_0 = \{\gamma \in \Gamma \,|\, \gamma|_C = \mathrm{id}_C\} \subseteq \Gamma\,.$$

Claim 1. The group Γ_0 *is torsion free.*

Proof of claim 1. Let $\gamma_0 \neq \mathrm{id}$ be an element of finite order in Γ_0. The finite cyclic group $\langle \gamma_0 \rangle$ is reductive. Hence the fixed point locus $(\mathbb{A}^2)^{\gamma_0}$ is smooth by the Luna Étale Slice Theorem [35] (see also [44, §6]). Thus an irreducible component $C = C_{a,b}$ of this locus must be smooth as well. However, under our assumptions the curve $C_{a,\,b}$ is singular. □

An element $\gamma \in \Gamma$ sends the polynomial $q(x, y) = y^a - x^b$ to $\lambda_\gamma q$, where $\lambda_\gamma \in \mathbb{G}_m$. Letting $\psi(\gamma) = \lambda_\gamma$ yields a character $\psi \colon \Gamma \to \mathbb{G}_m$ of Γ. The following claim is immediate.

Claim 2. $\Gamma = \mathbb{T}_{a,b} \cdot \Gamma_1$, *where* $\Gamma_1 = \ker(\psi)$.

Claim 3. The kernel $\Gamma_1 = \ker(\psi)$ *is a torsion group. Furthermore, there is a positive integer* N *such that the orders of all elements in* Γ_1 *divide* N.

Proof of claim 3. The group Γ_1 acts on every fiber $C_\alpha = \{q(x, y) = \alpha\}$ of q. Since for $\alpha \neq 0$ the affine curve C_α has positive genus, its automorphism group is finite of order, say, N. For any $\alpha \neq 0 \neq \beta$ the curves C_α and C_β are isomorphic. So their automorphism groups are isomorphic, too. Since $\gamma^N|_{C_\alpha} = \mathrm{id}_{C_\alpha}$ for every $\gamma \in \Gamma_1$ and $\alpha \neq 0$, we have $\gamma^N = \mathrm{id}$. □

Claim 4. $\Gamma_0 = \{\mathrm{id}\}$.

Proof of claim 4. According to claim 2 we have $\Gamma_0 \subseteq \mathbb{T}_{a,b} \cdot \Gamma_1$. Writing an element $\gamma_0 \in \Gamma_0$ as $\gamma_0 = \gamma_{a,b}(t) \circ \gamma_1$, from $\gamma_0|_C = \mathrm{id}_C$ we obtain $\gamma_1^{-1}|_C = \gamma_{a,b}(t)|_C$. Hence $\mathrm{id}_C = \gamma_1^{-N}|_C = \gamma_{a,b}(t^N)|_C$. It follows that $t^N = 1$.

Since $\Gamma_0 \cap \Gamma_1 = \{\mathrm{id}\}$ the map $\psi|_{\Gamma_0} : \Gamma_0 \to \mathbb{G}_m$ is injective. So $\psi(\gamma_0) = \psi(\gamma_{a,b}(t))$ has finite order dividing N. Due to claim 1 we can conclude that $\Gamma_0 = \{\mathrm{id}\}$. □

Claim 5. $\Gamma = \mathbb{T}_{a,b}$.

Proof of claim 5. For any $\gamma \in \Gamma$ there exists $t \in \mathbb{C}^\times$ such that $\gamma|C = \gamma_{a,b}(t)|C$. Hence $\gamma \circ \gamma_{a,b}^{-1}(t) \in \Gamma_0 = \{\mathrm{id}\}$ and so $\gamma = \gamma_{a,b}(t) \in \mathbb{T}_{a,b}$. □

This ends the proof of proposition 2.12.

Using propositions 2.8 and 2.12 we describe in propositions 2.13 and 2.14 below the structure of the stabilizer $\mathrm{Stab}(C)$ for reduced (but possibly reducible) acyclic plane curves C of the remaining types (VI) and (V), respectively.

Proposition 2.13. *Let C be an acyclic curve of type (VI) given by equation (2), where $r \geq 1$ and $\gcd(a,b) = 1$. If $\min\{a,b\} > 1$ then $\mathrm{Stab}(C)$ is a quasitorus of rank 1 contained in the maximal torus $\mathbb{T}$.*

Proof. Let $C^i = \{y^a - \kappa_i x^b = 0\}$, $i = 1, \ldots, r$, be the irreducible components of the curve C. Clearly $\mathbb{T}_{a,b} \subseteq \mathrm{Stab}(C)$. If $r = 1$ then by proposition 2.12 $\mathrm{Stab}(C) = \mathbb{T}_{a,b}$. Suppose that $r > 1$. Consider a finite abelian group $H = \mathrm{Stab}(C) \cap \mathbb{T}_{1,0}$. We claim that $\mathrm{Stab}(C) = H \cdot \mathbb{T}_{a,b} \subseteq \mathbb{T}$. Hence this is a quasitorus of rank one, as stated. Indeed, if $\delta \in \mathrm{Stab}(C) \setminus \mathbb{T}_{a,b}$ then $\delta(C^1) = C^i$ for some $i > 1$. If $h \in \mathbb{T}_{1,0}$ is such that $h(C^i) = C^1$ then $\gamma = h \circ \delta \in \mathrm{Stab}(C^1) = \mathbb{T}_{a,b}$. Hence $h = \gamma \circ \delta^{-1} \in H$ and so $\delta = h^{-1} \circ \gamma \in H \cdot \mathbb{T}_{a,b}$. Now the claim follows. This ends the proof. □

Proposition 2.14. *Let C be an acyclic curve C of type (V) given by equation (2), where $r \geq 1$ and $\varepsilon_y + r \geq 2$. If $a = 1 < b$ then $\mathrm{Stab}(C)$ is a quasitorus of rank 1 conjugated in the group $\mathrm{Aut}(\mathbb{A}^2)$ to a subgroup of the torus $\mathbb{T}$.*

Proof. If $\varepsilon_x = 1$ the proof is easy and can be left to the reader (cf. corollary 2.10(b)). Thus we may restrict to the case $\varepsilon_x = 0$. By our assumptions C is reducible and all components C^i of C are smooth and mutually tangent at the origin (we let here $C^0 = C_y$ if $\varepsilon_y = 1$). Consider the pencil $\mathcal{L} = \{C_\mu\}_{\mu \in \mathbb{P}^1}$, where $C_\mu = \{y - \mu x^b = 0\}$ for $\mu \neq \infty$ and $C_\infty = C_x$.

Claim 1. The pencil $\mathcal{L} \setminus \{C_x\}$ is stable under the action of the group $\mathrm{Stab}(C)$ on $\mathbb{A}^2$.

Proof of claim 1. The unique singular point $\bar{0} \in C$ is fixed under the action of $\mathrm{Stab}(C)$. Furthermore, for every $g \in \mathrm{Stab}(C)$ and every $\mu \in \mathbb{C}$, either

$C_\mu \subseteq C$ or $g(C_\mu) \cap C = \{\bar{0}\}$. In the latter case, letting $B_\mu = g(C_\mu)$ it follows that the restriction $(y - \kappa_i x^b)|_{B_\mu}$ vanishes just at the origin. Hence in an affine coordinate, say, z in $B_\mu \simeq \mathbb{A}^1$ centered at the origin the latter function is a monomial $\lambda_i z^{\alpha_i}$, where $\lambda_i \in \mathbb{C}^\times$ and $\alpha_i \geq 1$. Therefore B_μ is a component of the curve

$$(y - \kappa_i x^b)^{\alpha_j} - \delta(y - \kappa_j x^b)^{\alpha_i} = 0, \quad \text{where} \quad \lambda_i^{\alpha_j} = \delta \lambda_j^{\alpha_i} .$$

Since B_μ is smooth one of the exponents α_i, α_j divides the other. Preservation of the local intersection indices under g implies that $i(B_\mu, C^i, \bar{0}) = b$ $\forall i$. It follows that actually $\alpha_i = \alpha_j$ $\forall i, j$. Finally B_μ coincides with a certain member $C_{\mu'} \in \mathcal{L}$, as claimed. □

Clearly $\nu(g) : \mathbb{A}^1 \to \mathbb{A}^1$, $\mu \mapsto \mu'$, is an affine transformation leaving invariant the set $K = \{\kappa_i\} \subseteq \mathbb{A}^1$ enriched by $\kappa_0 = 0$ in case where $\varepsilon_y = 1$ in (7). Let $\mathrm{Stab}(K)$ be the stabilizer of K in $\mathrm{Aut}(\mathbb{A}^1)$. The group $\mathrm{Stab}(K)$ fixes the center of gravity of K and so embeds in $\mathbb{G}_m$ onto a finite subgroup. The natural homomorphism $\varphi : \mathrm{Stab}(C) \to \mathrm{Stab}(K)$ fits in the exact sequence

$$1 \to \mathrm{Stab}_0(C) \to \mathrm{Stab}(C) \overset{\varphi}{\longmapsto} \mathrm{Stab}(K) \to \dots ,$$

where $\mathrm{Stab}_0(C) = \ker(\varphi) \subseteq \mathrm{Stab}(C)$ consists of the elements $g \in \mathrm{Aut}(\mathbb{A}^2)$ leaving invariant every component C^i of C.

Claim 2. $\mathrm{Stab}_0(C) = \mathbb{T}_{1,b}$.

Proof of claim 2. If $g \in \mathrm{Stab}_0(C)$ then $\varphi(g) = \mathrm{id}_K$. Since $|K| \geq 2$ it follows that $\nu(g) = \mathrm{id}_{\mathbb{A}^1}$ i.e., $g(C_\mu) = C_\mu$ $\forall \mu \in \mathbb{A}^1$. In particular, $g(C_0) = C_0$, where $C_0 = C_y$. By proposition 2.8 we have $g : (x, y) \mapsto (\alpha x + f(y), \beta y)$ for some $\alpha, \beta \in \mathbb{C}^\times$ and $f \in \mathbb{C}[y]$. The equality $g(C^1) = C^1$ implies that $f = 0$ and $\beta = \alpha^b$, that is, $g \in \mathbb{T}_{1,b}$. Now the claim follows. □

Thus $\mathrm{Stab}(C)$ is an extension of the one-torus $\mathbb{T}_{1,b}$ by a finite cyclic group. The proof can be completed due to the following

Claim 3. $\mathrm{Stab}(C)$ *is conjugated in* $\mathrm{Aut}(\mathbb{A}^2)$ *to a subgroup of the maximal torus* $\mathbb{T}$.

Proof of claim 3. For every $g \in \mathrm{Stab}(C)$ we have $g(C_x) = C_x$. Letting z_0 denote the center of gravity of K we consider an automorphism $\gamma \in \mathrm{Aut}(\mathbb{A}^2)$, $(x, y) \mapsto (x, y - z_0 x^b)$. It is easily seen that γ preserves the pencil $\mathcal{L}$, while $\nu(\gamma) : z \mapsto z - z_0$. Since $\nu(\gamma g \gamma^{-1}) = \nu(\gamma)\nu(g)\nu(\gamma)^{-1} : 0 \mapsto 0$ we obtain $\gamma g \gamma^{-1} : C_y \to C_y$ and $C_x \to C_x$. Hence $\gamma \circ g \circ \gamma^{-1} \in \mathbb{T}$ (see corollary 2.10(b)) and so $\gamma\, \mathrm{Stab}(C) \gamma^{-1} \subseteq \mathbb{T}$. □

However, for C as in proposition 2.14 the stabilizer $\mathrm{Stab}(C)$ is not necessarily contained in $\mathrm{GL}(2, \mathbb{C})$, as is seen in the following example.

Example 2.15. Let $C = \{y(y - x^b) = 0\}$, where $b > 1$, and let $g : (x, y) \mapsto (x, x^b - y)$. Then $g \in \mathrm{Stab}(C) \setminus \mathrm{GL}(2, \mathbb{C})$.

From corollary 2.10 and propositions 2.12-2.14 we deduce the following result.

Corollary 2.16. *The stabilizer* $\mathrm{Stab}(C)$ *of an acyclic plane curve* C *is abelian unless* C *is equivalent under the* $\mathrm{Aut}(\mathbb{A}^2)$*-action to a union of affine lines through the origin.*

3. Acyclic curves on affine toric surfaces

In this section we classify acyclic curves on affine toric surfaces, similarly as this is done for plane acyclic curves in theorems 1.1 and 1.3.

3.1. *Acyclic curves in the smooth locus*

We start with the curves that do not pass through the singular point.

Let d and e be coprime integers with $0 < e < d$, and let $\zeta \in \mathbb{C}^\times$ be a primitive root of unity of degree d. Consider an affine toric surface $X_{d,e} = \mathbb{A}^2/G_{d,e}$, where $G_{d,e} = \langle g \rangle$ is the cyclic group generated by an element $g = \begin{pmatrix} \zeta^e & 0 \\ 0 & \zeta \end{pmatrix} \in \mathrm{GL}(2, \mathbb{C})$. Let $Q = \pi(\bar{0})$ denote the unique singular point of $X_{d,e}$, where $\pi\colon \mathbb{A}^2 \to X_{d,e}$ is the quotient morphism and $\bar{0} = (0,0) \in \mathbb{A}^2$. We let $N(G_{d,e})$ denote the normalizer of the subgroup $G_{d,e}$ in $\mathrm{GL}(2, \mathbb{C})$ and by $\mathcal{N}(G_{d,e})$ that in the group $\mathrm{Aut}(\mathbb{A}^2)$.

Remark 3.1. By [17, Section 2.6], the surfaces $X_{d,e}$ and $X_{d',e'}$ are isomorphic if and only if $d = d'$ and either $e = e'$ or $ee' = 1 \mod d$. The latter two possibilities are related via the twist $\tau : (x, y) \mapsto (y, x)$ on $\mathbb{A}^2$.

Remark 3.2. Any affine toric variety X over $\mathbb{C}$ which does not have non-constant invertible functions admits a canonical realization as a quotient $X = \mathbb{A}^m /\!/ D$ for a linear action of a diagonalizable affine algebraic group D on $\mathbb{A}^m$, see [14]. Here m is the number of invariant prime divisors on X, the group of characters of D can be identified with the divisor class group of X, and the polynomial ring $\mathbb{C}[T_1, \ldots, T_m] = \mathcal{O}(\mathbb{A}^m)$ with the homogeneous coordinate ring, or the Cox ring of X. By [5, Theorem 5.1], any automorphism of X can be lifted to an automorphism of $\mathbb{A}^m$ which normalizes the D-action.

For a linear form l on $\mathbb{A}^2$ and for $c \in \mathbb{C}$ we let $C_l(c)$ denote the affine line $l = c$.

Theorem 3.3.

(a) *Up to the action of the automorphism group* $\mathrm{Aut}(X_{d,e})$*, any irreducible acyclic curve* $\mathcal{C}$ *on* $X_{d,e}$ *which does not pass through the singular point* $Q \in X_{d,e}$ *is equivalent either to* $\pi(C_x(1))$ *or to* $\pi(C_y(1))$*. In particular,* $\mathcal{C}$ *coincides with an orbit of a* $\mathbb{G}_a$*-action on* $X_{d,e}$*, and also with an orbit closure of a* $\mathbb{G}_m$*-action on* $X_{d,e}$*.*

(b) *The curves* $\pi(C_x(1))$ *and* $\pi(C_y(1))$ *are equivalent on* $X_{d,e}$ *if and only if* $e^2 \equiv 1 \mod d$*.*

Proof. To show (a) we let $\mathcal{C}$ be an irreducible acyclic curve on $X_{d,e}$ not passing through the singular point Q and $C = \pi^{-1}(\mathcal{C})$ be its total preimage in $\mathbb{A}^2$. The morphism π is finite of degree d, so any component C^i of C maps to $\mathcal{C}$ properly. Since the cyclic group $G_{d,e}$ acts freely on $\mathbb{A}^2 \setminus \{\bar{0}\}$ and $C \subseteq \mathbb{A}^2 \setminus \{\bar{0}\}$, the map $\pi|_{C^i} : C^i \to \mathcal{C}$ is a non-ramified covering. However, $\mathcal{C}$ being simply connected it does not admit any non-trivial covering. Therefore C has d disjoint irreducible components $C^1, \ldots, C^d$ mapped isomorphically onto $\mathcal{C}$ under π. Furthermore, the cyclic group $G_{d,e}$ acts simply transitively on the set $\{C^1, \ldots, C^d\}$.

Write a reduced defining equation of C^1 as $p - 1 = 0$, where $p \in \mathbb{C}[x, y]$. Any regular invertible function on a connected and simply connected variety is constant. Hence for every $i = 1, \ldots, d$ the restriction of $p|C^i$ is constant, say, κ_i i.e., $C^i \subseteq p^{-1}(\kappa_i)$. If C^1 were singular then by theorem 1.3 it would be equivalent to a curve $C_{a,b} = \{y^a - x^b = 0\}$, where $\min\{a, b\} > 1$. For $c \neq 0$ the Euler characteristic of the fiber $y^a - x^b = c$ is negative. Hence this fiber cannot carry a curve with Euler characteristic 1. This leads to a contradiction, because $d > 1$ by our assumption. Thus the curve C^1 is smooth.

It follows that every fiber of p is isomorphic to $\mathbb{A}^1$. Hence there is an automorphism $\delta \in \mathrm{Aut}(\mathbb{A}^2)$ sending the curves C^i to the lines $y = \kappa_i$ with distinct k_i, where $\kappa_1 = 1$. Moreover, we may suppose that $\delta(\bar{0}) = \bar{0}$. Letting $g' = \delta \circ g \circ \delta^{-1}$ we obtain $g'(\bar{0}) = \bar{0}$, $g'^d = g^d = \mathrm{id}$, and $g'(C_y) = C_y$ i.e., $g' \in \mathrm{Stab}(C_y) = \mathrm{Jonq}^+(\mathbb{A}^2)$, see proposition 2.8. Furthermore,

$$\rho(g') = dg'(\bar{0}) = d\delta(\bar{0}) \circ g \circ (d\delta(\bar{0}))^{-1} \in \mathbb{T},$$

where $\rho : \mathrm{Jonq}^+(\mathbb{A}^2) \to \mathbb{T}$ is the canonical surjection (see proposition 2.5). Hence the elements $g = \rho(g) \in \mathbb{T}$ and $\rho(g') \in \mathbb{T}$ are conjugated in $\mathrm{GL}(2, \mathbb{C})$ and so either $\rho(g') = \rho(g) = g$ or $\rho(g') = \tau g \tau$.

Since $\mathrm{ord}(g') = d > 1$ we have $g' \notin U^+$. It follows from lemma 2.1 and proposition 2.5 that $\mu^{-1} g' \mu = \rho(g')$ for a suitable $\mu \in U^+$. In the case where

$\rho(g') = g$ we obtain $\mu^{-1}\delta g\delta^{-1}\mu = g$, and in the case where $\rho(g') = \tau g\tau$ we get $\tau\mu^{-1}\delta g\delta^{-1}\mu\tau = g$. In the former case $\mu^{-1}\delta(C^1) = C_y(1)$, where $\mu^{-1}\delta \in \mathcal{N}(G_{d,e})$, while in the latter one $\tau\mu^{-1}\delta(C^1) = C_x(1)$, where again $\tau\mu^{-1}\delta \in \mathcal{N}(G_{d,e})$. Since the corresponding element normalizes the group $G_{d,\,e}$ it descends to an automorphism of the surface $X_{d,\,e}$ which sends $\mathcal{C}$ to the curve $\pi(C_y(1))$ in the former case and to $\pi(C_x(1))$ in the latter one.

Furthermore, $\mathcal{C}$ is an orbit of a $\mathbb{G}_a$-action on $X_{d,\,e}$ induced by a $\mathbb{G}_a$-action on $\mathbb{A}^2$ with C^1 as an orbit, which commutes with the $G_{d,\,e}$-action and is defined via

$$t.(x,y) = (x, y + tx^{e'}) \quad \text{if } \mathcal{C} = \pi(C_x(1)) \quad \text{and}$$

$$t.(x,y) = (x + ty^{e}, y) \quad \text{if } \mathcal{C} = \pi(C_y(1))\,,$$

respectively, where $t \in \mathbb{G}_a$. This shows (a).

To show (b) we assume first that $e^2 \equiv 1 \mod d$ (and so $e' = e$). Then the involution $\tau : (x,y) \mapsto (y,x)$ normalizes the subgroup $G_{d,\,e}$ and induces an automorphism of $X_{d,\,e}$ which interchanges the curves $\pi(C_x(1))$ and $\pi(C_y(1))$ (and also $\pi(C_x)$ and $\pi(C_y)$).

Conversely, assume that there is an automorphism of $X_{d,\,e}$ which sends $\pi(C_x(1))$ to $\pi(C_y(1))$. Due to remark 3.2 [d] this automorphism can be lifted to an automorphism of the Cox ring $\mathbb{C}[x,y]$ of the surface $X_{d,\,e}$. Hence there is an element $\gamma \in \mathcal{N}(G_{d,\,e})$ which sends $C_x(1)$ to $C_y(1)$ and, moreover, sends the variable y to x. By proposition 2.8(a) γ has the form $(x,y) \longmapsto (\alpha y + f(x), x)$. A direct computation shows that

$$\gamma^{-1} \circ g \circ \gamma : (x,y) \longmapsto (\zeta x + h(y), \zeta^e y)\,,$$

where $\gamma^{-1} \circ g \circ \gamma \in G_{d,\,e}$ because $\gamma \in \mathcal{N}(G_{d,\,e})$. It follows that $h = 0$ and $(\zeta^e)^e = \zeta$ and so $e^2 \equiv 1 \mod d$. Now the proof is completed. □

Corollary 3.4. *Every irreducible, acyclic curve $\mathcal{C}$ on $X_{d,\,e}$ not passing through the singular point $Q \in X_{d,\,e}$ belongs to a pencil $\mathcal{L}$ consisting of one-dimensional orbits of an effective $\mathbb{G}_a$-action on $X_{d,\,e}$ and a fixed point curve $\mathcal{C}_0$ passing through Q. The members of $\mathcal{L}$ different from $\mathcal{C}_0$ are equivalent under the $\mathbb{G}_m$-action on $X_{d,\,e}$ induced by the $\mathbb{T}_{1,1}$-action on $\mathbb{A}^2$.*

The union of several one-dimensional orbits of a $\mathbb{G}_a$-action on $X_{d,\,e}$ is a disconnected, simply connected curve. In fact every such curve arises in this way, as the reader can easily derive from the previous results.

[d] See also the proof of theorem 5.3 below for an alternative argument.

Corollary 3.5. *Every disconnected, simply connected curve $\mathcal{C}$ on $X_{d,e}$ is equivalent under the* $\mathrm{Aut}(X_{d,e})$*-action either to* $\bigcup_{i=1}^{r}\pi(C_x(\kappa_i))$*, or to* $\bigcup_{i=1}^{r}\pi(C_y(\kappa_i))$*, where $r \geq 2$ and $\kappa_1,\dots,\kappa_r \in \mathbb{A}^1$ are distinct.*

3.2. *Acyclic curves through the singular point*

In this subsection we describe the acyclic curves on a singular affine toric surface $X_{d,e}$ ($d > 1$) passing through the singular point Q. We let as before $C_{a,b} = \{y^a - x^b = 0\}$, where $a, b \geq 1$ and $\gcd(a,b) = 1$. In particular $C_{1,1} = \{x - y = 0\}$. We keep the notation $C_x = \{x = 0\}$ and $C_y = \{y = 0\}$.

Due to the following theorem, the set of all equivalence classes of irreducible, acyclic curves on $X_{d,e}$ through Q is countable. A similar fact in the smooth case $X_{1,1} = \mathbb{A}^2$ is well known, see theorem 1.3(a).

Theorem 3.6.

(a) *Up to the action of the automorphism group* $\mathrm{Aut}(X_{d,e})$*, every irreducible acyclic curve $\mathcal{C}$ on $X_{d,e}$ passing through the singular point $Q \in X_{d,e}$ is equivalent to one of the curves $\pi(C_{a,b})$, $\pi(C_x)$, or $\pi(C_y)$.*

(b) *The curves $\pi(C_x)$ and $\pi(C_y)$ are equivalent on $X_{d,e}$ if and only if $e^2 \equiv 1 \mod d$, if and only if the twist $\tau : (x,y) \longmapsto (y,x)$ descends to an automorphism of the surface $X_{d,e}$.*

Proof. (a) The curve $D = \pi^*(\mathcal{C}) \setminus \{\bar{0}\}$ is reduced and the projection $D \to \mathcal{C} \setminus \{Q\}$ is an unramified cyclic covering of degree d. It follows that the irreducible components $D^1,\dots,D^r$ of D are disjoint, and for every $i = 1,\dots,r$ the restriction $\pi|_{D^i} : D^i \to \mathcal{C} \setminus \{Q\}$ is an unramified cyclic covering. Furthermore, the cyclic group $G_{d,e}$ acts transitively on the set of these components i.e., the generator g of $G_{d,e}$ permutes them cyclically. In particular, r divides d and $D^i \simeq \mathbb{A}^1_* \ \forall i$, while the closures $C^i = D^i \cup \{\bar{0}\}$ and $C = D \cup \{\bar{0}\} = \pi^{-1}(\mathcal{C})$ are acyclic. Clearly, the curve C cannot have more than one singular point, hence it cannot be of type (III). According to the remaining types we distinguish the following cases.

Case 1: $r = 1$ and $C = C^1$ is a smooth acyclic curve of type (I). Then C^1 is $G_{d,e}$-stable and passes through the origin. Similarly as in the proof of theorem 3.3, one can show that in suitable new coordinates in $\mathbb{A}^2$ we have $C^1 = C_y$ and g acts diagonally either via $(x,y) \mapsto (\zeta^e x, \zeta y)$, or via $(x,y) \mapsto (\zeta^{e'} x, \zeta y)$. In the second case after transposition $(x,y) \mapsto (y,x)$ we obtain that $C^1 = C_x$ and g acts via $(x,y) \mapsto (\zeta x, \zeta^{e'} y)$, that is by an element of the cyclic group $G_{d,e}$. In any case up to the $\mathrm{Aut}(X_{d,\,e})$-action, the curve $\mathcal{C}$ is equivalent either to $\pi(C_x)$ or to $\pi(C_y)$.

Case 2: $r>1$ and C is an acyclic curve of type (II) or (IV) with an ordinary singularity at the origin. By theorem 1.3 a suitable automorphism $\gamma\in\mathrm{Aut}(\mathbb{A}^2)$ sends the reduced curve $C=\pi^*(\mathcal{C})$ to a union $C'=C'^1+\ldots+C'^r$ of affine lines through the origin given by equation

$$y(y-\kappa_2 x)\ldots(y-\kappa_r x)=0,\quad \text{where}\quad \kappa_i\in\mathbb{C}^\times\quad \text{are distinct}\,. \tag{16}$$

The curve C' is stable under the action on $\mathbb{A}^2$ of the cyclic group $\gamma G_{d,e}\gamma^{-1}=\langle g'\rangle$, where $g'=\gamma g\gamma^{-1}\in\mathrm{Aut}(\mathbb{A}^2)$. By corollary 2.9 $\mathrm{Stab}(C')\subseteq\mathrm{GL}(2,\mathbb{C})$, hence $g'\in\mathrm{GL}(2,\mathbb{C})$.

There exists an element $\delta\in\mathrm{GL}(2,\mathbb{C})$ such that $\delta g'\delta^{-1}=g$ is diagonal and acts via $(x,y)\mapsto(\zeta^e x,\zeta y)$. Since no component C^i of C is stable under g, the composition $\delta\gamma$ sends each C^i to a line through the origin different from a coordinate axis. Since all such lines are $\mathbb{T}$-equivalent, their images in the surface $X_{d,e}$ are also equivalent under the action on $X_{d,e}$ of the quotient torus $\mathbb{T}'=\mathbb{T}/G_{d,e}$.

The resulting automorphism $\delta\gamma$ from the centralizer of the subgroup $G_{d,e}$ in $\mathrm{Aut}(\mathbb{A}^2)$ rectifies C and sends C^1 to a line $\mathbb{T}$-equivalent to $C_{1,1}$. Consequently, the curve $\mathcal{C}$ on $X_{d,e}$ is equivalent to $\pi(C_{1,\,1})$ under the $\mathbb{T}'$-action on $X_{d,e}$ and the automorphism $\pi_*(\delta\gamma)\in\mathrm{Aut}(X_{d,e})$.

Case 3: $r>1$ and C is an acyclic curve of type (V) with smooth components C^i and a non-ordinary singularity at the origin. By theorem 1.3 in this case a suitable automorphism $\gamma\in\mathrm{Aut}(\mathbb{A}^2)$ sends the reduced curve $C=C^1+\ldots+C^r$ on $\mathbb{A}^2$ to a curve C' given by equation

$$x^{\varepsilon_x}y(y-\kappa_2x^b)\ldots(y-\kappa_rx^b)=0\quad\text{with distinct}\quad \kappa_i\in\mathbb{C}^\times\,, \tag{17}$$

where $b>1$ and $\varepsilon_x\in\{0,1\}$. The group $G'_{d,e}=\gamma G_{d,e}\gamma^{-1}\subseteq\mathrm{Aut}(\mathbb{A}^2)$ acts transitively on the set of components C'^i of C'. Since the $G'_{d,e}$-action preserves tangency we have $\varepsilon_x=0$. We may also suppose that $k_2=1$. Since $G'_{d,e}\subseteq\mathrm{Stab}(C')$, the singular point $\bar 0\in C'$ is fixed under the $G'_{d,e}$-action on $\mathbb{A}^2$.

There is an element $g'\in G'_{d,e}$ which sends the component $C'^2=\{y-x^b=0\}$ to $C'^1=C_y$. Since $h:(x,y)\mapsto(x,y-x^b)$ does the same, according to proposition 2.8(a) g' can be written as

$$g':(x,y)\longmapsto(\alpha x+f(y-x^b),\,\beta(y-x^b))$$

for some $\alpha,\beta\in\mathbb{C}^\times$ and $f\in\mathbb{C}[z]$. Hence g' maps the affine line $C'^1=C_y$ to the parameterized curve $\{(\alpha t+f(-t^b),-\beta t^b)\,|\,t\in\mathbb{A}^1\}$. However

$g'(C'^1) = C'^i$ for some $i \in \{2, \ldots, r\}$. It follows that $f = 0$. Therefore $g' : (x, y) \mapsto (\alpha x, \beta(y - x^b))$ and so

$$g'^k : (x, y) \mapsto \left(\alpha^k x, \beta^k y - \beta x^b(\beta^{k-1} + \alpha^b\beta^{k-2} + \ldots + \alpha^{b(k-1)})\right).$$

Since $g'^k = \mathrm{id}$ for some $k|d$, we have $\alpha^k = \beta^k = 1$ and $(\beta/\alpha^b)^{k-1} + \ldots + 1 = 0$. This implies that $\beta \neq \alpha^b$.

Thus the triangular automorphism

$$h' : (x, y) \mapsto \left(x, y + \frac{\beta}{\alpha^b - \beta} x^b\right)$$

is well defined and sends the curve C' to a new one $C'' = \sum_i C''^i$ given by a similar equation. Furthermore, h' conjugates g' with

$$g'' : (x, y) \longmapsto (\alpha x, \beta y),$$

where (α, β) can be written either as (ξ^e, ξ) or as $(\xi^{e'}, \xi)$ for a primitive dth root of unity ξ (recall that $ee' \equiv 1 \mod d$). In the latter case we apply additionally the transposition of coordinates to get $(\alpha, \beta) = (\xi, \xi^{e'})$. The composition $h'\gamma$ normalizes the group $G_{d,e}$ in $\mathrm{Aut}(\mathbb{A}^2)$ and sends C^1 to a member of the pencil $\mathcal{L}$, where as before

$$\mathcal{L} = \{y - \kappa x^b = 0\}_{\kappa \in \mathbb{C}} \cup \{bC_x\}.$$

However, every member of $\mathcal{L}$ different from the coordinate axes C_x, C_y is equivalent to the curve $C_{1,b}$ under the $\mathbb{T}$-action on $\mathbb{A}^2$. Finally $h'\gamma$ induces an automorphism θ of the quotient surface $X_{d,e}$ such that the image under θ of the curve $\mathcal{C} = \pi(C^1)$ is $\mathbb{T}'$-equivalent either to $\pi(C_y)$ or to $\pi(C_{1,b})$, as required.

Case 4: C is an acyclic curve of type (VI) with all the components C^i being singular. By theorem 1.3 a suitable automorphism $\gamma \in \mathrm{Aut}(\mathbb{A}^2)$ sends C to a curve C' given by equation (2), where $\varepsilon_x = \varepsilon_y = 0$, $a, b > 1$, and $\gcd(a, b) = 1$. We may also assume that $\kappa_1 = 1$ and so $C'^1 = C_{a,b}$. Let as before $g' = \gamma g \gamma^{-1}$. Since the $\mathbb{T}$-action is transitive on the members of the pencil

$$\mathcal{L} = \{y^a - \kappa x^b = 0\}_{\kappa \in \mathbb{C}} \cup \{bC_x\}$$

different from the coordinate axes C_x and C_y, there is an element $h \in \mathbb{T}$ such that $g'' = h \circ g'$ preserves the curve $C_{a,b}$. By proposition 2.12 $g'' \in \mathbb{T}$, hence also $g' \in \mathbb{T}$. Since the eigenvalues ζ^e and ζ of g and g' are the same, up to interchanging the coordinates we have $g' \in G_{d,e}$. Hence γ normalizes the group $G_{d,e}$. Reasoning as before we conclude that the curve $\mathcal{C}$ on the surface $X_{d,e}$ is equivalent to $\pi(C_{a,b})$, as stated. This shows (a). The proof of (b) goes in the same way as that of theorem 3.3(b) and so we leave it to the reader. □

Remarks 3.7. 1. If $e > 1$ then the curve $\pi^{-1}(\pi(C_x)) = C_x$ on $\mathbb{A}^2$ is irreducible, while $\pi^{-1}(\pi(C_{1,1}))$ is reducible. Hence these plane curves are not equivalent and so their images $\pi(C_x)$ and $\pi(C_{1,1})$ on the surface $X_{d,e}$ are not equivalent either (under the $\mathrm{Aut}(X_{d,e})$-action).

2. The following simple example shows that, in contrast with theorem 3.3, a curve on $X_{d,e}$ isomorphic to the affine line and passing through the singular point can be non-equivalent to the image of one of the curves $C_{1,1}$, C_x, and C_y.

Indeed, the curve $C_{2,3}$ in $\mathbb{A}^2$ is singular. Nonetheless, its image $\mathcal{C}$ in the affine toric surface $X_{5,4}$ passes through the singular point and is isomorphic to the affine line $\mathbb{A}^1$. Clearly, $\mathcal{C}$ is not equivalent in $X_{5,4}$ to $\pi(C_{1,1})$, $\pi(C_x)$, or $\pi(C_y)$.

3.3. *Acyclic curves as orbit closures*

Summarizing the results of the previous subsections we arrive at the following alternative description.

Theorem 3.8. *Let X be an affine toric surface over $\mathbb{C}$ with the acting torus T. Then every irreducible acyclic curve $\mathcal{C}$ on X coincides with the closure of a non-closed orbit of a regular $\mathbb{G}_m$-action on X.*[e] *Furthermore, up to an automorphism of X, such a curve $\mathcal{C}$ is the closure of a non-closed orbit of a subtorus of T.*

Proof. Let $\mathcal{C}$ be an irreducible acyclic curve on a toric surface X. If X is smooth then this is one of the surfaces $\mathbb{A}^1_* \times \mathbb{A}^1_*$, $\mathbb{A}^1 \times \mathbb{A}^1_*$, or $\mathbb{A}^2$. There is no acyclic curve on $X = \mathbb{A}^1_* \times \mathbb{A}^1_*$, and the only irreducible acyclic curves on $\mathbb{A}^1 \times \mathbb{A}^1_*$ are of the form $\mathbb{A}^1 \times \{\mathrm{pt}\}$. Hence our assertion holds for these surfaces. In the case of the affine plane $X \simeq \mathbb{A}^2$ the result follows from theorems 1.1 and 1.3.

By [17, §2.2], every singular affine toric surface is isomorphic to one of the surfaces $X_{d,e} = \mathbb{A}^2/G_{d,e}$, where $d > 1$. Finally in this case the result follows from theorems 3.3 and 3.6. □

[e] Clearly, such an orbit closure is acyclic.

Corollary 3.9. *Any irreducible acyclic curve on an affine toric surface has at most one singular point. If the surface is singular then this point coincides with the singular point of the surface.*

The following result generalizes theorem 1.1 of Abhyankar, Moh, and Suzuki.

Theorem 3.10. *Up to the action of the group* $\mathrm{Aut}(X_{d,e})$ *there are only finitely many different embeddings* $\mathbb{A}^1 \hookrightarrow X_{d,e}$.

Proof. It suffices to show that the smooth curves on $X_{d,\,e}$ of the form $\pi(C_{a,\,b})$ belong to a finite set of equivalence classes. Notice that the subalgebra $\mathbb{C}[x,y]^{G_{d,\,e}}$ of $G_{d,\,e}$-invariants is generated by the monomials $y^d, xy^{c_1}, \ldots, x^{d-1}y^{c_{d-1}}, x^d$, where $0 < c_k < d$ and $c_k + ke \equiv 0 \mod d$. These monomials define a closed embedding $X_{d,\,e} \hookrightarrow \mathbb{A}^{d+1}$. The image of the curve $\pi(C_{a,\,b})$ under this embedding is

$$(t^{db}, t^{a+c_1 b}, \ldots, t^{(d-1)a+c_{d-1}b}, t^{da}), \quad t \in \mathbb{C}\,.$$

This image is smooth if and only if one of the exponents, say, δ of our monomials coincides with the greatest common divisor of all the exponents. Since a and b are coprime $\delta | d$.

In the case where $\delta = ka + c_k b$ for some $k \geq 1$ we obtain $\delta = ka + c_k b \leq d$ and so $a + b \leq d$. The number of all possible such pairs (a, b) is finite.

If $\delta \in \{da, db\}$ then $a = 1$ or $b = 1$ because $\gcd(a, b) = 1$. Suppose for instance that $a = 1$ and $\delta = d$, the other case being similar. Thus

$$ek + c_k = 0 \mod d \quad \text{and} \quad k + bc_k \equiv 0 \mod d$$

for all possible values of $k \geq 1$. For $k = 1$ it follows that $eb \equiv 1 \mod d$. Hence the curve $C_{a,b} = C_{1,b}$ is stable under the $G_{d,e}$-action on $\mathbb{A}^2$. The automorphism

$$(x, y) \longmapsto (x, y - x^b)$$

commutes with the $G_{d,e}$-action and sends this curve to the axis C_y. Therefore for any $b \geq 1$ the curves $\pi(C_{1,b})$ and $\pi(C_y)$ on $X_{d,e}$ are equivalent. Now the proof is completed. □

Remark 3.11. Consider an affine toric variety X of dimension n. It is known (see [8]) that if an $(n-1)$-dimensional torus T' acts effectively on X, then T' is conjugate in the group $\mathrm{Aut}(X)$ to a subtorus of the acting torus T of X. An analogous result for tori of codimension ≥ 2 is unknown. We conclude section 3 by the following related problem.

Questions. Let X be an affine toric variety of dimension n with acting torus T. Assume that a closed subvariety Y of X admits in turn an action of an algebraic torus T' with an open orbit.

Is it true that Y can be realized as the orbit closure of a T'-action on X? Is, moreover, this T'-action on X conjugated to the action on X of a subtorus of the acting torus T?

This is indeed the case for $n = 2$ as follows from theorem 3.8.

3.4. *Reducible acyclic curves on affine toric surfaces*

Let us start with the following lemma.

Lemma 3.12. *Consider the affine toric surface $X = X_{d,e}$ with the quotient map $\pi : \mathbb{A}^2 \to X = \mathbb{A}^2/G_{d,e}$. If $\mathcal{C}$ is a reduced, simply connected curve on X, then the total transform $C = \pi^*(\mathcal{C})$ of $\mathcal{C}$ in $\mathbb{A}^2$ is also reduced and simply connected.*

Proof. If $d = 1$ i.e., $X \simeq \mathbb{A}^2$, or the curve $\mathcal{C}$ is irreducible, then the assertion follows by the same argument as in the proof of theorems 3.3 and 3.6. Assume further that $d > 1$ and $\mathcal{C}$ is reducible. Letting $Q = \pi(\bar{0}) \in X$ and $\pi^*(\mathcal{C}) = C = C^1 + \ldots + C^s$, where every irreducible component C^i of C is simply connected, we consider the following cases.

Case 1: Every irreducible component $\mathcal{C}^k$ of $\mathcal{C}$ passes through Q. Then any two such components meet only at Q, and it is easily seen that any two distinct components C^i and C^j of C also meet only at the origin. In this case C is connected and simply connected i.e., acyclic.

Case 2: There are two distinct intersecting components, say, $\mathcal{C}^k$ and $\mathcal{C}^l$ of $\mathcal{C}$ not passing through Q. According to theorem 3.3 (cf. also Corollaries 3.4 and 3.5), under the action of the normalizer $\mathcal{N}(G_{d,e})$ on $\mathbb{A}^2$ the total transform $\pi^*(\mathcal{C}^k)$ (or $\pi^*(\mathcal{C}^l)$) is equivalent to a union of d parallel lines, which are parallel either to C_x or to C_y and are cyclically permuted under the $G_{d,e}$-action. Furthermore, each component of $\pi^*(\mathcal{C}^k)$ meets every component of $\pi^*(\mathcal{C}^l)$ in d distinct points. All these d^2 intersection points must project to the unique intersection point $\mathcal{C}^k \cap \mathcal{C}^l$. Hence they should belong to a $G_{d,e}$-orbit, which is impossible. This contradiction shows that the components of $\mathcal{C}$ not passing through Q do not meet.

Case 3: There is just one component, say, $\mathcal{C}^1$ of $\mathcal{C}$ not passing through Q. The union of the other components of $\mathcal{C}$ can meet $\mathcal{C}^1$ in at most one point, and they meet each other at Q. Hence either $\mathcal{C}^1$ does not meet this union, or there is just one component, say, $\mathcal{C}^2$ of $\mathcal{C}$ passing through Q which meets $\mathcal{C}^1$. In the former case the reduced curve $C = \pi^*(\mathcal{C})$ is clearly simply connected, as stated. In the latter case we may assume as before that $\pi^*(\mathcal{C}^1)$ is a union of d lines parallel to a coordinate axis and cyclically permuted under the $\mathbb{G}_{d,e}$-action. Every component of the total transform $\pi^*(\mathcal{C}^2)$ is simply connected, passes through the origin, and meets one of these lines. Hence it meets all the parallel lines. If the curve $\pi^*(\mathcal{C}^2)$ is reducible then its irreducible components meet only at the origin, and meet one of the lines in $\pi^*(\mathcal{C}^1)$ in at least two distinct points. These points project in X to distinct smooth points. The latter contradicts the assumption of simply connectedness of the curve $\mathcal{C}$, because in this case we obtain a non-contractible cycle in $\mathcal{C}$. Hence the curve $\pi^*(\mathcal{C}^2)$ is irreducible and meets every line in $\pi^*(\mathcal{C}^1)$ in just one point, while the other curves $\pi^*(\mathcal{C}^k)$, $k \geq 3$, do not meet these lines at all. This shows that C is simply connected.

Case 4: There are two disjoint components, say, $\mathcal{C}^1$ and $\mathcal{C}^2$ of $\mathcal{C}$ not passing through Q. Since the total transforms $\pi^*(\mathcal{C}^1)$ and $\pi^*(\mathcal{C}^2)$ are disjoint, they can be simultaneously transformed into unions of lines parallel to the same coordinate axis. Hence every component, say, $\mathcal{C}^3$ of $\mathcal{C}$ passing through Q and meeting $\mathcal{C}^1$ meets also $\mathcal{C}^2$, and vice versa. Using the same argument as before it is easily seen that there could be at most one such component $\mathcal{C}^3$, and the total preimage $C = \pi^*(\mathcal{C})$ is simply connected, as required.

This ends the proof. □

Using this lemma, in the following theorem we give a description of all reduced, simply connected curves on affine toric surfaces.

Theorem 3.13. *Every reduced, acyclic curve $\mathcal{C}$ on $X = X_{d,e}$ is equivalent to a curve $\pi(C)$, where $C \subseteq \mathbb{A}^2$ is given by one of the equations*

$$(18) \qquad x\prod_{i=1}^{r}(y^a - \kappa_i x^b) = 0 \quad \textit{or} \quad y\prod_{i=1}^{r}(x^a - \kappa_i y^b) = 0, \quad \textit{where}$$

$$b \geq 0,\ a \geq 1, \quad \gcd(a,b) = 1\,,$$

and where $\kappa_i \in \mathbb{C}$ $(i = 1, \ldots, r)$ are pairwise distinct.

Proof. Indeed, by lemma 3.12 the reduced plane curve $C = \pi^*(\mathcal{C})$ is simply connected. However, by theorem 1.3 every reduced, acyclic curve in $\mathbb{A}^2$ is given in appropriate coordinates by equation (18), while every reduced, disconnected, simply connected affine plane curve is equivalent to a finite union of parallel lines. It remains to show that, up to a permutation of the symbols x and y, the corresponding coordinate change can be chosen in the normalizer $\mathcal{N}(G_{d,e})$ of the group $G_{d,e}$ in $\mathrm{Aut}(\mathbb{A}^2)$. The latter can be done in the same way as in the proof of theorems 3.3 and 3.6. We leave the details to the reader. □

4. Automorphism groups of affine toric surfaces

In this section we prove an analog of the Jung-van der Kulk theorem 1.4 for affine toric surfaces and study algebraic groups acting on such surfaces.

4.1. *Free amalgamated product structure*

Consider again an affine toric surface $X_{d,e} = \mathbb{A}^2/G_{d,e}$. We assume as usual that $1 \le e < d$, $\gcd(d,e) = 1$, and $G_{d,e} = \langle g \rangle$, where

$$g : (x,y) \longmapsto (\zeta^e x, \zeta y) \quad \text{with} \quad \zeta = \exp\left(\frac{2\pi i}{d}\right).$$

Notation 4.1. Let $G \subseteq \mathrm{GL}(2,\mathbb{C})$. Letting as before $N(G)$ denote the normalizer of G in the group $\mathrm{GL}(2,\mathbb{C})$ and $\mathcal{N}(G)$ that in the group $\mathrm{Aut}(\mathbb{A}^2)$, we abbreviate

$$N_{d,e} = N(G_{d,e}) \text{ and } \mathcal{N}_{d,e} = \mathcal{N}(G_{d,e}).$$

It is easily seen that

$$N_{d,e} = \begin{cases} \mathrm{GL}(2,\mathbb{C}) & \text{if } e = 1, \\ N(\mathbb{T}) = \langle \mathbb{T}, \tau \rangle & \text{if } e > 1 \text{ and } e^2 \equiv 1 \mod d, \\ \mathbb{T} & \text{otherwise,} \end{cases}$$

where $\tau : (x,y) \longmapsto (y,x)$ is a twist and $\mathbb{T}$ stands for the maximal torus in $\mathrm{GL}(2,\mathbb{C})$ consisting of the diagonal matrices. We let $B^{\pm}$ denote the Borel subgroup of all upper (lower, respectively) triangular matrices in $\mathrm{GL}(2,\mathbb{C})$. Consider the subgroups

$$N^{\pm}_{d,e} = N_{d,e} \cap \mathrm{Jonq}^{\pm}(\mathbb{A}^2) = \begin{cases} B^{\pm} & \text{if } e = 1, \\ \mathbb{T} & \text{otherwise} \end{cases} \tag{19}$$

and

$$\mathcal{N}^{\pm}_{d,e} = \mathcal{N}_{d,e} \cap \mathrm{Jonq}^{\pm}(\mathbb{A}^2).$$

The latter subgroups are described in lemma 4.5 below. Notice that

$$\mathcal{N}^{+}_{d,e} \cap \mathcal{N}^{-}_{d,e} = \mathrm{Jonq}^{+}(\mathbb{A}^2) \cap \mathrm{Jonq}^{-}(\mathbb{A}^2) = \mathbb{T}. \tag{20}$$

With this notation we can state an analog of theorem 1.4 by Jung and van der Kulk.

Theorem 4.2. *If* $e^2 \not\equiv 1 \mod d$ *then*

$$\mathrm{Aut}(X_{d,e}) \simeq \mathcal{N}^+_{d,e}/G_{d,e} *_{\mathbb{T}/G_{d,e}} \mathcal{N}^-_{d,e}/G_{d,e}\,, \tag{21}$$

while for $e^2 \equiv 1 \mod d$ *we have*

$$\mathrm{Aut}(X_{d,e}) \simeq \mathcal{N}^+_{d,e}/G_{d,e} *_{\mathcal{N}^+_{d,e}/G_{d,e}} N_{d,e}/G_{d,e}\,. \tag{22}$$

There should be possible to derive this theorem by using the techniques elaborated by Danilov and Gizatullin [19]. However, we prefer a direct approach through an equivariant version 4.8 of the Abhyankar-Moh-Suzuki theorem. Within this approach theorem 4.2 is an immediate consequence of lemma 4.3 and proposition 4.4 below.

Lemma 4.3. *There is an isomorphism*

$$\mathrm{Aut}(X_{d,e}) \simeq \mathcal{N}_{d,e}/G_{d,e}\,.$$

Proof. The affine plane $\mathbb{A}^2$ can be viewed as the spectrum of a Cox ring of the toric surface $X_{d,e}$, see [14] or [3, I.6.1]. Hence every automorphism $\Phi \in \mathrm{Aut}(X_{d,e})$ can be lifted (in a non-unique way) to an element $\varphi \in \mathcal{N}_{d,e}$.[f] This yields an exact sequence (see [5, Thm. 5.1])

$$1 \to G_{d,e} \to \mathcal{N}_{d,e} \to \mathrm{Aut}(X_{d,e}) \to 1\,,$$

as claimed. □

Proposition 4.4. *If* $e^2 \not\equiv 1 \mod d$ *then*

$$\mathcal{N}_{d,e} \simeq \mathcal{N}^+_{d,e} *_{\mathbb{T}} \mathcal{N}^-_{d,e}\,, \tag{23}$$

while for $e^2 \equiv 1 \mod d$

$$\mathcal{N}_{d,e} \simeq \mathcal{N}^+_{d,e} *_{N^+_{d,e}} N_{d,e}\,. \tag{24}$$

The proof is done in lemmas 4.5–4.14 below.

Theorem 4.2 follows now from lemma 4.3 and proposition 4.4 due to the fact that the subgroup $G_{d,e}$ is normal in every group that participates in (23) and (24). Indeed, this can be seen directly or, alternatively, derived

[f] Alternatively, this follows from the monodromy theorem, see the proof of theorem 5.3 below.

as a simple consequence of a theorem by Cohen [12] on preservation of the free amalgamated product structure in the quotient.

Recall (see remark 2.2) that the polynomial ring $A = \mathbb{C}[t]$ possesses a $\mathbb{Z}/d\mathbb{Z}$-grading

$$A = \bigoplus_{i=0}^{d-1} A_{d,i}, \quad \text{where} \quad A_{d,i} = t^i\mathbb{C}[t^d]\,.$$

In terms of this grading the normalizer $\mathcal{N}^{\pm}_{d,e}$ admits the following description.

Lemma 4.5.

(a) *The group $\mathcal{N}^{+}_{d,e}$ ($\mathcal{N}^{-}_{d,e}$, respectively) consists of all de Jonqières transformations φ^+ as in (7) (φ^- as in (8), respectively) with $f \in A_{d,e}$ ($f \in A_{d,e'}$, respectively).*

(b) *The subgroup $\mathcal{N}^{\pm}_{d,e}$ is the centralizer of $G_{d,e}$ in the group* $\mathrm{Jonq}^{\pm}(\mathbb{A}^2)$.

Proof. We stick to the plus-case, the proof in the other one being similar. We have

$$\varphi^+ \circ g \circ (\varphi^+)^{-1} : (x,y) \longmapsto \left(\zeta^e x + f(\zeta\frac{y}{\beta}) - \zeta^e f(\frac{y}{\beta}), \zeta y\right)\,.$$

Hence $\varphi^+ \circ g \circ (\varphi^+)^{-1} \in G_{d,e}$ if and only if $f(\zeta t) = \zeta^e f(t)$, if and only if $f \in A_{d,e}$. This shows (a). In the latter case $\varphi^+ \circ g \circ (\varphi^+)^{-1} = g$, so (b) follows. □

For a pair of polynomials $\varphi = (u,v)$ we let $\deg\varphi = \max\{\deg u, \deg v\}$. The following result is an immediate consequence of lemma 4.5.

Lemma 4.6. *Assume as before that $1 \le e < d$ and $\gcd(d,e) = 1$. Then the following hold.*

(a) $\varphi(\bar{0}) = \bar{0}\ \forall \varphi \in \mathcal{N}^{\pm}_{d,e}$.

(b) $\mathcal{N}^{\pm}_{d,e} \cap \mathrm{Aff}(\mathbb{A}^2) = N^{\pm}_{d,e} = \begin{cases} B^{\pm}, & e = 1, \\ \mathbb{T}, & e > 1. \end{cases}$

(c) *Let $\varphi^{\pm}$ be as in (7) and (8), respectively. Assume that $\varphi^{\pm} \notin \mathbb{T}$. Then $\deg\varphi^+ \ge e$ and $\deg\varphi^- \ge e'$, where $1 \le e' < d$ and $ee' \equiv 1$* mod d.

The next two lemmas provide a $G_{d,e}$-equivariant version of the Abhyankar-Moh-Suzuki theorem 1.1.

Lemma 4.7. *Let C be a smooth, polynomial curve in $\mathbb{A}^2$ parameterized via $t \longmapsto (u(t), v(t))$, where $u, v \in t\mathbb{C}[t]$. If $G_{d,e} \subseteq \mathrm{Stab}(C)$ then either $(u, v) \in A_{d,e} \times A_{d,1}$ or $(u, v) \in A_{d,1} \times A_{d,e'}$.*

Proof. The tangent vector to C at the origin is $w = (u'(0), v'(0)) \in \mathbb{A}^2$. Since it is stable under the tangent $G_{d,e}$-action then either $g(w) = \zeta^e w$ or $g(w) = \zeta w$. Thus $g|C : C \to C$ acts either via $t \longmapsto \zeta^e t$ or via $t \longmapsto \zeta t$. In the former case

$$g \circ (u, v)(t) = (\zeta^e u(t), \zeta v(t)) = (u(\zeta^e t), v(\zeta^e t)),$$

and in the latter one

$$g \circ (u, v)(t) = (\zeta^e u(t), \zeta v(t)) = (u(\zeta t), v(\zeta t)).$$

Now the assertion follows. □

Lemma 4.8. *For a curve C as in lemma 4.7 there is an automorphism $\varphi \in \langle \mathcal{N}^+_{d,e}, \mathcal{N}^-_{d,e} \rangle$ which sends C to one of the coordinate axes C_x and C_y.*

Proof. If $u = 0$ or $v = 0$ there is nothing to prove. Thus we may suppose that $\deg u \geq \deg v > 0$. By the Abhyankar-Moh-Suzuki theorem 1.1 we have $\deg u = n \deg v$ for some $n \in \mathbb{N}$. By virtue of lemma 4.7 either $(\deg u, \deg v) \equiv (e, 1) \mod d$ or $(\deg u, \deg v) \equiv (1, e') \mod d$. In both cases it follows that $n \equiv e \mod d$ and so $\varphi_1 \in \mathcal{N}^+_{d,e}$, where $\varphi_1 : (x, y) \longmapsto (x - cy^n, y)$ (see lemma 4.5(a)). We have $\varphi_1(u, v) = (u_1, v_1) = (u - cv^n, v)$. So we can choose $c \in \mathbb{C}^\times$ in such a way that $\deg u_1 < \deg u$. We can continue this procedure recursively until we reach one of the pairs $(u_s, v_s) = (\alpha t, 0)$ or $(u_s, v_s) = (0, \beta t)$, where $\alpha, \beta \in \mathbb{C}^\times$. Then the product $\varphi = \varphi_s \circ \ldots \circ \varphi_1$ is a required automorphism. □

Lemma 4.9. *For any $\varphi \in \mathcal{N}_{d,e}$ we have $\varphi(\bar{0}) = \bar{0}$ and*

$$G_{d,e} \subseteq \mathrm{Stab}(\varphi(C_x)) \cap \mathrm{Stab}(\varphi(C_y)). \tag{25}$$

In particular,

$$\mathcal{N}_{d,e} \cap \mathrm{Aff}(\mathbb{A}^2) = N_{d,e}. \tag{26}$$

Proof. Since φ normalizes the subgroup $G_{d,e} = \langle g \rangle$ we have $\varphi^{-1} \circ g \circ \varphi = g^k$ for some $k \in \mathbb{N}$ and so $g \circ \varphi = \varphi \circ g^k$. Hence $g(\varphi(C_x)) = \varphi(g^k(C_x)) = \varphi(C_x)$ and, similarly, $g(\varphi(C_y)) = \varphi(C_y)$. This yields (25). From

$$g(\varphi(\bar{0})) = \varphi(g^k(\bar{0})) = \varphi(\bar{0})$$

we deduce that $\varphi(\bar{0}) \in (\mathbb{A}^2)^g = \{\bar{0}\}$ i.e. $\varphi(\bar{0}) = \bar{0}$. Now the last assertion follows easily. □

Lemma 4.10. *If* $e^2 \not\equiv 1 \mod d$ *then*

$$\mathcal{N}_{d,e} = \langle \mathcal{N}^+_{d,e}, \mathcal{N}^-_{d,e} \rangle \,, \tag{27}$$

while for $e^2 \equiv 1 \mod d$,

$$\mathcal{N}_{d,e} = \langle \mathcal{N}^+_{d,e}, N_{d,e} \rangle = \langle \mathcal{N}^+_{d,e}, \tau \rangle \,. \tag{28}$$

Proof. For $\varphi \in \mathcal{N}_{d,e}$ we let $C = \varphi^{-1}(C_y)$. By lemma 4.9 (see (25)) the cyclic group $G_{d,e}$ stabilizes C and $\bar{0} \in C$. By virtue of lemma 4.8 there is an automorphism $\psi \in \langle \mathcal{N}^+_{d,e}, \mathcal{N}^-_{d,e} \rangle$ which sends C to one of the coordinate axes C_x, C_y. Letting $\gamma = \psi \circ \varphi^{-1} \in \mathcal{N}_{d,e}$ we get $\gamma(C_y) = \psi(C) \in \{C_x, C_y\}$. If $\gamma(C_y) = C_x$ then the images of the coordinate axes $\pi(C_x)$ and $\pi(C_y)$ are equivalent in the surface $X_{d,e}$. According to theorem 3.6 in this case $e^2 \equiv 1 \mod d$ and $\tau \in N_{d,e}$.

Thus $\gamma(C_y) = C_y$ (i.e. $\gamma \in \mathrm{Stab}(C_y)$) if $e^2 \not\equiv 1 \mod d$. By proposition 2.8 in this case

$$\mathrm{Stab}(C_y) \cap \mathcal{N}_{d,e} = \mathrm{Jonq}^+(\mathbb{A}^2) \cap \mathcal{N}_{d,e} = \mathcal{N}^+_{d,e} \,.$$

Hence $\varphi \in \langle \mathcal{N}^+_{d,e}, \mathcal{N}^-_{d,e} \rangle$ and so (27) follows.

Assume further that $e^2 \equiv 1 \mod d$ and $\gamma(C_y) = C_x$. Then $\tau \circ \gamma(C_y) = C_y$ and so $\tau \circ \gamma \in \mathcal{N}^+_{d,e}$ and $\gamma \in \langle \mathcal{N}^+_{d,e}, \tau \rangle$. It follows that

$$\varphi = \gamma^{-1} \circ \psi \in \langle \mathcal{N}^+_{d,e}, \mathcal{N}^-_{d,e}, \tau \rangle = \langle \mathcal{N}^+_{d,e}, \tau \rangle = \langle \mathcal{N}^+_{d,e}, N_{d,e} \rangle \,.$$

Now the proof is completed. □

We need the following analog of lemma 4.1 in [28], see also [50, Theorem 5.3.1] and [52, Lemma 1.9]. For the reader's convenience we provide a short argument.

Lemma 4.11. *Let as before* $1 \le e < d$, *where* $\gcd(d, e) = 1$. *Consider an automorphism* $\varphi \in \mathrm{Aut}(\mathbb{A}^2)$ *with components* $u, v \in \mathbb{C}[x, y]$, *written as an alternating product*

$$\varphi = \varphi_s \cdot \ldots \cdot \varphi_1 \quad \textit{with} \quad \varphi_i \in \mathcal{N}^{\pm}_{d,e} \setminus N^{\pm}_{d,e} \quad \textit{and} \tag{29}$$

$$\varphi_{i+1} \in \mathcal{N}^{\mp}_{d,e} \setminus N^{\mp}_{d,e}, \quad i = 1, \ldots, s-1,$$

where $s \geq 1$. *Then*

$$\deg u > \deg v \quad \textit{if} \quad \varphi_s \in \mathcal{N}^{+}_{d,e} \quad \textit{and} \quad \deg u < \deg v \quad \textit{if} \quad \varphi_s \in \mathcal{N}^{-}_{d,e}.$$

In both cases

$$\deg \varphi := \max\{\deg u, \deg v\} = \prod_{i=1}^{s} \deg \varphi_i . \tag{30}$$

Proof. Both assertions are evidently true if $s = 1$. Letting $s > 1$ we assume by induction that they hold for the product $\psi = \varphi_{s-1} \cdot \ldots \cdot \varphi_1 = (\tilde{u}, \tilde{v})$. Thus

$$\deg \psi = \prod_{i=1}^{s-1} \deg \varphi_i , \tag{31}$$

$\deg \tilde{u} < \deg \tilde{v}$ if $\varphi_{s-1} \in \mathcal{N}^{-}_{d,e}$ (i.e. $\varphi_s \in \mathcal{N}^{+}_{d,e}$), and $\deg \tilde{u} > \deg \tilde{v}$ otherwise. In the former case the induction step goes as follows (the proof in the latter case is similar). By lemma 4.5(a) we can write φ_s as in (7), where $f \in A_{d,e}$ and $\deg f \geq 2$. Letting

$$\varphi : (x, y) \longmapsto (u, v) = (\alpha\tilde{u} + f(\tilde{v}), \beta\tilde{v})$$

from (31) we obtain

$$\deg u = \deg f(\tilde{v}) = \deg f \cdot \deg \tilde{v} = \deg \varphi_s \cdot \deg \psi$$

$$= \prod_{i=1}^{s} \deg \varphi_i \geq 2 \deg \tilde{v} > \deg \tilde{v} = \deg v .$$

This ends the proof. □

Now we can deduce the first part of proposition 4.4.

Lemma 4.12. *If* $e > 1$ *then*

$$\langle \mathcal{N}^{+}_{d,e}, \mathcal{N}^{-}_{d,e} \rangle \simeq \mathcal{N}^{+}_{d,e} *_{\mathbb{T}} \mathcal{N}^{-}_{d,e} . \tag{32}$$

In particular, if $e^2 \not\equiv 1 \mod d$ *then*

$$\mathcal{N}_{d,e} \simeq \mathcal{N}^{+}_{d,e} *_{\mathbb{T}} \mathcal{N}^{-}_{d,e} . \tag{33}$$

Proof. The second assertion follows from the first by virtue of lemma 4.10. To show the first one we recall that for $e > 1$

$$\mathcal{N}^+_{d,e} \cap \mathcal{N}^-_{d,e} = \mathbb{T} = N^+_{d,e} = N^-_{d,e}\,,$$

see (19) and (20). By a standard procedure (see [33, Ch. IX, §35, (6)]) any element $\Phi \in \langle \mathcal{N}^+_{d,e}, \mathcal{N}^-_{d,e}\rangle$ can be written in the form

$$\Phi = t\varphi = t\varphi_s \cdot \ldots \cdot \varphi_1\,,$$

where $t \in \mathbb{T}$, $s \geq 0$, and for $s > 0$ the factors $\varphi_i \notin \mathbb{T}$ are as in (29). If $s > 0$ then by (30) we obtain

$$\deg \Phi = \deg \varphi = \prod_{i=1}^{s} \deg \varphi_i > 1\,.$$

Hence $\Phi = \mathrm{id}$ if and only if $s = 0$ and $t = \mathrm{id}$. Thus there is no non-trivial relation in the group $\langle \mathcal{N}^+_{d,e}, \mathcal{N}^-_{d,e}\rangle$ between elements of the generating subgroups $\mathcal{N}^\pm_{d,e}$ and so (32) holds (cf. [52, §13]). □

To finish the proof of proposition 4.4 we need the following auxiliary result from the combinatorial group theory due to Hanna Neumann ([41, Corollary 8.11]).

Theorem 4.13. *In the amalgamated free product $G = A *_C B$ with the unified subgroup $C = A \cap B$, consider two subgroups $\tilde{A} \subseteq A$ and $\tilde{B} \subseteq B$, and let $\tilde{G} = \langle \tilde{A}, \tilde{B}\rangle$. Assume that $\tilde{A} \cap C = \tilde{C} = \tilde{B} \cap C$. Then $\tilde{G} = \tilde{A} *_{\tilde{C}} \tilde{B}$.*

The next lemma proves the second part of proposition 4.4.

Lemma 4.14. *For $e^2 \equiv 1 \mod d$ we have*

$$\mathcal{N}_{d,e} \simeq \mathcal{N}^+_{d,e} *_{N^+_{d,e}} N_{d,e}\,. \tag{34}$$

In particular, for $e = 1$

$$\mathcal{N}_{d,1} \simeq \mathcal{N}^+_{d,1} *_{B^+} \mathrm{GL}(2,\mathbb{C}), \quad \textit{where} \quad B^+ = \mathcal{N}^+_{d,e} \cap \mathrm{GL}(2,\mathbb{C})\,, \tag{35}$$

while if $e > 1$ and $e^2 \equiv 1 \mod d$ then

$$\mathcal{N}_{d,e} \simeq \mathcal{N}^+_{d,e} *_{\mathbb{T}} N(\mathbb{T}), \quad \textit{where} \quad \mathbb{T} = \mathcal{N}^+_{d,e} \cap N(\mathbb{T})\,. \tag{36}$$

Proof. We assume in the sequel that $e^2 \equiv 1 \mod d$. Let us note first that (35) and (36) are formal consequences of (34) since

$$N_{d,e} = \begin{cases} \mathrm{GL}(2,\mathbb{C}) & \text{if } e = 1 \\ N(\mathbb{T}) & \text{if } e > 1 \end{cases} \quad \text{and} \quad N^+_{d,e} = \begin{cases} B^+ & \text{if } e = 1 \\ \mathbb{T} & \text{if } e > 1\,. \end{cases}$$

Let us show (34). It follows from our definitions and lemma 4.9 that

$$(37) \qquad \mathcal{N}^+_{d,e} \cap \mathrm{Aff}^+(\mathbb{A}^2) = N_{d,e} \cap \mathrm{Aff}^+(\mathbb{A}^2) = N^+_{d,e}\,.$$

Letting

$$A = \mathrm{JONQ}^+(\mathbb{A}^2), \quad B = \mathrm{Aff}(\mathbb{A}^2), \quad \text{and} \quad C = A \cap B = \mathrm{Aff}^+(\mathbb{A}^2)$$

by the Jung-van der Kulk theorem 1.4 we obtain $G = \mathrm{Aut}(\mathbb{A}^2) = A *_C B$. Letting further

$$\tilde{A} = \mathcal{N}^+_{d,e}, \quad \tilde{B} = N_{d,e}, \quad \text{and} \quad \tilde{C} = N^+_{d,e} = \tilde{A} \cap \tilde{B}$$

we see by (37) that $\tilde{A} \cap C = \tilde{B} \cap C = \tilde{C}$. Now (34) follows by applying Hanna Neumann's theorem 4.13. □

4.2. *Algebraic groups actions on affine toric surfaces*

We can deduce now the following analog of the Kambayashi-Wright theorem 1.6 for affine toric surfaces.

Theorem 4.15. *Let $G \subseteq \mathrm{Aut}(X_{d,e})$ be an algebraic group acting on an affine toric surface $X_{d,e}$, where as before $1 \le e < d$ and $\gcd(d,e) = 1$.*

(a) If $e^2 \not\equiv 1 \mod d$ then G is conjugate in the group $\mathrm{Aut}(X_{d,e})$ to a subgroup of one of the groups $\mathcal{N}^+_{d,e}/G_{d,e}$ and $\mathcal{N}^-_{d,e}/G_{d,e}$.
(b) If $e^2 \equiv 1 \mod d$ and $e > 1$ then G is conjugate to a subgroup of one of the groups $\mathcal{N}^+_{d,e}/G_{d,e}$ and $N(\mathbb{T})/G_{d,e}$.
(c) If $e = 1$ then G is conjugate to a subgroup of one of the groups $\mathcal{N}^+_{d,e}/G_{d,e}$ and $\mathrm{GL}(2,\mathbb{C})/G_{d,e}$.

Proof. Consider the canonical surjection

$$\pi_* : \mathcal{N}_{d,e} \to \mathcal{N}_{d,e}/G_{d,e} \simeq \mathrm{Aut}(X_{d,e})$$

(see lemma 4.3). The algebraic group $\tilde{G} = \pi_*^{-1}(G)$ has bounded degree. Under the assumption of (a) we can conclude that the length $s = \mathrm{length}(\varphi)$ in (29) is uniformly bounded for all $\varphi \in \tilde{G}\backslash\mathbb{T}$. The same holds for $\pi_*(\varphi) \in G$ with respect to the free amalgamated product structure (21) as in theorem 4.2. Now (a) follows by Serre's theorem [46] (cf. subsection 1.2). Due to (34)–(36) a similar argument applies also in the remaining cases (b) and (c). □

Since unipotent elements form an abelian subgroup in $\mathcal{N}^\pm_{d,e}$ and a maximal unipotent subgroup in $\mathrm{GL}(2,\mathbb{C})$ is abelian, we obtain the following corollary.

Corollary 4.16. *Any unipotent algebraic subgroup $U \subseteq \mathrm{Aut}(X_{d,e})$ is abelian.*

For a reductive group acting on $X_{d,e}$, the following analog of proposition 2.5 holds.

Theorem 4.17. *Let $G \subseteq \mathrm{Aut}(X_{d,e})$ be a reductive algebraic group.*

(a) *If $e^2 \not\equiv 1 \mod d$ then G is conjugate in the group $\mathrm{Aut}(X_{d,e})$ to a subgroup of the torus $\mathbb{T}/G_{d,e}$.*
(b) *If $e^2 \equiv 1 \mod d$ and $e > 1$ then G is conjugate to a subgroup of the quotient $N(\mathbb{T})/G_{d,e}$.*
(c) *If $e = 1$ then G is conjugate to a subgroup of the quotient $\mathrm{GL}(2,\mathbb{C})/G_{d,e}$.*

Proof. Clearly the group $\tilde{G} = \pi_*^{-1}(G) \subseteq \mathcal{N}_{d,e}$ is reductive. By theorem 4.15 we may assume that $\tilde{G}$ is a subgroup of one of the corresponding factors. It suffices to restrict to the case where $\tilde{G} \subseteq \mathcal{N}^{\pm}_{d,e}$, since in the other case the assertions are evidently true. In the former case by proposition 2.5 the group $\tilde{G}$ is abelian and conjugate to a subgroup of the torus $\mathbb{T}$ via an element $\mu \in U^{\pm}$. We claim that such an element μ can be chosen within the subgroup $U^{\pm} \cap \mathcal{N}^{\pm}_{d,e}$. Let us show this assuming that $\tilde{G} \subseteq \mathcal{N}^{+}_{d,e}$, the other case being similar. Indeed, consider an element $\varphi^+ \in \tilde{G} \subseteq \mathcal{N}^{+}_{d,e}$ as in (7). By lemma 4.5 we have $f = \sum_{m\geq 0} a_m y^m \in A_{d,e}$, that is, $a_m = 0$ $\forall m \not\equiv e \mod d$. Hence in (12) we can also choose $g = \sum_{m\geq 0} b_m y^m \in \mathbb{k}[y]$ so that $b_m = 0$ $\forall m \not\equiv e \mod d$ and so $g \in A_{d,e}$ too. Again by lemma 4.5 the latter ensures that $\mu \in U^+ \cap \mathcal{N}^{+}_{d,e}$. Since $\tilde{G}$ is abelian, by virtue of lemma 2.4 such an element μ can be found simultaneously for all $\varphi^+ \in \tilde{G}$. Since μ normalizes the cyclic group $G_{d,e}$ it descends to an automorphism $\bar{\varphi} \in \mathrm{Aut}(X_{d,e})$ that conjugates G to a subgroup of the torus $\mathbb{T}/G_{d,e}$. The proof is completed. □

The following simple example clarifies case (c) above.

Example 4.18. The surface $X_{d,1}$ is a Veronese cone i.e., the affine cone over a rational normal curve Γ_d of degree d in $\mathbb{P}^d$. The group $\mathrm{GL}(2,\mathbb{C})$ acts naturally on $X_{d,1}$ (inducing the action of $\mathrm{PGL}(2,\mathbb{C})$ on Γ_d) via the standard irreducible representation on the space of binary forms of degree d.

5. Acyclic curves and automorphism groups of non-toric quotient surfaces

In this section we classify acyclic curves on the quotient $X = \mathbb{A}^2/G$ of the affine plane by a nonabelian finite linear group G and describe the automorphism groups of such surfaces.

We assume in the sequel that the finite group $G \subseteq \mathrm{GL}(n, \mathbb{C})$ is small that is, does not contain any pseudoreflection. Recall that a *pseudoreflection* on $\mathbb{A}^n$ is a non-identical linear transformation of finite order fixing pointwise a hyperplane. By the Chevalley-Shephard-Todd Theorem, the quotient space $\mathbb{A}^n/G$ of a finite linear group $G \subseteq \mathrm{GL}(n)$ is isomorphic to $\mathbb{A}^n$ if and only if G is generated by pseudoreflections. Assuming that this is not the case and considering the normal subgroup $H \lhd G$ generated by all pseudoreflections, we can decompose the quotient morphism $\pi\colon \mathbb{A}^n \to \mathbb{A}^n/G$ into a two-step factorization

$$\mathbb{A}^n \to \mathbb{A}^n/H \simeq \mathbb{A}^n \to X = \mathbb{A}^n/G\,.$$

For $n = 2$ we obtain in this way a presentation $X \simeq \mathbb{A}^2/(G/H)$, where the small linear group G/H acts on $\mathbb{A}^2$ freely off the origin. Thus we may assume, without loss of generality, that G is small. Under this assumption $X = \mathbb{A}^2/G$ is a normal affine surface with a unique singular point $Q = \pi(\bar{0})$, and the quotient morphism $\pi : \mathbb{A}^2 \to X$ is unramified outside this point.

It is well known that any finite subgroup of the group $\mathrm{SL}(2, \mathbb{C})$ either is cyclic, or is a binary dihedral (respectively, binary tetrahedral, octahedral, icosahedral) group, see e.g. [13] or [30]. The finite subgroups of $\mathrm{GL}(2, \mathbb{C})$ are cyclic extensions of these groups; we refer to [42] for their description.

For $n = 2$ every small abelian group G is conjugate to a cyclic group $G_{d,e}$ and so $X = \mathbb{A}^2/G$ is a toric surface. If G is small but nonabelian then X is non-toric. In the next theorem we examine this alternative possibility.

Theorem 5.1. *Let $X = \mathbb{A}^2/G$, where $G \subseteq \mathrm{GL}_2(\mathbb{C})$ is a nonabelian, finite, small group. Then any irreducible acyclic curve $\mathcal{C}$ in X is the image $\pi(L)$ of an affine line $L \subseteq \mathbb{A}^2$ passing through the origin.*

Proof. Let $\mathcal{C}$ be an irreducible acyclic curve on X. Assume first that $\mathcal{C}$ does not pass through the singular point $Q = \pi(\bar{0}) \in X$. Arguing as in the proof of theorem 3.3 we get

$$\pi^*(\mathcal{C}) = C^1 + \ldots + C^d\,,$$

where $d = |G|$, the irreducible components C^i are disjoint, $\pi : C^i \to \mathcal{C} \simeq \mathbb{A}^1$ is an isomorphism for all i, and the group G acts transitively on the set of these components. Since $d > 1$, in suitable new coordinates (x, y) on $\mathbb{A}^2$ the curves C^i are given by equations $y = \kappa_i$ with $\kappa_i \in \mathbb{C}^\times$. Any $g \in G$ sends y to $\lambda_g y$, where $g \mapsto \lambda_g$ defines a character $\psi\colon G \to \mathbb{G}_m$. The G-action on the set of components being simply transitive, ψ is injective and so G must be abelian, contrary to our assumption. This contradiction shows that $Q \in \mathcal{C}$.

Consider the reduced acyclic plane curve $C = \pi^*(\mathcal{C}) \subseteq \mathbb{A}^2$ passing through the origin. If C were smooth at the origin then the finite linear group G preserving C would preserve also the tangent line to C at the origin. So by Maschke's Theorem G should be abelian, a contradiction. Thus $\bar{0} \in C$ is a singular point.

By corollary 2.16 the stabilizer $\mathrm{Stab}(C) \supseteq G$ is abelian unless C is equivalent to a union of $r \geq 2$ affine lines through the origin. Since G is assumed to be nonabelian, the latter condition holds indeed.

We claim that C is in fact a union of lines, which yields the assertion. Indeed, consider the irreducible decomposition $C = C^1 + \ldots + C^r$, where $r \geq 2$. Notice that every component C^i of C has a unique place at infinity. Since for every $j \neq i$, $C^j = g(C^i)$ for some $g \in G$, where g is linear, we have $\deg(C^j) = \deg(C^i) = \delta$. Assume to the contrary that $\delta > 1$. Since G is nonabelian, by Maschke's Theorem the tautological representation $G \hookrightarrow \mathrm{GL}(2, \mathbb{C})$ is irreducible. Hence the induced G-action on $\mathbb{P}^1 = \mathbb{P}(\mathbb{A}^2)$ has no fixed point. It follows that for some pair of indices $i \neq j$, the points at infinity of the projective curves $\overline{C^i}$ and $\overline{C^j} = \overline{g(C^i)}$ are distinct. By Bezout Theorem these curves meet in $\delta^2 > 1$ points in $\mathbb{P}^2$. Since all these points are situated in the affine part $\mathbb{A}^2$ and the intersection index of C^i and C^j at the origin equals 1, these curves must have extra intersection points in the affine part. This contradicts the fact that C is simply connected. This contradiction ends the proof. □

The following corollary is immediate. Part (a) is an analog of theorem 3.8.

Corollary 5.2. *Under the assumptions of theorem 5.1 the following hold.*

(a) *The only irreducible acyclic curves in X are the closures of one-dimensional orbits of the $\mathbb{G}_m$-action on X by homotheties.*
(b) *The only simply connected curves in X are the images of finite unions of affine lines through the origin in $\mathbb{A}^2$.*

Using theorem 5.1 we can deduce the following description of the automorphism group of X, and as well an information on the equivalence classes of irreducible acyclic curves in X.

Theorem 5.3. *Let as before $X = \mathbb{A}^2/G$, where G is a nonabelian small finite subgroup of* $\mathrm{GL}(2,\mathbb{C})$. *Then the following hold.*

(a) $\mathrm{Aut}(X) \simeq N(G)/G$, *where the normalizer $N(G)$ of G in* $\mathrm{GL}(2,\mathbb{C})$ *is a finite extension of the one-torus $\mathbb{T}_{1,1} = Z(\mathrm{GL}(2,\mathbb{C}))$. Consequently,* $\mathrm{Aut}(X)$ *is a finite extension of the one-torus $\mathbb{T}_{1,1}/Z(G)$.*

(b) Let $\mathcal{S}$ denote the set of all irreducible acyclic curves $\mathcal{C} = \pi(L)$ on X. Then the $\mathrm{Aut}(X)$*-action on $\mathcal{S}$ has finite orbits, and the orbit space $\mathcal{S}/$* $\mathrm{Aut}(X)$ *is a rational curve.*

Proof. (a) We claim that every automorphism α of the quotient surface $X = \mathbb{A}^2/G$ comes from an automorphism of $\mathbb{A}^2$ normalizing the group G. Indeed, since G is small, the restriction $\pi|_{\mathbb{A}^2\setminus\{\bar{0}\}} : \mathbb{A}^2 \setminus \{\bar{0}\} \to X \setminus \{Q\}$ is an unramified Galois cover with the Galois group G. Since the surface $\mathbb{A}^2\setminus\{\bar{0}\}$ is simply connected, this is a universal cover. Furthermore, $\alpha \in \mathrm{Aut}(X)$ fixes the singular point Q and induces an automorphism of the smooth locus $X \setminus \{Q\}$. By the monodromy theorem, both compositions $\alpha \circ \pi$, $\alpha^{-1} \circ \pi$: $\mathbb{A}^2 \setminus \{\bar{0}\} \to X \setminus \{Q\}$ can be lifted to holomorphic maps $\tilde{\alpha}, \widetilde{\alpha^{-1}} : \mathbb{A}^2 \setminus \{\bar{0}\} \to \mathbb{A}^2 \setminus \{\bar{0}\}$ in such a way that $\tilde{\alpha} \circ \widetilde{\alpha^{-1}} = \mathrm{id}_{\mathbb{A}^2\setminus\{\bar{0}\}}$. According to the Hartogs Principle, the holomorphic automorphism $\tilde{\alpha}$ of $\mathbb{A}^2 \setminus \{\bar{0}\}$ extends to such an automorphism of the whole plane $\mathbb{A}^2$. Let us show that $\tilde{\alpha}$ is (bi)regular. The induced homomorphism $\tilde{\alpha}^* : \mathbb{C}[x,y] \to \mathrm{Hol}(\mathbb{A}^2)$ into the algebra of entire holomorphic functions in two variables sends the ring of invariants $\mathbb{C}[x,y]^G$ to the polynomial ring $\mathbb{C}[x,y]$. Hence the entire holomorphic functions $f = \tilde{\alpha}^*(x)$ and $g = \tilde{\alpha}^*(y)$ are integral over $\mathbb{C}[x,y]$. Consequently, they are of polynomial growth and so are polynomials. Since $\tilde{\alpha}$ covers α it belongs to the normalizer $\mathcal{N}(G)$ of G in the group $\mathrm{Aut}(\mathbb{A}^2)$. This proves our claim.

By theorem 5.1 the image $\alpha(\mathcal{C})$ of an irreducible acyclic curve $\mathcal{C} = \pi(L)$ on X is again such a curve. Hence any lift $\tilde{\alpha} \in \mathrm{Aut}(\mathbb{A}^2)$ of α preserves the collection of lines through the origin. Similarly as in the proof of corollary 2.9, it is easily seen that an automorphism with the latter property is linear i.e., $\tilde{\alpha} \in \mathrm{GL}(2,\mathbb{C})$. Hence $\tilde{\alpha} \in N(G)$. So finally $\mathrm{Aut}(X) = N(G)/G$. The proof of the remaining assertions is easy and can be left to the reader. □

Remarks 5.4. 1. Let as before $\mathcal{N}(G)$ denote the normalizer of G in the full automorphism group $\mathrm{Aut}(\mathbb{A}^2)$. The quotient group $\mathcal{N}(G)/G$ embeds into $\mathrm{Aut}(X)$. By virtue of theorem 5.3 it follows that $\mathcal{N}(G) = N(G)$ for any nonabelian small subgroup $G \subseteq \mathrm{GL}(2,\mathbb{C})$.

2. In contrast, for a toric surface $X_{d,e} = \mathbb{A}^2/G_{d,e}$ the group $\mathrm{Aut}(X_{d,e})$ acts infinitely transitively on the smooth locus X_{reg} i.e., m-transitively for

every $m \geq 1$, see [6]. In particular, the group $\mathrm{Aut}(X)$ is infinite dimensional. The equivalence class under the $\mathrm{Aut}(X)$-action of any curve $\mathcal{C}$ on X is infinite dimensional, too. Indeed, given an arbitrary set K of m distinct points on X_{reg}, due to the m-transitivity we can interpolate K by an image of $\mathcal{C}$ under a suitable automorphism (cf. corollary 4.18 in [4]).

Example 5.5. Consider the quaternion group

$$Q_8 = \left\{ \pm I_2, \pm \begin{pmatrix} i & 0 \\ 0 & -i \end{pmatrix}, \pm \begin{pmatrix} 0 & 1 \\ -1 & 0 \end{pmatrix}, \pm \begin{pmatrix} 0 & i \\ i & 0 \end{pmatrix} \right\} \subseteq \mathrm{GL}(2, \mathbb{C}).$$

This is a non-splittable central extension of the Klein four-group $V_4 \simeq \mathbb{Z}/2\mathbb{Z} \times \mathbb{Z}/2\mathbb{Z}$ by the center $\{\pm I_2\}$ of Q_8. The algebra of invariants $\mathbb{C}[x,y]^{Q_8}$ is generated by the homogeneous polynomials

$$f_1 = x^4 + y^4, \quad f_2 = x^2y^2, \quad \text{and} \quad f_3 = x^5y - xy^5$$

satisfying a relation of degree 12. The map $X = \mathbb{A}^2/G \to \mathbb{A}^3$ defined by these polynomials identifies X with a surface of degree 12 in $\mathbb{A}^3$ given by a quasihomogeneous polynomial equation with weights $(2,2,3)$. By theorem 5.1 every irreducible acyclic curve $\mathcal{C}$ on X is the image of an affine line L on $\mathbb{A}^2$ passing through the origin. Such a curve $\mathcal{C}$ is smooth if and only if the polynomial f_3 vanishes on L, if and only if $L = \mathbb{C}v$, where v is one of the vectors

$$(1,0), \quad (0,1), \quad (1,1), \quad (1,-1), \quad (1,i), \quad (1,-i). \tag{38}$$

Thus $\mathcal{C}$ coincides with one of the corresponding affine lines on X passing through the singular point Q. In particular, these three lines are the only affine lines on X in $\mathbb{A}^3$, and the image of any embedding $\mathbb{A}^1 \hookrightarrow X$ coincides with one of them. These lines are equivalent under the action on X of the automorphism group $\mathrm{Aut}(X) = N(Q_8)/Q_8$. Indeed, the vectors (38) belong to the same orbit of the normalizer $N(Q_8) \subseteq \mathrm{GL}(2,\mathbb{C})$.

References

1. S.S. Abhyankar, T.T. Moh. Embeddings of the line in the plane. J. Reine Angew. Math. 276 (1975), 148–166.
2. R.C. Alperin. Homology of the group of automorphisms of $\Bbbk[x,y]$. J. Pure Appl. Algebra 15 (1979), 109–115.
3. I.V. Arzhantsev, U. Derenthal, J. Hausen, and A. Laface. Cox rings. arXiv:1003.4229, see also the authors' webpages.
4. I.V. Arzhantsev, H. Flenner, S. Kaliman, F. Kutzschebauch, and M. Zaidenberg. Flexible varieties and automorphism groups. Duke Math. J. 162, no. 4 (2013), 767–823.

5. I.V. Arzhantsev, S.A. Gaifullin. Cox rings, semigroups and automorphisms of affine algebraic varieties. Sbornik Math. 201 (2010), no. 1, 1–21.
6. I.V. Arzhantsev, K. Kuyumzhiyan, and M. Zaidenberg. Flag varieties, toric varieties, and suspensions: three instances of infinite transitivity. Sbornik Math. 203 (2012), no. 7, 923–949.
7. T. Asanuma. Nonlinearizable algebraic group actions on $\mathbb{A}^n$. J. Algebra 166 (1994), 72–79.
8. F. Berchtold, J. Hausen. Demushkin's theorem in codimension one. Math. Z. 244 (2003), no. 4, 697–703.
9. A. Bialynicki-Birula. Remarks on the action of an algebraic torus on k^n. I, Bull. Acad. Polon. Sci. Sér. Sci. Math. Astronom. Pkys. 14 (1966), 177–181; II, *ibid.* 15 (1967), 123–125.
10. J. Chadzyński, K.T. Krasiński. On a formula for the geometric degree and Jung theorem. Univers. Iagellonicae Acta Mathem. 28 (1991), 81–84.
11. N.V. Chau. A simple proof of Jungs theorem on polynomial automorphisms of $\mathbb{C}^2$. Acta Math. Vietnam. 28 (2003), 209–214.
12. D.E. Cohen. A topological proof in group theory. Proc. Cambridge Philos. Soc. 59 (1963), 277–282.
13. J.H. Conway, D.A. Smith. On Quaternions and Octonions. Natick, Massachusetts: AK Peters, Ltd, 2003.
14. D.A. Cox. The homogeneous coordinate ring of a toric variety. J. Alg. Geom. 4 (1995), 17–50.
15. W. Dicks. Automorphisms of the polynomial ring in two variables. Publ. Sec. Math. Univ. Autonoma Barcelona 27 (1983), 155–162.
16. J. Fernández de Bobadilla. A new geometric proof of Jung's theorem on factorisation of automorphisms of $\mathbb{C}^2$. Proc. Amer. Math. Soc. 133 (2005), 15–19.
17. W. Fulton. Introduction to toric varieties. Annals of Math. Studies 131, Princeton University Press, Princeton, NJ, 1993.
18. M. Furushima. Finite groups of polynomial automorphisms in $\mathbb{C}^n$. Tohoku Math. J. (2) 35 (1983), 415–424.
19. M.H. Gizatullin, V.I. Danilov. Automorphisms of affine surfaces. I. Izv. Akad. Nauk SSSR Ser. Mat. 39 (1975), 523–565.
20. R.V. Gurjar. A new proof of Suzuki's formula. Proc. Indian Acad. Sci. Math. Sci. 107 (1997), 237–242.
21. S.M. Gusein-Zade, I. Luengo, and A. Melle-Hernández. On generating series of classes of equivariant Hilbert schemes of fat points. Mosc. Math. J. 10 (2010), 593–602.
22. A. Gutwirth. An inequality for certain pencils of plane curves. Proc. Amer. Math. Soc. 12 (1961), 631–638.
23. A. Gutwirth. The action of an algebraic torus on the affine plane. Trans. Amer. Math. Soc. 105 (1962), 407–414.
24. J.E. Humphreys. Linear algebraic groups. Graduate Texts in Mathematics 21, Springer-Verlag, 1998.
25. T. Igarashi. Finite Subgroups of the Automorphism Group of the Affine Plane. Master's thesis, Osaka University, 1977.

26. H. Jung. Über ganze birationale Transformationen der Ebene. J. Reine Angew. Math. 184 (1942), 161–174.
27. S. Kaliman, M. Koras, L. Makar-Limanov, and P. Russell. $\mathbf{C}^*$-actions on $\mathbf{C}^3$ are linearizable. Electron. Res. Announc. Amer. Math. Soc. 3 (1997), 63–71.
28. T. Kambayashi. On the absence of nontrivial separable forms of the affine plane. J. Algebra 35 (1975), 449–456.
29. T. Kambayashi. Automorphism group of a polynomial ring and algebraic group action on an affine space. J. Algebra 60 (1979), 439–451.
30. A.I. Kostrikin. Introduction to algebra. Universitext. Springer-Verlag, New York-Berlin, 1982.
31. H. Kraft. Challenging problems on affine n-space. Séminaire Bourbaki, Vol. 1994/95. Astérisque 237 (1996), Exp. No. 802, 5, 295–317.
32. H. Kraft, G. Schwarz. Finite automorphisms of affine N-space. Automorphisms of affine spaces. Curaçao, 1994, 55–66, Kluwer Acad. Publ., Dordrecht, 1995.
33. A.G. Kurosh. The theory of groups. Vol. 1, 2. Chelsea Publishing Co., New York 1960.
34. S. Lamy. Une preuve géométrique du théorème de Jung. Enseign. Math. (2) 48 (2002), 291–315.
35. D. Luna. Slices étales. Bull. Soc. Math. France 33 (1973), 81–105.
36. L.G. Makar-Limanov. On automorphisms of certain algebras. Candidates Dissertation, Moscow State University, 1970. (In Russian.)
37. J.H. McKay, S.S. Wang. An elementary proof of the automorphism theorem for the polynomial ring in two variables. J. Pure Appl. Algebra 52 (1988), 91–102.
38. M. Miyanishi. Open algebraic surfaces. CRM Monograph Series, 12. American Mathematical Society, Providence, RI, 2001.
39. M. Miyanishi, T. Nomura. Finite group scheme actions on the affine plane. J. Pure Appl. Algebra 71 (1991), 249–264.
40. M. Nagata. On automorphism group of $k[x, y]$. Kinokuniya Book Store, Tokyo, 1972.
41. H. Neumann. Generalized free products with amalgamated subgroups. Amer. J. Math. 70 (1948), 590–625.
42. K.A. Nguyen, M. van der Put, and J. Top. Algebraic subgroups of $\mathrm{GL}_2(\mathbb{C})$. Indag. Math. (N.S.) 19 (2008), 287–297.
43. V.L. Popov. On polynomial automorphisms of affine spaces. Izv. Math. 65 (2001), 569–587.
44. V.L. Popov, E.B. Vinberg. Invariant Theory. In: Algebraic Geometry IV, A.N. Parshin, I.R. Shafarevich (eds), Berlin, Heidelberg, New York: Springer-Verlag, 1994.
45. R. Rentschler. Opérations du groupe additif sur le plan affine. C. R. Acad. Sci. Paris Sér. A-B 267 (1968), 384–387.
46. J.-P. Serre. Trees. Springer Monographs in Mathematics. Springer-Verlag, Berlin, 2003.
47. M. Suzuki. Propriétés topologiques des polynômes de deux variables complexes, et automorphismes algébriques de l'espace $\mathbb{C}^2$. J. Math. Soc. Japan 26 (1974), 241–257.

48. A. Van den Essen. Polynomial automorphisms and the Jacobian conjecture. Progress in Mathematics, 190. Birkhauser Verlag, Basel, 2000.
49. W. van der Kulk. On polynomial rings in two variables. Nieuw Arch. Wiskunde (3) 1 (1953), 33–41.
50. D. Wright. Algebras which resemble symmetric algebras. Ph.D. Thesis, Columbia Univ., New York, 1975.
51. D. Wright. Abelian subgroups of $\mathrm{Aut}_k(k[X, Y])$ and applications to actions on the affine plane. Illinois J. Math. 23 (1979), 579–634.
52. D. Wright. Polynomial automorphism groups. Polynomial automorphisms and related topics, 1–19, Publishing House for Science and Technology, Hanoi, 2007.
53. M.G. Zaidenberg. Rational actions of the group $\mathbb{C}^*$ on $\mathbb{C}^2$, their quasi-invariants and algebraic curves in $\mathbb{C}^2$ with Euler characteristic 1. Soviet Math. Dokl. 31 (1985), 57–60.
54. M.G. Zaidenberg. Affine lines on $\mathbb{Q}$-homology plans and group actions. Transform. Groups 11 (2006), 725–735.
55. M.G. Zaidenberg, V.Y. Lin. An irreducible simply connected algebraic curve in $\mathbb{A}^2$ is equivalent to a quasi-homogeneous curve. Soviet Math. Dokl. 28 (1983), 200–204.

Hirzebruch surfaces and compactifications of $\mathbb{C}^2$

M. Furushima* and A. Ishida**

Department of Mathematics, Kumamoto University,
Kumamoto, Kurokami 2-39-1, Japan
**E-mail: wagami@sci.kumamoto-u.ac.jp*
***E-mail: ishidarakkuma.0927@gmail.com*
http://www.sci.kumamoto-u.ac.jp/math/index-e.html

In this paper, we shall prove that there exists a reduced curve $C^+ = C_1^+ \cup C_2^+$ on the Hirzebruch surface $\mathbb{F}_m$ $(m \geq 0)$ such that (i) $\mathbb{F}_m - C^+ \cong \mathbb{C}^2$, (ii) C_1^+ is a smooth rational curve and C_2^+ has a cusp singularity at $\mathrm{p} = C_1^+ \cap C_2^+ = \mathrm{Sing}C^+$. Moreover we shall examine about the relation of the compactification of $\mathbb{C}^2$ with Abhyankar-Moh-Suzuki on the linearization of lines on the affine plane.

Keywords: Hirzebruch surfaces, compactification, rational surfaces.

1. Introduction

Let (M, C) be a compactification of $\mathbb{C}^2$, that is, M be a connected smooth projective algebraic surface and C a closed algebraic curve in M such that $M - C$ is isomorphic to $\mathbb{C}^2$. Set $C = \bigcup_{i=1}^{r} C_i$, where C_i is an irreducible curve $(1 \leq i \leq r)$. Then we have:

(1.1) M is rational (cf. Kodaira [4]).
(1.2) $\mathrm{H}^1(C;\mathbb{Z}) \cong \mathrm{H}^1(M;\mathbb{Z}) \cong \mathrm{H}^3(M;\mathbb{Z}) \cong \mathrm{H}^3(C;\mathbb{Z}) = 0$, in particular, each component C_i $(1 \leq i \leq r)$ is a simply connected rational curve.
(1.3) $\mathrm{H}^2(M;\mathbb{Z}) \cong \mathrm{H}^2(C;\mathbb{Z})$, hence, the second Betti number $\mathrm{b}_2(M) = \mathrm{b}_2(C) = r$.
(1.4) Pic $M \cong \oplus_{i=1}^{r} \mathbb{Z}\mathcal{O}_M(C_i)$.

If $r = 1$, then M is isomorphic to $\mathbb{P}^2$ and C is a line on $\mathbb{P}^2$ by Remmert-Van de Ven [6]. If $r = 2$, then M is isomorphic to the Hirzebruch surface $\mathbb{F}_m$ of some degree $m \geq 0$ and C consists of two irreducible components C_1 and C_2. Let us denote by Sing$C$ the singular locus of C. Then we obtain Sing$C = C_1 \cap C_2 = \{p\}$ (one point) by Brenton [2]. In particular, if $C = C_1 \cup C_2$ is *minimal normal* according to Morrow [5], then the dual graph $\Gamma(C)$ of C looks like:

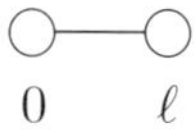

$(\ell \neq -1,\ |\ell| \geq m)$

However, in the case of $r = 2$, $C = C_1 \cup C_2$ is not necessarily minimal normal. In fact, Brenton [2] constructed the following example.

Theorem 1.1 (Brenton). *There exists a compactification* $(\mathbb{F}_1, C^{(1)})$ *of* $\mathbb{C}^2$ *such that*

(1) $C^{(1)} = C_{11} \cup C_{12}$, *where* C_{11} *is a smooth rational curve and* C_{12} *is a rational curve with a cusp singularity at* $\mathrm{p} = C_{11} \cap C_{12}$. *In particular, one has the multiplicity* $\mathrm{mult}_{\mathrm{p}}\, C_{12} = 2$.
(2) $(C_{11}^2)_{\mathbb{F}_1} = 1$, $(C_{12}^2)_{\mathbb{F}_1} = 8$, *and* $(C_{11} \cdot C_{12})_{\mathbb{F}_1} = 3$.

The compactification $(\mathbb{F}_1, C^{(1)})$ is based on the minimal normal compactification with the dual graph:

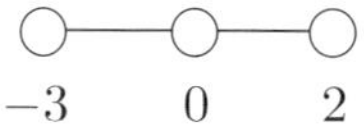

(see Table (c) in [5]).

On the other hand, applying the Abhyankar-Moh-Suzuki's theorem ([1],[7]) on the algebraic embeddings of the complex line $\mathbb{C}$ into the complex plane $\mathbb{C}^2$, we can construct new compactifications $(\mathbb{F}_1, C^{(n)})$ $(n \geq 1)$ containing $(\mathbb{F}_1, C^{(1)})$ as a special case.

Theorem 1.2. *There exists a curve* $C^{(n)} = C_{n1} \cup C_{n2}$ *in the Hirzebruch surface* $\mathbb{F}_1$ *such that*

(1) $\mathbb{F}_1 - C^{(n)} \cong \mathbb{C}^2$.
(2) In Pic $\mathbb{F}_1 \cong \mathbb{Z}[\Sigma_1] \oplus \mathbb{Z}[F]$, *we have a linear equivalence:*

$$C_{n1} \sim \Sigma_1 + F, \quad C_{n2} \sim (n-1)\Sigma_1 + nF,$$

where Σ_1 *is the minimal section and* F *a general fiber in* $\mathbb{F}_1$.
(3) C_{n1} *is a smooth rational curve and* C_{n2} *is a singular rational curve with one cusp at* $\mathrm{p} = C_{n1} \cap C_{n2}$. *In particular,* $\mathrm{mult}_{\mathrm{p}} C_{n2} = n - 1$.
(4) $(C_{n1}^2)_{\mathbb{F}_1} = 1$, $(C_{n2}^2)_{\mathbb{F}_1} = n^2 - 1$, $(C_{n1} \cdot C_{n2})_{\mathbb{F}_1} = n$.

Specifically, the Brenton's example is the case of $n = 3$.

Next, one can also construct more general examples based on the minimal normal compactification with the dual graph:

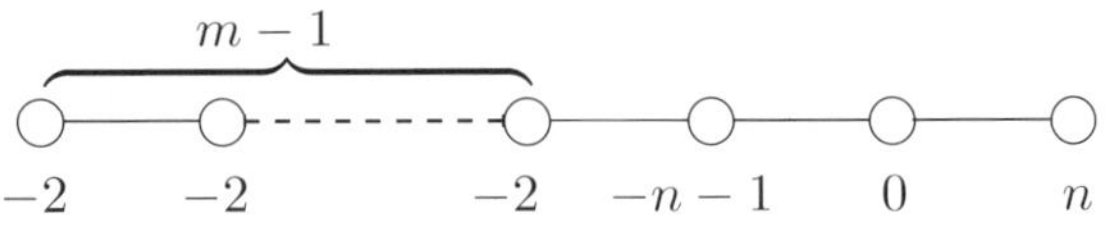

(see Table (d) in [5]).

Theorem 1.3. *There exists a curve* $C^+ = C_1^+ \cup C_2^+$ *in the Hirzebruch surafce* $M^+ = \mathbb{F}_m$ $(m \geq 0)$ *such that*

(i) $M^+ - C^+ \cong \mathbb{F}_m - C^+ \cong \mathbb{C}^2$.
(ii) $C_1^+ \cong \mathbb{P}^1$ *and* C_2^+ *is a rational curve with one cusp singularity* p *with* $\mathrm{mult}_{\mathrm{p}} C_2^+ = n$.
(iii) $C_1^+ \cap C_2^+ = \{\mathrm{p}\}$.
(iv) *There exists a fiber* F_0 *such that* $C_1^+ \cap C_2^+ \cap F_0 = \{\mathrm{p}\}$.
(iv) $C_1^+ \sim \Sigma_m^0 + mF \sim \Sigma_m^\infty$ *and* $C_2^+ \sim n\Sigma_m^0 + (nm+1)F$ *in* $\mathrm{Pic}\ \mathbb{F}_m$, *where* Σ_m^0 *is the minimal section and* F *is a general fiber of* $\mathbb{F}_m$.

Particularly, the Brenton's example is the case where $m = 1$, $n = 2$.

Moreover, using the above example $(\mathbb{F}_m, C^+)$, we shall explain the relation between the Abhyankar-Moh-Suzuki's theorem and the Hirzebruch surface $\mathbb{F}_m$ as a compactification of $\mathbb{C}^2$. Finally we shall propose some related problems.

2. A proof of Theorem 1.2

We need the following two easy lemmas.

Lemma 2.1. *Let* M_1, M_2 *be smooth complex surfaces and* $\varphi : M_1 \to M_2$ *be a biholomorphic mapping. Take a point* $\mathrm{p}_1 \in M_1$ *and set* $\mathrm{p}_2 = \varphi(\mathrm{p}_1)$. *Let* $\pi_i : \mathrm{B}_{\mathrm{p}_i}(M_i) \to M_i$ $(i = 1, 2)$ *be the blow-up of* M_i *at the point* p_i. *Then there exists a biholomorphic mapping* $\widehat{\varphi} : \mathrm{B}_{\mathrm{p}_1}(M_1) \to \mathrm{B}_{\mathrm{p}_2}(M_2)$ *such that* $\pi_2 \circ \widehat{\varphi} = \varphi \circ \pi_1$.

Lemma 2.2. *Let* $\pi : \mathrm{B}_{\mathrm{p}}(\mathbb{C}^2) \to \mathbb{C}^2$ *be the blow-up of* $\mathbb{C}^2$ *at a point* $\mathrm{p} \in \mathbb{C}^2$. *Let* ℓ *be a linear line in* $\mathbb{C}^2$ *passing through the point* p *and* $\overline{\ell}$ *the proper transform of* ℓ *in* $B_{\mathrm{p}}(\mathbb{C}^2)$. *Then* $\mathrm{B}_{\mathrm{p}}(\mathbb{C}^2) - \overline{\ell} \cong \mathbb{C}^2$.

We shall prove Theorem 1.2 as shown below. Let us first consider the following algebraic curve of degree n:

$$\Delta := \{(z_0 : z_1 : z_2) \in \mathbb{P}^2 : z_2^{n-1} z_1 = z_0^n\}.$$

In the affine part $U := \{z_2 \neq 0\} \cong \mathbb{C}^2$ with the affine coordinates

$$x = \frac{z_0}{z_2}, \ y = \frac{z_1}{z_2},$$

one has $\Delta_0 := \Delta \cap U = \{y = x^n\}$. Then the following coordinates transformation

$$s = x, \ t = y - x^n$$

gives the biholomorphic mapping $\varphi : U \to W := \mathbb{C}^2$ with $\varphi(\Delta_0) = \ell_0 = \{t = 0\}$, where (s, t) is a coordinate system of $W = \mathbb{C}^2$. Let $\mathrm{B}_{\mathbf{o}}(U) \to U$ (resp. $\mathrm{B}_{\mathbf{o}}(W) \to W$) be the blow-up of U (resp. W) at the point $\mathbf{o} = (0, 0) \in U$ (resp. $\mathbf{o} = \varphi(\mathbf{o}) \in W$) and $\overline{\Delta}_0$ (resp. $\overline{\ell}_0$) the proper transform of Δ_0 (resp. ℓ_0) in $\mathrm{B}_{\mathbf{o}}(U)$ (resp. in $\mathrm{B}_{\mathbf{o}}(W)$).

Claim 2.1. $\mathrm{B}_{\mathbf{o}}(U) - \overline{\Delta}_0 \cong \mathbb{C}^2$.

In fact, by Lemma 2.1, one has isomorphisms $\mathrm{B}_{\mathbf{o}}(U) \cong \mathrm{B}_{\mathbf{o}}(W)$ and $\mathrm{B}_{\mathbf{o}}(U) - \overline{\Delta}_0 \cong \mathrm{B}_{\mathbf{o}}(W) - \overline{\ell}_0$. By Lemma 2.2, we have the claim.

Claim 2.2. $\mathbb{F}_1 - C^{(n)} \cong \mathbb{C}^2$.

In fact, let $\pi : M = \mathrm{B}_{\mathrm{p}}(\mathbb{P}^2) \cong \mathbb{F}_1 \longrightarrow \mathbb{P}^2$ be the blow-up of $\mathbb{P}^2$ at the point $\mathrm{p} = (0 : 0 : 1)$ with the exceptional set $\Sigma \cong \mathbb{P}^1$ with $\Sigma^2 = -1$. Let $C_{n,1}$ and $C_{n,2}$ be proper transformations of the line $\{z_2 = 0\}$ and Δ in $\mathbb{F}_1$ respectively, and set $C^{(n)} = C_{n,1} \cup C_{n,2}$. Since $\mathbb{F}_1 - (C_{n,1} \cup C_{n,2}) \cong \mathrm{B}_{\mathbf{o}}(U) - \overline{\Delta}_0$, by Claim 2.1, we have the claim.

By construction, one can easily verify that

$$C_{n,1} \sim \Sigma_1 + F$$
$$C_{n,2} \sim (n-1)\Sigma_1 + nF$$

in $\mathrm{Pic}\, \mathbb{F}_1 \cong \mathbb{Z}[F] \oplus \mathbb{Z}[\Sigma_1]$, where F (resp. Σ_1) is a fiber (the minimal section) of $\mathbb{F}_1$. An easy computation yields that

$$C_{n,1}^2 = 1, \ \ C_{n,2}^2 = n^2 - 1, \ \ C_{n,1} \cap C_{n,2} = \{\mathrm{p}\}, \ \ (C_{n,1} \cdot C_{n,2})_{\mathbb{F}_1} = n.$$

We remark that the Brenton's example is the case of $n = 3$. This completes the proof of Theorem 1.2.

Remark 1. In general, let Δ_0 be a smooth algebraic curve of degree n in $\mathbb{C}^2$ which is homeomorphic to the affine line $\mathbb{C}$. Then according to

Abhyankar-Moh-Suzuki's theorem ([1], [7]), there exists a polynomial automorphism $\varphi : \mathbb{C}^2 \to \mathbb{C}^2$ such that $\varphi(\Delta_0)$ is linear. Let Δ be the closure of Δ_0 in $\mathbb{P}^2$ and $\ell_\infty \subset \mathbb{P}^2$ the line with $\mathbb{P}^2 - \ell_\infty \cong \mathbb{C}^2$. Then one has $\Delta \cap \ell_\infty = \{p_\infty\}$ and $\mathrm{mult}_{p_\infty} \Delta \leq n-1$. Let $\mathrm{B_p}(\mathbb{P}^2) \cong \mathbb{F}_1$ be the blow-up of $\mathbb{P}^2$ at a point $\mathrm{p} \in \Delta_0$ and C_1 (resp. C_2) the proper transform of Δ (resp. ℓ_∞) in $\mathbb{F}_1$. We put $C = C_1 \cup C_2$. Then one has easily

$$C_1 \sim \Sigma_1 + F$$
$$C_2 \sim (n-1)\Sigma_1 + nF$$

in Pic $\mathbb{F}_1 \cong \mathbb{Z}[\Sigma_1] \oplus \mathbb{Z}[F]$ and $\mathbb{F}_1 - C \cong \mathbb{C}^2$. This gives a general construction of non-minimal normal compactifications of $\mathbb{C}^2$ with the second Betti number equal to two.

3. A proof of Theorem 1.3

Let us first consider the minimal normal compactification (M, C) of $\mathbb{C}^2$ with the weighted dual graph $\Gamma(C) = \Gamma^{(m)}(C; n)$ (see Theorem 9 in [5]):

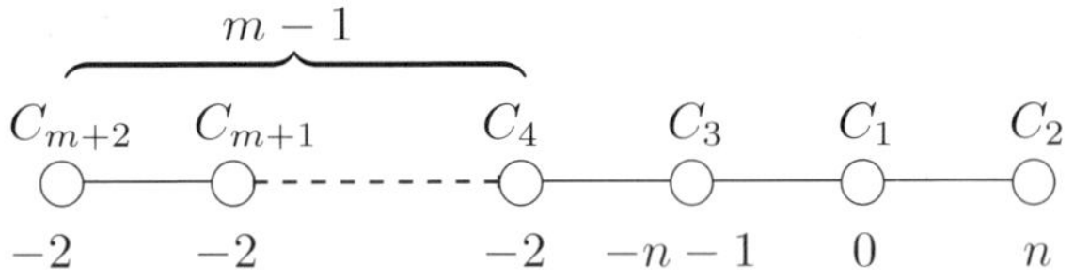

Figure 1

Let C_1 be the component of C which corresponds to the vertex with the weight $C_1^2 = 0$ in the graph $\Gamma(C)$. Then there exists a surjective morphism $\pi : M \longrightarrow \mathbb{P}^1$ which has C_1 as a regular fiber $\pi^{-1}(0)$ and C_2, C_3 the sections of $M \overset{\pi}{\to} \mathbb{P}^1$. One sees that the curve $\bigcup_{i=1}^{m-1} C_{i+3}$ $(m > 1)$ is contained in a singular fiber $T_0 := \pi^{-1}(\infty)$. Then one can show that T_0 is a unique singular fiber. In fact, let $T_0, T_1, \ldots, T_k$ be all the singular fibers of $M \overset{pi}{\to} \mathbb{P}^1$ and assume that $k > 0$. Let us denote by $1 + \alpha_i$ (resp. δ_i) the number of irreducible components of T_i (resp. the number of irreducible components of T_i which are not contained in C). Since $\sum_{i=0}^{k}(1 + \alpha_i - \delta_i) + 3 = \mathrm{b}_2(C) = \mathrm{b}_2(M) = 2 + \sum_{i=0}^{k} \alpha_i$, one has

$\sum_{i=0}^{k}(\delta_i - 1) = 1$. The singular fiber T_i $(i > 0)$ contains no irreducible components of C. This implies that $\delta_i = 1 + \alpha_i \geq 2$ $(i > 0)$, in particular, one has $\delta_0 = 1$. Thus we have $T_0 = \bigcup_{i=1}^{m} C_{i+3}$, where the component C_{m+3} is the (-1)-curve, which is absurd. Therefore T_0 is a unique singular fiber.

Then one sees that the dual graph of the singular fiber $T_0 = \pi^{-1}(\infty)$ looks like:

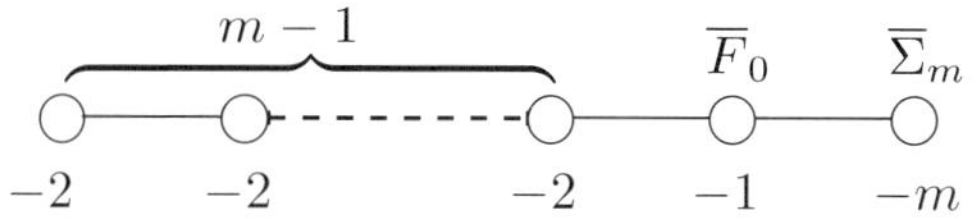

where $\overline{\Sigma}_m$ (resp. $\overline{F}_0$) is a component of the singular fiber $T_0 = \pi^{-1}(\infty)$ with the self-intersection number $-m$ (resp. -1). The picture of the fibration $\pi : M \to \mathbb{P}^1$ over $\mathbb{P}^1$ looks like Figure 2.

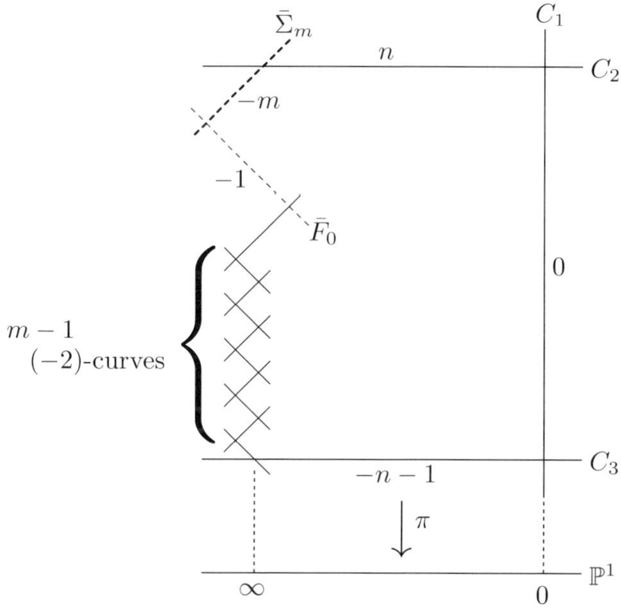

Figure 2

Take a general point $p \in C_1$. By the successive blow-ups of infinitely near points lying over p (cf. pp. 278–280 in [3]), the graph $\Gamma(C)$ changes to the graph $\Gamma(\overline{C})$ as shown below:

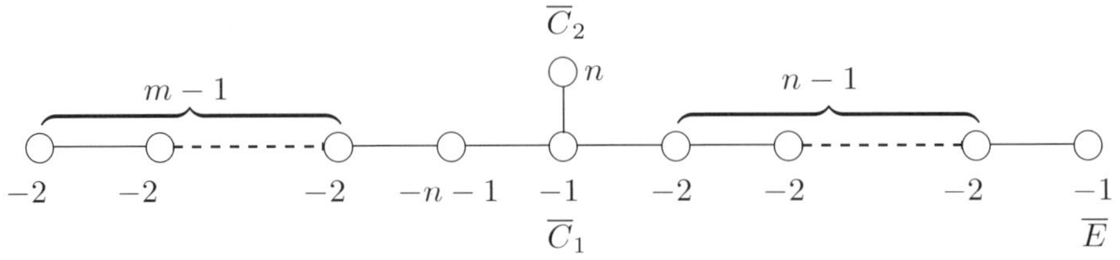

Then we obtain a new compactification $(\overline{M}, \overline{C})$ of $\mathbb{C}^2$ and a birational morphism $\phi : \overline{M} \to M$ with $\phi(\overline{C}) = C$.

The exceptional curves $\overline{C}_* \subset \overline{C}$ corresponding to the subgraph $\Gamma(\overline{C}_*)$:

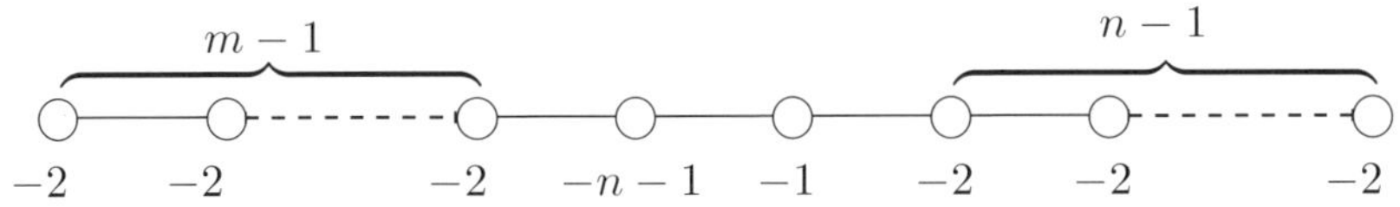

can be contracted to a smooth point, that is, there exists a birational morphism $\varphi : \overline{M} \longrightarrow M^+$ onto a smooth projective surface M^+ such that $\overline{M} - \overline{C}_* \cong M^+ - \mathrm{p}$, where $\mathrm{p} = \varphi(\overline{C}_*)$. We set $C_1^+ = \varphi_* \overline{E}$, $C_2^+ = \varphi_* \overline{C}_2$ and $C^+ = C_1^+ \cup C_2^+$. Then we have $M^+ - C^+ \cong \overline{M} - \overline{C} \cong \mathbb{C}^2$. Since $\mathrm{b}_2(M^+) = 2$, M^+ is isomorphic to some Hirzebruch surface. We put $F_0 = \varphi_* \overline{F}_0$ and $\Sigma_m^0 = \varphi_* \overline{\Sigma}_m$. Then, by construction, one sees the following:

(1) $M^+ \cong \mathbb{F}_m$ and $\mathrm{Pic}\, \mathbb{F}_m \cong \mathbb{Z}[\Sigma_m^0] \oplus \mathbb{Z}[F_0]$.
(2) $(\Sigma_m^0)^2 = -m$, $F_0^2 = 0$.
(3) $(C_1^+)^2 = m$, $(C_2^+)^2 = n(mn+2)$.
(4) $(C_1^+ \cdot \Sigma_m^0) = 0$, $C_1^+ \cap F_0 = C_2^+ \cap F_0 = C_1^+ \cap C_2^+ = \{\mathrm{p}\}$, in particular, $(C_2^+ \cdot F_0) = n$.
(5) $C_1^+ \sim \Sigma_m^\infty$ and C_2^+ has a cusp singularity at p with $\mathrm{mult}_{\mathrm{p}} C_2^+ = n$.
(6) $C_1^+ \sim \Sigma_m^0 + mF_0 \sim \Sigma_m^\infty$ and $C_2^+ \sim n\Sigma_m^0 + (nm+1)F_0$.

This completes the proof.

Remark 2. The Brenton's example is obtained by setting $m = 1$, $n = 2$.

4. Abhyankar-Moh-Suzuki's theorem

Let us recall the minimal normal compactification (M, C) of $\mathbb{C}^2$ with the dual graph $\Gamma(C) = \Gamma^{(m)}(C; n)$ as before. Let C_1 and C_2 be the components of C corresponding to the vertices with the weights $C_1^2 = 0$ and $C_2^2 = n$ respectively. Blowing up of M at the point $\mathrm{p}_0 = C_1 \cap C_2$ and blowing down the proper transform of C_1, which is the (-1) curve, we have a minimal normal compactification $(M_1, C^{(1)})$ with the graph $\Gamma^{(m)}(C^{(1)}; n-1)$. Repeating this process n-times, one gets a minimal normal compactification $(M_n, C^{(n)})$ with the graph $\Gamma^{(m)}(C^{(n)}; 0)$ shown below:

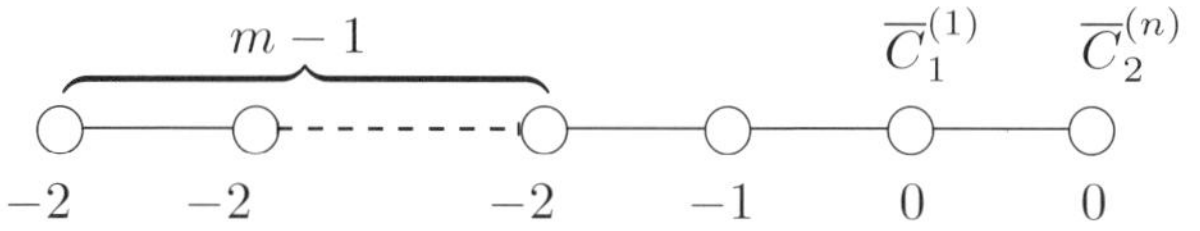

and a birational map $\nu : M \dashrightarrow M_n$. Blowing down the exceptional curves corresponding to the following subgraph:

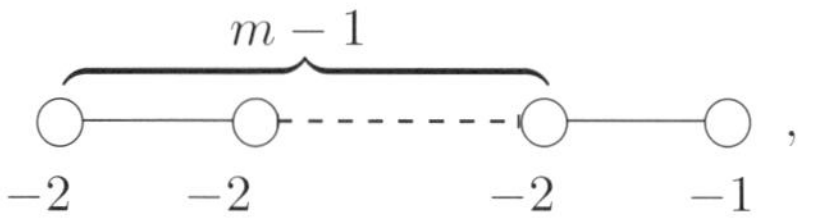

we obtain a smooth projective surface M^* and a surjective morphism $\mu : M_n \to M^*$. We put $C_1^* = \mu_* \overline{C}_1^{(n)}$, $C_2^* = \mu_* \overline{C}_2^{(n)}$ and $C^* = C_1^* \cup C_2^*$. Then one sees that $M^* \cong \mathbb{F}_m$, C_1^* is a section with $(C_1^*)^2 = m$ and C_2^* is a fiber of $\mathbb{F}_m$. Let $\sigma : M^+ = \mathbb{F}_m \dashrightarrow \mathbb{F}_m = M^*$ be the composition of birational maps

$$\mathbb{F}_m \cong M^* \xleftarrow{\mu} M_n \overset{\nu}{\dashleftarrow} M \xleftarrow{\phi} \overline{M} \xrightarrow{\varphi} M^+ \cong \mathbb{F}_m,$$

that is, $\sigma := \mu \circ \nu \circ \phi \circ \varphi^{-1}$.

Then we have an isomorphism

$$\mathbb{F}_m - C^+ \overset{\sigma}{\cong} \mathbb{F}_m - C^* \cong \mathbb{C}^2.$$

We put $F_0^* = (\mu \circ \nu)_* \overline{F_0}$ and $\Sigma_m^* = (\mu \circ \nu)_* \overline{\Sigma}_m$. Then F_0^* is a fiber and Σ_m^* the minimal section in $M^* = \mathbb{F}_m$. In particular, one has

$$\sigma(C_2^+) = C_2^*.$$

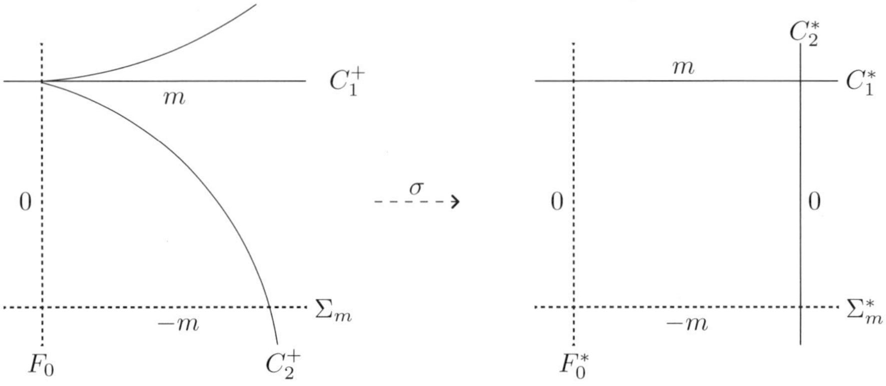

Figure 3

By construction, one has $\mathbb{F}_m - (F_0 \cup C_1^+) \cong \mathbb{C}^2 \cong \mathbb{F}_m - (F_0^* \cup C_1^*)$. Furthermore, the birational automorphism σ induces an algebraic automorphism

$$\sigma_0 : \mathbb{F}_m - (F_0 \cup C_1^+) \to \mathbb{F}_m - (F_0^* \cup C_1^*).$$

We set

$$A := C_2^+ \cap \Big(\mathbb{F}_m - (C_1^+ \cup F_0)\Big) \subset \mathbb{C}^2 \quad \text{and} \quad L := C_2^* \cap \Big(\mathbb{F}_m - (C_1^* \cup F_0)\Big) \subset \mathbb{C}^2.$$

Then A is a smooth algebraic curve in $\mathbb{C}^2$ with $A \cong \mathbb{C}$ and L is a linear algebraic curve in $\mathbb{C}^2$. Since $\sigma(C_2^+) = C_2^*$, one has $\sigma_0(A) = L$. This shows that the affine algebraic curve $A \cong \mathbb{C}$ is linearizable by the algebraic automorphism σ_0 of $\mathbb{C}^2$. This gives a proof of Abhyankar-Moh-Suzuki's theorem to the above affine algebraic curve $A \cong \mathbb{C}$.

Theorem 4.1. *Let $A \hookrightarrow \mathbb{C}^2$ be a smooth algebraic curve, C_1^+ the closure of A in the Hirzebruch surface $\mathbb{F}_m$ as a compactification of $\mathbb{C}^2$ and C_2^+ the section of $\mathbb{F}_m$ with $(C_2^+)^2 = m \geq 0$. We set $C^+ = C_1^+ \cup C_2^+$. Assume that*

(1) $\mathbb{F}_m - C^+ \cong \mathbb{C}^2$

(2) $(\mathbb{F}_m, C^+)$ is birationally equivalent to the minimal normal compactification (M, C) of $\mathbb{C}^2$ with the dual graph $\Gamma(C) = \Gamma^{(m)}(C; n)$ as shown before, that is, there exists a birational map $\tau : \mathbb{F}_m \dashrightarrow M$ such that the restriction $\tau_0 : \mathbb{F}_m - C^+ \to M - C$ is biregular.

Then there exists an algebraic automorphism σ of $\mathbb{C}^2$ such that $\sigma(A)$ is linear.

We remark that Theorem 4.1 will suggest a possibility of a birational geometric proof of Abhyankar-Moh-Suzuki's theorem.

Finally we shall propose the following:

Problem 4.1. *Let* C *be an ample divisor on the Hirzebruch surface* $\mathbb{F}_m$. *Suppose that the affine part* $V = \mathbb{F}_m - C$ *is the integral homology 2-cell, that is,* $\mathrm{H}_i(V;\mathbb{Z}) = 0$ *for* $i > 0$. *Then is* V *isomorphic to* $\mathbb{C}^2$*?*

Problem 4.2. *Determine the structure of the Hirzebruch surface* $\mathbb{F}_m$ *as a compactification of* $\mathbb{C}^2$.

References

1. S. Abhyankar and T. Moh, *Embeddings of the line in the plane, J. Reine Angew. Math.* **276**, 148–166 (1975).
2. L. Brenton, *A note on compactifications of* $\mathbb{C}^2$, *Math. Ann.* **206**, 303–310 (1973).
3. S. Iitaka, *Algebraic Geometry: An Introduction to Birational Geometry of Algebraic Varieties*, G.T.M. **76**, Springer-Verlag, New York, Heidelberg, Berlin, 1982.
4. K. Kodaira, *Holomorphic mappings of polydiscs into compact complex manifolds, J. Differential Geometry* **6**, 33–46 (1971).
5. J. Morrow, *Minimal normal compactifications of* $\mathbb{C}^2$, *Proceedings of the Conference on Complex Analysis, Rice University Studies* **59**, 97–112 (1973).
6. R. Remmert and A. Van de Ven, *Zwei Zätze über die komplex-projective Ebene, Nieuw Arch, Wisk.(3)* **8**, 147–157 (1960).
7. M. Suzuki, *Propriétés topologiques des polynômes de deux variables complexes, et automorphismes algébriques de l'espace* $\mathbb{C}^2$, *J. Math. Soc. Japan* **26**, 241–257 (1974).
8. M. Suzuki, *Compactifications of* $\mathbb{C} \times \mathbb{C}^*$ *and* $(\mathbb{C}^*)^2$, *Tôhoku Math. J.(2)* **31**, 453–468 (1979).

Cyclic multiple planes, branched covers of S^n and a result of D. L. Goldsmith

R.V. Gurjar

School of Mathematics, Tata Institute of Fundamental Research, Homi-Bhabha Road, Mumbai 400005, India
E-mail: gurjar@math.tifr.res.in

Dedicated to Professor Masayoshi Miyanishi on the occasion of his 70th birthday

We will prove two essential generalizations of O. Zariski's well-known result about irregularity of cyclic multiple planes. We will also provide a correct proof of a general result of D. L. Goldsmith. In our proofs we will use P. A. Smith's theory of finite group actions on simplicial complexes.

1. Introduction

In [12] O. Zariski proved the following well-known result.

(1) Irregularity of cyclic multiple planes. *Let $f(x, y)$ be an irreducible polynomial with complex coefficients. Let p be a prime number and $n = p^l$ for some integer $l \geq 1$. Then the first Betti number of a smooth projective model, say $\widetilde{S}$, of the affine surface $S := \{z^n - f = 0\}$ is 0.*

In this paper, by the first Betti number $b_1(T)$ of a simplicial complex T (or a topological manifold T) we mean the rank of the singular homology group $H_1(T; \mathbf{Z})$. It is easy to see that if $H_1(T; \mathbf{Z}/(p)) = (0)$ for a prime p then $b_1(T) = 0$. The converse is not true in general since $H_1(T; \mathbf{Z})$ can be a finite p-group so that $b_1(T) = 0$ but $H_1(T; \mathbf{Z}/(p))$ is non-trivial.

The following classical result from Knot Theory is an analogue of Theorem 1 ([3]).

2010 *Mathematics Subject Classification*: 14J80, 57M12, 57Q45.

(2) Branched covers of Knots. *Let* $S^m \subset S^{m+2}$ *be a codimension two tame Knot. Let* $n = p^l$ *be a prime power and let* $M \to S^{m+2}$ *be a Galois cover of degree* n *ramified totally over* S^m *and unramified outside* S^m. *Then* $b_1(M) = 0$.

In [5], D. L. Goldsmith proved a general "going up" result.

Let $\mathbf{Z}_p$ denote the cyclic group $\mathbf{Z}/(p)$ of prime order p.

We will prove a slightly stronger result than the one stated by Goldsmith.

(3) Goldsmith's going up theorem. *Let* $n = p^l$ *be a prime power. Suppose* (K, L) *is a simplicial pair of finite dimension with* K *a connected complex. Let* $\widetilde{K} \to K$ *be a connected regular topological covering of degree* n *and let* $\widetilde{L}$ *be the inverse image of* L *in* $\widetilde{K}$. *If* $H_i(K, L; \mathbf{Z}_p) = (0)$ *for all* $i \leq r$ *for some integer* $r \geq 0$ *then* $H_i(\widetilde{K}, \widetilde{L}; \mathbf{Z}_p) = (0)$ *for all* $i \leq r$.

It is easy to deduce (2) from (3) (see §4). Unfortunately, the proof of (3) in [5] appears to be incorrect (see, §5).

Nowadays it is customary to prove (1), (2), (3) using Knot-theoretic ideas arising from the famous short exact sequence of certain chain complexes corresponding to an infinite cyclic cover due to J. Milnor [7]. See for example, [6]. There is also a cohomological method for dealing with generalizations of (1). See, [2].

The aim of this paper is to give different proofs of (1), (2), (3) and also to prove two stronger results below. The author does not know if this stronger form is known to experts. We prove all these results by using arguments and results from the classical Smith Theory. We will deduce (2) from (3) and correct the proof of (3). We will implicitly use the classical result that given any complex algebraic variety X and a closed subvariety Y of X there is a triangulation of X such that Y is a subcomplex. See, [4].

Our main result is the following generalization of Zariski's theorem.

Theorem. *Let* X *be a normal irreducible quasiprojective variety/*$\mathbf{C}$ *of dimension* ≥ 2 *and let* D *be a connected reduced divisor in* X *(which may be empty). Assume that* $H_1(X; \mathbf{Z}_p) = 0$. *Let* $\pi : \overline{X} \to X$ *be a quotient* *morphism from a normal irreducible variety $\overline{X}$ for a cyclic group of order $n = p^l$ acting on $\overline{X}$ for some prime p, which is totally ramified over D and unramified outside D. Then $H_1(\overline{X}; \mathbf{Z}_p) = 0$.*

In the situation of Zariski's result, the affine surface $\{z^{p^l} - f = 0\}$ (and its resolution of singularities) is a Zariski-open subset of the corresponding branched cover of $\mathbf{P}^2$ (resp., a resolution of singularities of the branched cover of $\mathbf{P}^2$). Hence by Lemma 3 in §2, we see that our Theorem is stronger than Zariski's result.

The next two corollaries are both stronger than Zariski's result since Zariski assumed that $f(x, y)$ is an irreducible polynomial and his conclusion was about the first Betti number of a smooth projective model of $\{z^{p^l} - f = 0\}$. See, Lemma 3 in §2.

Corollary 1. *Let $f(x, y)$ be a reduced (possibly reducible) polynomial with complex coefficients. Let p be a prime number and $n = p^l$ for some integer $l \geq 1$. Assume that $C := \{f = 0\}$ is connected. Then the affine surface $S := \{z^n - f = 0\}$ has trivial first Betti number.*

Corollary 2. *Let $f(x, y)$ be an irreducible polynomial with complex coefficients and $n = p^l$. Then the first Betti number of a resolution of singularities S' of the surface S in Corollary 1 is 0.*

The author feels that it is Corollary 2 which is really behind Zariski's result, and not the assumption of smooth projective model in Zariski's result, since $b_1(S') \geq b_1(\widetilde{S})$. (See, Lemma 3 in §2).

Algebraists will find the next result a useful consequence of Corollary 1.

Corollary 3. *Let $f(x, y)$ be a reduced (possibly reducible) polynomial over $\mathbf{C}$ such that $\{f = 0\}$ is connected. Let $n = p^l$ be a prime power. Then the only units in the affine domain $\mathbf{C}[X, Y, Z]/(Z^{p^l} - f(X, Y))$ are non-zero constants in $\mathbf{C}$.*

Remarks. (1) The Corollaries are valid for affine varieties of the form $\{x_{m+1}^n - f(x_1, x_2, \ldots, x_m) = 0\}$, where $m \geq 2$, $n = p^l$. We have stated them for surfaces for the sake of simplicity.

(2) If f is reducible and C is singular then a resolution of singularities $\widetilde{S}$ of S can have $b_1 > 0$. See, §6 for an example. Compare this with Corollary 2 below.
(3) If n is not a prime power then Corollary 1 is not valid, even if $\{f = 0\}$ is smooth and irreducible. See [11] for an example.
(4) If C is not connected then Corollary 1 is false in general. See, §6 for an example.

Our proofs are surprisingly short and easy, once we use the basic Smith sequences and their properties. The reader will not find any Alexander polynomials, or their properties in any arguments! The author feels that this proof will make Zariski's important result more accessible to algebraic geometers, and because of our main Theorem the applicability of Zariski's result becomes wider.

2. Preliminaries

We will recall some terminology and results from Smith Theory. A good reference for this is G. Bredon's book [1, Chapter III].

Let G be a finite group which acts by simplicial homeomorphisms on a finite dimensional simplicial complex K. We say that K is a *regular G-complex* if the following condition (B) is satisfied.

(B) For any subgroup H of G the following holds:

If $h_0, h_1, \ldots, h_m \in H$ and $(\nu_0, \nu_1, \ldots, \nu_m)$ is a simplex in K such that $(h_0\nu_0, h_1\nu_1, \ldots, h_m\nu_m)$ is also a simplex in K then there exists an $h \in H$ such that $h\nu_i = h_i\nu_i$ for all i.

It is clear that G acts simplicially on the successive barycenric subdivisions $K', K'', \ldots$ of K. It can be easily proved that K'' is a regular G-complex and (B) is equivalent to either of the following two conditions on K''. See, [1].

(A) For any $g \in G$ and any simplex $s \subset K''$, $s \cap g(s)$ is pointwise fixed by g.

(A′) If a vertex ν of K'' and $g\nu$ belong to the same simplex of K'' then $\nu = g\nu$.

By passing to K'' we will assume that K is a regular G-complex. It is easy to see that the quotient space K/G inherits a natural simplicial structure such that $K \to K/G$ is a simplicial map. It is also clear that the fixed point set K^G is a subcomplex of K.

Now we assume that $G = \mathbf{Z}_p$. Assume that K is a regular G-complex and $L \subset K$ a subcomplex which is G-stable and also a regular G-complex. We write G multiplicatively. Clearly K^G embeds in K/G under the map $K \to K/G$. Let $\mathbf{Z}_p(G)$ be the group ring and g a fixed generator of G. We define two elements in $\mathbf{Z}_p(G)$ as follows: $\sigma := 1+g+\cdots+g^{p-1}$ and $\tau := 1-g$. We have $\sigma\tau = 0 = \tau\sigma$. One checks easily that $\sigma = \tau^{p-1}$. If $\rho := \tau^i$, we write $\overline{\rho} = \tau^{p-i}$. Then $\tau = \overline{\sigma}$, $\sigma = \overline{\tau}$. The chain complex $C(K, L; G)$ will be written simply as $C(K, L)$. This is a graded module over $\mathbf{Z}_p(G)$.

Now we state the basic result due to P.A. Smith, which (and the corresponding long exact homology sequence) we call the *Smith sequence.*

Proposition. *For each* $\rho = \tau^i$, $1 \le i \le p-1$, *there is a short exact sequence*

$$(0) \to \overline{\rho}C(K, L) \oplus C(K^G, L^G) \xrightarrow{j} C(K, L) \xrightarrow{\rho} \rho C(K, L) \to (0).$$

Here j is induced by inclusions of the two direct summands.

For $\rho = \tau^i$, $1 \le i \le p-1$, we write $H^\rho(K, L)$ for the homology of the complex $\rho C(K, L)$. We will need the following useful results. Write $K^* := K/G$ for simplicity.

Lemma 1. *$H^\sigma(K, L)$ is naturally isomorphic to $H(K^*, K^G \cup L^*)$.*

The other groups $H^\rho(K, L)$ do not have such a nice interpretation. The proof of the Proposition implies the following exact sequence. See, the discussion before 3.8, page 125, in [1].

Lemma 2. $(0) \to \sigma C(K, L) \xrightarrow{incl} \tau^j C(K, L) \xrightarrow{\tau} \tau^{j+1} C(K, L) \to (0)$ *for* $1 \le j \le p-1$.

Corollary. *With the notation of Lemma 2, if for some i we have $H_i^{\sigma}(K,L) = (0)$ then we get the inequalities*

$$dim\ H_i^{\tau^j}(K,L) \leq dim\ H_i^{\tau^{j+1}}(K,L)$$

for $1 \leq j \leq p-1$. In particular, $dim\ H_i^{\tau}(K,L) \leq dim\ H_i^{\tau^{p-1}}(K,L) = dim\ H_i^{\sigma}(K,L) = 0$.

Here *dim* is dimension as $\mathbf{Z}_p$ vector space. The above Corollary is a crucial observation in our proofs of the Theorem, and Goldsmith's result. The Corollary is proved easily by considering the long exact sequence corresponding to the short exact sequence in Lemma 2.

Using Čech homology P. A. Smith extended the results described above to a wide class of topological spaces. For details, see [1, Chapter III].

We will be repeatedly using the following general result about topology of algebraic varieties. It is certainly well-known to the experts.

Lemma 3. *Let $Y \subset Z$ be an inclusion of a non-empty Zariski-open subset Y into a normal irreducible variety Z. Then the natural homomorphism $\pi_1(Y) \to \pi_1(Z)$ is surjective. In particular, the natural homomorphism $H_1(Y;\Gamma) \to H_1(Z;\Gamma)$ is also surjective for any abelian coefficient group Γ. Thus, $b_1(Y) \geq b_1(Z)$.*

Remark. Very briefly, Lemma 3 follows from the fact that a normal complex analytic variety V is locally connected, i.e. for any point $p \in V$ and any small neighborhood N of p in V and any proper closed analyic subvariety S of V, the complement $N - S$ is path-connected. This implies that any loop in V is homologous to a loop not meeting S.

3. Proof of the Theorem

Let G be the Galois group of order p^l acting on $\overline{X}$ such that $\overline{X}/G \cong X$. We will prove the result by induction on l.

Since G contains a normal subgroup of order p^{l-1}, say H, we consider $\overline{X}/H$ with the induced morphism $\overline{X}/H \to X$ of degree p with Galois group $\mathbf{Z}_p$. Clearly this map is totaly ramified over D and unramified outside D. This implies that $\overline{X}/H$ is normal. Hence by an easy induction it is enough to prove that $H_1(\overline{X}/H;\mathbf{Z}_p) = 0$. Now we can assume that $l = 1$.

We use the Smith sequence in the Proposition for $G = \mathbf{Z}_p$, $K = \overline{X}$, $L = \pi^{-1}(D) \subset \overline{X}$. We will simply write $H_i(K, L)$ for homology with $\mathbf{Z}_p$-coefficients. Note that $K^G = L^G = L$.

$$(0) \to \tau C(K, L) \xrightarrow{i} C(K, L) \xrightarrow{\sigma} \sigma C(K, L) \to (0).$$

This gives the long exact sequence

$$\to H_1^\tau(K, L) \to H_1(K, L) \to H_1^\sigma(K, L) \to .$$

By Lemma 1, $H_1^\sigma(K, L) \cong H_1(X, D)$. The relative homology sequence for the pair (X, D) has the following terms

$$H_1(D) \to H_1(X) \to H_1(X, D) \to H_0(D) \to .$$

Since $H_1(X, \mathbf{Z}_p) = 0$ and D is connected by our assumptions, we see that $H_1(X, D) = (0)$. Now by Corollary to Lemma 2, we get $H_1^\tau(K, L) = (0)$. This implies that $H_1(K, L) = (0)$.

By a similar argument we get $H_1(K) = (0)$ as follows. We apply the Smith sequence to $L = \phi$. We first get $H_1^\sigma(K) = H_1(X, K^G) = (0)$. Next, this implies that $H_1^\tau(K) = (0)$, and finally $H_1(K) = (0)$.

This completes the proof of the Theorem.

Proof of Corollary 1.

Since C is totally ramified for the map $(x, y, z) \to (x, y)$ its inverse image in $\{z^{p^l} - f = 0\}$ is connected for each l. Clearly, the surface $\{z^n - f = 0\}$ is normal since f is reduced. Hence Corollary 1 follows immediately from the Theorem above.

Proof of Corollary 2.

We let $X_0 := \mathbf{C}^2 - Sing\ C$, $C_0 := C - Sing\ C$. Then $b_1(X_0) = 0$ and C_0 is connected. The inverse image of X_0 in $\{z^{p^l} - f = 0\}$, say $\widetilde{X_0}$, is smooth, irreducible. By the Theorem $H_1(\widetilde{X_0}) = (0)$. Since $\widetilde{X_0}$ is a Zariski-open subset of $\widetilde{S}$ we have a surjection $H_1(\widetilde{X_0}) \to H_1(\widetilde{S})$ by Lemma 3. Now Corollary 2 follows.

Proof of Corollary 3.

By our assumptions the affine surface $S := \{z^n - f = 0\}$ is irreducible and normal. By Corollary 1, $H_1(S; \mathbf{Z}_p) = (0)$. This easily implies that the first Betti number of S is 0. A non-constant unit, say u, in the coordinate ring, R, of S would have given finite unramified connected abelian coverings of S of arbitrarily large degrees by considering $R[u^{1/N}]$ for large N and taking the normalization of the corresponding affine surface. This is not possible since $b_1(S) = 0$.

Proof of (1). Zariski's theorem.

With the notation of the proof of Corollary 2, $\widetilde{X_0}$ is a Zariski-open subset of a smooth projective model $\widetilde{X}$ of $\{z^{p^l} - f = 0\}$. Hence there is a surjection $H_1(\widetilde{X_0}) \to H_1(\widetilde{X})$. Thus $b_1(\widetilde{X}) = 0$.

This proves Zariski's theorem.

4. Branched covers of S^n

We will prove (2) assuming (3) in the Introduction.

For simplicity, we assume that $S^m \subset S^{m+2}$ is a differentiable embedding. Let $p \in S^m$ and $p \in D \subset S^{m+2}$ a small $(m+2)$-disc with center p. It is easy to see that $D - S^m$ is homeomorphic to $S^1 \times D_0$, where D_0 is an open $(m+1)$-disc. It is easy to see that the induced maps $H_i(D - S^m) \to H_i(S^{m+2} - S^m)$ are isomorphisms, even with $\mathbf{Z}$-coefficients. In fact, $\pi_1(D - S^m; \mathbf{Z}) \cong \mathbf{Z}$, generated by a small loop $\gamma \subset D - S^m$ going around S^m and all higher homology groups are trivial. Except for the assertion about π_1, the same is true about $S^{m+2} - S^m$. Let $M \to S^{m+2}$ be a cyclic cover of degree p^l, branched precisely over S^m. By abuse of notation we write the inverse image of S^m in M again by S^m. Let $\widetilde{D - S^m}$ be the inverse image of $D - S^m$ in M. By (3), the induced maps (with $\mathbf{Z}_p$-coefficients) $H_i(\widetilde{D - S^m}) \to H_i(M - S^m)$ are isomorphisms. As remarked above, we have $\pi_1(\widetilde{D - S^m}) \cong \mathbf{Z}$ (being a subgroup of index n in $\pi_1(D - S^m)$) and hence $H_1(M - S^m; \mathbf{Z}_p) \cong \mathbf{Z}_p$. A generator of this group is a small loop around S^m, lying in $M - S^m$. From this we deduce easily that $H_1(M) = (0)$. Therefore $H_1(M; \mathbf{Z})$ is finite.

This proves that (2) is a consequence of (3).

5. Goldsmith's result

First we will point out where the mistake occurs in the proof of Goldsmith. In the proof of Theorem 2.1 in [5], which is our result (3), the author states that the sequence

$$(0) \to C_*(\widetilde{K}, \widetilde{L}) \xrightarrow{1+g+\cdots+g^{p^l-1}} C_*(\widetilde{K}, \widetilde{L}) \overset{g-1}{\to} C_*(\widetilde{K}, \widetilde{L})$$

is exact. But any chain of the form $gc - c$ for any chain c is in the kernel of the map given by $1 + g + \cdots + g^{p^l-1}$.

Now we will use the Smith sequence and give another proof of (3) in the Introduction.

By an easy induction on l, we can reduce to the case $l = 1$. As in the proof of the Theorem we have the inequalities

$$dim\ H_i^\tau(\widetilde{K}, \widetilde{L}) \leq dim\ H_i^\sigma(\widetilde{K}, \widetilde{L})$$

for $0 \leq i \leq r$, using the assumption that $K^G = \phi$ ($\widetilde{K} \to K$ is a regular topological covering). But $H_i^\sigma(\widetilde{K}, \widetilde{L}) \cong H_i(K, L)$ for all i, which are trivial for $i \leq r$ by assumption. Hence $H_i^\tau(\widetilde{K}, \widetilde{L}) = (0)$ for all $i \leq r$. Now the exact sequence

$$H_i^\tau(\widetilde{K}, \widetilde{L}) \to H_i(\widetilde{K}, \widetilde{L}) \to H_i^\sigma(\widetilde{K}, \widetilde{L})$$

implies that $H_i(\widetilde{K}, \widetilde{L}) = (0)$ for all $i \leq r$.

This completes the proof of Goldsmith's result.

Examples.

(1) Consider the surface $S := \{z^6 + y^3 + x^2 = 0\}$. Taking $p = 2$, by Corollary 1 we have $H_1(S; \mathbf{Q}) = (0)$. In fact, S admits a good $\mathbf{C}^*$-action so that S is topologically contractible. By resolving the singularity of S we obtain a smooth quasi-projective surface S' with an $\mathbf{A}^1$-fibration $S' \to E$, where E is an ellitic curve. This implies that $b_1(S'; \mathbf{Q})$ has rank at least 2.

Note that the curves $z^6 + x^2$, $z^6 + y^3$ are both reducible. This example illustrates the necessity of the assumptions of irreducibility of $f(x, y)$ and that n is a prime power in Zariski's theorem and Corollary 2.

(2) In [8], [9] and [11] examples are given of an irreducible polynomial $f(x, y)$ such that the affine curve $\{f = 0\}$ is smooth and the irregularity of the smooth affine surface $\{z^6 - f = 0\}$ is > 1.

(3) The affine curve $\{z^2 - x(x-1) = 0\}$ is isomorphic to $\mathbf{C}^* = \mathbf{C} - \{\textit{one point}\}$, which has Betti number 1. The same equation defines a smooth affine surface in $\mathbf{C}^3$ which is a cylinder over the above curve. Hence the first Betti number of the cylinder is also 1. This shows that the assumption of connectedness of C in Corollary 1 is necessary, even when n is a prime power.

References

1. G. Bredon, Introduction to Compact Transformation Groups. Academic Press, 1972.
2. H. Esnault, Fibre de Milnor d'un Cône sur une courbe plane singulière, Invent. Math., **68** (1982), no. 3, 477-496.
3. L. Goeritz, Die Bettische Zahlen Der Zyklischen Überlagerungsraume Der Knotenaussenraume, Amer. J. Math., **56** (1934), no. 1-4, 194-198.
4. B. Giesecke, Simpliziale Zerlegung abzählbarer analytische Räume, Math. Zeit., **83** (1964), 177-213.
5. D.L. Goldsmith, A Linking Invariant of Classical Link Concordance, Knot Theory. Proceedings, Plans-sur-Bex, Switzerland 1977, Lecture Notes in Math. 685, 135-170.
6. A. Libgober, Alexander polynomials of plane algebraic curves and cyclic multiple planes, Duke Math. J., **49** (1982), 833-851.
7. J. Milnor, Infinite Cyclic Coverings, Conference on the Topology of Manifolds at Michigan State University, 1967, Boston, Prindle, Weber & Schmidt (1969), 115-133.
8. M. Oka and D.T. Pho, Classification of sextics of torus type, Tokyo J. Math. **25** (2002), 399-433.
9. D.T. Pho, Classification of singularities of torus curves of type (2,3), Kodai Math. J., **24** (2001), 259-284.
10. F. Sakai, On the irregularity of cyclic coverings of the projective plane, Contemp. Math., **162** (1994), 359-369.
11. H. Tokunaga, Some examples of Zariski pairs arising from certain elliptic K3 surfaces, II: Degtyarev's conjecture, Math. Z., **230** (1999), 389-400.
12. O. Zariski, On the irregularity of cyclic multiple planes, Ann. of Math., **32** (1931), 485-511.

$\mathbb{A}^1_*$-fibrations on affine threefolds

R.V. Gurjar

School of Mathematics, Tata Institute for Fundamental Research,
400005 Homi Bhabha Road, Mumbai, India
E-mail: gurjar@math.tifr.res.in

M. Koras

Institute of Mathematics, Warsaw University,
ul. Banacha 2, Warsaw, Poland
E-mail: koras@mimuw.edu.pl

K. Masuda

School of Science and Technology, Kwansei Gakuin University,
Hyogo 669-1337, Japan
E-mail: kayo@kwansei.ac.jp

M. Miyanishi

Research Center for Mathematical Sciences, Kwansei Gakuin University,
Hyogo 669-1337, Japan
E-mail: miyanisi@kwansei.ac.jp

P. Russell

Department of Mathematics and Statistics, McGill University,
Montreal, 805 Sherbooke St. West, Canada
E-mail: russell@math.mcgill.ca

The main theme of the present article is $\mathbb{A}^1_*$-fibrations defined on affine threefolds. The difference between $\mathbb{A}^1_*$-fibration and the quotient morphism by a G_m-action is more essential than in the case of an $\mathbb{A}^1$-fibration and the quotient morphism by a G_a-action. We consider necessary (and partly sufficient) conditions under which a given $\mathbb{A}^1_*$-fibration becomes the quotient morphism by a G_m-action. Then we consider flat $\mathbb{A}^1_*$-fibrations which are expected to be surjective, but this turns out to be not the case by an example of Winkelmann [47]. This example gives also a quasi-finite endomorphism of $\mathbb{A}^2$ which is not surjective [14]. We consider then the structure of a smooth affine threefold which has a flat $\mathbb{A}^1$-fibration or a flat $\mathbb{A}^1_*$-fibration. More precisely, we consider affine threefolds with the additional condition that they are contractible.

Keywords: $\mathbb{A}^1_*$-fibration, G_m-action, contractible affine threefold, affine modification, $\mathbb{A}^1$-fibration, G_a-action.

Introduction

In affine algebraic geometry, our knowledge on affine algebraic surfaces is fairly rich with various methods of studying them. Meanwhile, knowledge of affine threefolds is very limited. It is partly because strong geometric approaches are not available or still under development.

A possible geometric approach is to limit ourselves to the case where the affine threefolds in consideration have fibrations by surfaces or curves, say the affine plane $\mathbb{A}^2$ or the affine line $\mathbb{A}^1$ or have group actions which give the quotient morphisms. To be more concrete, $\mathbb{A}^2$-fibrations were observed to give characterizations of the affine 3-space $\mathbb{A}^3$ (see [20, 35, 37]). Meanwhile, the quotient morphism $q : Y \to Y//G_a$ for an affine threefold has been considered by many people including P. Bonnet, Sh. Kaliman, D. Finston, J.K. Deveney et al. See [9] for the references.

In [9], some of the present authors considered $\mathbb{A}^1$-fibrations and quotient morphisms for an action of G_a. There, a main point is that an $\mathbb{A}^1$-fibration is factored by a quotient morphism by G_a and the study is essentially reduced to the quotient morphism by G_a. In this article, we consider $\mathbb{A}^1_*$-fibrations and quotient morphisms by G_m. Contrary to the case of $\mathbb{A}^1$-fibrations, $\mathbb{A}^1_*$-fibrations are not necessarily factored by a quotient morphism by G_m. The possibility to factor depends on the nature of the singular fibers of the $\mathbb{A}^1_*$-fibration (see Theorem 2.20). The quotient morphism by G_m provides us a fertile ground for research, and the $\mathbb{A}^1_*$-fibrations seem to be even more fertile, but mysterious. For example, singular fibers of $\mathbb{A}^1_*$-fibration are fully described in the surface case (see Lemma 2.2) but not completely in the case of threefolds, and the locus of singular fibers is shown to be a closed set with the unmixedness condition on singular fibers (cf. Lemma 2.11).

Our primary purpose in this article is to develop a theory of dealing with $\mathbb{A}^1_*$-fibrations by combining algebraic geometry with algebraic topology and commutative algebra and to apply it to elucidate the structure of such objects as homology (or contractible) threefolds. Thus the article contains known results as well as what we hope to be original ones.

Here we comment on the notation $\mathbb{A}^1_*$ for which the notation $\mathbb{C}^*$ is in more common use. But $\mathbb{C}^*$ is used to mean, for example, that an affine scheme Spec A has only constant invertible functions, i.e., $A^* = \mathbb{C}^*$. It is better to distinguish the curve $\mathbb{C}^*$ from the multiplicative group $\mathbb{C}^*$. Hence we use $\mathbb{A}^1_*$ for the curve and G_m for the group.

1. Preliminaries

We summarize various results which we make use of in the subsequent arguments. The ground field is always the complex number field $\mathbb{C}$. A *homology n-fold* is a smooth affine algebraic variety X of dimension n such that $H_i(X;\mathbb{Z}) = 0$ for every $i > 0$. If $H_i(X;\mathbb{Q}) = 0$ for every $i > 0$ instead, we call X a *$\mathbb{Q}$-homology n-fold.*

Lemma 1.1. *Let $X = \operatorname{Spec} A$ be a homology n-fold. Let (V, D) be a pair of a smooth projective variety V and a reduced effective divisor D on V such that X is a Zariski open set of V with $D = V - X$ and D is a divisor with simple normal crossings. Then the following assertions hold.*

(1) *X is factorial and $A^* = \mathbb{C}^*$.*
(2) $H^1(V, \mathcal{O}_V) = H^2(V, \mathcal{O}_V) = 0.$

Proof. Similar considerations can be found in [6]. To simplify the arguments, we treat the case $n = 3$. Consider the exact sequence of $\mathbb{Z}$-cohomology groups for the pair (V, D),

$$\begin{array}{llll} 0 \to H^0(V,D) \to H^0(V) & \to H^0(D) & \to H^1(V,D) \\ \to H^1(V) & \to H^1(D) & \to H^2(V,D) \to H^2(V) \\ \to H^2(D) & \to H^3(V,D) \to H^3(V) & \to H^3(D) \\ \to H^4(V,D) \to H^4(V) & \to H^4(D) & \to H^5(V,D) \\ \to H^5(V) & \to H^5(D). \end{array}$$

By Lefschetz duality, we have $H^i(V, D) \cong H_{6-i}(X)$ for $0 \le i \le 6$. Since X is a homology manifold, $H_i(X) = 0$ for $1 \le i \le 3$ and since X is affine, $H_i(X) = 0$ for $4 \le i \le 6$ (see [32, Theorem 7.1]). Hence we obtain the isomorphism $H^i(V) \cong H^i(D)$ for $0 \le i < 6$. Since $H^5(D) = 0$ as $\dim D = 2$, we have $H^5(V;\mathbb{Z}) = 0$. By Poincaré duality, we have $H_1(V;\mathbb{Z}) = 0$. Then the universal coefficient theorem [46, Theorem 3, p. 243] implies that $H^1(V;\mathbb{Z}) = 0$, and hence $H^1(V;\mathbb{C}) = 0$. The Hodge decomposition then implies that $H^1(V, \mathcal{O}_V) = 0$. Now consider an exact sequence

$$0 \longrightarrow \mathbb{Z} \longrightarrow \mathcal{O}_V \stackrel{\exp(2\pi i\ )}{\longrightarrow} \mathcal{O}_V^* \longrightarrow 0$$

and the associated exact sequence

$$H^1(V, \mathcal{O}_V) \to H^1(V, \mathcal{O}_V^*) \to H^2(V;\mathbb{Z}) \to H^2(V, \mathcal{O}_V). \tag{1}$$

On the other hand, consider the isomorphism $H^4(V;\mathbb{Z}) \cong H^4(D;\mathbb{Z})$. By the Mayer-Vietoris exact sequence, it follows that $H^4(D;\mathbb{Z})$ is a free abelian group generated by the classes of the irreducible components of D. The Poincaré duality and the universal coefficients theorem implies

$$H^4(V;\mathbb{Z}) \cong H_2(V;\mathbb{Z}) \cong H^2(V;\mathbb{Z}),$$

where we note that $H_1(V;\mathbb{Z}) = 0$. Hence we know that $H^2(V;\mathbb{Z})$ is a free abelian group generated by the first Chern classes of the irreducible components of D. Since all these classes are algebraic, we conclude that $H^2(V;\mathbb{Z}) = H^{1,1}(V) \cap H^2(V;\mathbb{C})$. Hence $H^2(V,\mathcal{O}_V) = 0$.

Now, by (1) above, we know that

$$\mathrm{Pic}\,(V) \cong H^2(V;\mathbb{Z}) \cong H^4(D;\mathbb{Z}).$$

$\mathrm{Pic}\,(X)$ is isomorphic to the quotient group of $\mathrm{Pic}\,(V)$ modulo the subgroup generated by the classes of irreducible components of D. Hence $\mathrm{Pic}\,(X) = 0$, and X is factorial. Since there are no linear equivalence relations among the irreducible components of D, it follows that $A^* = \mathbb{C}^*$. □

The same argument with the $\mathbb{Z}$-coefficients replaced by $\mathbb{Q}$-coefficients in the proof of Lemma 1.1 shows that *for a $\mathbb{Q}$-homology n-fold X it holds that*

(1) *X is $\mathbb{Q}$-factorial, i.e., $\mathrm{Pic}\,(X)$ is a finite abelian group, and $A^* = \mathbb{C}^*$.*
(2) $H^1(V,\mathcal{O}_V) = H^2(V,\mathcal{O}_V) = 0$.

The following result is an important consequence of a result of Hamm [13] that an affine variety of dimension n defined over $\mathbb{C}$ has the homotopy type of a CW complex of real dimension n.

Lemma 1.2. *Let X be an affine variety of dimension n. Then $H_n(X;\mathbb{Z})$ is a free abelian group of finite rank. Furthermore, $H^i(X;\mathbb{Z}) = 0$ for $i > n$.*

Proof. The vanishing of $H^i(X;\mathbb{Z})$ follows from the first assertion and the vanishing of $H_i(X;\mathbb{Z})$ for $i > n$. By the universal coefficient theorem, we have

$$H^i(X:\mathbb{Z}) = \mathrm{Hom}(H_i(X;\mathbb{Z}),\mathbb{Z}) \oplus \mathrm{Ext}^1(H_{i-1}(X;\mathbb{Z}),\mathbb{Z}).$$

If $i = n+1$, $H^{n+1}(X;\mathbb{Z}) = 0$ because $H_n(X;\mathbb{Z})$ is torsion free. For $i > n+1$, the result is clear because $H_i(X;\mathbb{Z}) = H_{i-1}(X;\mathbb{Z}) = 0$. □

The following result (see [41, Lemma 1.5] for the proof) is frequently used below.

Lemma 1.3. *Let $f : Y \to X$ be a dominant morphism of algebraic varieties such that the general fibers are irreducible. Then the natural homomorphism $\pi_1(Y) \to \pi_1(X)$ is surjective.*

As an extension of the above argument, one can give a different proof of the following result in [33, Theorem 1]. In the original proof, one has to use, in a crucial step of the proof, a rather difficult result that a smooth, quasi-affine surface X of log Kodaira dimension $-\infty$ is either affine-ruled or contains $\mathbb{A}^2/\Gamma$ as an open set so that $X - \mathbb{A}^2/\Gamma$ is a disjoint union of affine lines, where Γ is a small finite subgroup of $\mathrm{GL}(2, \mathbb{C})$ (cf. the proof of [33]).

Theorem 1.4. *Let the additive group scheme G_a act nontrivially on the affine 3-space $\mathbb{A}^3$. Then $\mathbb{A}^3//G_a \cong \mathbb{A}^2$.*

For the proof, we need two lemmas.

Lemma 1.5. *Let Y be a smooth contractible affine threefold with a nontrivial G_a-action and let $X = Y//G_a$. Let V be a normal projective surface containing X as an open set such that V is smooth along $D := V - X$ and D is a divisor of simple normal crossings. Let $\rho : V' \to V$ be a minimal resolution of singularities of X. Let $p_g(V')$ be the geometric genus of the surface V' which is a birational invariant independent of the choice of V'. Then the following assertions hold.*

(1) *Both X and X° are simply connected, where X° is the smooth part of X.*
(2) *$H_1(X; \mathbb{Z}) = 0$. Further, if $p_g(V') = 0$ and $H^1(D; \mathbb{Z}) = 0$, then $H_2(X; \mathbb{Z}) = 0$ and hence X is contractible.*
(3) *X is smooth under the additional assumptions in the assertion (2).*

Proof. (1) Write $Y = \mathrm{Spec}\, B$ and $X = \mathrm{Spec}\, A$. Then B is factorial by Lemma 1.1 and hence A is factorial, too. So, the singular locus of X is a finite set of quotient singular points [9, Lemma 3.4]. Hence the quotient morphism $q : Y \to X$ has no fiber components of dimension two and $Y - q^{-1}(X^\circ)$ has dimension ≤ 1. Hence $\pi_1(q^{-1}(X^\circ)) = \pi_1(Y) = (1)$. By Lemma 1.3, both X and X° are simply connected.

(2) Let X' be the inverse image $\rho^{-1}(X)$ and let E be the exceptional locus of ρ which is a divisor with simple normal crossings. Let $F = D + E$.

Then we have an exact sequence of cohomology groups

$$H^3(V', F; \mathbb{Z}) \to H^3(V'; \mathbb{Z}) \to H^3(F : \mathbb{Z})$$

where $H^3(F;\mathbb{Z}) = 0$ and $H^3(V', F; \mathbb{Z}) \cong H_1(X^\circ; \mathbb{Z})$ by the Lefschetz duality and $H_1(X^\circ;\mathbb{Z}) = 0$ because $\pi_1(X^\circ) = 1$. Hence $H_1(V';\mathbb{Z}) \cong H^3(V';\mathbb{Z}) = 0$. As in the proof of Lemma 1.1, this implies that $H^1(V';\mathbb{C}) = 0$ and $H^1(V', \mathcal{O}_{V'}) = 0$. Hence we have an exact sequence

$$0 \to H^1(V', \mathcal{O}^*_{V'}) \to H^2(V'; \mathbb{Z}) \to H^2(V', \mathcal{O}_{V'}).$$

Since $H^2(V', \mathcal{O}_{V'}) = 0$ by the hypothesis, we have $\text{Pic}\,(V') \cong H^1(V', \mathcal{O}^*_{V'}) \cong H^2(V';\mathbb{Z})$. Namely the group $H^2(V';\mathbb{Z})$ of topological 2-cocycles is generated by algebraic classes of divisors. Since X is factorial, $\text{Pic}\,(V')$ is generated by the classes of irreducible components of F. Since $H^2(F;\mathbb{Z})$ is a free abelian group generated by the classes of irreducible components of F, the natural homomorphism $H^2(V';\mathbb{Z}) \to H^2(F;\mathbb{Z})$ is an isomorphism.

Consider the long exact sequence associated to the pair (V', F)

$$H^1(V';\mathbb{Z}) \to H^1(F;\mathbb{Z}) \to H^2(V', F; \mathbb{Z}) \to H^2(V';\mathbb{Z}) \to H^2(F;\mathbb{Z}),$$

where $H^1(V';\mathbb{Z}) = 0$ as $H_1(V';\mathbb{Z}) = 0$ and $H^2(V';\mathbb{Z}) \to H^2(F;\mathbb{Z})$ is an isomorphism. Hence $H^1(F;\mathbb{Z}) \cong H^2(V', F; \mathbb{Z})$. Since the divisors D and E are disjoint from each other, $H^1(F;\mathbb{Z}) \cong H^1(D;\mathbb{Z}) \oplus H^1(E;\mathbb{Z})$, where $H^1(D;\mathbb{Z}) = 0$ by the hypothesis and $H^1(E;\mathbb{Z}) = 0$ because X has at worst quotient singular points. So, $H^1(F;\mathbb{Z}) = 0$. By the Lefschetz duality, $H^2(V, F; \mathbb{Z}) \cong H_2(X^\circ;\mathbb{Z})$. It follows that $H_2(X^\circ;\mathbb{Z}) = 0$.

Now let $\{P_1, \ldots, P_r\}$ be the set of singular points of X. Choose a closed neighborhood T_i of P_i for every i so that $T_i \cap T_j = \emptyset$ if $i \neq j$. Let ∂T_i be the boundary of T_i and let $\partial T = \partial T_1 \cup \cdots \cup \partial T_r$. By [40, Proof of Lemma 2.2], we have an exact sequence

$$0 \to H_2(X^\circ;\mathbb{Z}) \to H_2(X;\mathbb{Z}) \to H_1(\partial T;\mathbb{Z}) \to H_1(X^\circ;\mathbb{Z}),$$

where $H_1(X^\circ;\mathbb{Z}) = 0$ as $\pi_1(X^\circ) = 1$ and $H_1(\partial T;\mathbb{Z})$ is a finite abelian group because $\pi_1(\partial T_i)$ is a finite group as the local fundamental group $\pi_{1,P_i}(X)$ of the quotient singular point P_i. Since $H_2(X^\circ;\mathbb{Z}) = 0$ as shown above, $H_2(X;\mathbb{Z})$ is a finite abelian group. Since $H_2(X;\mathbb{Z})$ is torsion free by Lemma 1.2, it follows that $H_2(X;\mathbb{Z}) = 0$.

(3) Since $\pi_1(X) = \pi_1(X^\circ) = 1$ and X is contractible, X is smooth by the so-called *affine Mumford theorem* [8, Theorem 3.6] which we state below. □

Lemma 1.6. *Let X be a normal affine surface such that X is contractible and the smooth part X° of X is simply connected. Then X is smooth.*

Now the different proof of Theorem 1.4 is given as follows. With the notations in Lemma 1.5, let $Y = \mathbb{A}^3$. Choose a general linear hyperplane L in such a way that the quotient morphism q restricted on L is a dominant morphism to $X := \mathbb{A}^3//G_a$ and L meets the inverse image $q^{-1}(\mathrm{Sing}X)$ in finitely many points. Then it follows that X is rational and $\overline{\kappa}(X^\circ) = -\infty$. This implies that $p_g(V') = 0$ and $H^1(D;\mathbb{Z}) = 0$ with the notations in Lemma 1.5. Hence X is smooth. It is clear that A is factorial, $A^* = \mathbb{C}^*$ and $\overline{\kappa}(X) = -\infty$. By a characterization of the affine plane, X is isomorphic to $\mathbb{A}^2$.

Remark 1.7 To be accurate, we have to use the result in [33, p. 49] mentioned before Theorem 1.4 in the course of the proof of Lemma 1.6. Hence it is very desirable to give a new proof of Lemma 1.6 which is more topological and not depending on the result used in [33].

In [18, Theorem 2.7], Kaliman and Saveliev state a result stronger than Lemma 1.5. They state that *for every nontrivial G_a-action on a smooth contractible affine algebraic threefold Y, the quotient $X = Y//G_a$ is a smooth contractible affine surface.*

2. $\mathbb{A}^1_*$-fibration

Let $p : Y \to X$ be a dominant morphism of algebraic varieties. We call p an $\mathbb{A}^1$-(resp. $\mathbb{A}^1_*$-) fibration if general fibers of p are isomorphic to $\mathbb{A}^1$ (resp. $\mathbb{A}^1_*$). Here $\mathbb{A}^1_*$ is the affine line with one point deleted. A *singular* fiber of p is a fiber which is not scheme-theoretically isomorphic to $\mathbb{A}^1$ (resp. $\mathbb{A}^1_*$). We say that an $\mathbb{A}^1_*$-fibration is *untwisted* (resp. *twisted*) if the generic fiber $Y_K := Y \times_X \mathrm{Spec}\, K$ has two K-rational places at infinity (resp. if Y_K has one non K-rational place at infinity), where K is the function field of X over $\mathbb{C}$. In the untwisted case, Y_K is isomorphic to $\mathbb{A}^1_{*,K} := \mathrm{Spec}\, K[t, t^{-1}]$, while Y_K is a non-trivial K-form of $\mathbb{A}^1_{*,K}$ and Pic $(Y_K) \cong \mathbb{Z}/2\mathbb{Z}$ in the twisted case. If Y is a factorial affine variety, then any $\mathbb{A}^1_*$-fibration on Y is untwisted because Y_K is then factorial. In fact, if $Y = \mathrm{Spec}\, B$ and $X = \mathrm{Spec}\, A$ then B is factorial and hence the quotient ring $B \otimes_A K$ is factorial. Since $Y_K = \mathrm{Spec}\, B \otimes_A K$, Y_K is factorial. In the present article, an $\mathbb{A}^1_*$-fibration is *always assumed to be untwisted.*

If the additive group scheme G_a (resp. the multiplicative group scheme G_m) acts on an affine threefold Y then the algebraic quotient $Y//G_a$ (resp. $Y//G_m$) exists. Namely, if $Y = \mathrm{Spec}\, B$ and X is the quotient of Y by G_a (resp. G_m), then the invariant subring $A = B^{G_a}$ (resp. $A = B^{G_m}$) is the coordinate ring of X and the quotient morphism $q : Y \to X$ is given by the inclusion $A \hookrightarrow B$. Note that A is finitely generated over $\mathbb{C}$ by a lemma of Zariski in the case of G_a [49] and by the well-known result on reductive group actions in the case of G_m. In the case of G_m-actions, we assume *unless otherwise mentioned* that $\dim A = \dim B - 1$. Hence q has relative dimension one. In the case of a G_a-action, the quotient morphism $q : Y \to X$ is an $\mathbb{A}^1$-fibration which is not necessarily surjective, but in the case of a G_m-action, q is surjective and either an $\mathbb{A}^1$-fibration or an $\mathbb{A}^1_*$-fibration according as general orbits admit fixed points or not. If the quotient morphism $q : Y \to X$ is an $\mathbb{A}^1_*$-fibration, it is untwisted by a theorem of Rosenlicht [43]. We recall the following fundamental result on the quotients under reductive algebraic groups [2]. When we consider the quotient morphism by G_a (or G_m), we denote it by q in most cases, but an $\mathbb{A}^1$-fibration (or $\mathbb{A}^1_*$-fibration) by p.

Lemma 2.1. *Let G be a connected reductive algebraic group acting on a smooth affine algebraic variety Y and let $q : Y \to X$ be the quotient morphism. Then, for two points P_1, P_2, we have $q(P_1) = q(P_2)$ if and only if $\overline{GP_1} \cap \overline{GP_2} \neq \emptyset$. Hence for any point Q of X, the fiber $q^{-1}(Q)$ contains a unique closed orbit.*

In order to distinguish the case of surfaces from the case of threefolds, we use the notation like $p : X \to C$ in the case of surfaces and retain the notation $p : Y \to X$ for the case of threefolds or varieties in general.

Lemma 2.2. *Let X be a normal affine surface and let $p : X \to C$ be an $\mathbb{A}^1_*$-fibration. Then the following assertions hold.*

(1) *A singular fiber F of p is written in the form $F = \Gamma + \Delta$, where $\Gamma = m\mathbb{A}^1_*, \emptyset$ or $m\mathbb{A}^1 + n\mathbb{A}^1$ with two $\mathbb{A}^1$'s meeting in one point which may be a cyclic quotient singular point of X, $\Gamma \cap \Delta = \emptyset$ and Δ is a disjoint union of affine lines with multiplicities, each of which may have a unique cyclic singular quotient point of X. Each type of a singular fiber is realizable.*
(2) *If p is the quotient morphism of a G_m-action on X and p is an $\mathbb{A}^1_*$-fibration, the part Δ is absent in a singular fiber F. Hence F is either $m\mathbb{A}^1_*$ or $m\mathbb{A}^1 + n\mathbb{A}^1$ in the above list.*

Proof. (1) In [36, Lemma 4] and [48, Proposition 5.1 (b)], the case where X is a smooth affine surface is treated. In [40, Lemma 2.9], singular fibers of an $\mathbb{A}^1_*$-fibration are classified, where singularities of the surface X are assumed to be quotient singularities, but one can drop this condition and make the same argument by assuming simply that X is normal. The normal case is also treated in [4, Proposition 3.8 and Theorem 4.18].

(2) By the assertion (1) and Lemma 2.1, the only possible cases of a singular fiber of the quotient morphism p are the two cases listed in the statement and the case where $F_{\rm red}$ is irreducible and isomorphic to $\mathbb{A}^1$ which may contain a unique cyclic quotient singularity. We show that the last case does not occur. Suppose that $F_{\rm red} \cong \mathbb{A}^1$. Then F contains a unique fixed point P. Suppose first that X is smooth at P. Look at the induced tangential representation of G_m on $T_{X,P}$ which is diagonalizable and has weights $-a, b$ with $a, b > 0$. Then it follows that F is locally near P a union of two irreducible components meeting in P. This is a contradiction. Suppose that X is singular at P. Let $\sigma : \widetilde{X} \to X$ be a minimal resolution of singularity at P. Then the inverse image $\sigma^{-1}(F_{\rm red})$ is the proper transform G of $F_{\rm red}$ and a linear chain of $\mathbb{P}^1$'s with G meeting one of the terminal components of the linear chain. Since the G_m-ation on X lifts to $\widetilde{X}$, each irreducible component of $\sigma^{-1}(F_{\rm red})$ is G_m-stable and the other terminal component, say $\widetilde{F}$, of the linear chain has an isolated fixed point, say $\widetilde{P}$. Note that we obtain an affine surface from $\widetilde{X}$ by removing all components of $\sigma^{-1}(F_{\rm red})$ except for the component $\widetilde{F}$. Now looking at the induced tangential representation of G_m on $T_{\widetilde{X},\widetilde{P}}$, we have a contradiction as in the above smooth case. So, the case $F_{\rm red} \cong \mathbb{A}^1$ does not occur. □

Special attention has to be paid in the case where the G_m-quotient morphism $q : Y \to X$ is an $\mathbb{A}^1$-fibration. We consider the surface case first and then treat the case of threefolds.

Lemma 2.3. *Let X be a normal affine surface with a G_m-action. Suppose that the quotient morphism $\rho : X \to C$ is an $\mathbb{A}^1$-fibration, where C is a normal affine curve. Then the following assertions hold.*

(1) *There exists a closed curve Γ on X such that the restriction of q onto Γ induces an isomorphism between Γ and C, i.e., Γ is a cross-section of ρ.*
(2) *Suppose further that X is smooth. Then every fiber of q is reduced and isomorphic to $\mathbb{A}^1$. Hence X is a line bundle over C.*
(3) *Suppose that X is smooth and C is rational. Then X is isomorphic to a direct product $C \times \mathbb{A}^1$.*

Proof. (1) Let $\overline{\rho} : \overline{X} \to \overline{C}$ be a G_m-equivariant completion of $\rho : X \to C$. Namely, $\overline{X}$ is a normal projective surface containing X as an open set, $\overline{C}$ is the smooth completion of C. We may assume that $\overline{X}$ is smooth along $\overline{X} \setminus X$, ρ extends to a morphism $\overline{\rho}$ and the G_m-action extends to $\overline{X}$ so that $\overline{\rho}$ is a $\mathbb{P}^1$-fibration. Since ρ is an $\mathbb{A}^1$-fibration, there exists an open set U of C such that $\rho^{-1}(U) \cong U \times \mathbb{A}^1$ with G_m acting on $\mathbb{A}^1$ in a natural fashion. Let Γ_0 be the fixed point locus in $\rho^{-1}(U)$ and let Γ (resp. $\overline{\Gamma}$) be the closure of Γ_0 in X (resp. $\overline{X}$). Then the restriction of $\overline{\rho}$ onto $\overline{\Gamma}$ is a birational morphism onto $\overline{C}$, hence an isomorphism. Thus $\overline{\Gamma}$ is a cross-section of $\overline{\rho}$. On the other hand, $\overline{\rho}$ has another cross-section $\overline{\Gamma}_\infty$ lying outside X. Since ρ is an $\mathbb{A}^1$-fibration, every fiber of ρ is a disjoint union of the affine lines, hence it consists of a single affine line by Lemma 2.1. Suppose that for a point $\alpha \in C$, the fiber F_α does not meet Γ. Then $\overline{\Gamma}$ meets the fiber $\overline{F}_\alpha := \overline{\rho}^{-1}(\alpha)$ outside F_α, and F_α has an isolated G_m-fixed point P. This leads to a contradiction if we argue as in the proof of Lemma 2.2. Namely, if X is smooth at P, then consider the induced tangential representation at P. Otherwise, consider a minimal resolution of singularity at P and a G_m-fixed point appearing in one of the terminal components of the exceptional locus which does not meet the proper transform of F_α. Thus Γ meets every fiber of ρ and is smooth.

(2) With the above notations, suppose that X has a singular point on the fiber F_α. Since G_m acts transitively on $F_\alpha - F_\alpha \cap \Gamma$, X has cyclic singularity at the point $P_\alpha := F_\alpha \cap \Gamma$. Let $\sigma : \widetilde{X} \to \overline{X}$ be the minimal resolution of singularities and let $\Delta = \sigma^{-1}(P_\alpha)$. The $\mathbb{P}^1$-fibration $\overline{\rho} : \overline{X} \to \overline{C}$ lifts to a $\mathbb{P}^1$-fibration $\widetilde{\rho} : \widetilde{X} \to \overline{C}$. The proper transform $\widetilde{\Gamma}$ of $\overline{\Gamma}$ is a cross-section of $\widetilde{\rho}$. The fiber $\widetilde{F}_\alpha$ of $\widetilde{\rho}$ corresponding to $\overline{F}_\alpha$ contains the linear chain Δ in such a way that one terminal component, say G, meets $\widetilde{\Gamma}$ and the other terminal component meets the proper transform of F_α. Further, all the components except for the terminal component G meeting $\widetilde{\Gamma}$ can be contracted smoothly. In fact, we can replace the fiber F_α by $G - \{Q\}$ without losing the affineness of X, where Q is the point where G meets the adjacent component of $\widetilde{F}_\alpha$. If X is smooth along F_α, it is reduced. If X is smooth, the $\mathbb{A}^1$-fibration ρ has no singular fibers. Hence X is an $\mathbb{A}^1$-bundle over C [9, Lemma 1.15]. Since the two cross-sections $\overline{\Gamma}$ and $\overline{\Gamma}_\infty$ are disjoint from each other over C, it follows that X is a line bundle over C.

(3) Since Pic $(C) = 0$, every line bundle is trivial. □

The proof of the assertion (2) implies that every normal affine surface with a G_m-action and an $\mathbb{A}^1$-fibration as the quotient morphism is constructed from a line bundle over C by a succession of blowing-ups with centers on the zero section and contractions of linear chains. The following result is well-known.

Lemma 2.4. *Let $Y = \text{Spec } B$ be a smooth affine threefold with a G_m-action and let $q : Y \to X := \text{Spec } A$ be the quotient morphism. Assume that q is an $\mathbb{A}^1$-fibration and that q is equi-dimensional. Then the following assertions hold.*

(1) *We may assume that B is positively graded, that is, B is a graded A-algebra indexed by $\mathbb{Z}_{\geq 0}$.*
(2) *A is normal and factorially closed in B.*
(3) *$q : Y \to X$ has a cross-section S. Namely, each fiber of q has dimension one and meets S in one point transversally.*
(4) *X is smooth, and Y is a line bundle over X.*
(5) *If Y is factorial, then the equi-dimensionality condition is automatically satisfied and Y is isomorphic to a direct product $X \times \mathbb{A}^1$ with G_m acting on the factor $\mathbb{A}^1$ in a natural fashion.*

Proof. (1) By [9, Lemma 1.2], there exists a nonzero element $a \in A$ such that $B[a^{-1}] = A[a^{-1}][u]$, where u is an element of B algebraically independent over the quotient field $Q(A)$. In fact, u is a variable on a general fiber of q. We may assume that the G_m-action on the fiber is given by ${}^t u = tu$. Note that the G_m-action gives a $\mathbb{Z}$-graded ring structure on B. Let b be a homogeneous element of B. We can write $a^r b = f(u)$, where $f(u) = a_0 u^m + a_1 u^{m-1} + \cdots + a_m \in A[u]$ with $a_0 \neq 0$. Then we have

$$a^r({}^t b) = f(tu) = a_0 t^m u^m + a_1 t^{m-1} u^{m-1} + \cdots + a_m.$$

Hence $a^r b = a_0 u^m$ and b has degree m. This implies that every homogeneous element of B has degree ≥ 0.

(2) The normality of A is well-known [27, p. 100], and the factorial closedness follows from the assertion (1).

(3) Let S_0 be the closed set in the open set $q^{-1}(D(a))$ defined by $u = 0$. Hence S_0 is the locus of the fixed points in the fibers $q^{-1}(Q)$ when $Q \in D(a)$. Let S be the closure of S_0 in Y. Then S is a rational cross-section of q. We shall show that S meets every fiber of q in one point. Let P be a point of X, let C be a general irreducible curve on X passing through P and let $Z = Y \times_X C$. Then $q_C : Z \to C$, the projection onto C, is an $\mathbb{A}^1$-fibration. Let $\nu_Z : \widetilde{Z} \to Z$ and $\nu_C : \widetilde{C} \to C$ be the normalizations of Z and C respectively. Then there exists an $\mathbb{A}^1$-fibration $\rho : \widetilde{Z} \to \widetilde{C}$ such that $\nu_C \cdot \rho = q_C \cdot \nu_Z$. The normal affine surface $\widetilde{Z}$ has the induced G_m-action and ρ is the quotient morphism. Choosing the open set $D(a)$ small enough,

we may assume that $C \cap D(a)$ is a non-empty set contained in the smooth part of C. Let $\Gamma_0 = S_0 \cap Z$ and let $\widetilde{\Gamma}$ be the closure of Γ_0 in $\widetilde{Z}$. Let $\widetilde{P}$ be a point of $\widetilde{C}$ lying over P. Let $\widetilde{F} = \rho^{-1}(\widetilde{P})$ and $F = q^{-1}(P)$. By Lemma 2.3, $\widetilde{\Gamma}$ meets $\widetilde{F}$ in a single point. This implies that S meets F in a point, say R. By [9, Theorem 1.15 and Lemma 3.5] and Lemma 2.1, F consists of one irreducible component which is contractible and smooth outside the point R because it admits a non-trivial G_m-action with a fixed point R. Since Y is smooth, the local intersection multiplicity $i(S, F; R) = 1$. This implies that F is reduced and smooth also at R. Hence F is isomorphic to $\mathbb{A}^1$.

(4) The morphism $q : Y \to X$ induces a quasi-finite birational morphism $q|_S : S \to X$. Then it is an isomorphism by Zariski's main theorem. Since S is smooth by the argument in the proof of the assertion (3), X is also smooth. Furthermore, every fiber of q is an affine line. Hence Y is an $\mathbb{A}^1$-bundle over X. Since a smooth completion $\overline{q} : \overline{Y} \to \overline{X}$ is a $\mathbb{P}^1$-fibration with two cross-sections (the zero section and the infinity section) which are disjoint over X, Y is a line bundle over X.

(5) By [9, Lemma 1.10], q does not have codimension one fiber components. Since X is factorial, Y is isomorphic to $X \times \mathbb{A}^1$. This fact follows from the proof of the assertion (1). In fact, one can take the element u to be a prime element of B. Then, the relation $a^r b = a_0 u^m$ for a homogeneous element b of degree m implies that $b = cu^m$ for a certain element $c \in A$. Hence $B = A[u]$. □

In [9, Theorem 1.4], it is shown that any $\mathbb{A}^1$-fibration $p : Y \to X$ on an affine threefold Y is factored by the quotient morphism $q : Y \to Z$ by a certain G_a action on Y. Meanwhile, this is not the case with an $\mathbb{A}^1_*$-fibration. Examples can be easily produced by using Lemma 2.2, (1).

A homology threefold Y is a *contractible threefold* if it is topologically contractible. A homology threefold is contractible if and only if $\pi_1(Y) = 1$. We consider a G_m-action on such a threefold.

Lemma 2.5. *Let G_m act nontrivially on a $\mathbb{Q}$-homology threefold Y. Then the fixed point locus Y^{G_m} is a non-empty connected closed subset of X. Furthermore, $H_i(Y^{G_m}; \mathbb{Z}/p\mathbb{Z}) = 0$ for every $i > 0$ and for infinitely many prime numbers p.*

Proof. Since $H_i(Y; \mathbb{Z})$ is a finite group, there exist infinitely many primes p such that Y is $\mathbb{Z}/p\mathbb{Z}$-acyclic, i.e., $H_i(Y; \mathbb{Z}/p\mathbb{Z}) = 0$ for every $i > 0$. We choose such a prime p with the additional property that p does not divide any weight of the induced tangential G_m-action on $T_{Y,Q}$ for every fixed point

Q. We take a point $Q \in Y^{G_m}$. By [24, Theorem 1], there exists a G_m-stable open neighborhood U of Q in the Euclidean topology and algebraic functions x, y, z around the point Q which form a system of analytic coordinates on U. The G_m-action is given by ${}^t(x,y,z) = (t^a x, t^b y, t^c z)$ with integers a, b, c, where $p \nmid \alpha$ for any $\alpha \in \{a,b,c\} \setminus \{0\}$. Then $U \cap Y^{G_m} = U \cap Y^{\mathbb{Z}/p\mathbb{Z}}$. Since $Y^{G_m} \subseteq Y^{\mathbb{Z}/p\mathbb{Z}}$, it follows that Y^{G_m} is a connected component of $Y^{\mathbb{Z}/p\mathbb{Z}}$. By [5], we have the inequality

$$\sum_i \dim H_i(Y^{\mathbb{Z}/p\mathbb{Z}}; \mathbb{Z}/p\mathbb{Z}) \le \sum_i \dim H_i(Y; \mathbb{Z}/p\mathbb{Z}) = 1.$$

This implies that $Y^{\mathbb{Z}/p\mathbb{Z}}$ is connected and equal to Y^{G_m}. Hence Y^{G_m} is connected and $H_i(Y^{G_m}; \mathbb{Z}/p\mathbb{Z}) = 0$ for every $i > 0$. □

We have the following result.

Theorem 2.6. *Let G_m act nontrivially on a $\mathbb{Q}$-homology threefold Y and let $q : Y \to X$ be the quotient morphism. Suppose that q has relative dimension one. Then the following assertions hold.*

(1) *Y^{G_m} is $\mathbb{Q}$-acyclic. If Y is a homology threefold, then Y^{G_m} is $\mathbb{Z}$-acyclic.*
(2) *If $\dim Y^{G_m} = 2$ then $Y^{G_m} \cong X$ and Y is a line bundle over X. If Y is a homology threefold, then $Y \cong X \times \mathbb{A}^1$.*
(3) *If $\dim Y^{G_m} = 1$ then $Y^{G_m} \cong \mathbb{A}^1$ and $q|_{Y^{G_m}} : Y^{G_m} \to X$ is a closed embedding. If Y is contractible and q is equi-dimensional, then X is a smooth contractible surface of log Kodaira dimension $-\infty$ or 1.*
(4) *If $\dim Y^{G_m} = 0$ then the G_m-action on Y is hyperbolic, i.e., the tangential representation on $T_{Y,Q}$ for the unique fixed point Q has mixed weights, e.g., either $a_1 < 0, a_2 > 0$ and $a_3 > 0$, or $a_1 > 0, a_2 < 0$ and $a_3 < 0$. In this case, there is an irreducible component of codimension one contained in the fiber of q passing through the point Q.*

Proof. (1) The $\mathbb{Q}$-acyclicity is verified below case by case according to $\dim Y^{G_m}$. When X is a homology threefold, the acyclicity of Y^{G_m} follows from the Smith theory [44, §22.3].

(2) By Lemma 2.1, Y^{G_m} lies horizontally to the morphism q and the restriction $q|_{Y^{G_m}} : Y^{G_m} \to X$ is a bijection (hence a birational morphism). In particular, Y^{G_m} is irreducible. Further, $q : Y \to X$ is an $\mathbb{A}^1$-fibration. Since Y is $\mathbb{Q}$-factorial by a remark after Lemma 1.1, q is equi-dimensional

(cf. [9, Lemma 1.14]). Then Lemma 2.4 implies that X is smooth and $q : Y \to X$ is a line bundle. By Zariski's main theorem, we conclude that $Y^{G_m} \cong X$. Furthermore, since Y is contractible to X, it follows that Y^{G_m} is $\mathbb{Q}$-acyclic (resp. $\mathbb{Z}$-acyclic) if Y is a $\mathbb{Q}$-homology threefold (resp. $\mathbb{Z}$-homology threefold). If Y is a homology threefold, then X is factorial and a line bundle over X is trivial. Hence $Y \cong X \times \mathbb{A}^1$.

(3) Y^{G_m} is a connected curve. Further, Y^{G_m} is smooth. In fact, the smoothness of the fixed point locus is a well-known fact for a reductive algebraic group action on a smooth affine variety (cf. [44])[a]. Hence Y^{G_m} is irreducible. Then Y^{G_m} is an affine line because $H_1(Y^{G_m}; \mathbb{Z}/p\mathbb{Z}) = 0$. By Lemma 2.1, q induces a closed embedding of Y^{G_m} into X [b]. If Y is contractible, X is contractible by [28, Theorem B]. By Lemma 1.3, $\pi_1(X^\circ) = (1)$, where X° is the smooth part of X. Then X is smooth by Lemma 1.6. So, X is a smooth contractible surface containing a curve isomorphic to $\mathbb{A}^1$. Hence X has log Kodaira dimension $-\infty$ or 1 (cf. [10, 48]).

(4) The induced G_m-action on the tangent space $T_{Y,Q}$ of the unique fixed point Q must have mixed weights. Otherwise, $\dim Y^{G_m} > 0$ near the point Q. This is a contradiction. To prove the last assertion, we have only to show it when Y is identified with the tangent linear 3-space $T_{Y,Q}$. Then the G_m-action is diagonalized and ${}^t(x, y, z) = (t^{a_1}x, t^{a_2}y, t^{a_3}z)$ with respect to a suitable system of coordinates (x, y, z). Then the invariant elements, viewed as elements in $\Gamma(Y, \mathcal{O}_Y)$, are divisible by x. Hence the fiber F of q passing through Q contains an irreducible component $\{x = 0\}$ which is a hypersurface of Y. □

Question 2.7. In the case (3) above, is Y isomorphic G_m-equivariantly to a direct product $Z \times Y^{G_m}$, where Z is a smooth affine surface with G_m

---

[a]In the case where G is a connected reductive algebraic group acting on a smooth affine variety Y, the G-action near a fixed point Q is locally analytically G-equivalent to a linear representation [24, 30]. In the case of a linear representation of G on the affine space $\mathbb{C}^n$, the fixed point locus near the origin is a linear subspace and it is G-equivariantly a direct summand of $\mathbb{C}^n$ by the complete reducibility of G. The smoothness of the fixed point locus Y^G near the point Q follows from this observation.

[b]Let $Q \in Y^{G_m}$. In view of [24], there exists a system of local (analytic) coordinates $\{x, y, z\}$ at Q such that G_m acts as ${}^t(x, y, z) = (x, t^{-a}y, t^b z)$ with $ab > 0$ and $y = z = 0$ defining the curve Y^{G_m} near Q. Then the quoteint surface X at the point $P = q(Q)$ has a system of local analytic coordinates $\{x, x^{b'}y^{a'}\}$ with $a' = a/d, b' = b/d$ and $d = \gcd(a, b)$. Then the curve $q(Y^{G_m})$ is defined by $x^{a'}y^{b'} = 0$ near P and X is smooth near the curve $q(Y^{G_m})$. Hence Y^{G_m} and $q(Y^{G_m})$ are locally isomorphic at Q and P. Since $q|_{Y^{G_m}} : Y^{G_m} \to X$ is injective, it is a closed embedding. Furthermore, the fiber of q through the point Q is a cross $a'\mathbb{A}^1 + b'\mathbb{A}^1$. See the definition after Corollary 2.8.

acting on it? If this is the case, q is the direct product $q = q' \times Y^{G_m}$ of the quotient morphism $q' : Z \to C$, where $C = Z//G_m$.

We discuss this question in Section 4 in detail. Theorem 2.6 has the following consequence.

Corollary 2.8. *Let the notations and the assumptions be the same as in Theorem 2.6. If the quotient morphism $q : Y \to X$ has equi-dimension one, then the fixed point locus Y^{G_m} has positive dimension.*

Now we consider an $\mathbb{A}^1_*$-fibration $p : Y \to X$, where Y is a smooth affine threefold and X is a normal affine surface. For a point P, the scheme-theoretic fiber $Y \times_X \operatorname{Spec} k(P)$ is denoted by $p^{-1}(P)$ or F_P, where $k(P)$ is the residue field of P in X. A fiber F_P of p is called *singular* if F_P is not isomorphic to $\mathbb{A}^1_*$ over $k(P)$. The set of points $P \in X$ such that F_P is singular is denoted by $\operatorname{Sing}^{st}(p)$ and called the *strict singular locus* or the *degeneracy locus* of p. For a technical reason (cf. Lemma 2.13 and a remark below it), we define the *singular locus* $\operatorname{Sing}(p)$ as the union of $\operatorname{Sing}^{st}(p)$ and the set $\operatorname{Sing}(X)$ of singular points. A singular fiber F_P is called a *cross* (resp. *tube*) [c] if it has the form $F_P \cong m\mathbb{A}^1 + n\mathbb{A}^1$ (resp. $m\mathbb{A}^1_*$). When we speak of a cross on a threefold or a surface which is smooth at the intersection point Q of two lines, we assume that two affine lines meet each other transversally at Q. By abuse of the terminology, we call a singular fiber F on a normal surface a cross even when the surface has a cyclic quotient singularity at the intersection point Q and the proper transforms of the two lines on the minimal resolution of singularity form a linear chain together with the exceptional locus.

We first recall a result of Bhatwadekar-Dutta [1, Theorem 3.11].

Lemma 2.9. *Let R be a Noetherian normal domain and let A be a finitely generated flat R-algebra such that*

(1) *The generic fiber $K \otimes_R A$ is a Laurent polynomial ring $K[T, T^{-1}]$ in one variable over $K = Q(R)$.*
(2) *For each prime ideal $\mathfrak{p}$ of R of height one, the fiber* $\operatorname{Spec} A \otimes_R k(\mathfrak{p})$ *is geometrically integral but is not $\mathbb{A}^1$ over $k(\mathfrak{p})$.*

Then there exists an invertible ideal I in R such that A is a $\mathbb{Z}$-graded R-algebra isomorphic to the R-subalgebra $R[IT, I^{-1}T^{-1}]$ of $K[T, T^{-1}]$. In

---

[c] The naming of cross is apparent. The fiber $m\mathbb{A}^1_*$ is a projective line with two end points lacking and thickened with multiplicity. It looks like a tube.

particular, Spec A *is locally* $\mathbb{A}^1_*$ *and hence an* $\mathbb{A}^1_*$*-fibration over* Spec R.

In geometric terms, a weaker version of Lemma 2.9 in the case of dimension three is stated as follows.

Lemma 2.10. *Let* $p : Y \to X$ *be an* $\mathbb{A}^1_*$*-fibration with a smooth affine threefold* Y *and let* P *be a closed point of* X*. Suppose that the following conditions are satisfied.*

(1) *There is an open neighborhood* U *of* P *such that* U *is smooth and* p *is equi-dimensional over* U.
(2) *Every fiber* $F_{P'}$ *of* p *for* $P' \in U \setminus \{P\}$ *is isomorphic to* $\mathbb{A}^1_*$.

Then the fiber F_P *is isomorphic to* $\mathbb{A}^1_*$ *and* $p : p^{-1}(U) \to U$ *is an* $\mathbb{A}^1_*$*-bundle.*

Proof. By replacing X by a smaller affine open neighborhood of P contained in U, we may assume that X is smooth, p is equi-dimensional and $\mathrm{Sing}(p) \cap (X \setminus \{P\}) = \emptyset$. Then p is flat. Further, there exists a $\mathbb{P}^1$-fibration $\overline{p} : \overline{Y} \to X$ such that Y is an open set of $\overline{Y}$ and $\overline{p}|_Y = p$. Since the $\mathbb{A}^1_*$-fibration p is assumed to be untwisted, the generic fiber $\overline{Y}_K$ has two K-rational points ξ_1, ξ_2, where K is the function field of X over $\mathbb{C}$. Let S_i be the closure of ξ_i in $\overline{Y}$ for $i = 1, 2$. By the assumption (2) above, S_1, S_2 are cross-sections over $X \setminus \{P\}$. Namely, $p^{-1}(X \setminus \{P\})$ is an $\mathbb{A}^1_*$-bundle. Hence $p^{-1}(X \setminus \{P\}) \cong \mathcal{S}pec\,(\mathcal{O}_{X\setminus\{P\}}[\mathcal{L}', {\mathcal{L}'}^{-1}])$, where $\mathcal{L}'$ is an invertible sheaf on $X \setminus \{P\}$. Since X is smooth and P is a point of codimension two, $\mathcal{L}'$ is extended to an invertible sheaf $\mathcal{L}$ on X. Since Y is affine and $\mathrm{codim}\,_Y p^{-1}(P) = 2$, it follows that $Y \cong \mathcal{S}pec\,(\mathcal{O}_X[\mathcal{L}, \mathcal{L}^{-1}])$. So, $p : Y \to X$ is an $\mathbb{A}^1_*$-bundle and $F_P \cong \mathbb{A}^1_*$. □

Let F_P be a fiber of an $\mathbb{A}^1_*$-fibration $p : Y \to X$ with $P \in X$. Let C be a general smooth curve on X passing through P. Let Z be the normalization of $Y \times_X C$. Then p induces an $\mathbb{A}^1_*$-fibration $p_C : Z \to C$ and the fiber $\widetilde{F}_P$ over the point P is a finite covering of the fiber F_P (in the sense that the normalization morphism induces a finite morphism $\widetilde{F}_P \to F_P$. In view of Lemma 2.2, we say that the $\mathbb{A}^1_*$-fibration p is *unmixed* if $\widetilde{F}_P$ for every $P \in X$ is $\mathbb{A}^1_*$, $\mathbb{A}^1 + \mathbb{A}^1$, or a disjoint union of $\mathbb{A}^1$'s when taken with reduced structure. This definiton does not depend on the choice of the curve C.

Lemma 2.11. *Let* $p : Y \to X$ *be an unmixed* $\mathbb{A}^1_*$*-fibration such that* Y *and* X *are smooth. Then the singular locus* $\mathrm{Sing}(p)$ *is a closed subset of* X *of codimension one.*

Proof. Lemma 2.9 or Lemma 2.10 implies that every irreducible component of the closure $\overline{\mathrm{Sing}(p)}$ of $\mathrm{Sing}(p)$ has codimension one.

So, we have only to show that $\mathrm{Sing}(p)$ is a closed set. It suffices to show that if the fiber F_P over $P \in X$ is $\mathbb{A}^1_*$ then there exists an open neighborhood U of P such that $F_{P'} \cong \mathbb{A}^1_*$ for every $P' \in U$. Let C be a curve passing through P. If C is not a component of $\overline{\mathrm{Sing}(p)}$, then the fiber of p over the generic point of C is $\mathbb{A}^1_*$ and hence geometrically integral. Suppose that C is a component of $\overline{\mathrm{Sing}(p)}$. The unmixedness condition on p implies that for a general point P' of C, the fiber $F_{P'}$ is either irreducible and dominated by $\mathbb{A}^1_*$ (case (i)), or each irreducible component of $F_{P'}$ is dominated by a contractible curve (case (ii)). Suppose that the case (ii) occurs. Then there exists a normal affine surface $\widetilde{Z}$ with an $\mathbb{A}^1$-fibration $\widetilde{p} : \widetilde{Z} \to \widetilde{C}$ satisfying the following commutative diagram

$$\begin{array}{ccc} \widetilde{Z} & \xrightarrow{\ \mu\ } & Y \times_X C \\ \downarrow{\scriptstyle \widetilde{p}} & & \downarrow{\scriptstyle p_C} \\ \widetilde{C} & \xrightarrow{\ \nu\ } & C \end{array}$$

where μ is a quasi-finite morphism and ν is the normalization morphism. Since any singular fiber of an $\mathbb{A}^1$-fibration on a normal affine surface is a disjoint union of the affine lines, it turns out that the fiber F_P, which is a fiber of p_C in the above diagram, is dominated by the affine line, which is a connected component of the fiber of $\widetilde{p}$. This is a contradiction, and the case (ii) cannot occur. Now removing from X all irreducible components of $\overline{\mathrm{Sing}(p)}$ for whose general points the case (ii) occurs and replacing Y by the inverse image of the open set of X thus obtained, we may assume that the case (i) occurs for general points of the irreducible components of $\overline{\mathrm{Sing}(p)}$. Let

$$dp : \mathcal{T}_Y \to p^*\mathcal{T}_X$$

be the tangential homomorphism of the tangent bundles on Y and X. Let $\mathcal{C}$ be the cokernel of dp. Then $\mathcal{C}$ is a coherent $\mathcal{O}_Y$-Module. Hence $\mathrm{Supp}\,(\mathcal{C})$ is a closed set T such that $T = p^{-1}(p(T))$ and $T = \overline{\mathrm{Sing}(p)}$. Hence the point P belongs to $p(T)$. However, dp is everywhere surjective on the fiber F_P and $F_P \not\subset \mathrm{Supp}\,(\mathcal{C})$. This is a contradiction.

By the above argument, every irreducible curve C through the point P has the property that the general fibers are isomorphic to $\mathbb{A}^1_*$. Hence the fiber of p over the generic point of C is geometrically integral. By Lemma 2.9, we know that $B \otimes_A \mathcal{O}_P \cong \mathcal{O}_P[x, x^{-1}]$. This implies that there exists an open neighborhood U of P satisfying the required property. $\square$

In the proofs of Lemmas 2.9, 2.10 and 2.11, the flatness condition on the morphism p should not be overlooked. Hence the assumption that X be normal instead of being smooth does not seem to be sufficient for the conclusion. So we ask the following question.

Question 2.12. Let $p : Y \to X$ be an $\mathbb{A}^1_*$-fibration with a smooth affine threefold Y. Is a point $P \in X$ smooth if the fiber F_P is isomorphic to $\mathbb{A}^1_*$?

The following is a partial answer to this question.

Lemma 2.13. *Let Y be a smooth affine threefold with an effective G_m-action and let $q : Y \to X$ be the quotient morphism. Suppose that the G_m-action is fixed point free. Then X has at worst cyclic quotient singular points.*

Proof. Suppose that P is a singular point of X. Let Q be a point of Y such that $q(Q) = P$ and let G_Q be the isotropy group at Q which is a finite cyclic group. Then, by Luna's étale slice theorem [30], there exists an affine subvariety V of Y with a G_Q-action and an étale morphism $\varphi/G_Q : V/G_Q \to X$ such that $Q \in V$, V is smooth at Q and $(\varphi/G_Q)(Q) = P$. This implies that X has at worst cyclic quotient singularity at P. Since the ground field is $\mathbb{C}$, we can work with an analytic slice instead of an étale slice. $\square$

We can easily show that if X is smooth at P then the fiber F_P has multiplicity equal to $m = |G_Q|$ near the point Q. In fact, if V is an analytic slice with coordinates x, y, the G_Q-action on V is given by ${}^\zeta(x, y) = (\zeta x, \zeta^b y)$ with a generator ζ of G_Q which is identified with an m-th primitive root of unity, where $0 \le b < m$. Then P is singular if and only if $b > 0$. Hence $b = 0$ if P is smooth. This implies that (x^m, y) is a local system of parameters of X at P. Hence the fiber F_P has multiplicity m. If X is singular at P, the fiber F_P is a multiple fiber. In fact, with the above notations, $\mathfrak{m}_{X,P}\widehat{\mathcal{O}}_{Y,Q}$ is generated by

$$\{x^m, y^m\} \cup \{x^s y^n \mid bn + s \equiv 0 \pmod{m},\ 0 < s < m,\ 0 < n < m\}.$$

Since $\widehat{\mathcal{O}}_{Y,Q} = \mathbb{C}[[x, y, z]]$ with a fiber coordinate z of F_P, it follows that $\mathfrak{m}_{X,P}\widehat{\mathcal{O}}_{Y,Q}$ is not a radical ideal. Hence the fiber F_P is not reduced near the point Q.

The singular locus $\mathrm{Sing}(p)$ may consist of a single point as shown in the following example.

Example 2.14. Let $\mathbb{A}^3 \to \mathbb{A}^2$ be the morphism defined by $(x, y, z) \mapsto (xy, x^2(xz + 1))$, where (x, y, z) is a system of coordinates of $\mathbb{A}^3$. Then p is an $\mathbb{A}^1_*$-fibration and $\mathrm{Sing}(p) = \{(0, 0)\}$. In fact, set $\alpha = xy,\ \beta = x^2(xz+1)$. Then $p^{-1}(\alpha, \beta) = \{(\alpha y^{-1}, y, \alpha^{-3}y(\beta y^2 - \alpha^2) \mid y \in \mathbb{C}^*\} \cong \mathbb{A}^1_*$ if $\alpha \neq 0$, $p^{-1}(0, \beta) = \{y = 0,\ x^2(xz + 1) = \beta\} \cong \mathbb{A}^1_*$ if $\alpha = 0,\ \beta \neq 0$ and $p^{-1}(0, 0) = \{x = 0\} \cup \{y = xz + 1 = 0\} \cong \mathbb{A}^2 \cup \mathbb{A}^1_*$ if $\alpha = \beta = 0$. □

The following example shows that a disjoint union of affine lines may appear as a singular fiber of an $\mathbb{A}^1_*$-fibration.

Example 2.15. Let $p : \mathbb{A}^3 \to \mathbb{A}^2$ be the morphism defined by $(x, y, z) \mapsto (xy, x^2(yz + 1))$. Then p is an $\mathbb{A}^1_*$-fibration such that $\mathrm{Sing}(p) \cong \mathbb{A}^1$ and the fibers are given as follows. Set $\alpha = xy,\ \beta = x^2(yz + 1)$. Then $p^{-1}(\alpha, \beta) = \{(\alpha y^{-1}, y, \alpha^{-2}(\beta y - \alpha^2 y^{-1}) \mid y \in \mathbb{C}^*\} \cong \mathbb{A}^1_*$ if $\alpha \neq 0$, $p^{-1}(0, \beta) = \{y = 0,\ x^2 = \beta\} \cong \mathbb{A}^1(x = \sqrt{\beta}, y = 0) \sqcup \mathbb{A}^1(x = -\sqrt{\beta}, y = 0)$ if $\alpha = 0,\ \beta \neq 0$ and $p^{-1}(0, 0) = \{x = 0\} \cong \mathbb{A}^2$ if $\alpha = \beta = 0$. □

Winkelmann [47] kindly communicated us of the following example of a flat $\mathbb{A}^1_*$-fibration $p : \mathbb{A}^3 \to \mathbb{A}^2$ which is not surjective.

Proposition 2.16. *Let $f(x) = \prod_{i=1}^n (x - i)$ and let $p : \mathbb{A}^3 \to \mathbb{A}^2$ be the morphism defined by $(x, y, z) \mapsto (x + zx + zyf(x), x + yf(x))$. Then the following assertions hold.*

(1) *A general fiber is isomorphic to $\mathbb{A}^1_{n*}$, where $\mathbb{A}^1_{n*}$ is the affine line minus n points. Hence p is an $\mathbb{A}^1_*$-fibration if $n = 1$.*
(2) $\mathrm{Im}\,(p) = \mathbb{A}^2 \setminus S$, *where $S = \{(k, 0) \mid 1 \leq k \leq n\}$.*
(3) *p is equi-dimensional. Hence p is flat.*
(4) *The singular locus* $\mathrm{Sing}(p)$, *that is the locus of points P of the base $\mathbb{A}^2$ over which $p^{-1}(P) \not\cong \mathbb{A}^1_{n*}$, is the union of $\{(\alpha, 0) \mid \alpha \in \mathbb{C}\}$ and $\bigcup_{k=1}^n \{(\alpha, k) \mid \alpha \in \mathbb{C}\}$. Hence* $\mathrm{Sing}(p)$ *consists of disjoint $n+1$ affine lines. For any $\alpha \in \mathbb{C} - \{1, \ldots, n\}$, the fiber $p^{-1}(\alpha, 0) \cong \mathbb{A}^1$. For any $k \in \{1, \ldots, n\}$ and $\alpha \in \mathbb{C}$, the fiber $p^{-1}(\alpha, k) \cong \mathbb{A}^1 \cup \mathbb{A}^1_{(n-1)*}$.*

Proof. (1) Set $\alpha = x + z(x + yf(x))$ and $\beta = x + yf(x)$. Then $\alpha = x + z\beta$. Hence if $\beta \neq 0$, we have $z = (\alpha - x)/\beta$. Further, if $\beta \notin \{1, 2, \ldots, n\}$, then the equation $\beta = x + yf(x)$ is solved as $y = (\beta - x)/f(x)$. In fact, if (x, y) is a solution of $\beta = x + yf(x)$ and $f(x) = 0$, then $x = \beta$ and $\beta \in \{1, 2, \ldots, n\}$, a contradiction. Hence $p^{-1}(\alpha, \beta) \cong \mathbb{A}^1 - \{1, 2, \ldots, n\} \cong \mathbb{A}^1_{n*}$.

(2) Suppose $\beta = 0$. Then $x = \alpha$ and z is free if $\alpha = x + z\beta$ has a solution. Further, $0 = \alpha + yf(\alpha)$, which implies $\alpha = 0$ if $f(\alpha) = 0$. But this is impossible. So, $p^{-1}(k, 0) = \emptyset$ if $k \in \{1, 2, \ldots, n\}$. If $f(\alpha) \neq 0$, y is solved as $y = -\alpha/f(\alpha)$. So, $p^{-1}(\alpha, 0) = \{(\alpha, -\alpha/f(\alpha), z) \mid z \in \mathbb{C}\} \cong \mathbb{A}^1$. Suppose $\beta \in \{1, 2, \ldots, n\}$. Then $z = (\alpha - x)/\beta$ and the equation $x + yf(x) = \beta$ is written as $(x - \beta)(yg(x) + 1) = 0$, where $f(x) = g(x)(x - \beta)$. If $x = \beta$ then y is free. If $yg(x) + 1 = 0$ then $y = -1/g(x)$. This implies that $p^{-1}(\alpha, \beta) = \mathbb{A}^1 \cup \mathbb{A}^1_{(n-1)*}$ with two components meeting in a single point $(\beta, -1/g(\beta), (\alpha - \beta)/\beta)$ in the (x, y, z)-coordinates. By the above reasoning, we know that $\mathrm{Im}\,(p) = \mathbb{A}^2 \setminus S$ as stated above. The assertions (3) and (4) are also shown in the above argument. □

If we restrict the morphism $p : \mathbb{A}^3 \to \mathbb{A}^2$ in the above proposition to a general linear plane in $\mathbb{A}^3$, we obtain an example of an endomorphism $\varphi : \mathbb{A}^2 \to \mathbb{A}^2$ which is not surjective. Such examples have been constructed by Jelonek [14].

We further study the singular fibers of $\mathbb{A}^1_*$-fibrations.

Example 2.17. Let $\widetilde{B} = \mathbb{C}[x, y, z, z^{-1}]$ be the Laurent polynomial ring in z over a polynomial ring $\widetilde{A} = \mathbb{C}[x, y]$. Let G_m and a cyclic group G of order n act on $\widetilde{B}$ by

$$^t(x, y, z) = (x, y, tz) \quad \text{and} \quad ^\zeta(x, y, z) = (x, \zeta y, \zeta^d z),$$

where ζ is a primitive n-th root of 1, G is identified with the subgroup $\langle \zeta \rangle$ of G_m generated by ζ and d is a positive integer with $\gcd(d, n) = 1$. Then the following assertions hold.

(1) Let B (resp. A) be the G-invariant subring of $\widetilde{B}$ (resp. $\widetilde{A}$) and let d' be a positive integer such that $dd' \equiv 1 \pmod{n}$. Then $B = \mathbb{C}[x, y^n, y/z^{d'}, z^n, z^{-n}] = \mathbb{C}[x, \eta, U, T, T^{-1}]/(U^n = \eta T^{-d'})$ and $A = \mathbb{C}[x, \eta]$, where $\eta = y^n$, $U = y/z^{d'}$ and $T = z^n$.
(2) Let $Y = \mathrm{Spec}\, B$, $X = \mathrm{Spec}\,(A)$ and $q : Y \to X$ be the morphism induced by the inclusion $A \hookrightarrow B$. Then q is the quotient morphism of Y by the G_m-action given by $^t(x, \eta, U, T) = (x, \eta, t^{-d'}U, t^n T)$.
(3) The singular locus $\mathrm{Sing}(q)$ is the line $\eta = 0$ on X, and the singular fibers over $\mathrm{Sing}(q)$ are of the form $n\mathbb{A}^1_*$.

□

In the above example, the G-action on $\widetilde{B}$ commutes with the G_m-action. Hence the G_m-action descends onto Y. Write $dd' = 1 + cn$. Then $^t(U^d T^c) =$

$t^{-1}(U^d T^c)$. Hence the isotropy subgroup is trivial (resp. G) if $U \neq 0$ (resp. $U = 0$).

Lemma 2.18. *Let $p : Y \to X$ be an $\mathbb{A}^1_*$-fibration satisfying the following conditions.*

(1) *Y and X are smooth, p is equi-dimensional and* $\mathrm{Sing}(p)$ *is a smooth irreducible curve, say C.*
(2) *There exists a positive integer $n > 1$ such that the fiber F_P over every point $P \in C$ is of the form $n\mathbb{A}^1_*$.*
(3) *Either X is factorial, or there exists an invertible sheaf $\mathcal{L}$ on X such that $\mathcal{L}^{\otimes n} \cong \mathcal{O}_X(-C)$.*

Then there exists a cyclic covering $\mu : \widetilde{X} \to X$ of order n ramifying totally over C such that the normalization $\widetilde{Y}$ of $Y \times_X \widetilde{X}$ is an $\mathbb{A}^1_*$-bundle over $\widetilde{X}$, where $\nu : \widetilde{Y} \to Y \times_X \widetilde{X}$ is the normalization morphism. Further, there exists a G_m-action on $\widetilde{Y}$ such that the projection $\widetilde{q} : \widetilde{Y} \xrightarrow{\mu} Y \times_X \widetilde{X} \to \widetilde{X}$ is the quotient morphism by the G_m-action. The quotient morphism $\widetilde{q}$ commutes with the cyclic covering group G, and hence descends down to the quotient morphism $q : Y \to X$ by the induced G_m-action which coincides with the given $\mathbb{A}^1_*$-fibration.

Proof. Write $X = \mathrm{Spec}\, A$. If X is factorial, let $f \in A$ define the curve C and let $\widetilde{X} = \mathrm{Spec}\, A[\xi]/(\xi^n - f)$. If $\mathcal{O}_X(-C) \cong \mathcal{L}^{\otimes n}$, let $\widetilde{X} = \mathcal{S}pec\,_X \bigoplus_{i=0}^{n-1} \mathcal{L}^{\otimes i}$. Then the morphism $\mu : \widetilde{X} \to X$ is a cyclic covering of order n totally ramifying over the curve C.

Let $\mathcal{U} = \{U_i\}_{i \in I}$ be an affine open covering of $p^{-1}(C)$ and let η_i be an element of $\Gamma(U_i, \mathcal{O}_Y)$ such that $\eta_i = 0$ defines $p^{-1}(C)_{\mathrm{red}}\,|_{U_i}$ for each i. We consider the case where C is defined by $f = 0$. The other case can be treated in a similar fashion. We can write $f = u_i \eta_i^n$ for $u_i \in \Gamma(U_i, \mathcal{O}_Y^*)$. Then $\widetilde{Y}$ over U_i is defined as $\mathcal{S}pec\,_{U_i} \mathcal{O}_{U_i}[\xi/\eta_i]$, where $(\xi/\eta_i)^n = u_i$. This implies that the morphism $\rho : \widetilde{Y} \xrightarrow{\nu} Y \times_X \widetilde{X} \xrightarrow{p_1} Y$ with the first projection p_1 is a finite étale morphism, whence $\widetilde{Y}$ is smooth. It is clear that $\widetilde{q} : \widetilde{Y} \to \widetilde{X}$ is an $\mathbb{A}^1_*$-fibration. Since ρ is finite and étale, $\widetilde{q}^{-1}(\widetilde{Q})$ with $\widetilde{Q} \in \mu^{-1}(C)_{\mathrm{red}}$ is a reduced $\mathbb{A}^1_*$ or a disjoint union of reduced $\mathbb{A}^1_*$'s. Let D be a general smooth curve on $\widetilde{X}$ passing through $\widetilde{Q}$ and let $\widetilde{Z}$ be the normalization of $\widetilde{q}^{-1}(D)$. Then the canonical morphism $\overline{q} : \widetilde{Z} \to D$ is an $\mathbb{A}^1_*$-fibration on a normal affine surface $\widetilde{Z}$ and the fiber $\overline{q}^{-1}(\widetilde{Q})$ maps surjectively onto the fiber $\widetilde{q}^{-1}(\widetilde{Q})$. By Lemma 2.2, $\overline{q}^{-1}(\widetilde{D})$ does not have two or more irreducible components which surject onto $\mathbb{A}^1_*$. Hence $\widetilde{q}^{-1}(\widetilde{Q})$ consists of a single reduced component isomorphic to $\mathbb{A}^1_*$. This implies that $\widetilde{Y}$ is an $\mathbb{A}^1_*$-bundle over $\widetilde{X}$.

An $\mathbb{A}^1_*$-bundle has a standard G_m-action along the fibers. Namely, if $\widetilde{q}^{-1}(V_j) \cong V_j \times \mathrm{Spec}\, \mathbb{C}[\tau_j, \tau_j^{-1}]$ for an open covering $\mathcal{V} = \{V_j\}_{j \in J}$ of $\widetilde{X}$, then G_m acts as ${}^t\tau_j = t\tau_j$. This action clearly commutes with the action of the cyclic covering group G. Note that $\widetilde{Y}/G \cong Y$. Hence the G_m-action on $\widetilde{Y}$ descends down to Y, and gives rise to a G_m-action on Y. The quotient morphism $q : Y \to Y//G_m$ coincides with $p : Y \to X$ because $q = p$ on the open set $p^{-1}(X \setminus C)$. □

A G_m-action of a smooth affine threefold is called *equi-dimensional* if the quotient morphism $q : Y \to X$ is equi-dimensional.

Lemma 2.19. *Let Y be a smooth affine threefold with a G_m-action and let $q : Y \to X$ be the quotient morphism with $X = Y//G_m$. Suppose that q has equi-dimension one, q is an $\mathbb{A}^1_*$-fibration and X is smooth. Then a singular fiber of q is either a tube or a cross.*

Proof. Let F_P be a singular fiber F of q. Let C be a smooth irreducible curve on X such that $P \in C$. Let $Z = Y \times_X C$ and let $\nu : \widetilde{Z} \to Z$ be the normalization morphism. Then the projection $\rho : Z \to C$ and $\widetilde{\rho} = \rho \cdot \nu : \widetilde{Z} \to C$ are the quotient morphisms by the induced G_m-actions on Z and $\widetilde{Z}$. The fiber $\rho^{-1}(P)$ is the fiber F_P and is the image of $\widetilde{F}_P = \widetilde{\rho}^{-1}(P)$ by ν. By Lemma 2.2, the fiber $\widetilde{F}_P$ is either a tube or a cross. If $\widetilde{F}_P$ is a tube, then F_P is also a tube because ν restricted onto $\widetilde{F}_P$ commutes with the G_m-action. Suppose that $\widetilde{F}_P$ is a cross. Since the intersection point $\widetilde{Q}$ of two affine lines of $\widetilde{F}_P$ is a fixed point and $\nu_{\widetilde{F}_P}$ is surjective, either F_P consists of two components meeting in a point $Q = \nu(\widetilde{Q})$ or F_P is a contractible curve with Q a fixed point and $F_P - \{Q\}$ a G_m-orbit. In the latter case, two branches of a cross are folded into a single curve. But this is impossible because the G_m-action on a cross viewed near the point $\widetilde{Q}$ has weights $-a, b$ respectively on two branches with $ab > 0$. So, F_P consists of two branches meeting in one point Q. Consider the induced representation of G_m on the tangent space $T_{Y,Q}$ at the fixed point Q. We can write it as ${}^t(x, y, z) = (x, t^{-a}y, t^b z)$ for a suitable system of local coordinates $\{x, y, z\}$. Then the fiber F_P is given by $xy = 0$ locally at Q. Hence two branches of F_P meet transversally at Q, and F_P is a cross. □

We can prove a *converse* of this result.

Theorem 2.20. *Let $p : Y \to X$ be an $\mathbb{A}^1_*$-fibration on a smooth affine threefold Y. Suppose that X is normal, p is equi-dimensional and the singular fibers are tubes or crosses over the points of* $\mathrm{Sing}(p)$ *except for a* *finite set of points. Then there exists an equi-dimensional G_m-action on Y such that the quotient morphism $q : Y \to Y//G_m$ coincides with the given $\mathbb{A}^1_*$-fibration p. Hence X has at worst cyclic quotient singularities.*

Proof. Since $\mathrm{Sing}(X)$ is a finite set, if a G_m-action is constructed on $p^{-1}(X \setminus \mathrm{Sing}(X))$ in such a way that the quotient morphism coincides with p restricted on $p^{-1}(X \setminus \mathrm{Sing}(X))$, then the G_m-action extends to Y by Hartog's theorem and the quotient morphism coincides with p. Since, as shown below, the construction of a G_m-action is local over $X \setminus \mathrm{Sing}(X)$, we may restrict ourselves to an affine open set of $X \setminus \mathrm{Sing}(X)$ and assume that X is smooth from the beginning.

By Lemmas 2.9 and 2.10, the singular locus $\mathrm{Sing}(p)$ has no isolated points. Let $\mathrm{Sing}(p) = C_1 \cup \cdots \cup C_r$ be the irreducible decomposition. Let

$$X_i = X \setminus (C_1 \cup \cdots \cup \check{C_i} \cup \cdots \cup C_r)$$

for $1 \le i \le r$, let $Y_i = p^{-1}(X_i)$ and let $p_i = p|_{Y_i} : Y_i \to X_i$. Then Y_i and X_i are affine and the $\mathbb{A}^1_*$-fibration p_i has an irreducible singular locus, say C_i by abuse of the notation. If C_i has singular points, we replace Y_i and X_i first by $p_i^{-1}(X \setminus \mathrm{Sing}(C_i))$ and $X \setminus \mathrm{Sing}(C_i)$ respectively and then by $p_i^{-1}(U_{i\lambda})$ and $U_{i\lambda}$ respectively, where $\{U_{i\lambda}\}_{\lambda \in \Lambda}$ is an affine open covering of X_i. If there exist a family of equi-dimensional G_m-action $\sigma_i : G_m \times Y_i \to Y_i$ which induce the standard multiplication $\tau \mapsto t\tau$ for a variable on a general fiber τ, then the actions $\{\sigma_i\}_{i=1}^r$ patch together and define an equi-dimensional G_m-action $\sigma : G_m \times Y \to X$. By Hartog's theorem, this G_m-action is also extended over the fibers lying over the singular points of $\mathrm{Sing}(p)$. In fact, the union of the fibers over the singular points of $\mathrm{Sing}(p)$ has codimension greater than one.

So, we assume that $\mathrm{Sing}(p)$ is an irreducible smooth curve C. By shrinking X again to a smaller open set, we may assume that the curve C is principal, i.e., defined by a single equation $f = 0$. If the singular fibers over the points of C are tubes, the existence of a G_m-action follows from Lemma 2.18. Suppose that the fibers over C are crosses. Then $p^{-1}(C) = E_1 \cup E_2$ with irreducible component E_1, E_2. Then $Y \setminus E_1$ and $Y \setminus E_2$ are affine and the fibration p restricted to $Y \setminus E_1$ and $Y \setminus E_2$ have tubes over the curve C. Hence Lemma 2.18 implies that there exist G_m-actions on $Y \setminus E_1$ and $Y \setminus E_2$ and that they coincides on the general fibers of p. Hence they patch together and define a G_m-action on Y.

The last assertion follows from Lemma 2.13. □

For examples with tubes or crosses as the singular fibers, we refer to

Example 2.17 for tubes and Lemmas 4.1 and 4.4 below for crosses. In view of Lemma 2.2, we have a satisfactory description on singular fibers of $\mathbb{A}^1_*$-fibrations on normal affine surfaces. If Z is a normal affine surface with an $\mathbb{A}^1_*$-fibration $\rho : Z \to C$, then $p = \rho \times C' : Z \times C' \to C \times C'$ gives an $\mathbb{A}^1_*$-fibration on a normal affine threefold $Z \times C'$, where C' is a smooth affine curve. Hence the same singular fibers as in the surface case appear in the product threefold case. But we can say much less in general. Let $p : Y \to X$ be an $\mathbb{A}^1_*$-fibration, where Y is a smooth affine threefold and X is a smooth affine surface. Let F_P be the singular fiber of p lying over a point $P \in X$. Let C be a smooth curve on X through P and let Z be the normalization of $Y \times_X C$. Then the induced morphism $p_C : Z \to C$ is an $\mathbb{A}^1_*$-fibration. Hence the fiber F_P has a finite covering $\widetilde{F}_P \to F_P$, where $\widetilde{F}_P$ is a fiber of p_C and hence has the form as described in Lemma 2.2. We do not know exactly what the singular fiber F_P itself looks like.

Concerning the coexistence of tubes and crosses in the quotient morphism $q : Y \to X$ by a G_m-action, we have the following result.

Lemma 2.21. *Let Y be a homology threefold with an effective, equi-dimensional G_m-action and let $q : Y \to X$ be the quotient morphism. Suppose that q is an $\mathbb{A}^1_*$-fibration, X is smooth and* $\dim Y^{G_m} = 1$. *Then there are no tubes as singular fibers of q.*

Proof. By the assertion (3) of Theorem 2.6, $Y^{G_m} \cong \mathbb{A}^1$ and $q|_{Y^{G_m}} : Y^{G_m} \to X$ is a closed embedding. By Lemma 2.19, any fiber through a point of Y^{G_m} is a cross. Let $F = F_P$ be a tube of multiplicity $m > 1$. Let p be a prime factor of m and let $\Gamma = \mathbb{Z}/p\mathbb{Z}$. Consider the induced Γ-action on Y as Γ is a subgroup of G_m. Then F is contained in the Γ-fixed point locus Y^Γ (see the argument after the proof of Lemma 2.13). Let Q be a point on F. Then the induced tangential representation of Γ on $T_{Y,Q}$ is written as

$$^\zeta(x, y, z) = (x, \zeta^a y, \zeta^b z), \quad 0 \le a < p, \ 1 \le b < p,$$

where ζ is a primitive p-th root of unity and x is a coordinate with the tangential direction of the fiber. If $a > 0, \ b > 0$, then the point $P = q(Q)$ is a singular point of X (see the argument after the proof of Lemma 2.13). Hence $a = 0$. This implies that the component of Y^Γ containing F has dimension two. By the Smith theory, the locus Y^Γ is a connected closed set. Hence there is a point Q' of Y^{G_m} such that the fiber $F'_{P'}$ through Q' is contained in Y^Γ. Then we can write $F'_{P'} = m_1\mathbb{A}^1 + m_2\mathbb{A}^2$ with $p \mid m_i$ for $i = 1, 2$. But this is impossible. In fact, let C be a smooth curve on X passing through the point P' and set $Z = Y \times_X C$. Then Z has an induced

G_m-action. Let $\widetilde{Z}$ be the normalization of $Z \times_C \widetilde{C}$, where $\widetilde{C} \to C$ is a finite covering of order p totally ramifying over the point P'. Then there is an induced G_m-action on $\widetilde{Z}$ such that the induced morphism $\widetilde{q} : \widetilde{Z} \to \widetilde{C}$ is the quotient morphism. The fiber $\widetilde{q}^{-1}(\widetilde{P}')$ with $\widetilde{P}'$ a point of $\widetilde{C}$ lying over P' consists of p-copies of the cross $\mathbb{A}^1 + \mathbb{A}^1$ when taken with reduced structure. But this is impossible. □

Finally in this section, we discuss Question 2.12. In fact, we prove a more general result.

Lemma 2.22. *Let $p : Y \to X$ be a dominant morphism from an algebraic variety of dimension $n + r$ to an algebraic variety X of dimension n. Let P be a point of X and let Q be a point of Y such that $p(Q) = P$. Suppose that Y is smooth at Q and the fiber $F := p^{-1}(P)$ is reduced, r-dimensional and smooth at Q. Then P is smooth at X.*

Proof. Let $(R, \mathfrak{m})$ (resp. $(S, \mathfrak{M})$) be the local ring of X (resp. Y) at P (resp. Q). Since the fiber F is defined by $\mathfrak{m}S$ at Q and since $S/\mathfrak{m}S$ is a regular local ring as Q is a smooth point of F, it follows that $\mathfrak{M} = \mathfrak{m}S + (z_{n+1}, \ldots, z_{n+r})S$ for elements $z_{n+1}, \ldots, z_{n+r} \in \mathfrak{M}$. Since $(S, \mathfrak{M})$ is a regular local ring of dimension $n + r$, we find n elements $z'_1, \ldots, z'_n$ of $\mathfrak{m}$ with images $z_1, \ldots, z_n$ in S such that $\{z_1, \ldots, z_n, z_{n+1}, \ldots, z_{n+r}\}$ is a regular system of parameters of $(S, \mathfrak{M})$. Since the completion $(\widehat{S}, \widehat{\mathfrak{M}})$ is isomorphic to $\mathbb{C}[[z_1, \ldots, z_n, z_{n+1}, \ldots, z_{n+r}]]$, we can express any $h \in \widehat{S}$ as a formal power series in $z_{n+1}, \ldots, z_{n+r}$ with coefficients in $\mathbb{C}[[z_1, \ldots, z_n]]$,

$$h = \sum_{\vec{i}=0}^{\infty} \alpha_{\vec{i}} z^{\vec{i}}, \quad \alpha_{\vec{i}} \in \mathbb{C}[[z_1, \ldots, z_n]],$$

where $z^{\vec{i}} = z_{h+1}^{i_{n+1}} \cdots z_{n+r}^{i_{n+r}}$ for $\vec{i} = (i_{n+1}, \ldots, i_{n+r})$. We shall show that the completion $(\widehat{R}, \widehat{\mathfrak{m}})$ is isomorphic to the formal power series ring $\mathbb{C}[[z_1, \ldots, z_n]]$. We consider the local complete intersection

$$H = \{z_j = 0 \mid n + 1 \leq j \leq n + r\}$$

in Y near the point Q as a section transversal to the fiber F at the point Q. We may assume that the restriction $p|_H : H \to X$ is quasi-finite. Hence the injective local ring homomorphism $R \to S \to S/zS$ induces an injective local homomorphism

$$(\widehat{R}, \widehat{\mathfrak{m}}) \to (\widehat{S}, \widehat{\mathfrak{M}}) \to (\widehat{S}/z\widehat{S}, \widehat{\mathfrak{M}}/z\widehat{S}),$$

because $(\widehat{S}/(z_{n+1},\ldots,z_{n+r})\widehat{S}, \widehat{\mathfrak{M}}/(z_{n+1},\ldots,z_{n+r})\widehat{S})$ is viewed as the completion of the local ring $\mathcal{O}_{H,Q}$. Sending z_i to z'_i, $i=1,\ldots,n$, we obtain a homomorphism

$$\mathbb{C}[[z_1,\ldots,z_n]] \longrightarrow \widehat{R},$$

which gives an isomorphism when composed with the mapping $\widehat{R} \to (\widehat{S}/z\widehat{S})$. Let h' be an element of $\widehat{R}$ and h its image in $\widehat{S}$. With h expressed as a power series as above, we find that h' and the image in $\widehat{R}$ of $\alpha_{\vec{0}} \in \mathbb{C}[[z_1,\ldots,z_n]]$ have the same image in $\widehat{\mathfrak{M}}/z\widehat{S}$ and hence coincide. Hence $\widehat{R} \cong \mathbb{C}[[z_1,\ldots,z_n]]$. This implies that X is smooth at P. □

3. Homology threefolds with $\mathbb{A}^1$-fibrations

In [9], it is shown that an $\mathbb{A}^1$-fibration $p: Y \to X$ from a smooth affine threefold to a normal surface has a factorization

$$p: Y \xrightarrow{q} \widetilde{X} \xrightarrow{\sigma} X,$$

where $q: Y \to \widetilde{X}$ is the quotient morphism by a G_a-action and $\sigma: \widetilde{X} \to X$ is the birational morphism such that $\widetilde{A} = \Gamma(\widetilde{X}, \mathcal{O}_{\widetilde{X}})$ is the factorial closure of $A = \Gamma(X, \mathcal{O}_X)$ in $B = \Gamma(Y, \mathcal{O}_Y)$. Namely,

$$\widetilde{A} = \{b \in B \mid b \text{ is a factor of an element } a \in A\}.$$

Thus we may look into the quotient morphism $q: Y \to X$ by a G_a-action. We set $\mathrm{Sing}(q) = \{P \in X \mid F_P \not\cong \mathbb{A}^1\}$, where F_P is the fiber over $P \in X$, and call it the *singular locus* of q.

Lemma 3.1. *Let $q: Y \to X$ be the quotient morphism of a smooth affine threefold Y with respect to a G_a-action. Suppose that q is equi-dimensional. Then the following assertions hold.*

(1) *If a fiber F_P has an irreducible component which is reduced in F_P, then the point P is smooth in X.*
(2) $\mathrm{Sing}(q)$ *is a closed set.*

Proof. (1) The assertion follows from Lemma 2.22.

(2) Since q has equi-dimension one, the fixed point locus Y^{G_a} consists of fiber components [9, Corollary 3.2]. Then every fiber of q is a disjoint union of contractible curves [9, Lemma 3.5]. Further, a contractible irreducible component is isomorphic to $\mathbb{A}^1$ if it contains a non-fixed point, or is contained in Y^{G_a}.

It suffices to show that given a fiber F_P isomorphic to $\mathbb{A}^1$ there exists an open neighborhood U of P in X with $F_{P'} \cong \mathbb{A}^1$ for all $P' \in U$. The subsequent proof is similar to the one for G_m-actions (cf. Lemma 2.11). Let C be an irreducible curve on X through P and let Z be the normalization of $Y \times_X C$. Suppose that the fiber of q over a general point of C consists of m copies of $\mathbb{A}^1$, where $m > 1$. Then Z has an induced G_a-action such that

$$\begin{array}{ccc} Z & \xrightarrow{\mu} & Y \times_X C \\ \downarrow{\scriptstyle \widetilde{q}} & & \downarrow{\scriptstyle q} \\ \widetilde{C} & \xrightarrow{\nu} & C \end{array}$$

where $\widetilde{q}$ is the quotient morphism and $\widetilde{C}$ is the normalization of C in Z. The morphism $\nu : \widetilde{C} \to C$ is a finite covering of degree m (the Stein factorization), which is ramified over the point P. It then follows that the fiber F_P is non-reduced, and this is a contradiction.

Consider the closure $\overline{\mathrm{Sing}(q)}$ and remove from X all the irreducible components of $\overline{\mathrm{Sing}(q)}$ over which a general fiber of q consists of more than one irreducible component. Thus we may assume that all singular fibers of q are of the form $m\mathbb{A}^1$ with $m > 1$. Note that P is a smooth point by the assertion (1). Replacing X by an affine open neighborhood U of P and accordingly Y by the inverse image $q^{-1}(U)$, we may assume that X is smooth. Consider the tangential homomorphism of the tangent bundles

$$dq : \mathcal{T}_Y \to q^*\mathcal{T}_X$$

and let $\mathcal{C}$ be the cokernel of dq. Then $\mathcal{C}$ is a coherent $\mathcal{O}_Y$-module. The support $T = \mathrm{Supp}\,(\mathcal{C})$ is a closed set such that $T = q^{-1}(q(T))$. If $P \in \overline{\mathrm{Sing}(q)}$, then $F_P \subset T$, but $F_P \cap T = \emptyset$. This is a contradiction. Then, by Dutta [3, Theorem], there exists an open neighborhood U of P such that $q^{-1}(U)$ is an $\mathbb{A}^1$-bundle over U. Hence $\mathrm{Sing}(q)$ is a closed set. □

Concerning a G_a-action, we ask the following

Question 3.2. Let Y be an affine variety with a G_a-action. Suppose that the algebraic quotient $X = Y//G_a$ exists as an affine variety, i.e., the G_a-invariant subring of $\Gamma(Y, \mathcal{O}_Y)$ is an affine domain. Let Q be a point of Y with trivial isotropy group, i.e., a point which is not G_a-fixed. Is then the fiber F_P of the quotient morphism $q : Y \to X$ passing through the point Q reduced, where $P = q(Q)$?

The answer is negative and an example is given by an affine pseudo-plane [9, Corollary 3.16]. See also [39, Lemma 2.4].

Example 3.3. Let C be the smooth conic in $\mathbb{P}^2 = \mathrm{Proj}\ \mathbb{C}[X_0, X_1, X_2]$ defined by $X_2^2 = X_0X_1$. Let L be the tangent line of C at the point $(X_0, X_1, X_2) = (0, 1, 0)$. Hence L is defined by $X_0 = 0$. Let Λ be the linear pencil generated by C and $2L$. Then the G_a-action on $\mathbb{P}^2$ defined by

$$^tX_0 = X_0, \quad ^tX_1 = X_1 + 2X_2t + X_0t^2, \quad ^tX_2 = X_2 + X_0t$$

has a unique fixed point Q and preserves the members of the pencil Λ. Let X be the complement of C in $\mathbb{P}^2$. Then Λ defines an $\mathbb{A}^1$-fibration $q : X \to \mathbb{A}^1$ which turns out to be the quotient morphism of the induced G_a-action on X. Although the G_a-action has no fixed points, q has a multiple fiber 2ℓ which comes from the member $2L$ of Λ. The affine surface X is an affine pseudo-plane and its universal covering is a Danielewski surface. □

This is the case for every affine pseudo-plane $q : X \to Z$ of type (m, r) with an integer $r \geq 1$ such that $r \equiv 1 \pmod m$. For the definition of an affine pseudo-plane of type (m, r), see [31] where the type is denoted by (d, r) instead of (m, r). In particular, $q : X \to Z$ is an $\mathbb{A}^1$-fibration with a unique singular fiber F_0 of multiplicity $m > 1$, i.e., $F_0 = m\mathbb{A}^1$ and $Z \cong \mathbb{A}^1$. X is denoted by $X(m, r)$.

Lemma 3.4. *Let $q : X \to Z$ be an affine pseudo-plane of type (m, r) with $r \equiv 1 \pmod m$. Then q is given by a fixed-point free G_a-action.*

Proof. Let $\widetilde{X}(m, r)$ be the universal covering of $X(m, r)$. Let $H(m) = \mathbb{Z}/m\mathbb{Z}$ be the covering group which is identified with the m-th roots of unity. By [31, Lemma 2.6], $\widetilde{X}(m, r)$ is isomorphic to a hypersurface in $\mathbb{A}^3 = \mathrm{Spec}\ \mathbb{C}[x, y, z]$ defined by

$$x^rz + (y^m + a_1xy^{m-1} + \cdots + a_{m-1}x^{m-1}y + a_mx^m) = 1, \quad a_i \in \mathbb{C}.$$

The Galois group $H(m)$ acts as

$$^\lambda(x, y, z) = (\lambda x, \lambda y, \lambda^{-r}z), \quad \lambda \in H(m).$$

The projection $(x, y, z) \mapsto x$ defines an $\mathbb{A}^1$-fibration $\widetilde{q} : \widetilde{X}(m, r) \to \widetilde{Z}$, where $\widetilde{Z} \cong \mathbb{A}^1$. Further, there is a G_a-action on $\widetilde{X}(m, r)$ defined by

$$\begin{aligned}
&^t(x, y, z) = \\
&\quad (x, y + tx^r, z - x^{-r}\{((y + tx^r)^m + a_1x(y + tx^r)^{m-1} + \cdots + a_mx^m) \\
&\qquad - (y^m + a_1xy^{m-1} + \cdots + a_{m-1}x^{m-1}y + a_mx^m)\}).
\end{aligned}$$

Then it follows that

(1) the fiber $\widetilde{F}_0$ of $\widetilde{q}$ over the point $x = 0$ is a disjoint union of m-copies of the affine line,
(2) the G_a-action preserves the fibration $\widetilde{q}$,
(3) the G_a-action preserves and acts non-trivially on each connected component of $\widetilde{F}_0$,
(4) if $r \equiv 1 \pmod{m}$ then the $H(m)$-action commutes with the G_a-action.

Hence the G_a-action descends to a G_a-action on $X(m, r)$ which has no fixed points. □

We consider, however, a result implied by the assumptions in Question 3.2. We need some auxiliarly results. Let $B = \Gamma(Y, \mathcal{O}_Y)$ and let $A = \Gamma(X, \mathcal{O}_X)$. Let δ be a locally nilpotent derivation on B which corresponds to the given G_a-action. Further, let $\mathfrak{m}$ be the maximal ideal of A corresponding to the point P. First we recall the following result in [39, Theorem 3.3].

Lemma 3.5. *Let B be an affine domain over $\mathbb{C}$ and let δ be a locally nilpotent derivation on B. Let $A = \operatorname{Ker} \delta$. Suppose that $B/\mathfrak{m}B$ is an integral domain over $\mathbb{C}$ of dimension one and that the associated G_a-action on* Spec $B/\mathfrak{m}B$ *has no fixed points. Then the following assertions hold.*

(1) *For any integer $n > 0$, there exists an element $z_n \in B$ such that $B/\mathfrak{m}^n B = R_n[z_n]$, where R_n is an Artin local ring and we denote the residue class of z_n in $B/\mathfrak{m}^n B$ by the same letter.*
(2) *For $m > n$, we have a natural exact sequence*

$$0 \to \mathfrak{m}^n R_m \to R_m \xrightarrow{\theta_{nm}} R_n \to 0,$$

where θ_{nm} is a local homomorphism. Let $\widehat{R} = \varprojlim_n R_n$. Then $\widehat{R}$ is a complete local ring.

Let $\mathcal{O}$ be the local ring $\mathcal{O}_{X,P}$ and denote the ideal $\mathfrak{m}\mathcal{O}$ by the same letter $\mathfrak{m}$. Let $\widehat{\mathcal{O}}$ be the $\mathfrak{m}$-adic completion of $\mathcal{O}$. Then $B \otimes_A \mathcal{O}$ is an affine domain over $\mathcal{O}$ with the associated locally nilpotent derivation δ such that δ is nonzero and $\operatorname{Ker} \delta = \mathcal{O}$. Further *we assume the following condition (H) that* $(\mathfrak{m}^n(B \otimes_A \mathcal{O})) \cap A = \mathfrak{m}^n$ *for every* $n > 0$. This condition is satisfied if $B \otimes_A \mathcal{O}$ is $\mathcal{O}$-flat (see [11, Lemma 1.4]).

Proposition 3.6. *With the above notations and assumptions, the following assertions hold.*

(1) $(B \otimes_A \mathcal{O})/\mathfrak{m}^n(B \otimes_A \mathcal{O}) \cong B/\mathfrak{m}^n B = R_n[z_n]$.

(2) *The projective limit* $\widehat{B} = \varprojlim_n (B \otimes_A \mathcal{O})/\mathfrak{m}^n(B \otimes_A \mathcal{O})$ *contains* $B \otimes_A \mathcal{O}$ *as a subring. Further,* $\widehat{B}$ *has a derivation* $\widehat{\delta}$ *and an element* $\widehat{z}$ *such that* $\widehat{\delta}\,|_{B\otimes_A\mathcal{O}} = \delta$, $\operatorname{Ker}\widehat{\delta} = \widehat{R}$ *and* $\widehat{\delta}(\widehat{z}) = 1$. *The ring* $\widehat{B}$ *itself is a subring of a formal power series ring* $\widehat{R}[[\widehat{z}]]$.

(3) *Let* $\widehat{B}_{\rm ln}$ *be the subring of* $\widehat{B}$ *consisting of elements for which* $\widehat{\delta}$ *is locally nilpotent. Then* $\widehat{B}_{\rm ln} = \widehat{R}[\widehat{z}]$ *and* $B \otimes_A \mathcal{O}$ *is a subring of* $\widehat{B}_{\rm ln}$.

Proof. (1) The assertion follows from Lemma 3.5.

(2) It suffices to show that if b is an element of $\bigcap_n \mathfrak{m}^n(B \otimes_A \mathcal{O})$ then $b = 0$. Suppose that $b \neq 0$. Then there exists an integer $r \geq 0$ such that $\delta^r(b) \neq 0$ and $\delta^{r+1}(b) = 0$. Then $\delta^r(b) \in \operatorname{Ker}\delta = \mathcal{O}$. Since $\delta(\mathfrak{m}^n(B \otimes_A \mathcal{O})) \subseteq \mathfrak{m}^n(B \otimes_A \mathcal{O})$ for every $n > 0$, by the condition (H), we have $\delta^r(b) \in (\bigcap_n \mathfrak{m}^n(B \otimes_A \mathcal{O})) \cap \mathcal{O} = \bigcap_n \mathfrak{m}^n$, which is zero by Krull's intersection theorem. Hence $\delta^r(b) = 0$. This is a contradiction. Hence $B \otimes_A \mathcal{O}$ is a subring of $\widehat{B}$. The sequence $\{z_n\}_{n\geq 1}$ is a Cauchy sequence because $z_m - z_n \in \mathfrak{m}^n(B \otimes_A \mathcal{O})$ for $m > n$. Hence it converges to an element $\widehat{z}$ of $\widehat{B}$. The derivation δ extends to a derivation $\widehat{\delta}$. Since we may assume that $\delta(z_n) \equiv 1 \pmod{\mathfrak{m}^n(B \otimes_A \mathcal{O})}$, it follows that $\widehat{\delta}(\widehat{z}) = 1$ in $\widehat{B}$. Let b_n be an element of $B \otimes_A \mathcal{O}$. Since $(B \otimes_A \mathcal{O})/\mathfrak{m}^n(B \otimes_A \mathcal{O}) \cong R_n[z_n]$ and $\widehat{z} - z_n \in \mathfrak{m}^n\widehat{B}$, there exists a polynomial $f_n(\widehat{z}) \in \widehat{R}[\widehat{z}]$ such that $b_n - f_n(\widehat{z}) \in \mathfrak{m}^n\widehat{B}$. This implies that a Cauchy sequence $\{b_n\}_{n\geq 0}$ in $B \otimes_A \mathcal{O}$ is approximated by a Cauchy sequence $\{f_n(\widehat{z})\}_{n\geq 0}$ in $\widehat{B}$. So, $\widehat{B}$ is a subring of $\widehat{R}[[\widehat{z}]]$.

(3) Let b be an element of $B \otimes_A \mathcal{O}$. Write

$$b = \alpha_0 + \alpha_1\widehat{z} + \cdots + \alpha_i\widehat{z}^i + \cdots = \sum_{i=0}^{\infty} \alpha_i\widehat{z}^i$$

as an element of $\widehat{R}[[\widehat{z}]]$. If this is an infinite series, the module $M(b)$ generated over $\widehat{\mathcal{O}}$ by $\widehat{\delta}^i(b)$ for $i \geq 0$ is not finitely generated over $\widehat{\mathcal{O}}$ because it is not finitely generated over $\widehat{R}$. Meanwhile, $\widehat{\delta}^i(b) = \delta^i(b)$ for every $i \geq 0$ and $M(b)$ is finitely generated over $\widehat{\mathcal{O}}$ because δ is locally nilpotent. Hence $b \in \widehat{R}[\widehat{z}]$. This implies that $B \otimes_A \mathcal{O}$ is a subring of $\widehat{R}[\widehat{z}]$. It is clear that $\widehat{B}_{\rm ln} = \widehat{R}[\widehat{z}]$. □

We recall the following result of Kaliman-Saveliev [18, Corollary 2.8].

Lemma 3.7. *Let* Y *be a smooth contractible affine threefold with a fixed-point free* G_a*-action. Then* Y *is isomorphic to* $X \times \mathbb{A}^1$ *with* G_a *acting on the second factor, where* $X = Y//G_a$.

This result inspires us a challenging problem.

Question 3.8. Let Y be a homology threefold with a fixed-point free G_a-action. Is Y G_a-equivariantly isomorphic to a product $X \times \mathbb{A}^1$ with G_a acting on the second factor, where $X = Y//G_a$?

If Y is replaced by a $\mathbb{Q}$-homology threefold, the answer is negative.

Remark 3.9. Let $q : X \to Z$ be an affine pseudo-plane as in Lemma 3.4. Then X is a $\mathbb{Q}$-homology plane. Let $Y = X \times \mathbb{A}^1$ with G_a acting on X and trivially on the second factor. Then Y is a $\mathbb{Q}$-homology threefold, $Y//G_a = Z \times \mathbb{A}^1$ and the quotient morphism is $q_Y = q \times \mathrm{id}_{\mathbb{A}^1} : X \times \mathbb{A}^1 \to Z \times \mathbb{A}^1$. Hence $\mathrm{Sing}(q_Y) \cong \mathbb{A}^1$.

4. Contractible affine threefolds with $\mathbb{A}^1_*$-fibrations

We first discuss Question 2.7.

Lemma 4.1. *Let X be a homology plane of log Kodaira dimension $-\infty$ or 1 and let C be a curve on X isomorphic to $\mathbb{A}^1$. Let $V = X \times \mathbb{A}^1$ with a G_m-action induced from the standard action on the $\mathbb{A}^1$-factor, i.e., ${}^t(Q, x) = (Q, tx)$, where $Q \in X$ and x is a coordinate of $\mathbb{A}^1$. Let $\sigma : W \to V$ be the blowing-up of V with center $C \times (0)$ which is identified with C. Let $Y = W \setminus \sigma'(X \times (0))$, where $\sigma'(X \times (0))$ is the proper transform of $X \times (0)$. Then the following assertions hold.*

(1) *Y is a homology threefold with an induced G_m-action such that the quotient morphism $q : Y \to Y//G_m$ has the quotient space $Y//G_m \cong X$. If X is contractible, the threefold Y is also contractible.*
(2) *The singular locus $\mathrm{Sing}(q)$ is the curve C. For every point P of C, the fiber $q^{-1}(P)$ consists of two affine lines meeting in one point. Thus we have the situation treated in Theorem 2.6, (3).*
(3) *If $\overline{\kappa}(X) = 1$, then $\overline{\kappa}(Y) = 1$. Hence Y cannot be written as $Y \cong Z \times Y^{G_m}$, i.e., the answer to the question 2.7 is negative.*
(4) *If $\overline{\kappa}(X) = -\infty$, i.e., $X \cong \mathbb{A}^2$, then $Y \cong Z \times Y^{G_m}$ with Z a smooth affine surface with a G_m-action.*

Proof. (1) Set $D = X \times (0)$, $L = C \times \mathbb{A}^1$, $L' = \sigma'(L)$ and $E' = \sigma^{-1}(C) \setminus \sigma'(D)$. Then $\sigma : (Y, E') \to (V, D)$ is an affine modification with $\sigma(E') = C$ (see [19]). Write $X = \mathrm{Spec}\, A$. Then $A[x]$ is the coordinate ring of V. Since A is factorial by Lemma 1.1, the curve C is defined by an element ξ of A.

The hypersurface $D = X \times (0)$ in V is defined by $x = 0$. Let I be the ideal of $A[x]$ generated by ξ and x. Then Y has the coordinate ring $\Sigma_{I,x}(A[x])$ which is the affine modification of $A[x]$ along (x) with center I. Clearly, Y is a smooth affine threefold. By [19, Proposition 3.1 and Theorem 3.1], it follows that Y is a homology threefold and is contractible provided so is X. Let $q_0 : V \to X$ be the projection to X, which is in fact the quotient morphism by the G_m-action such that $V^{G_m} = D$. By the above process, the G_m-action is inherited on Y and the quotient morphism $q : Y \to Y//G_m$ is induced by q_0.

(2) Meanwhile, in passing from V to Y, the fiber over a point $Q \in X \setminus C$ loses the point $(Q, 0)$ in D and becomes isomorphic to $\mathbb{A}^1_*$. The fiber $q^{-1}(Q)$ is a cross $\mathbb{A}^1 + \mathbb{A}^1$ with two $\mathbb{A}^1$ meeting in the point $q^{-1} \cap (L' \cap E')$. Hence $E' \cap L' = Y^{G_m}$ and $\mathrm{Sing}(q) = C$.

(3) Suppose that $\overline{\kappa}(X) = 1$. In fact, there is a unique affine line lying on X. Since the general fibers of q are isomorphic to $\mathbb{A}^1_*$, we have an inequality $\overline{\kappa}(Y) \geq \overline{\kappa}(X) + \overline{\kappa}(F)$ by Kawamata [23], where F is a general fiber of q. Hence $\overline{\kappa}(Y) \geq 1$. Furthermore, X itself has an $\mathbb{A}^1_*$-fibration $\pi : X \to B$ such that C is a fiber of π and $\pi^{-1}(U) \cong U \times \mathbb{A}^1_*$, where U is an open set of B contained in $\mathbb{A}^1_{2*}$. By the above construction, $q^{-1}(\pi^{-1}(U)) \cong \pi^{-1}(U) \times \mathbb{A}^1_* \cong U \times \mathbb{A}^1_* \times \mathbb{A}^1_*$. Hence $\overline{\kappa}(Y) \leq \overline{\kappa}(U \times \mathbb{A}^1_* \times \mathbb{A}^1_*) = 1$. This implies that $\overline{\kappa}(Y) = 1$. If $Y \cong Z \times \mathbb{A}^1$ as inquired in Question 2.7, then it would follow that $\overline{\kappa}(Y) = -\infty$. But this is not the case.

(4) Suppose that $\overline{\kappa}(X) = -\infty$. Then $X \cong \mathbb{A}^2$, and the affine line C is chosen to be a coordinate line by AMS theorem. Namely, there exists a system of coordinates (ξ, η) of X such that C is defined by $\xi = 0$. The affine modification $\Sigma_{I,x}(A[x])$ is equal to $\mathbb{C}[x, \xi/x, \eta]$. Set $y = \xi/x$ and $R = \mathbb{C}[x, y]$. Then the induced G_m-action on $\Sigma_{I,x}(A[x])$ is given by ${}^t(x, y, \eta) = (tx, t^{-1}y, \eta)$. Hence the threefold Y is isomorphic to $Z \times \mathbb{A}^1$, where $Z = \mathrm{Spec}\, R \cong \mathbb{A}^2$ and $\mathbb{A}^1 = \mathrm{Spec}\, \mathbb{C}[\eta]$. So, the answer to Question 2.7 is affirmative. □

Remark 4.2. The quotient morphism $q : Y \to X$ has crosses $\mathbb{A}^1 + \mathbb{A}^1$ with each multiplicity one as the singular fibers over $\mathrm{Sing}(q) \cong \mathbb{A}^1$. The locus of intersection points of crosses is the fixed point locus $\Gamma = Y^{G_m}$. There are two embedded affine planes Z_1, Z_2 meeting transversally along Γ. If Y is a homology threefold, Z_1, Z_2 are defined by $f_1 = 0, f_2 = 0$. Hence Γ is defined by the ideal $I_1 = (f_1, f_2)$ of $B_1 = \Gamma(Y, \mathcal{O}_Y)$. The affine modification $B_2 = \Sigma_{I_1, f_1}(B)$ or $B'_2 = \Sigma_{I, f_2}(B)$ gives rise to a smooth affine threefold $Y_2 = \mathrm{Spec}\, B_2$ or $Y'_2 = \mathrm{Spec}\, B'_2$ with an equi-dimensional G_m-

action, which gives the quotient morphism $q_2 : Y_2 \to X$ or $q_2' : Y_2' \to X$. Both Y_2 and Y_2' are homology threefolds (resp. contractible threefolds) if Y is a homology threefold (resp. contractible threefold). A difference between Y_2 (or Y_2') and Y is that the crosses have multiplicities. Namely, Y_2 (resp. Y_2') has crosses $2\mathbb{A}^1 + \mathbb{A}^1$ (resp. $\mathbb{A}^1 + 2\mathbb{A}^1$). We can repeat this process to produce homology threefolds or contractible threefolds which have crosses with higher multiplicities. If a cross is written as $m\mathbb{A}^1 + n\mathbb{A}^1$, then $\gcd(m, n) = 1$ (see the argument in the last part of the proof of Lemma 2.21).

Another remark to Lemma 4.1 is the following.

Remark 4.3. Take a $\mathbb{Q}$-homology plane X instead of a homology plane in the construction of Y in Lemma 4.1. Then we can consider such $\mathbb{Q}$-homology planes with log Kodaira dimension $-\infty, 0$ and 1. We take an embedded line C in X and blow up the center $C \times (0)$ in $V = X \times \mathbb{A}^1$. By the same construction, we obtain a smooth affine threefold Y with an equidimensional G_m-action. The threefold Y has the same homology groups as X. Hence Y is a $\mathbb{Q}$-homology threefold, which is not of the product type $Z \times \mathbb{A}^1$ with a $\mathbb{Q}$-homology plane Z provided $\overline{\kappa}(X) \neq -\infty$. For the embedded lines in the case of log Kodaira dimension 0, see [12, Theorem] for a complete classification.

In Lemma 4.1, the center of blowing-up is $C \times (0)$, where C is an embedded line in a homology plane, and the resulting homology threefold has log Kodaira dimension at most one. We can apply a process similar to the one used in Lemma 4.1 with a point as the center to obtain the first assertion of the following lemma. This is first constructed in [26, Example 3.7].

Lemma 4.4. (1) *Let X be a homology plane and let P_0 be a point of X. Let $Y_0 = X \times \mathbb{A}^1$ with G_m acting trivially on X and on $\mathbb{A}^1$ with weight -1. Let $Q_0 = P_0 \times (0)$. Blow up the point Q_0 and remove the proper transform of $X \times (0)$ to obtain a smooth affine threefold Y. Then Y is a homology threefold with a G_m-action such that the quotient morphism $q : Y \to X$ is induced by the first projection $p_1 : Y_0 \to X$, the fixed point locus Y^{G_m} consists of the unique point which is the intersection point of the proper transform of $P_0 \times \mathbb{A}^1$ with the exceptional surface of the blowing-up (whence the G_m-action on Y is* hyperbolic*) and $\overline{\kappa}(Y) = \overline{\kappa}(X)$. If X is contractible, so is Y.*

(2) *Let Y be a smooth homology threefold with a hyperbolic G_m-action. Let Q_0 be the unique fixed point and let Z_0 be the two-dimensional fiber component of the quotient morphism $q : Y \to X$, where $Q_0 \in Z_0$ and $X = Y//G_m$ (cf. Theorem 2.6). Let Y' be the affine modification $\Sigma_{Q_0, Z_0}(Y)$ which is the blowing-up of Y at the center Q_0 with the proper transform of Z_0 deleted off. Then Y' is a smooth homology threefold. If Y is contactible, then so is Y'.*

The statement and the proof depend on [19, §3].

Example 4.5. Let Y be the Koras-Russell threefold $x + x^2y + z^2 + t^3 = 0$. Then a hyperbolic G_m-action on Y is given by

$$^{\lambda}(x, y, z, t) = (\lambda^6 x, \lambda^{-6} y, \lambda^3 z, \lambda^2 t), \quad \lambda \in \mathbb{C}^*.$$

The fixed point Q_0 is $(0, 0, 0, 0)$, and the two-dimensional fiber component Z_0 is defined by $y = 0$. Let B be the coordinate ring of Y and let $I = (x, y, z, t)$ which is the maximal ideal of Q_0. The affine modification $B' = \Sigma_{I,y}(B)$, which is the coordinate ring of Y', is given as

$$B' = \mathbb{C}[x', y, z', t'], \quad x' = \frac{x}{y}, \; z' = \frac{z}{y}, \; t' = \frac{t}{y}.$$

Hence Y' is a hypersurface $x' + {x'}^2 y^2 + {z'}^2 y + {t'}^3 y^2 = 0$ and the hyperbolic G_m-action is given by

$$^{\lambda}(x', y, z', t') = (\lambda^{12} x', \lambda^{-6} y, \lambda^9 z', \lambda^8 t'), \quad \lambda \in \mathbb{C}^*.$$

We can further repeat the affine modifications of the same kind to Y', Y'' etc.

The above Koras-Russell threefold and its affine modifications are examples of smooth contractible threefolds Y with a hyperbolic G_m-action such that the quotient X is isomorphic to that of the tangent space T_{Q_0} at the unique fixed point Q_0 of Y by the induced tangential representation. In [26, Theorem 4.1] a description of all such threefolds is given.

In [17] the Makar-Limanov invariants $\mathrm{ML}(Y)$ of such threefolds are computed. To apply this result the equation for Y' has to be brought into a standard form that exhibits Y' as a cyclic cover of $\mathbb{A}^3$. To this end let $B'' = B'^{\omega} = \mathbb{C}[x', y, \zeta, t']$ with a square root ω of the unity and $\zeta = {z'}^2$. Put

$$x'' = -({x'}^2 y + \zeta + {t'}^3 y).$$

Then $x' = x''y$ and we see that $B'' = \mathbb{C}[x'', y, t']$ is a polynomial ring in three variables and

$$x'' + x''^2 y^3 + \zeta + t'^3 y = 0.$$

Hence it follows that $B' = \mathbb{C}[x'', y, z', t']$ with a defining equation

$$x'' + x''^2 y^3 + z'^2 + t'^3 y = 0.$$

It now follows from [17, Theorem 8.4] that $\mathrm{ML}(Y) = \mathbb{C}[x]$ and $\mathrm{ML}(Y') = B'$. □

In order to show that a fixed point exists under a G_m-action on a $\mathbb{Q}$-homology threefold Y, we used the Smith theory and its variant. The following result without using the Smith theory is of some interest.

Lemma 4.6. *Let Y be a smooth affine variety with an effective G_m-action. Suppose that there are no fixed points. Then the Euler number $e(Y)$ of Y is zero.*

Proof. Let $q : Y \to X$ be the quotient morphism. Since there are no fixed points, every fiber is isomorphic to $\mathbb{A}^1_*$ when taken with reduced structure. The general fibers of q are reduced $\mathbb{A}^1_*$ and special fibers are multiple A^1_*. We work with the complex analytic topology. Considering the isotropy groups of the fibers, there exists a descending chain of closed subsets of X

$$F_0 \supset F_1 \supset \cdots \supset F_i \supset F_{i+1} \supset \cdots$$

such that $F_0 = X$ and the isotropy group of the fiber over a point of $F_i - F_{i+1}$ is constant, say G_i. Then $F_i - F_{i+1}$ is covered by open sets $\{U_{i\lambda}\}_{\lambda \in \Lambda_i}$ such that $q^{-1}(U_{i\lambda})_{\mathrm{red}} \cong G_m \times_{G_i} V_{i\lambda}$, where $V_{i\lambda}$ is a suitable slice (cf. [30, Théorème du slice étale et Remarque 3°]). Hence $q^{-1}(F_i)_{\mathrm{red}}$ is a $\mathbb{C}^*$-bundle over the open set $F_i - F_{i+1}$. This implies that the Euler number $e(Y)$ is zero. □

Lemma 4.6 implies that any non-trivial G_m-action on a smooth affine variety Y has a fixed point if $e(Y) \neq 0$. Furthermore, considering the induced tangential representation at a fixed point, we know that the fixed point locus Y^{G_m} is smooth. However, we do not know if the fixed point locus Y^{G_m} is connected. The following example shows that the connectedness fails in general.

Example 4.7. Let $X = \mathbb{A}^2$ and let $Y_0 = X \times \mathbb{A}^1$ which has a standard G_m-action on the factor $\mathbb{A}^1$ with the point (0) as a fixed point. Choose two parallel lines ℓ_1, ℓ_2 on X, and let $Z_i = \ell_i \times \mathbb{A}^1$ for $i = 1, 2$. Let $\sigma : W \to Y_0$ be the blowing-ups with centers $\ell_1 \times (0)$ and $\ell_2 \times (0)$ and let Y be W with the proper transform of $X \times (0)$ removed. Then the restriction of $p_1 \cdot \sigma$ onto Y gives a morphism $q : Y \to X$ which is eventually the quotient morphism by the G_m-action on Y induced by the one on Y_0, where p_1 is the projection $Y_0 \to X$. The singular fibers of q are crosses over $\ell_1 \cup \ell_2$. The locus of intersection points of crosses is the fixed point locus Y^{G_m}. Hence Y^{G_m} is not connected. The Euler number $e(Y)$ of the threefold Y is two. □

In Section 2, we observed a G_m-action on a $\mathbb{Q}$-homology threefold Y such that the quotient morphism $q : Y \to X$ has relative dimension one. The following result deals with the case of q having relative dimension two.

Proposition 4.8. *Let Y be a $\mathbb{Q}$-homology threefold with a G_m-action. Suppose that the quotient morphism has equi-dimension two. Let $q : Y \to C$ be the quotient morphism, where C is a smooth affine curve. Then each fiber is isomorphic to $\mathbb{A}^2$, and C is isomorphic to $\mathbb{A}^1$. Hence Y is isomorphic to $\mathbb{A}^3$.*

Proof. Let F be a fiber of q. Since $\dim F = 2$, there is a unique fixed point Q such that the closure of every orbit passes through Q. The locus Γ of points Q is the fixed point locus Y^{G_m} and $q|_\Gamma : \Gamma \to C$ is a bijection. Hence it is an isomorphism. Since Y is smooth and the local intersection multiplicity $i(F, \Gamma; Q) = 1$, it follows that F is smooth near Q. Hence F itself is smooth. Then it is easy to show that F is isomorphic to $\mathbb{A}^2$ with an elliptic G_m-action. By a theorem of Sathaye [45], $q : Y \to C$ is an $\mathbb{A}^2$-bundle over C. Since Y is then contractable to C, the curve C is $\mathbb{Q}$-acyclic. Hence $C \cong \mathbb{A}^1$ and q is necessarily trivial. □

Question 4.9. Let Y be a $\mathbb{Q}$-homology n-fold with a G_m-action. Suppose that the quotient morphism has equi-dimension $n - 1$. Is Y isomorphic to the affine space $\mathbb{A}^n$?

In fact, one can show that each fiber of the quotient morphism $q : Y \to C$ is isomorphic to $\mathbb{A}^{n-1}$ with the induced elliptic G_m-action. If q is locally trivial in the Zariski topology, C is $\mathbb{Q}$-acyclic. Hence $C \cong \mathbb{A}^1$ and $Y \cong \mathbb{A}^n$. If Y is a homology n-fold, Question 4.9 has a positive answer which is a theorem of Kraft-Shwarz [29, Theorem 5].

Remark 4.10. In [25], a general result has been proved. Namely, let Y be a homology n-fold. Suppose that Y is dominated by an affine space and Y is endowed with an effective action of $T = G_m^{n-2}$ such that $\dim Y^T > 0$. Then Y is T-equivariantly isomorphic to the affine space $\mathbb{A}^n$ with a linear action of T. Hence a contractible threefold Y having a non-hyperbolic G_m-action is G_m-equivariantly isomorphic to A^3 provided Y is dominated by an affine space.

The following result deals with more general cases of the quotient morphism having equi-dimension one.

Lemma 4.11. *Let Y be a smooth affine threefold with a G_m-action. Suppose that the quotient morphism $q : Y \to X$ has equi-dimension one. Then the following assertions hold.*

(1) *If $\dim Y^{G_m} = 2$, then Y^{G_m} is isomorphic to X and hence Y^{G_m} is connected. The morphism q defines a line bundle over X.*
(2) *If $\dim Y^{G_m} = 1$ and $e(Y) > 0$, then Y^{G_m} is smooth and consists of connected components $\Gamma_1, \ldots, \Gamma_r$, one of which is isomorphic to $\mathbb{A}^1$. Further, if Y^{G_m} is connected, then $e(Y) = 1$ and the quotient surface X is a normal affine surface with an embedded line and at worst cyclic quotient singularities.*

Proof. (1) Since q does not contain a fiber component of dimension two, Y^{G_m} lies horizontally to the fibration q. Hence each fiber contains a unique fixed point. This implies that each fiber is isomorphic to $\mathbb{A}^1$. Considering the tangent space $T_{Y,Q}$ and the induced tangential representation for each $Q \in Y^{G_m}$, we know that Y^{G_m} is smooth and isomorphic to X. Namely, q is an $\mathbb{A}^1$-fibration with all reduced fibers isomorphic to $\mathbb{A}^1$ and has two cross-sections Y^{G_m} and a section at infinity. Hence q is in fact a line bundle.

(2) The morphism q is then an $\mathbb{A}^1_*$-fibration and Y^{G_m} is a smooth curve. Let $Y^{G_m} = \Gamma_1 \sqcup \cdots \sqcup \Gamma_r$ be the decomposition into connected components. Let $\overline{\Gamma}_i$ be the smooth completion of Γ_i. Let g_i be the genus of $\overline{\Gamma_i}$ and let n_i be the number of points in $\overline{\Gamma}_i \setminus \Gamma_i$. Note that G_m acts on $q^{-1}(X - q(Y^{G_m}))$ without fixed points. Hence, by Lemma 4.6, the Euler number of $q^{-1}(X - q(Y^{G_m}))$ is zero. Note that $q^{-1}(q(\Gamma_i))$, taken with reduced structure, is a union of two $\mathbb{A}^1$-bundles over $q(\Gamma_i)$ meeting transversally along the section Γ_i. In fact, since the fiber $q^{-1}(P)$ over a point $P \in q(\Gamma_i)$ is a cross with each branch meeting Γ_i transversally in one point, $q^{-1}(q(\Gamma_i))$ consists of two irreducible components $W_i^{(1)}$ and $W_i^{(2)}$, each of which has an $\mathbb{A}^1$-fibration with a cross-section Γ_i. Hence $W_i^{(1)}$ and $W_2^{(2)}$ are $\mathbb{A}^1$ bundles over $q(\Gamma_i)$ meeting transversally along Γ_i. Hence the Euler number of $q^{-1}(q(\Gamma_i))$ is

equal to $2 - 2g_i - n_i$. This observation yields a relation

$$0 < e(Y) = \sum_{i=1}^{r}(2 - 2g_i - n_i).$$

Since Γ_i is an affine curve, we have $n_i \geq 1$. If none of $\Gamma_1, \ldots, \Gamma_r$ is isomorphic to $\mathbb{A}^1$, then the right side of the above equality is less than or equal to zero, which is a contradiction. Hence one of them is isomorphic to $\mathbb{A}^1$. If Y^{G_m} is connected, then $Y^{G_m} = \Gamma_1 \cong \mathbb{A}^1$. It follows from the footnotes in the proof of Theorem 2.6 that q induces a closed embedding of Y^{G_m} into X and that X is smooth near $q(Y^{G_m})$. By Lemma 2.13, X has at worst cyclic quotient singularities in the open set $X \setminus (q(Y^{G_m}))$. □

Remark 4.12. Given a smooth affine surface X_0, we can produce a smooth affine surface X by a half-point attachment [38, p. 233] which contains X_0 as an open set and has an embedded affine line C. By the same procedure as in Lemma 4.1, we take a product $X \times \mathbb{A}^1$ and apply the affine modification of $X \times \mathbb{A}^1$ with center $C \times (0)$ in $X \times (0)$. The resulting threefold is a smooth affine threefold Y with a G_m-action such that the quotient morphism $q : Y \to Y//G_m$, where $X \cong Y//G_m$ and $Y^{G_m} \cong C$. Further, $\overline{\kappa}(Y) = \overline{\kappa}(X)$.

References

1. S.M. Bhatwadekar and A.K. Dutta, On $\mathbb{A}^*$-fibrations, J. Pure and Applied Algebra **149** (2000), 1–14.
2. A. Borel, Linear algebraic groups, Second edition, Graduate Texts in Mathematics **126**, Springer-Verlag, New York, 1991.
3. A.K. Dutta, On $\mathbb{A}^1$-bundles of affine morphisms, J. Math. Kyoto Univ. **35** (3) (1995) 377–385.
4. H. Flenner and M. Zaidenberg, Normal affine surfaces with $\mathbb{C}^*$-actions, Osaka J. Math. **40** (2003), no. 4, 981–1009.
5. E.E. Floyd, On periodic maps and the Euler characteristics of associated spaces, Trans. Amer. Math. Soc. **72** (1952), 138–147.
6. T. Fujita, On the topology of noncomplete algebraic surfaces, J. Fac. Sci. Univ. Tokyo Sect. IA Math. **29** (1982), no. 3, 503–566.
7. R.V. Gurjar, Two-dimensional quotients of $\mathbb{C}^n$ are isomorphic to $\mathbb{C}^2/\Gamma$, Transf. Groups, 12 (2007), no. 1, 117–125.
8. R.V. Gurjar, M. Koras, M. Miyanishi and P. Russell, Affine normal surfaces with simply-connected smooth locus, Math. Ann. Published online: 06 July, 2011.
9. R.V. Gurjar, K. Masuda and M. Miyanishi, $\mathbb{A}^1$-fibrations on affine threefolds, J. Pure and Applied Algebra, **216** (2012), 296–313.
10. R.V. Gurjar and M. Miyanishi, Affine lines on logarithmic **Q**-homology planes, Math. Ann. 294 (1992), 463-482.

11. R.V. Gurjar and M. Miyanishi, Pure subalgebras of polynomial rings, preprint.
12. R.V. Gurjar and A.J. Parameswaran, Affine lines on $\mathbb{Q}$-homology planes, J. Math. Kyoto Univ. **35** (1995), no. 1, 63–77.
13. H.A. Hamm, Zur Homotopietyp Steinscher Räume, J. Reine Angew. Math. **338** (1983), 121–135.
14. Z. Jelonek, A number of points in the set $\mathbb{C}^2 \setminus F(\mathbb{C}^2)$, Bull. Polish Acad. Sci. Math. 47 (1999), no. 3, 257–261.
15. Sh. Kaliman, Free $\mathbb{C}_+$-actions on $\mathbb{C}^3$ are translations, Invent. Math. **156** (2004), no. 1, 163–173.
16. Sh. Kaliman, Actions of $\mathbb{C}^*$ and $\mathbb{C}_+$ on affine algebraic varieties, Algebraic geometry-Seattle 2005, Part 2, 629–654, Proc. Sympos. Pure Math., **80**, Part 2, Amer. Math. Soc., Providence, RI, 2009.
17. Sh. Kaliman and L. Makar-Limanov, On the Russell-Koras contractible threefolds, J. Algebraic Geom. **6** (1997), no. 2, 247–268.
18. Sh. Kaliman and N. Saveliev, $\mathbb{C}_+$-actions on contractible threefolds, Michigan Math. J. **52** (2004), no. 3, 619–625.
19. Sh. Kaliman and M. Zaidenberg, Affine modifications and affine hypersurfaces with a very transitive automorphism group, Transf. Groups **4** (1999), 53–95.
20. Sh. Kaliman and M. Zaidenberg, Families of affine planes: the existence of a cylinder, Michigan Math. J. **49** (2001), no. 2, 353–367.
21. T. Kambayashi and P. Russell, On linearizing algebraic torus actions, J. pure and applied algebra **23** (1982), 243–250.
22. Y. Kawamata, On the classification of non-complete algebraic surfaces, Algebraic Geometry, Lecture Notes in Math., vol. 732, Springer, Berlin, 1979, 215-232.
23. Y. Kawamata, Addition formula of logarithmic Kodaira dimensions for morphisms of relative dimension one, Proc. of the Internat. Symposium on Algebraic Geometry (Kyoto Univ., Kyoto, 1977), pp. 207–217, Kinokuniya Book Store, Tokyo, 1978.
24. M. Koras, Linearization of reductive group actions, in *Group Actions and Vector Fields*, Lecture Notes in Math. **956** (1982), 92–98.
25. M. Koras and P. Russell, Codimension 2 torus actions on affine n-space, *Group actions and invariant theory* (Montreal, PQ, 1988), 103–110, CMS Conf. Proc. **10**, Amer. Math. Soc., Providence, RI, 1989.
26. M. Koras, and P. Russell, Contractible threefolds and $\mathbb{C}^*$-actions on $\mathbb{C}^3$, J. Algebraic Geom. **6** (1997), no. 4, 671–695.
27. H.P. Kraft, Geometrische Methoden in der Invariantentheorie, Aspects of Mathematics, D1. Friedr. Vieweg & Sohn, Braunschweig, 1984.
28. H.P. Kraft, T. Petrie and J.D. Randall, Quotient varieties. Adv. Math. 74 (1989), no. 2, 145–162.
29. H.P. Kraft and G.W. Schwarz, Reductive group actions with one-dimensional quotient, Inst. Hautes Etudes Sci. Publ. Math. No. **76** (1992), 1–97.
30. D. Luna, Slices étales, Bull. Soc. Math. France, Paris, Memoire **33**, Soc. Math. France, Paris, 1973.

31. K. Masuda and M. Miyanishi, Affine pseudo-planes and cancellation problem, Trans. Amer. Math. Soc. **357** (2005), 4867–4883.
32. J. Milnor, Morse theory, Ann. Math. Study **51**, Princeton Univ. Press, 1963.
33. M. Miyanishi, Normal affine subalgebras of a polynomial ring, Algebraic and Topological Theories - to the memory of Dr. Takehiko Miyata, Kinokuniya, 1985, 37-51.
34. M. Miyanishi, An algebric characterization of the affine plane, J. Math. Kyoto Univ. **15**, (1975), 169-184.
35. M. Miyanishi, An algebro-topological characterization of the affine space of dimension three, Amer. J. Math. **106** (1984), no. 6, 1469–1485.
36. M. Miyanishi, Etale endomorphisms of algebraic varieties, Osaka J. Math. **22** (1985), 345–364.
37. M. Miyanishi, Algebraic characterizations of the affine 3-space, Algebraic Geometry Seminar (Singapore, 1987), 53–67, World Sci. Publishing, Singapore, 1988.
38. M. Miyanishi, Open algebraic surfaces, Centre de Recherches Mathématiques, Vol **12**, Université de Montréal, Amer. Math. Soc., 2001.
39. M. Miyanishi, G_a-actions and completions, J. Algebra **319** (2008), 2845–2854.
40. M. Miyanishi and T. Sugie, Homology planes with quotient singularities, J. Math. Kyoto Univ. **31** (1991), no. 3, 755–788.
41. M.V. Nori, Zariski's conjecture and related problems. Ann. Sci. Ecole Norm. Sup. (4) 16 (1983), no. 2, 305–344.
42. K. Palka, Singular $\mathbb{Q}$-homology planes I, arXiv:0806.3110.
43. M. Rosenlicht, Some basic theorems on algebraic groups,. Amer. J. Math. **78** (1956), 401–443.
44. P. Russell, Gradings of polynomial rings, Algebraic geometry and its applications (C.L. Bajaj ed.), Springer, 1994.
45. A. Sathaye, Polynomial ring in two variables over a DVR: a criterion, Invent. Math. **74** (1983), no. 1, 159–168.
46. E.H. Spanier, Algebraic topology, Springer Verlag, 1966.
47. J. Winkelmann, A correspondence to P. Russell, August 23, 2011.
48. M. Zaidenberg, Isotrivial families of curves on affine surfaces and characterization of the affine plane, Izv. Akad. Nauk SSSR Ser. Mat. **51** (1987), no. 3, 534–567, 688; translation in Math. USSR-Izv. **30** (1988), no. 3, 503–532.
49. O. Zariski, Interprétations algébrico-géométriques du quatorzième problème de Hilbert, Bull.Sci. Math. **78** (1954), 155-168.

Acknowledgements

This article is a product of the discussions which we had at McGill University in August, 2011 at the occasion of the Workshop on Complex-Analytic and Algebraic Trends in the Geometry of Varieties sponsored by the Centre de Recherches Mathématiques, Montréal and held at Université de Quebec à Montréal (UQAM). We are very grateful to the Department of Mathematics and Statistics, McGill University, and Professor Steven Lu of UQAM for providing us this nice opportunity for discussions.

The second author was supported by a Polish grant NCN N201 608640. The third and fourth authors were respectively supported by Grant-in-Aid for Scientific Research (C) 22540059 and 21540055, JSPS. The fifth author was supported by a grant from NSERC, Canada.

Finally the authors appreciate various comments by the referees. As pointed out by one of them, $\mathbb{Q}$-homology threefold might as well be called $\mathbb{Q}$-homology 3-space.

Miyanishi's characterization of singularities appearing on $\mathbb{A}^1$-fibrations does not hold in higher dimensions

Takashi Kishimoto

Department of Mathematics, Faculty of Science,
Saitama University, Saitama 338-8570, Japan
E-mail: tkishimo@rimath.saitama-u.ac.jp

Dedicated to Professor Masayoshi Miyanishi on the occasion of his 70th birthday

In [Mi81, Theorems 1 and 2], Miyanishi characterizes two-dimensional singularities on normal affine surfaces equipped with $\mathbb{A}^1$-fibrations, more precisely, he proves in *loc. cit.* that any two-dimensional cyclic quotient singularity can be realized as that on a suitable normal affine surface possessing an $\mathbb{A}^1$-fibration and vice versa. In this paper, we shall observe that the category of cyclic quotient singularities are not large enough, in case of higher-dimension dim $\geqq 3$, to characterize those appearing on normal affine varieties admitting $\mathbb{A}^1$-fibrations and try to enlarge the extent of the class of such singularities. As a result, we show that any cyclic quotient singularity, any toric singularity, more generally, any singularity admitting an action of an algebraic torus $\mathbb{T}$ satisfying a certain condition about a σ-polyhedral divisor is realized in an affine variety X equipped with a $\mathbb{G}_a$-action such that the ring of invariants $\Gamma(\mathcal{O}_X)^{\mathbb{G}_a}$ is finitely generated. Furthermore, in case of toric singularities, such an X can be chosen for the Makar-Limanov invariant $\mathrm{ML}(\Gamma(\mathcal{O}_X))$ to be trivial.

Keywords: Singularities, $\mathbb{A}^1$-fibrations, $\mathbb{G}_a$-actions, $\mathbb{T}$-varieties.

1. Introduction

All varieties considered in this paper are defined over the field of complex numbers $\mathbb{C}$. As a principle in algebraic geometry, we understand that the existence of a family of rational curves on an algebraic variety imposes a certain restriction about the type of (isolated) singularities on it. For instance, as shown in [FlZa03a], it is known that if a Cohen-Macaulay normal quasi-projective variety X with an isolated singularity $x \in X$ admits a family of closed rational curves, say $\mathcal{F}$, which covers a Zariski open subset of X and members from $\mathcal{F}$ do not pass through x, then $x \in X$ is a rational singularity. Certainly, rational singularities form still a large class of singularities, but the existence of a more particular family of rational curves imposes more restrictive conditions about the type of singularities, for instance, the case of a family of mutually disjoint affine lines $\mathbb{A}^1$ on a normal affine surface X (in other word, X contains an $\mathbb{A}^1$-*cylinder*, see [Mi78, Chapter I]) (cf. [Mi81]). To state the result in [Mi81], we shall recall the notion of $\mathbb{A}^r$-*cylinders*. A Zariski open subset U on an algebraic variety X is said to be *an* $\mathbb{A}^r$-*cylinder* if U is isomorphic to $U_0 \times \mathbb{A}^r$, where U_0 is an algebraic variety of $\dim(U_0) = \dim(X) - r$. On the other hand, a surjective morphism $f : X \to Z$ is said to be *an* $\mathbb{A}^r$-*fibration* if a general (closed) fiber of f is isomorphic to $\mathbb{A}^r$ scheme-theoretically. We are mainly interested in the case of $r = 1$ in this paper. It is well known that if X is equipped with an $\mathbb{A}^1$-fibration, then X contains an $\mathbb{A}^1$-cylinder (cf. [KaMi78]). Conversely, even supposing that a normal affine algebraic variety X contains an $\mathbb{A}^1$-cylinder $U_0 \times \mathbb{A}^1 \cong U \subseteq X$, we do not know so far whether or not X admits an $\mathbb{A}^1$-fibration.[a] Meanwhile in case of $\dim(X) = 2$, the existence of an $\mathbb{A}^1$-fibration over a smooth curve Z on such an X is well known (see for instance [Mi81, §2]). However Z is not necessarily affine. For example, let us consider the affine surface $X = \mathbb{P}^1 \times \mathbb{P}^1 \backslash (\text{diagonal})$, where $\mathbb{A}^1$-cylinders arising from $\mathbb{A}^1$-fibrations on X obtained as restrictions of canonical projections of $\mathbb{P}^1 \times \mathbb{P}^1$. Anyway, the existence of an $\mathbb{A}^1$-cylinder on an affine algebraic surface X is equivalent to asking that X possesses an $\mathbb{A}^1$-fibration over a constructible set. The result of Miyanishi is then stated as follows:

Theorem 1.1. (cf. [Mi81, Theorems 1 and 2])

(1) *Let X be a normal affine surface which contains an $\mathbb{A}^1$-cylinder (or equivalently, which has an $\mathbb{A}^1$-fibration). Then X has at worst cyclic quotient singularities.*
(2) *Conversely, any two-dimensional cyclic quotient singularity is realized as that on a suitable normal affine surface containing an $\mathbb{A}^1$-cylinder (or equivalently, having an $\mathbb{A}^1$-fibration).*

Remark 1.1. The proof of Theorem 1.1 depends on a general theory of singular fibers about $\mathbb{P}^1$-fibrations on surfaces (cf. [Giz70], [Mi78, Chapter 2]) and a criterion about the contractibility of curves on surfaces (cf. [Ar62],

---

[a] In the particular case of $\mathrm{Pic}(X) \otimes_{\mathbb{Z}} \mathbb{Q} = (0)$, it follows that X admits a $\mathbb{G}_a$-action by applying [KPZ11, Proposition 3.1.5]. But even though if $\dim(X) \geqq 4$, the ring of invariants $\Gamma(\mathcal{O}_X)^{\mathbb{G}_a}$ is not necessarily finitely generated over $\mathbb{C}$. As a consequence, the extension to an $\mathbb{A}^1$-fibration on X whole is not guaranteed.

[Br68]). Note that the argument in [Mi81, Theorem 2] to obtain Theorem 1.1 (2) says implicitly that any two dimensional cyclic quotient singularity is embedded into a normal affine surface X with an $\mathbb{A}^1$-fibration $f : X \to Z$ over the affine line $Z \cong \mathbb{A}^1$ in such a way that f is yielded as a quotient of a suitable $\mathbb{G}_a$-action on X.[b] Meanwhile, there is not an information about the Makar-Limanov invariant $\mathrm{ML}(\Gamma(\mathcal{O}_X))$ of the coordinate ring of X, which is defined as the intersection of kernels of all LND's on $\Gamma(\mathcal{O}_X)$. But, by making use of the toric description of 2-dimensional cyclic quotient singularities, we can, in fact, choose X in such a way that $\mathrm{ML}(\Gamma(\mathcal{O}_X))$ is trivial, i.e., $\mathrm{ML}(\Gamma(\mathcal{O}_X)) = \mathbb{C}$ (see Proposition 3.1).

As Theorem 1.1 suggests, it is reasonable to believe that the type of singularities is further restricted once we admit an $\mathbb{A}^1$-fibration or an $\mathbb{A}^1$-cylinder. Thus we shall propose the following problem as a generalization of Theorem 1.1 for the case of higher-dimension:

Problem 1.1.

(1) *Let X be a normal affine variety. Supposing that X possesses an $\mathbb{A}^1$-fibration or an $\mathbb{A}^1$-cylinder, what kinds of singularities appear on X? Conversely,*
(2) *What kind of a germ of a singularity can be realized as a singularity of a normal affine algebraic variety admitting an $\mathbb{A}^1$-fibration or containing an $\mathbb{A}^1$-cylinder? More strongly,*
(3) *What kind of a germ of a singularity can be realized as a singularity of a normal affine algebraic variety equipped with a $\mathbb{G}_a$-action?*

Remark 1.2. Notice that for any normal affine algebraic variety Y, the product $X := Y \times \mathbb{A}^1$, which is normal, admits obviously an $\mathbb{A}^1$-cylinder. Thus it is hopeful to deal with affine algebraic varieties which are obtained as products with the affine spaces at the same time. This intent brings us to the notion of $\mathbb{T}$-varieties (see §2, Definition 2.3). Indeed, in terms of $\mathbb{T}$-varieties, the coordinate ring $A := \Gamma(\mathcal{O}_X) = \Gamma(\mathcal{O}_Y) \otimes_{\mathbb{C}} \mathbb{C}[t]$ of X is described in such a way that $A \cong A[Y, \mathcal{D}]$ with $\mathcal{D} := \sum_D \sigma\,[D]$ and $\sigma := \mathbb{Q}_{\geq 0} \subseteq N_{\mathbb{Q}} \cong \mathbb{Q}$, where D ranges over all prime divisors on Y (see Definitions 2.4, 2.5 and Theorem 2.1). Indeed, the $\mathbb{A}^1$-bundle $X = Y \times \mathbb{A}^1 \to Y$ is then obtained associated to a homogeneous LND on A of fiber type corresponding to σ as stated in Theorem 2.2 and Theorem 1.2 below.

[b]Notice that if an affine algebraic variety X admits a $\mathbb{G}_a$-action, then X contains an $\mathbb{A}^1$-cylinder (cf. [Mi78, Chapter 1]), but the existence of an $\mathbb{A}^1$-cylinder in X does not imply a $\mathbb{G}_a$-action on X, in general.

In view of Theorem 1.1, it seems to be natural to think that cyclic quotient singularities are those requested in Problem 1.1 even in case of higher dimension also. But, it follows, in fact, that the class of cyclic quotient singularities is not large enough to reply to questions in Problem 1.1 as seen in the following simple example:

Example 1.1. Let us consider the affine quadratic hypersurface $X = (XY - ZU = 0) \subseteq \mathbb{A}^4$, which possesses $\mathbb{A}^1$-fibrations. In fact, for instance, the restriction $\mathrm{pr}_{(X,Y)}|_X$ of the projection $\mathrm{pr}_{(X,Y)} : \mathbb{A}^4{}_{(X,Y,Z,U)} \to \mathbb{A}^2{}_{(X,Y)}$ onto X yields an $\mathbb{A}^1$-fibration. As the plane $(X = Z = 0) \subseteq X$ is not $\mathbb{Q}$-Cartier at the origin $(0,0,0,0) \in X$, it is not in particular a cyclic quotient singularity. On the other hand, it is classical and well known that X is a toric variety (see Example 2.4).

Indeed, we shall confirm the following result concerning Problem 1.1, (2), (3) in this article:

Theorem 1.2.

(1) *Any toric singularity (hence, in particular, any cyclic quotient sigularity) can be realized as a singular point on a normal affine variety X with at least* $\dim(X)$*'s $\mathbb{A}^1$-fibrations over normal affine varieties all of which are obtained as quotients of $\mathbb{G}_a$-actions on X such that* $\mathrm{ML}(\Gamma(\mathcal{O}_X)) = \mathbb{C}$. *More generally,*

(2) *A singularity embedded into an affine $\mathbb{T}$-variety* $\mathrm{Spec}\,(A[Z,\mathcal{D}])$, *where Z is a semiprojective variety, $\mathcal{D}$ is a proper σ-polyhedral divisor on Z (see §2, Definitions 2.3, 2.4 and 2.5), satisfying the following condition $(*)$ can be realized as a singular point on a normal affine variety X with an $\mathbb{A}^1$-fibration over a normal affine algebraic variety of dimension* $\dim(X) - 1$, *which is obtained as a quotient morphism of a $\mathbb{G}_a$-action:*

   $(*)$ *There exists a face ρ of σ of dimension* 1 *such that $\mathcal{D}(\mathbf{m})$ is big for any* $\mathbf{m} \in \mathrm{rel.int.}\,(\check{\sigma} \cap \rho^{\perp}) \cap M$.

Corollary 1.1. *There exist examples of singularities, which are not cyclic quotient, appearing on normal affine algebraic varieties admitting $\mathbb{A}^1$-fibrations.*[c]

[c]Example 1.1 is one of such examples. Another example is given in Example 3.2.

Remark 1.3.

(1) The class of singularities admitting an action of an algebraic torus $\mathbb{T}$ is too large in order to ensure the existence of $\mathbb{A}^1$-fibrations. For instance, let us observe the affine hypersurface of Fermat type $X := (x^d + y^d + z^d = 0) \subseteq \mathbb{A}^3$ with $d \geqq 3$. Notice that X admits a natural $\mathbb{T} = \mathbb{G}_m$-action, hence it is a $\mathbb{T}$-variety to guarantee that the coordinate ring $A = \Gamma(\mathcal{O}_X)$ of X is described as $A = A[Z, \mathcal{D}]$, where $Z := (x^d + y^d + z^d = 0) \subseteq \mathbb{P}^3$ and $\mathcal{D}$ is a certain proper σ-polyhedral divisor on Z with $\sigma = \mathbb{Q}_{\geqq 0} \subseteq N_{\mathbb{Q}} \cong \mathbb{Q}$ (see Theorem 2.1). Supposing that X admits an $\mathbb{A}^1$-fibration, the origin $\mathbf{o} = (0, 0, 0) \in X$ must be at worst a cyclic quotient singularity by virtue of Theorem 1.1. But this is not possible in case of $d \geqq 3$. Hence there does not exist any $\mathbb{G}_a$-action on X, which implies that $\mathcal{D}$ does not satisfy $(*)$ in Theorem 1.2 by taking Theorem 2.2 into account.

(2) There exists an example where a proper σ-polyhedral divisor $\mathcal{D}$ on Z does not satisfy $(*)$, hence there is no homogeneous LND of fiber type on $A[Z, \mathcal{D}]$ by Theorem 2.2, nevertheless there exist on the contrary homogeneous ones of horizontal type (see Definition 2.6). For instance, see Example 3.2 as such an example. As a consequence, $(*)$ is only a sufficient condition to ascertain the existence of a $\mathbb{G}_a$-action.

(3) Different from the case of homogeneous LND's of fiber type, it is worthwhile to notice that kernels of homogeneous LND's of horizontal type are not necessarily finitely generated (cf. Theorem 2.2 (2), [Ro90], [DaFr99], [Fr00]).

We shall state the scheme of this article. In §2, we prepare some definitions and known results from affine algebraic geometry and toric geometry. In particular, we shall mention briefly a correspondence between $\mathbb{G}_a$-actions on a given affine algebraic variety X and (homogeneous) locally nilpotent derivations on the coordinate ring $\Gamma(\mathcal{O}_X)$ though we believe that this stuff is well known for experts on affine algebraic geometry. Furthermore, we recall the notion of (affine) $\mathbb{T}$-varieties, which are by definition varieties equipped with an action of an algebraic torus $\mathbb{T}$. One of the features of affine $\mathbb{T}$-varieties are that they can be expressed by a combinatorial description in terms of polyhedral divisors (see Theorem 2.1), which is a generalization of a combinatorial one in terms of polyhedral cones in case of toric varieties. In §3, we show assertions stated in Theorem 1.2 and Corollary 1.1. One of the ingredients in this article lies in the description of locally nilpotent derivations on coordinate rings of $\mathbb{T}$-varieties, which are homogeneous with respect to $\mathbb{T}$-actions, due to Liendo (cf. [Li10a], [Li10b], see Theorem 2.2). Recall again that the technique in [Mi81] to obtain Theorem 1.1 depends on the general fact about singular fibers of $\mathbb{P}^1$-fibrations on surfaces and the contractibility criterion of curves, meanwhile, that used in this article is different from it. Thus (the proof of) Theorem 1.2 yields also an alternative proof for Theorem 1.1, (2) (see also Remark 3.1). In addition, we consider Problem 3.1 about singularities appearing on $\mathbb{A}^2$-fibrations.

2. Preliminaries

In this section, we recall some notions and facts from affine and toric geometry. More precisely, we shall recall the correspondence of (homogeneous) locally nilpotent derivations and $\mathbb{G}_a$-actions (which are compatible with an action of an algebraic torus $\mathbb{T}$), a description of $\mathbb{T}$-varieties in terms of proper σ-polyhedral divisors due to Altmann-Hausen (see Theorem 2.1) and the result of Liendo (see Theorem 2.2). We think that some of stuffs treated here are familiar to experts in affine algebraic geometry, notwithstanding, we shall prepare them because of their importance in this article. First of all, we begin with locally nilpotent derivations:

Definition 2.1. Let A be a finitely generated $\mathbb{C}$-algebra, which is an integral domain. A derivation Δ on A is said to be a *locally nilpotent derivation* (LND, for short) if for any element $a \in A$ there exists $N(a) \in \mathbb{N}$ such that $\Delta^n(a) = 0$ for $\forall n \geqq N(a)$. Furthermore, in the case where A is graded by $\mathbb{Z}^n$,[d] say $A = \bigoplus_{\mathbf{d} \in \mathbb{Z}^n} A_{\mathbf{d}}$, then a derivation Δ on A is said to be *homogeneous of degree* $\mathbf{e} \in \mathbb{Z}^n$ if $\Delta(A_{\mathbf{d}}) \subseteq A_{\mathbf{d}+\mathbf{e}}$ for all $\mathbf{d} \in \mathbb{Z}^n$.

Definition 2.2. Two LND's Δ_1 and Δ_2 on a finitely generated $\mathbb{C}$-algebra domain A is said to be *equivalent*, denoted by $\Delta_1 \sim \Delta_2$, if $\mathrm{Ker}\,(\Delta_1) = \mathrm{Ker}\,(\Delta_2)$. This is equivalent to being able to write $f_1\Delta_1 = f_2\Delta_2$ for some $f_1, f_2 \in \mathrm{Ker}(\Delta_1)$. Then we say also that $\mathbb{G}_a$-actions corresponding to Δ_1 and Δ_2 on $\mathrm{Spec}\,(A)$ are *equivalent* (see Remark 2.1 below).

Remark 2.1. Let A be a finitely generated $\mathbb{C}$-algebra domain, and let $X := \mathrm{Spec}\,(A)$ be the associated affine algebraic variety. Then it is well known that there exists a one to one correspondence between the set of LND's on A, say $\mathrm{LND}(A)$, and the set of $\mathbb{G}_a$-actions on X (cf. [Mi78],

[d]Equivalently, the variety $\mathrm{Spec}\,(A)$ admits an action of $\mathbb{G}_m^n$.

[Fr06]). More precisely, for $\Delta \in \mathrm{LND}(A)$ given, the corresponding $\mathbb{G}_a$-action on X is obtained associated to the co-action defined by:

$$\varphi_\Delta : A \ni a \mapsto \sum_{n \geqq 0} \frac{\Delta^n(a)}{n!} \otimes t^n \in A \otimes_{\mathbb{C}} \mathbb{C}[t],$$

where t is a coordinate on $\mathbb{G}_a$. If two mutually different LND's Δ_1 and Δ_2 on A are equivalent (see Definition 2.2), then the actions themselves of $\mathbb{G}_a$ associated to φ_{Δ_1} and φ_{Δ_2} are different, but the invariant rings coincide to each other. Hence their quotient morphisms coincide. In the case where X is equipped with an action by an algebraic torus $\mathbb{T} \cong \mathbb{G}_m^r$, or equivalently, its coordinate ring A is $\mathbb{Z}^r$-graded, to give a homogeneous LND on A is equivalent to saying that X possesses a $\mathbb{G}_a$-action that is normalized in $\mathrm{Aut}\,(X)$ by $\mathbb{T}$.

Example 2.1. Let A be the normal integral domain $\mathbb{C}[x,y,z]/(yz - x^3) = \mathbb{C}[\overline{x}, \overline{y}, \overline{z}]$, which yields a Du Val singularity of A_2-type at the origin of $X := \mathrm{Spec}\,(A) = (yz - x^3 = 0) \subseteq \mathbb{A}^3$. Notice that X is an affine toric variety of dimension 2, in other words, A is $\mathbb{Z}^2$-graded. Indeed, we see that $A = \bigoplus_{(p,q) \in \mathbb{Z}^2} A_{(p,q)}$, where

$$A_{(p,q)} := \left\{ f(\overline{x}, \overline{y}, \overline{z}) \in A \,\middle|\, (\lambda, \mu) \cdot f(\overline{x}, \overline{y}, \overline{z}) = \lambda^p \mu^q f(\overline{x}, \overline{y}, \overline{z}),\ \forall\, (\lambda, \mu) \in \mathbb{G}_m^2 \right\},$$

where $(\lambda, \mu) \in \mathbb{G}_m^2$ acts in such a way that $(\lambda, \mu) \cdot (\overline{x}, \overline{y}, \overline{z}) = (\lambda \overline{x}, \mu \overline{y}, \lambda^3 \mu^{-1} \overline{z})$. Let Δ_1 and Δ_2 be LND's on A defined as in the following fashion:

$$\Delta_1(\overline{x}) = \overline{y},\ \Delta_1(\overline{y}) = 0,\ \Delta_1(\overline{z}) = 3\overline{x}^2, \quad \Delta_2(\overline{x}) = \overline{z},\ \Delta_2(\overline{y}) = 3\overline{x}^2,\ \Delta_2(\overline{z}) = 0.$$

It is then straightforward to confirm that both of Δ_1 and Δ_2 are, in fact, homogeneous LND's on A of degree $(-1, 1)$ and $(2, -1)$, respectively, therefore these give rise to two $\mathbb{G}_a$-actions on X, which are normalized by the action of $\mathbb{G}_m^2$ in $\mathrm{Aut}\,(X)$. Indeed, we can verify that these two LND's Δ_1 and Δ_2 exhaust homogeneous LND's on A with respect to the $\mathbb{Z}^2$-grading up to the equivalence (cf. Theorem 2.2).

In general, for a given normal affine algebraic variety X, it is not reasonable to determine all LND's even up to equivalence. Notwithstanding, in the case that X is a $\mathbb{T}$-variety (see Definition 2.3 below), the description of homogeneous LND's of fiber type (see Definition 2.6) on the coordinate ring $\Gamma(\mathcal{O}_X)$ is realized in an explicit manner in terms of σ-polyhedral divisors and extremal rays of σ (cf. [Li10a], [Li10b]). In order to state the results in [Li10a], [Li10b] exactly, we need some preparation of terminologies.

Definition 2.3. Let $\mathbb{T}$ be an algebraic torus of $\dim(\mathbb{T}) = r$, i.e., $\mathbb{T} \cong \mathbb{G}_m^r$. A normal affine algebraic variety X is said to be a $\mathbb{T}$-*variety* if X admits an effective action of $\mathbb{T}$. This is equivalent to saying that the coordinate ring $A = \Gamma(\mathcal{O}_X)$ of X is effectively $\mathbb{Z}^r$-graded, namely A can be written in such a way that $A = \bigoplus_{\mathbf{m} \in \mathbb{Z}^r} A_{\mathbf{m}}$ and the set $\{\mathbf{m} \in \mathbb{Z}^r \mid A_{\mathbf{m}} \neq 0\}$ is not contained in a proper sublattice of $\mathbb{Z}^r$.

Remark 2.2. It is clear that an affine toric variety is at the same time a $\mathbb{T}$-variety. Indeed, an affine $\mathbb{T}$-variety X is toric in case of $\dim(\mathbb{T}) = \dim(X)$. But it is not difficult to construct an example of $\mathbb{T}$-varieties which are never toric (see Example 2.2). On the other hand, it is possible that a fixed $\mathbb{T}$-variety $X = \mathrm{Spec}(A)$ becomes a toric variety simultaneously by changing the grading of A. For instance, the affine plane $\mathbb{A}^2 = \mathrm{Spec}(\mathbb{C}[s,t])$ is a toric variety if we define an action of $\mathbb{G}_m^2$ on $\mathbb{C}[s,t]$ by $(\lambda,\mu)\cdot(s,t) = (\lambda s, \mu t)$, where $(\lambda,\mu) \in \mathbb{G}_m^2$. Meanwhile, by defining $\mathbb{T} = \mathbb{G}_m$-action on $\mathbb{C}[s,t]$ in such a way that $\lambda\cdot(s,t) = (\lambda s, t)$ (resp. $\lambda\cdot(s,t) = (\lambda s, \lambda t)$), it has a structure as a $\mathbb{T}$-variety, in fact, $\mathbb{C}[s,t]$ is then $\mathbb{Z}$-graded as $\mathbb{C}[s,t] = \bigoplus_{m \geqq 0} \mathbb{C}[t]s^m$ (resp. $\mathbb{C}[s,t] = \bigoplus_{d \geqq 0} \mathbb{C}[s,t]_{(d)}$, where $\mathbb{C}[s,t]_{(d)}$ is a $\mathbb{C}$-vector space spanned by homogeneous polynomials in (s,t) of usual degree d).

Example 2.2. (cf. [FlZa03b]) Let $C_0 = \mathrm{Spec}(A_0)$ be a smooth affine algebraic curve with $K_0 := \mathrm{Frac}(A_0)$ the function field. Let $a \in C_0$ be a point and let us consider the $\mathbb{Q}$-divisor $D(a) := -(e/d)[a]$ on C_0, where $d, e \geqq 1$ and $\gcd(d,e) = 1$. Then:

$$A := \bigoplus_{m \geqq 0} H^0(C_0, \lfloor mD(a) \rfloor)s^m \subseteq K_0[s]$$

is a finitely generated $\mathbb{C}$-algebra domain of dimension 2 which is effectively $\mathbb{Z}$-graded, namely, $X := \mathrm{Spec}(A)$ is a $\mathbb{T}$-variety with $\mathbb{T} = \mathbb{G}_m$.[e] Let us denote by $\iota : C_0 \hookrightarrow X$ the inclusion morphism associated with the projection $A \ni \bigoplus_{m \geqq 0} a_m s^m \mapsto a_0 \in A_0$. On the other hand, letting us denote by $\pi : X \to C_0$ the morphism associated with the inclusion $A_0 \hookrightarrow A$, π is an $\mathbb{A}^1$-fibration having a section $\iota(C_0) \subseteq X$, which is a set of fixed points of the action of $\mathbb{T}$. Furthermore, π possesses the singular fiber $\pi^*(a)$ and the point $\iota(a) \in X$ on $\pi^*(a)$ is a cyclic quotient singularity of type (d, e'), where $0 \leqq e' < d$ is an integer such that $e \equiv e' \pmod{d}$ (cf. [FlZa03a, Proposition 3.8]), that is, a singularity which is obtained as the origin of a quotient of

[e]For a given $\mathbb{Q}$-divisor $D = \sum_i d_i D_i$ with D_i's prime divisors and $d_i \in \mathbb{Q}$, we denote by $\lfloor D \rfloor$ the *round down* of D, i.e., $\lfloor D \rfloor = \sum_i \lfloor d_i \rfloor D_i$, where $\lfloor d_i \rfloor$ is the integer satisfying $d_i - 1 < \lfloor d_i \rfloor \leqq d_i$.

$\mathbb{A}^2\,(s,t)$ by a cyclic group $\mathbb{Z}/d\mathbb{Z} \cong <\zeta>$ via $\zeta\cdot(s,t) = (\zeta s, \zeta^{e'} t)$, where ζ is a primitive e'-th root of unity. Notice that if C_0 is not rational, then X cannot admit any structure of a toric surface.

Example 2.2 demonstrates a principle to construct $\mathbb{T}$-varieties generally. In fact, the work [AlHa06] yields a complete description of $\mathbb{T}$-varieties of any dimension in terms of geometry of polyhedral cones. In order to state the results in [AlHa06], we shall recall several definitions:

Definition 2.4. Let N be a lattice of rank $(N) = r$ with the dual lattice $M := \mathrm{Hom}_{\mathbb{Z}}(N, \mathbb{Z})$. Let $\sigma \subseteq N_{\mathbb{Q}} := N \otimes_{\mathbb{Z}} \mathbb{Q}$ be a strongly convex polyhedral cone, i.e., a polyhedral cone in $N_{\mathbb{Q}}$ which does not contain any non-trivial vector space, let $\check{\sigma} \subseteq M_{\mathbb{Q}} := M \otimes_{\mathbb{Z}} \mathbb{Q}$ be the dual cone of σ, that is, $\check{\sigma} = \{\,\mathbf{m} \in M_{\mathbb{Q}} \mid\, <\mathbf{m},\mathbf{n}> \geqq 0 \ (\forall\, \mathbf{n} \in N_{\mathbb{Q}})\,\}$.

(1) A polyhedral domain in $N_{\mathbb{Q}}$ is called a *σ-tailed polyhedron* if it is expressed as the Minkowski sum $\sigma + \Delta$ of a bounded polyhedron Δ of $N_{\mathbb{Q}}$ and σ. We denote by $\mathrm{Pol}_{\sigma}(N_{\mathbb{Q}})$ the set of all σ-tailed polyhedra in $N_{\mathbb{Q}}$, which is equipped with a structure of an abelian semi-group via Minkowski sum. Notice that the neutral element of $\mathrm{Pol}_{\sigma}(N_{\mathbb{Q}})$ is σ.

(2) Let $\mathrm{CPL}_{\mathbb{Q}}(\check{\sigma})$ be the set of all piecewise linear $\mathbb{Q}$-valued function $h : \check{\sigma} \to \mathbb{Q}$ satisfying the following two properties:

$$h(\mathbf{m}+\mathbf{m}') \geqq h(\mathbf{m}) + h(\mathbf{m}') \quad \text{and} \quad h(\lambda\mathbf{m}) = \lambda h(\mathbf{m})$$

$$(\mathbf{m},\mathbf{m}' \in \check{\sigma},\ \lambda \in \mathbb{Q}_{\geqq 0}).$$

Note that the set $\mathrm{CPL}_{\mathbb{Q}}(\check{\sigma})$ with the usual addition of functions becomes an abelian semigroup with the neutral element 0.

Remark 2.3. As abelian semi-groups, $\mathrm{Pol}_{\sigma}(N_{\mathbb{Q}})$ and $\mathrm{CPL}_{\mathbb{Q}}(\check{\sigma})$ are isomorphic. Indeed, for a given $\Delta \in \mathrm{Pol}_{\sigma}(N_{\mathbb{Q}})$, let us define the $\mathbb{Q}$-valued function $h_{\Delta} \in \mathrm{CPL}_{\mathbb{Q}}(\check{\sigma})$ by:

$$h_{\Delta}(\mathbf{m}) := \min_{\mathbf{n}\in\Delta} <\mathbf{m},\mathbf{n}> .$$

Then the correspondence $\Delta \mapsto h_{\Delta}$ gives rise to an isomorphism between $\mathrm{Pol}_{\sigma}(N_{\mathbb{Q}})$ and $\mathrm{CPL}_{\mathbb{Q}}(\check{\sigma})$ (cf. [AlHa06]).

Definition 2.5. An algebraic variety Z is said to be *semi-projective* if Z is projective over an affine algebraic variety.[f] For a strongly convex polyhedral

[f]In particular, affine and projective varieties are semi-projective.

cone $\sigma \subseteq N_{\mathbb{Q}}$, we define a *$\sigma$-polyhedral divisor on Z* to be a formal sum $\mathcal{D} = \sum_D \Delta_D[D]$, where D ranges over all prime divisors on Z, $\Delta_D \in \mathrm{Pol}_\sigma(N_{\mathbb{Q}})$ and $\Delta_D = \sigma$ for all but finitely many D's. For each $\mathbf{m} \in \check{\sigma} \cap M$, we associate a $\mathbb{Q}$-divisor $\mathcal{D}(\mathbf{m})$ on Z as follows:

$$\mathcal{D}(\mathbf{m}) = \sum_D h_{\Delta_D}(\mathbf{m})[D],$$

where $h_{\Delta_D} : \check{\sigma} \to \mathbb{Q}$ is an element of $\mathrm{CPL}_{\mathbb{Q}}(\check{\sigma})$ defined as in Remark 2.3. Notice that $\mathcal{D}(\mathbf{m})$ is actually a $\mathbb{Q}$-divisor on Z as $\Delta_D \neq \sigma$ only for finitely many D's. Moreover, a σ-polyhedral divisor $\mathcal{D}$ is called *proper* if the following two conditions are satisfied:

(a) $\mathcal{D}(\mathbf{m})$ is semi-ample and $\mathbb{Q}$-Cartier for $\forall\, \mathbf{m} \in \check{\sigma} \cap M$,[g] and
(b) $\mathcal{D}(\mathbf{m})$ is big for $\forall\, \mathbf{m} \in \mathrm{rel.int}(\check{\sigma}) \cap M$.[h]

The significance to consider σ-polyhedral divisors can be summarized in the following fundamental result Theorem 2.1. In fact, Theorem 2.1 characterizes the coordinate rings of $\mathbb{T}$-varieties.

Theorem 2.1. (cf. [AlHa06]) *Let N be a lattice with the dual lattice M.*

(1) *For a normal semi-projective variety Z, a strongly convex polyhedral cone $\sigma \subseteq N_{\mathbb{Q}}$ and for a proper σ-polyhedral divisor $\mathcal{D}$ on Z,*

$$(\star)\quad A[Z,\mathcal{D}] := \bigoplus_{\mathbf{m}\in\check{\sigma}\cap M} H^0(Z, \lfloor \mathcal{D}(\mathbf{m}) \rfloor)\chi^{\mathbf{m}} \subseteq \mathbb{C}(Z)[\{\chi^{\mathbf{m}}\}_{\mathbf{m}\in\check{\sigma}\cap M}]$$

is a finitely generated normal effectively M-graded integral domain of dimension $\dim A[Z,\mathcal{D}] = \dim Z + \mathrm{rank}\,(M)$.[i]

(2) *Conversely, for any normal finitely generated effectively M-graded integral domain A, there exist a normal semi-projective variety Z, a strongly convex polyhedral cone $\sigma \subseteq N_{\mathbb{Q}}$ and a proper σ-polyhedral divisor $\mathcal{D}$ on Z such that A is isomorphic to $A[Z,\mathcal{D}]$ defined as in $(\star)$ of (1).*

By virtue of Theorem 2.1, the structure of $\mathbb{T}$-varieties can be interpreted in terms of proper σ-polyhedral divisors. For instance, $\mathbb{G}_a$-actions on a

[g] I.e., some multiple $n\mathcal{D}(\mathbf{m})$ $(n \in \mathbb{N})$ becomes a Cartier divisor and $|n\mathcal{D}(\mathbf{m})|$ yields a base point free linear system.

[h] We denote by $\mathrm{rel.int}(\check{\sigma})$ the interior of $(\check{\sigma})_{\mathbb{R}}$, which is the extension of $\check{\sigma} \subseteq M_{\mathbb{Q}}$ in $M_{\mathbb{R}} = M_{\mathbb{Q}} \otimes_{\mathbb{Q}} \mathbb{R}$, with respect to the usual Euclidian topology of $M_{\mathbb{R}}$.

[i] The algebra structure of $A[Z,\mathcal{D}]$ is determined by $\chi^{\mathbf{m}}\chi^{\mathbf{m}'} = \chi^{\mathbf{m}+\mathbf{m}'}$ for $\mathbf{m}, \mathbf{m}' \in \check{\sigma} \cap M$.

$\mathbb{T}$-variety Spec$(A[Z, \mathcal{D}])$, which are compatible with a $\mathbb{T}$-action and vertical with respect to the rational map Spec$(A[Z, \mathcal{D}]) \dashrightarrow Z$ associated to $\mathbb{C}(Z)(\{\chi^{\mathbf{m}}\}_{\mathbf{m}\in\check{\sigma}\cap M}) \supseteqq \mathbb{C}(Z)$, are translated into rays of σ, i.e., one-dimensional faces of σ, as seen in the works of A. Liendo [Li10a], [Li10b]. Before stating the results in [Li10a], [Li10b], we shall return to Example 2.2 by translating the homogeneous algebra A and a homogeneous LND on A yielding the $\mathbb{A}^1$-fibration $\pi : X \to C_0$ into a σ-polyhedral divisor and an extremal ray.

Example 2.3. Let $N \cong \mathbb{Z}$ be a lattice of rank 1 with $M := \mathrm{Hom}_{\mathbb{Z}}(N, \mathbb{Z})$ the dual lattice, let $\sigma := \mathbb{Q}_{\geqq 0}\mathbf{e} \subseteq N_{\mathbb{Q}}$ and let $\Delta := \big([-(e/d), 0] \cap \mathbb{Q}\big) + \sigma \in \mathrm{Pol}_{\sigma}(N_{\mathbb{Q}})$ be a σ-tailed polyhedron, where $d, e \geqq 1$. On the other hand, let $C_0 = \mathrm{Spec}\,(A_0)$ be a smooth affine algebraic curve and we take a point $a \in C_0$. Let us consider a proper σ-polyhedral divisor on C_0 (cf. Definition 2.5):

$$\mathcal{D} := \Delta\,[a] + \sum_{b\in C_0\setminus\{a\}} \sigma\,[b].$$

Then for $m\mathbf{e}^* \in \check{\sigma} \cap M = \mathbb{Z}_{\geqq 0}\mathbf{e}^*$ we have:

$$\mathcal{D}(m\mathbf{e}^*) = h_{\Delta}(m\mathbf{e}^*)[a] + \sum_{b\in C_0\setminus\{a\}} h_{\sigma}(m\mathbf{e}^*)\,[b] = h_{\Delta}(m\mathbf{e}^*)[a] = -\Big(\frac{e}{d}\Big)m[a],$$

to see:

$$\begin{aligned} A[C_0, \mathcal{D}] &= \bigoplus_{m\mathbf{e}^*\in\check{\sigma}\cap M} H^0(C_0, \lfloor \mathcal{D}(m\mathbf{e}^*)\rfloor)\chi^{m\mathbf{e}^*} \\ &\cong \bigoplus_{m\geqq 0} H^0(C_0, \lfloor mD(a)\rfloor)s^m = A \end{aligned}$$

as observed in Example 2.2. Hence the $\mathbb{T}$-variety X defined in Example 2.2 is isomorphic to Spec$\,(A[C_0, \mathcal{D}])$ with $\mathbb{T} \cong \mathbb{G}_m$. Notice that the $\mathbb{A}^1$-fibration $\pi : X \to C_0$ associated to $A_0 \hookrightarrow A \cong A[C_0, \mathcal{D}]$ is obtained, in other viewpoint, as the quotient of the $\mathbb{G}_a$-action on X corresponding to an LND δ on $A[C_0, \mathcal{D}]$ (up to equivalence) defined as follows:

$$\delta(f\chi^{m\mathbf{e}^*}) = mf\chi^{(m-1)\mathbf{e}^*} =< m\mathbf{e}^*, \mathbf{e} > f\chi^{m\mathbf{e}^*-\mathbf{e}^*},$$

$$f \in H^0(C_0, \lfloor \mathcal{D}(m\mathbf{e}^*)\rfloor) \quad (m \geqq 0).$$

Letting us denote by $\overline{\delta}$ the natural extension of δ to a derivation on $K_0[\{\chi^{\mathbf{m}}\}_{\mathbf{m}\in M}]$, it follows that $\overline{\delta}|_{K_0} = 0$, where $K_0 = \mathrm{Frac}(A_0)$ is the function field of C_0, in other words geometrically speaking, general fibers of π are contained in closures of general orbits of the $\mathbb{T}$-action on X.

The description of Example 2.3 brings us to the following definition about the type of homogeneous LND's on an algebra of the form $A[Z,\mathcal{D}]$.

Definition 2.6. Let δ be a homogeneous LND on an effectively M-graded integral domain $A := A[Z,\mathcal{D}] = \bigoplus_{m\in\check{\sigma}\cap M} H^0(Z,\lfloor\mathcal{D}(m)\rfloor)\chi^m$, where Z, $\sigma \subseteq N_{\mathbb{Q}}$ and $\mathcal{D}$ are as in Definition 2.5. By the Leibniz rule, we can extend δ to the quotient field $K := \mathrm{Frac}(A)$, which we denote by $\overline{\delta}$. It is easy to see that $K = K_0(M)$, where $K_0 := \mathbb{C}(Z)$ is the function field of Z and $K_0(M)$ is the quotient field of $K_0[M] = K_0[\{\chi^m\}_{m\in M}]$. We say that δ is *of fiber type* (resp. *of horizontal type*) if $\overline{\delta}|_{K_0} = 0$ (resp. $\overline{\delta}|_{K_0} \neq 0$). Geometrically speaking, in the case that δ is of fiber type, a general orbit of the $\mathbb{G}_a$-action associated to δ is contained in a closure of an orbit of the $\mathbb{T}$-action. (See for instance Example 3.2 as an example of a homogeneous LND of horizontal type.)

The results in [Li10b] concern how to describe homogeneous LND's on $A[Z,\mathcal{D}]$ of fiber type in terms of a geometry of $\sigma \subseteq N_{\mathbb{Q}}$ and the property of $\mathcal{D}(m)$.

Theorem 2.2. (cf. [Li10a]) *Let $A = A[Z,\mathcal{D}] = \bigoplus_{\mathbf{m}\in\check{\sigma}\cap M} H^0(Z,\lfloor\mathcal{D}(\mathbf{m})\rfloor)$ $\chi^{\mathbf{m}}$ be an effectively M-graded integral domain, where Z, $\sigma \subseteq N_{\mathbb{Q}}$ and $\mathcal{D}$ are as in Definition 2.5. We assume that $\sigma \neq \{0\}$. For a face $\rho = \mathbb{Q}_{\geqq 0}[\mathbf{u}] \prec \sigma$ of dimension 1 having $\mathbf{u} \in \rho$ as the primitive lattice vector along ρ, let us write $\tau := \check{\sigma} \cap \rho^{\perp} \prec \check{\sigma}$, let us denote by $\sigma_\rho \prec \sigma$ the cone spanned by all the rays of σ except for ρ, and let us set:*[j]

$$S_\rho := \left\{ \mathbf{m} \in \check{\sigma_\rho} \cap M \,\middle|\, <\mathbf{m},\mathbf{u}>= -1 \right\}.$$

For a given $\mathbf{e} \in S_\rho$, we shall define the following $\mathbb{Q}$-divisor $D_{\mathbf{e}}$ on Z:[k]

$$D_{\mathbf{e}} := \sum_D \max_{\mathbf{m}\in\check{\sigma}\cap M\setminus\tau\cap M} \{h_{\Delta_D}(\mathbf{m}) - h_{\Delta_D}(\mathbf{m}+\mathbf{e})\}[D].$$

Then the following assertions hold true:

(1) *The derivation $\delta_{\rho,\mathbf{e}}$ on the algebra $\mathbb{C}(Z)[\check{\sigma}\cap M] = \bigoplus_{\mathbf{m}\in\check{\sigma}\cap M} \mathbb{C}(Z)\chi^{\mathbf{m}}$ defined by $\delta_{\rho,\mathbf{e}}(\chi^{\mathbf{m}}) =<\mathbf{m},\mathbf{u}> \chi^{\mathbf{m}+\mathbf{e}}$ is a homogeneous LND on $\mathbb{C}(Z)[\check{\sigma}\cap M]$ of degree $\mathbf{e}$ such that $\mathrm{Ker}\,(\delta_{\rho,\mathbf{e}}) = \mathbb{C}(Z)[\tau\cap M] = \bigoplus_{\mathbf{m}\in\tau\cap M}\mathbb{C}(Z)\chi^{\mathbf{m}}$.*

[j]Notice that $S_\rho \neq \emptyset$.

[k]Note that $D_{\mathbf{e}}$ is well defined as $\Delta_D = \sigma$ for all but only finitely many D's.

(2) *Provided that* $H^0(Z, \lfloor -D_{\mathbf{e}} \rfloor) \neq 0$, *for any* $\varphi \in H^0(Z, \lfloor -D_{\mathbf{e}} \rfloor) \backslash \{0\}$, *the derivation* $\delta_{\rho,\mathbf{e},\varphi} := \varphi\delta_{\rho,\mathbf{e}}$ *on* $\mathbb{C}(Z)[\check{\sigma} \cap M]$ *becomes a homogeneous LND on* A *with the kernel* $\mathrm{Ker}\,(\delta_{\rho,\mathbf{e},\varphi}) = \bigoplus_{\mathbf{m} \in \tau \cap M} H^0(Z, \lfloor \mathcal{D}(\mathbf{m}) \rfloor)\chi^{\mathbf{m}}$. *In particular,* $\mathrm{Ker}\,(\delta_{\rho,\mathbf{e},\varphi})$ *is finitely generated as a* $\mathbb{C}$*-algebra.*

(3) $H^0(Z, \lfloor -D_{\mathbf{e}} \rfloor) \neq 0$ *for a suitable choice of* $\mathbf{e} \in S_\rho$ *if and only if* $\mathcal{D}(\mathbf{m})$ *is big for any* $\mathbf{m} \in \mathrm{rel.int.}\,(\tau) \cap M$.

(4) *There exists one-to-one correspondence between the set of homogeneous LND's of fiber type on* A *up to equivalence and the set of faces* $\rho \prec \sigma$ *of dimension one such that* $\mathcal{D}(\mathbf{m})$ *is big for* $\forall\, \mathbf{m} \in \mathrm{rel.int.}\,(\check{\sigma} \cap \rho^\perp) \cap M$. *Moreover, this correspondence is given by* $\rho \mapsto \delta_{\rho,\mathbf{e},\varphi}$.

Remark 2.4. Notice that if Z is of dimension zero, i.e., $X = \mathrm{Spec}\,(A)$ is an affine toric variety, then the condition mentioned in (2) and (3) of Theorem 2.2 to guarantee the existence of non-zero elements of $H^0(Z, \lfloor -D_{\mathbf{e}} \rfloor)$ is not necessary. In fact, as Z is a point, we do not have to consider a σ-polyhedral divisor $\mathcal{D}$ and the $\mathbb{Q}$-divisor $D_{\mathbf{e}}$. Furthermore, any homogeneous LND on X is then of fiber type automatically. Thence Theorem 2.2 yields a complete description of homogeneous LND's (up to equivalence) on the coordinate ring of a toric variety.

Example 2.4. As stated in Example 1.1, the affine quadratic hypersurface $(XY - ZU = 0) \subseteq \mathbb{A}^4$ is toric. Though this is a well known fact, we shall verify it below and describe all $\mathbb{G}_a$-actions on it, which are normalized by the action of an algebraic torus. Let us consider the following strongly convex cone in $N_{\mathbb{Q}} \cong \mathbb{Q}^3$:

$$\sigma := \mathbb{Q}_{\geqq 0}(0,1,0) + \mathbb{Q}_{\geqq 0}(0,0,1) + \mathbb{Q}_{\geqq 0}(-1,1,0) + \mathbb{Q}_{\geqq 0}(-1,0,1) \subseteq N_{\mathbb{Q}},$$

and let us look at four extremal rays of σ (i.e., faces of σ of dimension one):

$$\rho_1 := \mathbb{Q}_{\geqq 0}(0,1,0), \quad \rho_2 := \mathbb{Q}_{\geqq 0}(0,0,1),$$

$$\rho_3 := \mathbb{Q}_{\geqq 0}(-1,1,0), \quad \rho_4 := \mathbb{Q}_{\geqq 0}(-1,0,1) \prec \sigma.$$

It is easy to verify that $\check{\sigma} \cap M = \mathbb{Z}_{\geqq 0}(0,1,0) + \mathbb{Z}_{\geqq 0}(0,0,1) + \mathbb{Z}_{\geqq 0}(-1,0,0) + \mathbb{Z}_{\geqq 0}(1,1,1)$ and $A := \mathbb{C}[\check{\sigma} \cap M] = \mathbb{C}[\chi^{(0,1,0)}, \chi^{(0,0,1)}, \chi^{(-1,0,0)}, \chi^{(1,1,1)}] \cong \mathbb{C}[X,Y,Z,U]/(XY - ZU)$, where this isomorphism is yielded by $\chi^{(0,1,0)} \mapsto X, \chi^{(0,0,1)} \mapsto Y, \chi^{(-1,0,0)} \mapsto Z, \chi^{(1,1,1)} \mapsto U$. Letting us set $\tau_i := \check{\sigma} \cap \rho_i^\perp$, we see that:

$$\tau_1 \cap M = \mathbb{Z}_{\geqq 0}(0,0,1) + \mathbb{Z}_{\geqq 0}(-1,0,0), \; \tau_2 \cap M = \mathbb{Z}_{\geqq 0}(0,1,0) + \mathbb{Z}_{\geqq 0}(-1,0,0),$$

$$\tau_3 \cap M = \mathbb{Z}_{\geqq 0}(0,0,1) + \mathbb{Z}_{\geqq 0}(1,1,1),\ \tau_4 \cap M = \mathbb{Z}_{\geqq 0}(0,1,0) + \mathbb{Z}_{\geqq 0}(1,1,1).$$

With the notation in Theorem 2.2, we can ascertain by a simple calculation that:

$$\begin{aligned} S_{\rho_1} &= \{\,(a,-1,c) \in M \mid a \leqq -1,\ c \geqq 0\,\}, \\ S_{\rho_2} &= \{\,(a,b,-1) \in M \mid a \leqq -1,\ b \geqq 0\,\}, \\ S_{\rho_3} &= \{\,(a,a-1,c) \in M \mid a \geqq 1,\ c \geqq a\,\}, \\ S_{\rho_4} &= \{\,(a,b,a-1) \in M \mid b \geqq a \geqq 1\,\}, \end{aligned}$$

and we take $\mathbf{e}^{(i)} \in S_{\rho_i}$, for instance, as $\mathbf{e}^{(1)} := (-1,-1,0), \mathbf{e}^{(2)} := (-1,0,-1), \mathbf{e}^{(3)} := (1,0,1), \mathbf{e}^{(4)} := (1,1,0)$. Then $\delta_{\rho_i,\mathbf{e}^{(i)}}$ defined as in Theorem 2.2, (1) are described as:

$$\delta_{\rho_1,\mathbf{e}^{(1)}}(\chi^{(a,b,c)}) = b\chi^{(a-1,b-1,c)},\ \delta_{\rho_2,\mathbf{e}^{(2)}}(\chi^{(a,b,c)}) = c\chi^{(a-1,b,c-1)},$$

$$\begin{aligned} \delta_{\rho_3,\mathbf{e}^{(3)}}(\chi^{(a,b,c)}) &= (-a+b)\chi^{(a+1,b,c+1)}, \\ \delta_{\rho_4,\mathbf{e}^{(4)}}(\chi^{(a,b,c)}) &= (-a+c)\chi^{(a+1,b+1,c)}. \end{aligned}$$

According to Theorem 2.2, these four LND's exhaust all homogeneous LND's on A up to equivalence. Under the isomorphism $A \cong \mathbb{C}[X,Y,Z,U]/(XY - ZU)$, letting us denote by δ_i the LND on $\mathbb{C}[X,Y,Z,U]/(XY - ZU)$ corresponding to $\delta_{\rho_i,\mathbf{e}^{(i)}}$, we see that:

$$\begin{aligned} \delta_1 &= Z\frac{\partial}{\partial X} + Y\frac{\partial}{\partial U}, & \delta_2 &= Z\frac{\partial}{\partial Y} + X\frac{\partial}{\partial U}, \\ \delta_3 &= U\frac{\partial}{\partial X} + Y\frac{\partial}{\partial Z}, & \delta_4 &= U\frac{\partial}{\partial Y} + X\frac{\partial}{\partial Z}. \end{aligned}$$

3. Proof of Theorem 1.2

This section is devoted to the proof of Theorem 1.2. Before the proof, we shall look at the case of cyclic quotient singularities of dimension 2, which demonstrates the case of higher dimension also.

3.1.

As well known, in the case of dimension 2, the categories of toric singularities and cyclic quotient singularities coincide (cf. [CLS11, Chapter 10]). Let us consider a germ $(x \in X)$ of a surface singularity, which is a cyclic quotient of type (d,e) with $1 \leqq e \leqq d$ such that $\gcd(e,d) = 1$, i.e., analytically $(x \in X)$ is isomorphic to the origin of $\mathbb{A}^2/\mathbb{Z}_d(1,e)$, where the action of the cyclic group $\mathbb{Z}_d \cong <\zeta>$ on $\mathbb{A}^2 = \mathrm{Spec}\,(\mathbb{C}[x,y])$ is determined by $\zeta \cdot (x, y) = (\zeta x, \zeta^e y)$ with ζ a primitive d-th root of unity. Then the coordinate ring of $\mathbb{A}^2/\mathbb{Z}_d(1, e)$ is described as:

$$A := \mathbb{C}[\,\{x^a y^b\} \mid a, b \geqq 0,\ a + be \in d\mathbb{Z}\,].$$

On the other hand, A is realized in terms of toric algebra and homogeneous LND's on A are explicitly described as follows:

Proposition 3.1. *Let $\mathbf{e}_1$ and $\mathbf{e}_2$ be the standard generators of $\mathbb{Z}^2$ with the dual $\mathbf{e}_1^*$ and $\mathbf{e}_2^*$, respectively. Let us set $N := \mathbb{Z}\mathbf{e}_1 + \mathbb{Z}(\mathbf{e}_2/e) + \mathbb{Z}((\mathbf{e}_1 + \mathbf{e}_2)/d)$ with the dual $M := \mathrm{Hom}_{\mathbb{Z}}(N, \mathbb{Z})$, where $1 \leqq e \leqq d$ is coprime, i.e., $\gcd(e, d) = 1$, and p, q are integers such that $pe + qd = 1$. For a cone $\sigma := \mathbb{Q}_{\geqq 0}\mathbf{e}_1 + \mathbb{Q}_{\geqq 0}\mathbf{e}_2 \subseteq N_{\mathbb{Q}}$, we have the following:*

(1) $\check{\sigma} \cap M = \{\, a\mathbf{e}_1^* + eb\mathbf{e}_2^* \mid a, b \geqq 0,\ a + eb \in d\mathbb{Z}\,\}$.

(2) The $\mathbb{C}$-algebra

$$\mathbb{C}[\check{\sigma} \cap M] = \bigoplus_{\mathbf{m} \in \check{\sigma} \cap M} \mathbb{C}\chi^{\mathbf{m}} \cong \bigoplus_{a, b \geqq 0,\ a+eb \in d\mathbb{Z}} \mathbb{C}\chi^{(a,b)}$$

is isomorphic to A. More precisely, this isomorphism is given by the assignment $\chi^{(a,b)} \mapsto x^a y^b$.

(3) There exist mutually non-equivalent homogeneous LND's on $\mathbb{C}[\check{\sigma} \cap M]$, say δ_1 and δ_2, such that $\mathrm{Ker}(\delta_1) \cap \mathrm{Ker}(\delta_2) = \mathbb{C}$. More precisely, we can take such δ_i's as in the following fashion (up to equivalence):

$$\delta_1(\chi^{(a,b)}) = a\chi^{(a-1, b+p)}, \quad \delta_2(\chi^{(a,b)}) = b\chi^{(a+e, b-1)}.$$

(4) The LND ∂_i on A corresponding to δ_i under the isomorphism $\mathbb{C}[\check{\sigma} \cap M] \cong A$ in (2) is described as

$$\partial_1(x^a y^b) = a x^{a-1} y^{b+p}, \quad \partial_2(x^a y^b) = b x^{a+e} y^{b-1}.$$

In particular, we have $\mathrm{ML}(A) = \mathbb{C}$.

Proof. The assertions (1) and (2) are directly confirmed. On the other hand, (3) is a consequence of Theorem 2.2. For the convenience of readers, we shall demonstrate in an explicit manner. The cone $\sigma \subseteq N_{\mathbb{Q}}$ has two extremal rays $\rho_1 := \mathbb{Q}_{\geqq 0}[\mathbf{e}_1]$ and $\rho_2 := \mathbb{Q}_{\geqq 0}[(1/e)\mathbf{e}_2]$. For instance, let us look into a homogeneous LND corresponding to ρ_2. The set $S_{\rho_2} \subseteq M$ is, by definition (see Theorem 2.2), is described as:

$$S_{\rho_2} = \{\, a\mathbf{e}_1^* + eb\mathbf{e}_2^* \in M \mid a \geqq 0,\ b = -1\,\} = \{\, a\mathbf{e}_1^* - e\mathbf{e}_2^* \mid a \geqq 0,\ a - e \in d\mathbb{Z}\,\}.$$

Letting us choose $\mathbf{e}^{(2)} \in S_{\rho_2}$, for instance, as $\mathbf{e}^{(2)} := e\mathbf{e}_1^* - e\mathbf{e}_2^*$, the homogeneous LND $\delta_{\rho_2,\mathbf{e}^{(2)}}$ on $\mathbb{C}[\check{\sigma} \cap M]$ is then determined by:

$$\delta_{\rho_2,\mathbf{e}^{(2)}}(\chi^{(a,b)}) = \delta_{\rho_2,\mathbf{e}^{(2)}}(\chi^{a\mathbf{e}_1^*+be\mathbf{e}_2^*})$$

$$= < a\mathbf{e}_1^* + be\mathbf{e}_2^*, (1/e)\mathbf{e}_2 > \chi^{a\mathbf{e}_1^*+be\mathbf{e}_2^*+e\mathbf{e}_1^*+e\mathbf{e}_2^*} = b\chi^{(a+e,b-1)}.$$

By the same fashion, letting us take $\mathbf{e}^{(1)} \in S_{\rho_1}$ as $\mathbf{e}^{(1)} := -\mathbf{e}_1^* + pe\mathbf{e}_2^*$ for example, a homogeneous LND $\delta_{\rho_1,\mathbf{e}^{(1)}}$ on $\mathbb{C}[\check{\sigma} \cap M]$ is defined by:

$$\delta_{\rho_1,\mathbf{e}^{(1)}}(\chi^{(a,b)}) = \delta_{\rho_1,\mathbf{e}^{(1)}}(\chi^{a\mathbf{e}_1^*+be\mathbf{e}_2^*})$$

$$= < a\mathbf{e}_1^* + be\mathbf{e}_2^*, \mathbf{e}_1 > \chi^{a\mathbf{e}_1^*+be\mathbf{e}_2^*-\mathbf{e}_1^*+pe\mathbf{e}_2^*} = a\chi^{(a-1,b+p)},$$

as stated. Then by recalling that the isomorphism $\mathbb{C}[\check{\sigma} \cap M] \cong A$ is yielded by $\chi^{(a,b)} \mapsto x^a y^b$, we can see (4) readily. □

Remark 3.1. Proposition 3.1 gives rise to an alternative proof for Theorem 1.1, (2). In fact, it asserts that any cyclic quotient singularity of dimension two can be realized as that on a certain normal affine surface with trivial Makar-Limanov invariant. But historically this fact is implicitly found in the work of Gizatullin (cf. [Giz71]). Indeed, it is not difficult to verify that any non-degenerate affine toric surface admits a normal projective completion whose boundary is a chain of rational curves. Then by virtue of [Giz71] it admits at least two $\mathbb{A}^1$-fibrations over the affine line $\mathbb{A}^1$, which implies the triviality of its Makar-Limanov invariant. We would like to refer the readers to [Du04] for the precise statement and a self-contained argument which follows techniques from [Giz71].

3.2.

In what follows, we shall investigate in general the case of dimension $n := \dim \geqq 3$.

3.2.1.

At first we recall that cyclic quotient singularities are regarded as toric singularities as in Proposition 3.1. Let $\mathbb{A}^n/\mathbb{Z}_d(a_1, \cdots, a_n)$ be a cyclic quotient singularity, where the cyclic group $\mathbb{Z}_d \cong < \zeta >$ acts on coordinates $(x_1, \cdots, x_n)$ of $\mathbb{A}^n$ via $\zeta \cdot (x_1, \cdots, x_n) = (\zeta^{a_1}x_1, \cdots, \zeta^{a_n}x_n)$ with ζ a primitive d-th root of unity and $\gcd(a_1, \cdots, a_n, d) = 1$. We denote by A the coordinate ring of $\mathbb{A}^n/\mathbb{Z}_d(a_1, \cdots, a_n)$, namely,

$$A = \mathbb{C}[\{x_1^{b_1} \cdots x_n^{b_n}\} \mid b_i \geqq 0,\ b_1a_1 + \cdots + b_na_n \in d\mathbb{Z}].$$

As in Proposition 3.1, A is regarded as a toric algebra:

Lemma 3.1. *Let $\mathbf{e}_1, \cdots, \mathbf{e}_n$ be the standard generators of the lattice $\mathbb{Z}^n$ with the duals $\mathbf{e}_1^*, \cdots, \mathbf{e}_n^*$, respectively. Let us set $N := \mathbb{Z}(\mathbf{e}_1/a_1) + \cdots + \mathbb{Z}(\mathbf{e}_n/a_n) + \mathbb{Z}((\mathbf{e}_1 + \cdots + \mathbf{e}_n)/d)$ and let $M := \mathrm{Hom}_{\mathbb{Z}}(N, \mathbb{Z})$ be its dual lattice. For a cone $\sigma := \mathbb{Q}_{\geqq 0}\mathbf{e}_1 + \cdots + \mathbb{Q}_{\geqq 0}\mathbf{e}_n \subseteq N_{\mathbb{Q}}$, we have the following:*

(1) $\check{\sigma} \cap M = \{ b_1 a_1 \mathbf{e}_1^* + \cdots + b_n a_n \mathbf{e}_n^* \mid b_1, \cdots, b_n \geqq 0,\ b_1 a_1 + \cdots + b_n a_n \in d\mathbb{Z} \}$.

(2) The $\mathbb{C}$-algebra

$$\mathbb{C}[\check{\sigma} \cap M] = \bigoplus_{\mathbf{m} \in \check{\sigma} \cap M} \mathbb{C}\chi^{\mathbf{m}} \cong \bigoplus_{b_i \geqq 0,\ \sum_{i=1}^n b_i a_i \in d\mathbb{Z}} \mathbb{C}\chi^{(b_1, \cdots, b_n)}$$

is isomorphic to A. More precisely, this isomorphism is given by the assignement $\chi^{\sum_{i=1}^n b_i a_i \mathbf{e}_i^*} = \chi^{(b_1, \cdots, b_n)} \mapsto x_1^{b_1} \cdots x_n^{b_n}$.

(3) There exist n mutually non-equivalent homogeneous LND's on $\mathbb{C}[\check{\sigma} \cap M]$, say δ_i $(1 \leqq i \leqq n)$ such that $\bigcap_{i=1}^n \mathrm{Ker}\,(\delta_i) = \mathbb{C}$. In particular, $\mathrm{ML}(A) = \mathbb{C}$.

Proof. The assertions (1) and (2) can be done directly. The cone $\sigma \subseteq N_{\mathbb{Q}}$ possesses n-extremal rays $\rho_i := \mathbb{Q}_{\geqq 0}[(1/a_i)\mathbf{e}_i]$ with the corresponding faces of codimension 1, say $\tau_i := \check{\sigma} \cap \rho_i^{\perp}$ $(1 \leqq i \leqq n)$. After taking $\mathbf{e}^{(i)} \in S_{\rho_i}$ (see Theorem 2.2 for the notation), we obtain the LND $\delta_i := \delta_{\rho_i, \mathbf{e}^{(i)}}$ on $\mathbb{C}[\check{\sigma} \cap M]$ as in Theorem 2.2, (1). Note that δ_i's are mutualy non-equivalent, indeed, $\mathrm{Ker}\,(\delta_i) = \bigoplus_{\mathbf{m} \in \tau_i \cap M} \mathbb{C}\chi^{\mathbf{m}}$ and $\tau_i \neq \tau_j$ for $i \neq j$ to see $\mathrm{Ker}\,(\delta_i) \neq \mathrm{Ker}\,(\delta_j)$. It is clear that $\bigcap_{i-1}^n \tau_i = 0$ since σ is strongly convex, therefore we have $\bigcap_{i=1}^n \mathrm{Ker}\,(\delta_i) = \bigoplus_{\mathbf{m} \in (\cap_{i=1}^n \tau_i) \cap M} \mathbb{C}\chi^{\mathbf{m}} = \mathbb{C}$, as desired. Thus it follows, in particular, that $\mathrm{ML}(\mathbb{C}[\check{\sigma} \cap M]) = \mathrm{ML}(A) = \mathbb{C}$. □

Thence as a generalization of the result due to Miyanishi (cf. Theorem 1.1, (2)), any cyclic quotient singularity of any dimension can be embedded into an affine variety with a trivial Makar-Limanov invariant.

3.3.

In this subsection, we shall observe toric singularities. Let $(x \in X)$ be a germ of a toric singularity, namely, there exists a toric variety X associated to a suitable fan containing a point that is analytically isomorphic to $(x \in X)$. Since X is covered by affine toric open subsets, we may assume that X itself is an affine toric variety associated to a strongly convex polyhedral cone $\sigma \subseteq N_{\mathbb{Q}}$ with $n := \mathrm{rank}_{\mathbb{Z}}(N) = \dim(X)$. If σ is degenerate, i.e., $\dim_{\mathbb{Q}}(\sigma + (-\sigma)) < n$, then we take a suitable strongly convex non-degenerate polyhedral cone $\sigma_0 \subseteq N_{\mathbb{Q}}$ containing σ as a face. Then $\mathbb{C}[\check{\sigma} \cap M]$ is a localization of $\mathbb{C}[\check{\sigma_0} \cap M]$, thence X is an open subset of $\mathrm{Spec}(\mathbb{C}[\check{\sigma} \cap M])$. Hence we may and shall assume that X is a toric variety associated to a non-degenerate strongly convex cone $\sigma \subseteq N_{\mathbb{Q}}$. We denote by $\rho_i \prec \sigma$ $(1 \leqq i \leqq r)$ all extremal rays of σ, that is, faces of dimension 1 such that $\sum_{i=1}^{r} \rho_i = \sigma$. Note that $r \geqq n$. Then we can confirm the following assertion by the similar argument as in Lemma 3.1.

Lemma 3.2. *With the notation same as above, we have the following:*

(1) There exist exactly r $\mathbb{G}_a$-actions that are normalized in $\mathrm{Aut}(X)$ *by the $\mathbb{G}_m^n$-action on X (up to equivalence).*
(2) Each of such $\mathbb{G}_a$-actions on X is obtained corresponding to one of extremal rays $\rho_1, \cdots, \rho_r$ as in Theorem 2.2.
(3) $\mathrm{ML}(\Gamma(\mathcal{O}_X)) = \mathbb{C}$.

Proof. Let δ_i be a homogeneous LND on $\Gamma(\mathcal{O}_X)$ associated to an extremal ray ρ_i (see Theorem 2.2), and let us set $\tau_i := \check{\sigma} \cap \rho_i^{\perp}$ $(1 \leqq i \leqq r)$. These δ_i's are mutually non-equivalent with finitely generated kernels $\mathrm{Ker}(\delta_i) = \bigoplus_{\mathbf{m} \in \tau_i \cap M} \mathbb{C}\chi^{\mathbf{m}}$. Notice that every homogeneous LND on a toric algebra is of fiber type (cf. Definition 2.6 and Remark 2.4). Therefore any $\mathbb{G}_a$-action on X, which is compatible with an action of $\mathbb{G}_m^n$, is associated to one of $\rho_1, \cdots, \rho_r$ up to equivalence by Theorem 2.2. Thus the assertions (1) and (2) are confirmed. For $\chi^{\mathbf{m}} \in \bigcap_{i=1}^{r} \mathrm{Ker}(\delta_i)$, it satisfies $< \mathbf{m}, \rho_i >= 0$ for all $1 \leqq i \leqq r$. As $\rho_1, \cdots, \rho_r$ generate σ, it follows that $\mathbf{m}$ is zero on σ whole to deduce that $\mathbf{m} = 0$, i.e., $\chi^{\mathbf{m}} = 1$. Hence $\bigcap_{i=1}^{r} \mathrm{Ker}(\delta_i) = \mathbb{C}$, in particular, we show (3). □

Thus we complete the proof of Theorem 1.2, (1).

Example 3.1. The origin of the hypersurface $X = (XY - ZU = 0) \subseteq \mathbb{A}^4$ is a toric singularity. Indeed, the coordinate ring $\Gamma(\mathcal{O}_X)$ of X is represented as $\Gamma(\mathcal{O}_X) \cong \mathbb{C}[\check{\sigma} \cap M]$, where $\sigma \subseteq N_{\mathbb{Q}} \cong \mathbb{Q}^3$ is a strongly convex polyhedral cone (see Example 2.4). Since σ is spanned by 4 extremal rays, $\Gamma(\mathcal{O}_X)$ has exactly 4 homogeneous LND's $\delta_1, \cdots, \delta_4$ up to equivalence. More explicitly, these LND's are described up to equivalence as:

$$\delta_1 = Z\frac{\partial}{\partial X} + Y\frac{\partial}{\partial U}, \quad \delta_2 = Z\frac{\partial}{\partial Y} + X\frac{\partial}{\partial U},$$
$$\delta_3 = U\frac{\partial}{\partial X} + Y\frac{\partial}{\partial Z}, \quad \delta_4 = U\frac{\partial}{\partial Y} + X\frac{\partial}{\partial Z}.$$

Thus it is easy to ascertain $\mathrm{ML}(\Gamma(\mathcal{O}_X)) = \mathbb{C}$ actually.

3.4.

In this subsection, we consider a germ of singularity $(x \in X)$ admitting an action of algebraic torus $\mathbb{T} = \mathbb{G}_m^r$.[1] Since X is then a $\mathbb{T}$-variety, there exists a suitable semiprojective variety Z of $\dim(Z) = \dim(X) - r$, a strongly convex degenerate polyhedral cone $\sigma \subseteq N_{\mathbb{Q}}$ of $\mathrm{rank}_{\mathbb{Z}}(N) = r$ and a proper σ-polyhedral divisor $\mathcal{D}$ on Z such that the coordinate ring $\Gamma(\mathcal{O}_X)$ is described as $\Gamma(\mathcal{O}_X) \cong A[Z,\mathcal{D}] := \bigoplus_{\mathbf{m} \in \check{\sigma} \cap M} H^0(Z, \lfloor \mathcal{D}(\mathbf{m}) \rfloor)\chi^{\mathbf{m}}$ (cf. Theorem 2.1). Different from toric singularities, a $\mathbb{T}$-variety $X \cong \mathrm{Spec}\,(A[Z,\mathcal{D}])$ does not necessarily possess an $\mathbb{A}^1$-fibration (cf. Remark 1.3). In what follows, we suppose in addition the following condition $(*)$ about $\mathcal{D}$:

$(*)$ There exists a face ρ of σ of dimension 1 such that $\mathcal{D}(\mathbf{m})$ is big for any $\mathbf{m} \in \mathrm{rel.int.}\,(\check{\sigma} \cap \rho^{\perp}) \cap M$.

Under this condition $(*)$, we can construct a homogeneous LND $\delta_{\rho,\mathbf{e},\varphi}$ on $A[Z,\mathcal{D}]$ associated to ρ, an element $\mathbf{e} \in S_\rho$ and $\varphi \in H^0(Z, \lfloor -D_{\mathbf{e}} \rfloor)\backslash\{0\}$ with the finitely generated kernel by virtue of Theorem 2.2. Thus we obtain Theorem 1.2, (2).

Example 3.2. Let us investigate the following quadratic hypersurface:

$$X := (X^2 + YZ + UV = 0) \subseteq \mathbb{A}^5 = \mathrm{Spec}\,(\mathbb{C}[X,Y,Z,U,V]).$$

It seems that X is no longer a toric variety, however it is equipped with an action of $\mathbb{T} = \mathbb{G}_m^3$. In fact, we can define a $\mathbb{T}$-action, for instance, in such a way that:

$$(\mu_1,\mu_2,\mu_3)\cdot(X,Y,Z,U,V) = (\mu_1\mu_2\mu_3 X, \mu_1^2\mu_3^2 Y, \mu_2^2 Z, \mu_1^2\mu_2 U, \mu_2\mu_3^2 V),$$

$$(\mu_1,\mu_2,\mu_3) \in \mathbb{T}.$$

Then the coordinate ring $A := \Gamma(\mathcal{O}_X)$ is $(\mathbb{Z}_{\geqq 0})^3$-graded $A = \bigoplus_{(a,b,c)\in(\mathbb{Z}_{\geqq 0})^3} A_{(a,b,c)}$, where

$$A_{(a,b,c)} := \sum_{\left(\begin{smallmatrix} 1&2&0&2&0\\ 1&0&2&1&1\\ 1&2&0&0&2 \end{smallmatrix}\right)\left(\begin{smallmatrix} n_1\\ n_2\\ n_3\\ n_4\\ n_5 \end{smallmatrix}\right) = \left(\begin{smallmatrix} a\\ b\\ c \end{smallmatrix}\right)} \mathbb{C}\, X^{n_1}Y^{n_2}Z^{n_3}U^{n_4}V^{n_5}.$$

[1] In the case where the dimension r of $\mathbb{T}$ coincides with that of X, this is nothing but a toric singularity.

By the straightforward computation, we are able to see that

$$A_{(a,b,c)} \cong \Big(\bigoplus_{-\lfloor \min\{0,\alpha\}\rfloor \leqq n \leqq \lfloor \min\{\beta,\gamma\}\rfloor} \mathbb{C}\, t^n \Big)\, Z^{\alpha} U^{\beta} V^{\gamma},$$

with $t := YZU^{-1}V^{-1}$, $\alpha := -\frac{a}{4} + \frac{b}{2} - \frac{c}{4}$, $\beta := \frac{a}{2}$ and $\gamma := \frac{c}{2}$. Then it follows that:

$$A \cong \bigoplus_{(\alpha,\beta,\gamma)\in\check{\sigma}\cap M} H^0\big(\mathbb{P}^1, \lfloor \min\{0,\alpha\}\rfloor[0] + \lfloor \min\{\beta,\gamma\}\rfloor[\infty]\big)\, \chi^{(\alpha,\beta,\gamma)},$$

where $\sigma := \mathbb{Q}_{\geqq 0}(0,1,0) + \mathbb{Q}_{\geqq 0}(0,0,1) + \mathbb{Q}_{\geqq 0}(1,1,0) + \mathbb{Q}_{\geqq 0}(1,0,1) \subseteq N_{\mathbb{Q}} \cong \mathbb{Q}^3$ and the dual cone $\check{\sigma} = \mathbb{Q}_{\geqq 0}(1,0,0) + \mathbb{Q}_{\geqq 0}(0,1,0) + \mathbb{Q}_{\geqq 0}(0,0,1) + \mathbb{Q}_{\geqq 0}(-1,1,1) \subseteq M_{\mathbb{Q}}$ with the dual lattice $M := \mathrm{Hom}_{\mathbb{Z}}(N,\mathbb{Z})$ of N. Thus we have $A \cong A[\mathbb{P}^1, \mathcal{D}]$ with the notation as in Theorem 2.1, where $\mathcal{D}$ is a proper σ-polyhedral divisor on $\mathbb{P}^1$ defined by:

$$\mathcal{D} = (\Delta_0 + \sigma)\,[0] + (\Delta_\infty + \sigma)\,[\infty] + \sum_{z\in\mathbb{P}^1\setminus\{0,\infty\}} \sigma\,[z],$$

with

$$\Delta_0 := \big\{\, (\lambda,0,0) \,\big|\, \lambda \in [0,1]\cap\mathbb{Q} \,\big\} \quad \text{and} \quad \Delta_\infty = \big\{\, (0,1,0), (0,0,1) \,\big\}.$$

Since σ has four extremal rays, we have at most four homogeneous LND's *of fiber type* on $A[\mathbb{P}^1, \mathcal{D}]$ by Theorem 2.2. However, we can verify by the direct computation that any of these extremal rays does not satisfy the condition $(*)$, i.e., that requested in Theorem 2.2, (4).[m] As a consequence, there does not exist a homogeneous LND of fiber type on $A[\mathbb{P}^1, \mathcal{D}]$. Instead we can find homogeneous LND's *of horizontal type*, for example:

$$U\frac{\partial}{\partial Z} - Y\frac{\partial}{\partial V} \quad \text{and} \quad V\frac{\partial}{\partial Y} - Z\frac{\partial}{\partial U}$$

are homogeneous LND's on A of horizontal type of degree $(1,-1,1)$ and $(-1,1,-1)$, respectively. In fact, we can see that these two LND's do not eliminate $t = YZU^{-1}V^{-1}$ to confirm that they are of horizontal type. Notice that the divisor on X defined by $(X^2 + YZ = U = 0)$ is never $\mathbb{Q}$-Cartier at the origin $\mathbf{o} \in X$, in particular, $\mathbf{o} \in X$ cannot be a cyclic quotient singularity.

3.5.

Finally, we shall show Corollary 1.1.

[m]For instance, we can verify this fact by virtue of [Li10a, Remark 3.14].

3.5.1.

Let us consider Example 2.4, where we regard the quotient algebra $A = \mathbb{C}[X,Y,Z,U]/(XY - ZU)$ as a toric one by the following action of $\mathbb{G}_m^3$:

$$(\mu_1, \mu_2, \mu_3)\cdot(X,Y,Z,U) = (\mu_2 X, \mu_3 Y, \mu_1^{-1} Z, \mu_1\mu_2\mu_3 U), \quad (\mu_1,\mu_2,\mu_3) \in \mathbb{G}_m^3.$$

Indeed, we see that $A \cong \mathbb{C}[\check{\sigma} \cap M]$ with $\sigma = \mathbb{Q}_{\geqq 0}(0,1,0) + \mathbb{Q}_{\geqq 0}(0,0,1) + \mathbb{Q}_{\geqq 0}(-1,1,0) + \mathbb{Q}_{\geqq 0}(-1,0,1) \subseteq N_{\mathbb{Q}} \cong \mathbb{Q}^3$. According to Theorem 2.2, there exist 4 homogeneous LND's $\delta_1, \cdots, \delta_4$ up to equivalence corresponding to four extremal rays of σ as described in Example 2.4, so that $X = \mathrm{Spec}\,(A) = (XY - ZU = 0) \subseteq \mathbb{A}^4$ possesses four $\mathbb{G}_a$-actions normalized by the $\mathbb{G}_m^3$-action in $\mathrm{Aut}(X)$. On the other hand, the divisor on X defined by $(X = Z = 0)$ is not $\mathbb{Q}$-Cartier at the origin $\mathbf{o} \in X$. By noting that the local ring at a cyclic quotient singular point is $\mathbb{Q}$-factorial, i.e., a multiple of any height one ideal becomes principal, we know that $\mathbf{o} \in X$ can not be a cyclic quotient singularity.

3.5.2.

As another example to justify Corollary 1.1, let us look into Example 3.2. In this case also, the singular point of X, which is the origin $\mathbf{o} \in X$, can not be cyclic quotient, nevertheless, there exist homogeneous LND's on $A = \Gamma(\mathcal{O}_X)$ to deduce that X admits $\mathbb{G}_a$-actions which are normalized by the $\mathbb{G}_m^3$-action on it in $\mathrm{Aut}\,(X)$.

3.6.

3.6.1.

When we consider $\mathbb{G}_a^2$-actions instead of $\mathbb{G}_a$-actions, it seems to be reasonable to think that the type of appearing singularities are fairly restricted. Namely, we shall ask the following question:

Problem 3.1.

(1) *Let X be a normal affine variety. Supposing that X possesses an $\mathbb{A}^2$-fibration or an $\mathbb{A}^2$-cylinder, what kinds of singularities appear on X? Conversely,*
(2) *What kind of a germ of a singularity can be realized as a singularity of a normal affine algebraic variety admitting an $\mathbb{A}^2$-fibration or containing an $\mathbb{A}^2$-cylinder? More strongly,*
(3) *What kind of a germ of a singularity can be realized as a singularity of a normal affine algebraic variety equipped with a $\mathbb{G}_a^2$-action?*

Remark 3.2.

(1) Note that the possession of an $\mathbb{A}^2$-fibration on a normal affine algebraic variety X implies the existence of an $\mathbb{A}^2$-cylinder in X by virtue of [KaZa01, Theorem 0.1].

(2) Even in the case of $\mathbb{G}_a^2$-actions also, the category of cyclic quotient singularities is not large enough to respond to Problem 3.1, (3) even for dimension three. For instance, let us consider again Example 2.4, i.e., $X = (XY - ZU = 0) \subseteq \mathbb{A}^4$, whose origin is not a cyclic quotient singular point. On the other hand, as already seen there, $\delta_1, \cdots, \delta_4$ exhaust homogeneous LND's on $A = \Gamma(\mathcal{O}_X)$ up to equivalence. It is then straightforward to verify that $\delta_1 \circ \delta_2 = \delta_2 \circ \delta_1$, $\delta_1 \circ \delta_3 = \delta_3 \circ \delta_1$, $\delta_2 \circ \delta_4 = \delta_4 \circ \delta_2$ and $\delta_3 \circ \delta_4 = \delta_4 \circ \delta_3$, therefore X admits four $\mathbb{G}_a^2$-actions which are normalized by the action of $\mathbb{G}_m^3$ in $\mathrm{Aut}\,(X)$.

3.6.2.

In what follows, we shall investigate the case of dim = 3 concerning Problem 3.1 (1), so let us suppose that a normal affine threefold X possesses an $\mathbb{A}^2$-fibration over a smooth curve Z, say $\pi : X \to Z$. Since X is normal, the locus of singularities $\Gamma := \mathrm{Sing}(X)$ is at most of dimension one. Furthermore, Γ is vertical with respect to π. Indeed, otherwise, $\pi|_\Gamma : \Gamma \to Z$ is a quasi-finite morphism and a general fiber of π, which is isomorphic to $\mathbb{A}^2$ scheme-theoretically, intersects Γ at several points transversally. This implies that X is smooth at these intersection points, a contradiction. Thus each connected component of Γ is contained in the support of a fiber of π. If $\dim(\Gamma) = 0$, i.e., X has only isolated singularities, then we have the following:

Lemma 3.3. *Let x be a singular point of X, let us set $z := \pi(x)$ and let us denote by $F_z = \pi^*(z) \subseteq X$ the fiber of π over $z \in Z$. Suppose that X is $\mathbb{Q}$-factorial and F_z is reduced, irreducible and normal. Then the pair (X, F_z) has at worst plt singularities at x. In particular, $x \in X$ is a terminal singularity.*[n]

Proof. Letting A and R be respectively the coordinate rings of X and Z, and let us denote by $(\mathcal{O}, t\mathcal{O})$ the local ring at $z \in Z$ with $t \in \mathcal{O}$ a uniformisant. Then $A_\mathcal{O} := A \otimes_R \mathcal{O}$ is $\mathcal{O}$-flat. Indeed, since $tA_\mathcal{O}$ is a principal

[n]See [KM98, Chapter 2] for the definitions of terminal and plt singularities.

ideal of $A_{\mathcal{O}}$, it follows that any component of $\mathbb{V}(tA_{\mathcal{O}}) \subseteq \mathrm{Spec}(A_{\mathcal{O}})$ is of codimension one. As $A_{\mathcal{O}}$ is normal and $\mathcal{O}$ is regular, we know then that $A_{\mathcal{O}}$ is $\mathcal{O}$-flat.[o] Then, by the similar argument as in [Mi84, Lemma 2.3], it follows that F_z is $\mathbb{A}^1$-ruled. For the sake of readers, we shall explain it in more detail. As said in Remark 3.2 (1), $A \otimes_R K = A_{\mathcal{O}} \otimes_{\mathcal{O}} K$ has a non-trivial locally nilpotent K-derivation, say Δ, where $K := \mathrm{Frac}(R)$. Since $A_{\mathcal{O}}$ is finitely generated over $\mathcal{O}$, say $A_{\mathcal{O}} = \mathcal{O}[a_1, \cdots, a_r]$, there exists $N \in \mathbb{Z}$ such that $t^N \Delta(a_i) \in A_{\mathcal{O}}$ for $1 \leqq i \leqq r$. We choose N in such a way that N is the smallest among such integers. Then $\delta := t^N\Delta$ is a non-trivial locally nilpotent $\mathcal{O}$-derivation on $A_{\mathcal{O}}$ and it induces that on $\overline{A} := A_{\mathcal{O}}/tA_{\mathcal{O}}$, say $\overline{\delta}$, by setting:

$$\overline{\delta}\big(a \ (\mathrm{mod}\ tA_{\mathcal{O}})\big) = \delta(a) \ (\mathrm{mod}\ tA_{\mathcal{O}}) \qquad (a \in A_{\mathcal{O}}).$$

Thus $F_z = \mathrm{Spec}(\overline{A})$ is $\mathbb{A}^1$-ruled. Therefore, by Theorem 1.1 (1), F_z has only cyclic quotient singularities, in particular, klt singularities. Since $\dim(\Gamma) \leqq 0$ by hypothesis, X is Cartier in codimension two, i.e., any Weil divisor on X is Cartier except for at most finitely many (closed) points. Thence it follows that the pair (X, F_z) has at most plt singularities at x by virtue of Inversion of Adjunction (cf. [KM98, Theorem 5.50]). As F_z is a Cartier divisor on X, the fact that (X, F_z) is plt at x implies that X is terminal there as desired. □

Note that X is not necessarily terminal without the assumption that F_z is reduced as in the following example.

Example 3.3. Let us consider $X := \mathbb{A}^3(x, y, z)/\mathbb{Z}_d(1, 1, e)$ with $1 \leq e \leqq d$ such that $\gcd(e, d) = 1$, where the cyclic group $\mathbb{Z}_d \cong <\zeta>$ acts in such a way that:

$$\zeta \cdot (x, y, z) = (\zeta x, \zeta y, \zeta^e z), \quad \zeta := \exp\Big(\frac{2\pi\sqrt{-1}}{d}\Big).$$

It is clear that X is $\mathbb{Q}$-factorial. As seen in Lemma 3.1, X is a toric variety whose coordinate ring possesses exactly three homogeneous LND's up to equivalence, say $\delta_1, \delta_2, \delta_3$. More precisely, we can express them as follows:

$$\delta_1 = z^p \frac{\partial}{\partial x}, \ \delta_2 = z^p \frac{\partial}{\partial y}, \ \delta_3 = x^e \frac{\partial}{\partial z},$$

where $p > 0$, $q < 0$ are integers such that $pe + qd = 1$. Furthermore, we can see that $\delta_2 \circ \delta_3 = \delta_3 \circ \delta_2$, hence these give rise to an $\mathbb{A}^2$-fibration

[o]So far, we do not use the assumption that F_z is irreducible and reduced.

$\pi : X \to Z = \mathrm{Spec}\,(\mathbb{C}[x^d]) \cong \mathbb{A}^1$. The fiber of π over the point on Z defined by $x^d = 0$ is a multiple of a toric surface $\mathbb{A}^2\,(y,z)/\mathbb{Z}_d(1,e)$. On the other hand, the point of X, which is the image of the origin of $\mathbb{A}^3\,(x,y,z)$, is not terminal if $d \geqq e+2$.

Acknowledgements

The author would like to express his sincere gratitude to the referees for the useful comments and suggestions for the improvement of a readability of the article. The part of this work was done during the author's stay at l'Institut de Fourier (Grenoble) and l'Institut de Mathématiques de Bourgogne (Dijon). The author thanks these institutes for their hospitality and financial support. The author is supported by Grant-in-Aid for Scientific Research (No. 24740003) from JSPS.

References

AlHa06. K. Altmann and J. Hausen, *Polyhedral divisors and algebraic torus actions*, Math. Ann., 334 (2006), 557–607.

Ar62. M. Artin, *Some numerical criteria for contractibility of curves on algebraic surfaces*, Amer. J. Math., 84 (1962), 485–496.

Br68. E. Brieskorn, *Rationale singularitäten komplexer Flächen*, Invent. Math., 4 (1968), 336–358.

CLS11. D. Cox, J. Little and H. Schenck, *Toric varieties*, Graduate Studies in Math. Vol.214, American Mathematical Society, 2011.

DaFr99. D. Daigle and G. Freudenburg, *A counterexample to Hilbert's Fourteenth Problem in dimension five*, J. Algebra, 221 (1999), 528–535.

Du04. A. Dubouloz, *Completions of normal affine surfaces with a trivial Makar-Limanov invariant*, Michigan Math. J., 52 (2004), 289–308.

FlZa03a. H. Flenner and M. Zaidenberg, *Rational curves and rational singularities*, Math. Z., 244 (2003), 549–575.

FlZa03b. H. Flenner and M. Zaidenberg, *Normal affine surfaces with* $\mathbb{C}^*$*-actions*, Osaka J. Math., 40 (2003), 981–1009.

FlZa05. H. Flenner and M. Zaidenberg, *Locally nilpotent derivations on affine surfaces with a* $\mathbb{C}^*$*-action*, Osaka J. Math., 42 (2005), 931–974.

Fr00. G. Freudenburg, *A counterexample to Hilbert's Fourteenth Problem in dimension six*, Transformation Groups, 5 (2000), 61–71.

Fr06. G. Freudenburg, *Algebraic theory of locally nilpotent derivations*, Encyclopaedia Math. Sci. Vol.136, Springer, Berlin, 2006.

Fu93. W. Fulton, *Introduction to toric varieties*, Ann. of Math. Studies. Vol. 131, Princeton University Press, Princeton, New Jersey, 1993.

Giz70. M.H. Gizatullin, *On affine surfaces that can be completed by a nonsingular rational curve*, Izv. Akad. Nauk SSSR Ser. Math., 34 (1970), 778–802; Math. USSR-Izv., 4 (1970), 787–810.

Giz71. M.H. Gizatullin, *Quasihomogeneous affine surfaces*, Math. USSR-Izv., 5 (1971), 1057–1081.

KaZa01. S. Kaliman and M. Zaidenberg, *Families of affine planes: the existence of a cylinder*, Michigan Math. J., 49 (2001), 353–367.

KaMi78. T. Kambayashi and M. Miyanishi, *On flat fibrations by the affine line*, Illinois J. Math., 22 (1978), 662–671.

KPZ11. T. Kishimoto, Yu. Prokhorov, M. Zaidenberg, *Group actions on affine cones*, Affine Algebraic Geometry, The Russell Festschrift (Eds. D.Daigle, R. Ganong and M. Koras), CRM Proceedings & Lecture Note, Vol. 54, American Mathematical Society, 2011, 123–163.

KM98. J. Kollár and S. Mori, *Birational Geometry of algebraic varieties*, Cambridge Tracts in Mathematics. Vol. 134, Cambridge University Press, 1998.

Li10a. A. Liendo, *Affine $\mathbb{T}$-varieties of complexity one and locally nilpotent derivations*, Transformation Groups, 15 (2010), 389–425.

Li10b. A. Liendo, *$\mathbb{G}_a$-actions of fiber type on affine $\mathbb{T}$-varieties*, J. Algebra, 324 (2010), 3653–4665.

Mi78. M. Miyanishi, *Lectures on curves on rational and unirational surfaces*, Springer-Verlag (Berlin, Heidelberg, New York), 1978, Published for Tata Inst. Fund. Res., Bombay.

Mi81. M. Miyanishi, *Singularities on normal affine surfaces containing cylinder-like open sets*, J. Algebra, 68 (1981), 268–275.

Mi84. M. Miyanishi, *An algebro-topological characterization of the affine space of dimension three*, Amer. J. Math. 106 (1984), 1469–1486.

Ro90. P. Roberts, *An infinitely generated symbolic blow-up in a power series ring and a new counterexample to Hilbert's fourteenth problem*, J. Algebra, 132 (1990), 461–473.

A Galois counterexample to Hilbert's Fourteenth Problem in dimension three with rational coefficients

Ei Kobayashi and Shigeru Kuroda*

Department of Mathematics and Information Sciences,
Tokyo Metropolitan University,
1-1 Minami-Osawa, Hachioji,
Tokyo 192-0397, Japan
**E-mail: kuroda@tmu.ac.jp*

Dedicated to Professor Masayoshi Miyanishi on his seventieth birthday

We give a new counterexample to Hilbert's Fourteenth Problem in dimension three, and solve an open problem posed by Kuroda (2007).

Keywords: Hilbert's Fourteenth Problem.

1. Introduction

Let k be a field, $k[\mathbf{x}] := k[x_1, \ldots, x_n]$ the polynomial ring in n variables over k, where $n \in \mathbf{N}$, and $k(\mathbf{x})$ the field of fractions of $k[\mathbf{x}]$. Assume that L is a subfield of $k(\mathbf{x})$ containing k. Then, Hilbert's Fourteenth Problem asks whether the k-subalgebra $L \cap k[\mathbf{x}]$ of $k[\mathbf{x}]$ is always finitely generated. The first counterexample to this problem was given by Nagata [10] in 1958, where $n = 32$ and $\text{trans.deg}_k L = 4$. In 1990, Roberts [11] constructed a different type of counterexample with $\text{trans.deg}_k L = 6$ in the case of $n = 7$. To date, a variety of counterexamples to Hilbert's Fourteenth Problem have been constructed by generalizing those of Nagata and Roberts. Nagata's counterexample was generalized by Mukai [9], Steinberg [12] and Totaro [13], while Robert's counterexample was generalize by Kojima-Miyanishi [3], Daigle-Freudenburg [1] and Kuroda [4]. Kuroda also systematically constructed various counterexamples having their roots in Roberts (see the appendix of Kuroda [7] for detail). In 2004, he [5] gave one with $\text{trans.deg}_k L = 3$ in the case of $n = 3$. This is the "smallest"

2010 *Mathematics Subject Classification*: Primary 13F20; Secondary 13E15.

possible counterexample to Hilbert's Fourteenth Problem. Indeed, when $\text{trans.deg}_k L \leq 2$, the answer to the problem is affirmative due to Zariski [14]. This is also the first counterexample in the case where $k(\mathbf{x})/L$ is an algebraic extension. When $n = 3$, Kuroda [6] also gave a useful construction of counterexamples. As an application, he constructed one with $k(\mathbf{x})/L$ a Galois extension for some field $k \neq \mathbf{Q}$ of characteristic zero. On the other hand, for any field k of characteristic zero, and for any finite group G with $|G| \geq 2$, Kuroda [8] gave a counterexample L such that $k(\mathbf{x})/L$ is a Galois extension with Galois group isomorphic to G for sufficiently large $n \geq 4$. For example, he constructed such an example for $G = \mathbf{Z}/2\mathbf{Z}$ for each $n \geq 4$. In the case where $n = 3$ and $k = \mathbf{Q}$, however, no such a counterexample was found. So, Kuroda [6, Problem 1.5] posed the following problem.

Problem 1.1. *Assume that $n = 3$. Let L be a subfield of $\mathbf{Q}(\mathbf{x})$ such that $\mathbf{Q}(\mathbf{x})/L$ is a Galois extension. Is the $\mathbf{Q}$-subalgebra $L \cap \mathbf{Q}[\mathbf{x}]$ of $\mathbf{Q}[\mathbf{x}]$ always finitely generated?*

The purpose of this paper is to solve this problem in the negative. In what follows, we assume that $n = 3$ and k is of characteristic zero. For each $A = (a_{i,j}) \in GL(2, k)$, we define $\phi_A \in \text{Aut}_k k(\mathbf{x})$ by

$$\phi_A(x_i) = a_{1,i}x_1 + a_{2,i}x_2 \text{ for } i = 1, 2, \text{ and } \phi_A(x_3) = x_3.$$

Then, $GL(2, k) \ni A \mapsto \phi_A \in \text{Aut}_k k(\mathbf{x})$ is an injective homomorphism of groups. Hence, we may regard $GL(2, k)$ as a subgroup of $\text{Aut}_k k(\mathbf{x})$ by identifying A with ϕ_A. Let Γ be the subgroup of $GL(2, k)$ generated by

$$A = \begin{pmatrix} 0 & 1 \\ 1 & -1 \end{pmatrix} \quad \text{and} \quad B = \begin{pmatrix} 0 & 1 \\ 1 & 0 \end{pmatrix}.$$

For each $\epsilon \in \mathbf{Z}$, we define $\sigma_\epsilon \in \text{Aut}_k k(\mathbf{x})$ by

$$\sigma_\epsilon(x_1) = x_1 x_2^{\epsilon-1}, \quad \sigma_\epsilon(x_2) = x_2^{-1}, \quad \sigma_\epsilon(x_3) = x_2^{-1} + x_3, \tag{1}$$

and put $\Gamma_\epsilon = \sigma_\epsilon^{-1} \Gamma \sigma_\epsilon$.

The following is the main theorem of this paper, which gives a negative solution to Problem 1.1.

Theorem 1.2. *Assume that $n = 3$ and k is any field of characteristic zero. If $\epsilon \geq 4$, then the k-subalgebra $k(\mathbf{x})^{\Gamma_\epsilon} \cap k[\mathbf{x}]$ of $k[\mathbf{x}]$ is not finitely generated.*

We note that Γ is isomorphic to the symmetric group of degree three, and hence $|\Gamma| = 6$. Actually, since

$$A^2 = \begin{pmatrix} -1 & -1 \\ 1 & 0 \end{pmatrix} \quad \text{and} \quad AB = \begin{pmatrix} 1 & 0 \\ -1 & -1 \end{pmatrix},$$

we see that $A^3 = B^2 = E$ and $BAB = A^2$, where E is the 2×2 elementary matrix.

Theorem 1.2 is presented in the Master's thesis of the first author [2]. The proof given in this paper is a modification of that given in the thesis.

2. Invariant field

Set

$$f_1 = x_1^2 - x_1x_2 + x_2^2 \quad \text{and} \quad f_2 = x_1^3 - \frac{3}{2}x_1^2x_2 - \frac{3}{2}x_1x_2^2 + x_2^3.$$

The goal of this section is to prove the following lemma.

Lemma 2.1. *We have* $k(\mathbf{x})^\Gamma = k(f_1, f_2, x_3)$.

By definition, Γ is generated by A and B, and so by AB and B. Hence, f belongs to $k(\mathbf{x})^\Gamma$ if and only if $\phi_{AB}(f) = \phi_B(f) = f$ for each $f \in k(\mathbf{x})$. Observe that f_1 and f_2 are symmetric polynomials in x_1 and x_2 over k. Since $\phi_B(x_1) = x_2$ and $\phi_B(x_2) = x_1$, it follows that $\phi_B(f_i) = f_i$ for $i = 1, 2$. Since $\phi_{AB}(x_1) = x_1 - x_2$ and $\phi_{AB}(x_2) = -x_2$, we have

$$\phi_{AB}(2x_1 - x_2) = 2x_1 - x_2 \quad \text{and} \quad \phi_{AB}(x_2^2) = x_2^2.$$

Using $2x_1 - x_2$ and x_2^2, we can write

$$f_1 = \frac{1}{4}(2x_1 - x_2)^2 + \frac{3}{4}x_2^2 \quad \text{and} \quad f_2 = \frac{1}{8}(2x_1 - x_2)^3 - \frac{9}{8}(2x_1 - x_2)x_2^2.$$

Hence, $\phi_{AB}(f_i) = f_i$ holds for $i = 1, 2$. Thus, f_1 and f_2 belong to $k(\mathbf{x})^\Gamma$. Clearly, x_3 belongs to $k(\mathbf{x})^\Gamma$. Therefore, $K := k(f_1, f_2, x_3)$ is contained in $k(\mathbf{x})^\Gamma$. Consequently, we get

$$[k(\mathbf{x}) : K] \geq [k(\mathbf{x}) : k(\mathbf{x})^\Gamma] = |\Gamma| = 6.$$

To conclude $k(\mathbf{x})^\Gamma = K$, it suffices to prove that $[k(\mathbf{x}) : K] \leq 6$. Since f_1 and f_2 are symmetric polynomials in x_1 and x_2 over k, we see that K is contained in $L := k(x_1 + x_2, x_1x_2, x_3)$. Hence, we have

$$[k(\mathbf{x}) : K] = [k(\mathbf{x}) : L][L : K].$$

Because $[k(\mathbf{x}) : L] = 2$, we are reduced to proving that $[L : K] \leq 3$. Put $s = x_1 + x_2$. Then, $K(s)$ is contained in L. Since $x_1 + x_2 = s$, $x_1x_2 = s^2 - f_1$ and x_3 belong to $K(s)$, we know that L is contained in $K(s)$. Hence, L is equal to $K(s)$. A direct computation shows that

$$s^3 - 3f_1s + 2f_2 = (x_1^3 + 3x_1^2x_2 + 3x_1x_2^2 + x_2^3) - 3(x_1^3 + x_2^3) + 2\left(x_1^3 - \frac{3}{2}x_1^2x_2 - \frac{3}{2}x_1x_2^2 + x_2^3\right) = 0.$$

Hence, the degree of s over K is at most three. Thus, we have $[L : K] = [K(s) : K] \leq 3$. Therefore, we conclude that $k(\mathbf{x})^\Gamma = K$. This completes the proof of Lemma 2.1.

3. Kuroda's construction

In this section, we briefly review Kuroda's construction of counterexamples to Hilbert's Fourteenth Problem in the case of $n = 3$ (see Ref. [6] for details). Let δ_1 and δ_2 be natural numbers, and $\pi_1, \pi_2 \in k[z]$ polynomials in one variable over k. Assume that the four-tuple $\Delta := (\delta_1, \delta_2; \pi_1, \pi_2)$ satisfies the following conditions, where $\delta_0 := \gcd(\delta_1, \delta_2)$ and $\delta_i' := \delta_i/\delta_0$ for $i = 1, 2$.

(a) $\delta_1 < \delta_2$, and δ_2 is not divisible by δ_1.
(b) $\pi_1(0) = \pi_2(0) = 1$.
(c) π_1 or π_2 does not belong to the radical of the ideal

$$(\alpha_1 \pi_1^{\delta_2'} + \alpha_2 \pi_2^{\delta_1'})\bar{k}[z]$$

of $\bar{k}[z]$ for each $\alpha_1, \alpha_2 \in \bar{k}$, where $\bar{k}$ is an algebraic closure of k.

By (b), we have

$$\epsilon' := \max\{l \in \mathbf{Z} \mid \pi_1^{\delta_2'} - \pi_2^{\delta_1'} \in z^l k[z]\} \geq 1.$$

Take any $\epsilon \in \mathbf{Z}$ with

$$\epsilon\epsilon' \geq \frac{\delta_1\delta_2}{\delta_0} + 1 \quad \text{and} \quad \epsilon \geq \delta_0. \tag{2}$$

Then, we define L_Δ^ϵ to be the subfield of $k(\mathbf{x})$ generated by

$$\Pi_0 := x_2^{-\delta_0} + x_3 \quad \text{and} \quad \Pi_i := x_2^{-\delta_i}\pi_i(x_1 x_2^\epsilon) \quad \text{for} \quad i = 1, 2$$

over k.

With the notation and assumption above, the following theorem holds.

Theorem 3.1. (Kuroda [6]) *The k-subalgebra $L_\Delta^\epsilon \cap k[\mathbf{x}]$ of $k[\mathbf{x}]$ is not finitely generated.*

Let M_Δ be the subfield of $k(x_1, x_2)$ generated by $x_2^{\delta_i}\pi_i(x_1 x_2^{-1})$ for $i = 1, 2$ over k. Define $\sigma_\epsilon^{\delta_0} \in \mathrm{Aut}_k k(\mathbf{x})$ by

$$\sigma_\epsilon^{\delta_0}(x_1) = x_1 x_2^{\epsilon-1}, \quad \sigma_\epsilon^{\delta_0}(x_2) = x_2^{-1}, \quad \sigma_\epsilon^{\delta_0}(x_3) = -x_2^{\delta_0} + x_3.$$

Then, we remark that $\sigma_\epsilon^{\delta_0}(L_\Delta^\epsilon) = M_\Delta(x_3)$, since $\sigma_\epsilon^{\delta_0}(\Pi_0) = x_3$, and $\sigma_\epsilon^{\delta_0}(\Pi_i) = x_2^{\delta_i}\pi_i(x_1 x_2^{-1})$ for $i = 1, 2$.

4. Proof of Theorem 1.2

Let us complete the proof of Theorem 1.2. Define $\Delta = (\delta_1, \delta_2; \pi_1, \pi_2)$ by

$$\delta_1 = 2, \quad \delta_2 = 3, \quad \pi_1 = z^2 - z + 1, \quad \pi_2 = z^3 - \frac{3}{2}z^2 - \frac{3}{2}z + 1.$$

Then, we have $M_\Delta = k(f_1, f_2)$, since

$$x_2^{\delta_1}\pi_1(x_1x_2^{-1}) = x_1^2 - x_1x_2 + x_2^2 = f_1$$
$$x_2^{\delta_2}\pi_2(x_1x_2^{-1}) = x_1^3 - \frac{3}{2}x_1^2x_2 - \frac{3}{2}x_1x_2^2 + x_2^3 = f_2.$$

By virtue of Lemma 2.1, it follows that $M_\Delta(x_3) = k(f_1, f_2, x_3) = k(\mathbf{x})^\Gamma$. Since $\delta_0 = \gcd(2,3) = 1$, we have $\sigma_\epsilon^{\delta_0} = \sigma_\epsilon$ for each $\epsilon \in \mathbf{Z}$. Hence, we get

$$\sigma_\epsilon(L_\Delta^\epsilon) = \sigma_\epsilon^{\delta_0}(L_\Delta^\epsilon) = M_\Delta(x_3) = k(\mathbf{x})^\Gamma$$

as remarked after Theorem 3.1. This implies that

$$L_\Delta^\epsilon = k(\mathbf{x})^{\sigma_\epsilon^{-1}\Gamma\sigma_\epsilon} = k(\mathbf{x})^{\Gamma_\epsilon}.$$

Thus, it suffices to prove that $L_\Delta^\epsilon \cap k[\mathbf{x}]$ is not finitely generated over k when $\epsilon \geq 4$.

It is easy to see that Δ satisfies (a) and (b). We check (c). Since $\delta_0 = 1$, we have $\delta_i' = \delta_i$ for $i = 1, 2$. Suppose to the contrary that π_1 and π_2 belong to the radical I of the ideal

$$(\alpha_1\pi_1^3 + \alpha_2\pi_2^2)\bar{k}[z]$$

for some $\alpha_1, \alpha_2 \in \bar{k}$. Then, $\gcd(\pi_1, \pi_2)$ belongs to I. Since π_1 has no real root, while $\pi_2(-1) = \pi_2(1/2) = \pi_2(2) = 0$, we see that $\gcd(\pi_1, \pi_2) = 1$. Hence, I is the unit ideal of $\bar{k}[z]$. This implies that $\alpha_1\pi_1^3 + \alpha_2\pi_2^2 = \beta$ for some $\beta \in \bar{k} \setminus \{0\}$. Since $\pi_1(-1) = 3$ and $\pi_1(1/2) = 3/4$, we obtain $3^3\alpha_1 = \beta$ and $(3/4)^3\alpha_1 = \beta$ by substituting -1 and $1/2$ for z. This yields that $\alpha_1 = 0$, and so we get $\alpha_2\pi_2^2 = \beta$. Thus, $\alpha_2\pi_2^2$ belongs to $\bar{k} \setminus \{0\}$, a contradiction. Therefore, Δ satisfies (c).

We show that $\epsilon' = 2$. Put $\pi = \pi_1^3 - \pi_2^2$. Then, we have

$$\pi' = 3\pi_1'\pi_1^2 - 2\pi_2'\pi_2, \quad \pi'' = 3\pi_1''\pi_1^2 + 6(\pi_1')^2\pi_1 - 2\pi_2''\pi_2 - 2(\pi_2')^2.$$

Since $\pi_1' = 2z - 1$, $\pi_1'' = 2$, $\pi_2' = 3z^2 - 3z - 3/2$ and $\pi_2'' = 6z - 3$, we see that

$$\pi(0) = 1^3 - 1^2 = 0, \quad \pi'(0) = 3 \cdot (-1) \cdot 1^2 - 2 \cdot \left(-\frac{3}{2}\right) \cdot 1 = 0$$
$$\pi''(0) = 3 \cdot 2 \cdot 1^2 + 6 \cdot (-1)^2 \cdot 1 - 2 \cdot (-3) \cdot 1 - 2 \cdot \left(-\frac{3}{2}\right)^2 \neq 0.$$

Hence, π is divisible by z^2, but is not divisible by z^3. Thus, we conclude that $\epsilon' = 2$. Since $\delta_0 = 1$ and $\delta_1\delta_2/\delta_0 = 6$, it follows that $\epsilon \in \mathbf{Z}$ satisfies (2) if and only if $\epsilon \geq 4$. Therefore, if $\epsilon \geq 4$, then $L_\Delta^\epsilon \cap k[\mathbf{x}]$ is not finitely generated over k thanks to Theorem 3.1. This completes the proof of Theorem 1.2.

Acknowledgments

The second author is grateful to Prof. Akinari Hoshi (Rikkyo University) for informing him about the matrix group Γ. This work was supported in part by the Grant-in-Aid for Young Scientists (B) 21740026, The Ministry of Education, Culture, Sports, Science and Technology, Japan.

References

1. D. Daigle and G. Freudenburg, A counterexample to Hilbert's fourteenth problem in dimension 5, J. Algebra **221** (1999), 528–535.
2. E. Kobayashi, A Galois counterexample to Hilbert's Fourteenth Problem in dimension three with rational coefficients (Japanese), Master's thesis, Tokyo Metropolitan University, 2011.
3. H. Kojima and M. Miyanishi, On Roberts' counterexample to the fourteenth problem of Hilbert, J. Pure Appl. Algebra **122** (1997), 277–292.
4. S. Kuroda, A generalization of Roberts' counterexample to the fourteenth problem of Hilbert, Tohoku Math. J. **56** (2004), 501–522.
5. S. Kuroda, A counterexample to the Fourteenth Problem of Hilbert in dimension three, Michigan Math. J. **53** (2005), 123–132.
6. S. Kuroda, Hilbert's Fourteenth Problem and algebraic extensions, J. Algebra **309** (2007), 282–291.
7. S. Kuroda, Hilbert's fourteenth problem and algebraic extensions with an appendix on Roberts type counterexamples, Acta Math. Vietnam. **32** (2007), no. 2-3, 247–257.
8. S. Kuroda, Hilbert's Fourteenth Problem and invariant fields of finite groups, preprint.
9. S. Mukai, Counterexample to Hilbert's fourteenth problem for the 3-dimensional additive group, Preprint **1343**, Research Institute for Mathematical Sciences, Kyoto University, 2001.
10. M. Nagata, On the fourteenth problem of Hilbert, in Proceedings of the International Congress of Mathematicians, 1958, Cambridge Univ. Press, London, New York, 1960, 459–462.
11. P. Roberts, An infinitely generated symbolic blow-up in a power series ring and a new counterexample to Hilbert's fourteenth problem, J. Algebra **132** (1990), 461–473.
12. R. Steinberg, Nagata's example, in Algebraic Groups and Lie Groups, Austral. Math. Soc. Lect. Ser. **9**, Cambridge Univ. Press, 1997, 375–384.
13. B. Totaro, Hilbert's 14th problem over finite fields and a conjecture on the cone of curves, Compos. Math. **144** (2008), no. 5, 1176–1198.

14. O. Zariski, Interprétations algébrico-géométriques du quatorzième problème de Hilbert, Bull. Sci. Math. **78** (1954), 155–168.

Open algebraic surfaces of logarithmic Kodaira dimension one

Hideo Kojima

Department of Mathematics,
Faculty of Science, Niigata University,
8050 Ikarashininocho, Nishi-ku, Niigata 950-2181, Japan
E-mail: kojima@math.sc.niigata-u.ac.jp

Dedicated to Professor Masayoshi Miyanishi on the occasion of his 70th birthday

We give a structure theorem for open algebraic surfaces of logarithmic Kodaira dimension one in arbitrary characteristic. Moreover, by using the structure theorem, we give some results on logarithmic plurigenera of normal affine surfaces of logarithmic Kodaira dimension one.

Keywords: Open algebraic surface, logarithmic Kodaira dimension, logarithmic plurigenus.

0. Introduction

Let k be an algebraically closed field of characteristic $p \geq 0$, which we fix as a ground field.

Since Iitaka [4] introduced the notion of the logarithmic Kodaira dimension of an algebraic variety, classification theory of open algebraic surfaces in the case $p = 0$ has been developed by several mathematicians. In particular, Kawamata [7] gave structure theorems for open algebraic surfaces of non-negative logarithmic Kodaira dimension. Some of the results on open algebraic surfaces are valid also in the case $p > 0$. For example, the minimal model theory of open algebraic surfaces works also in the case $p > 0$ (see [12, Chapter 2], [13, Chapter 1]). Russell [14] and Miyanishi [11] studied open algebraic surfaces of logarithmic Kodaira dimension $-\infty$ in arbitrary characteristic. In [8], the author studied open rational surfaces of logarithmic Kodaira dimension zero with connected boundaries at infinity in arbitrary characteristic and classified the strongly minimal smooth affine surfaces of logarithmic Kodaira dimension zero (for the definition of strong minimality, see Section 1).

In the present article, we study open algebraic surfaces of logarithmic Kodaira dimension one in arbitrary characteristic. In Section 1, following [12, Chapter 2, Sections 3 and 4], we construct an almost minimal model and a strongly minimal model of a smooth open algebraic surface of non-negative logarithmic Kodaira dimension. In Section 2, we give a structure theorem for open algebraic surfaces of logarithmic Kodaira dimension one (cf. Theorem 2.1). Finally, in Section 3, by using the results in Section 2, we prove the following theorem.

Theorem 0.1. *Let S be a normal affine surface with $\overline{\kappa}(S - \operatorname{Sing} S) = 1$. Then the following assertions hold true:*

(1) $\overline{P}_8(S - \operatorname{Sing} S) > 0$ *or* $\overline{P}_{12}(S - \operatorname{Sing} S) > 0$. *In particular,* $\overline{P}_{24}(S - \operatorname{Sing} S) > 0$.
(2) *If* $\operatorname{char}(k) \neq 2$, *then* $\overline{P}_4(S - \operatorname{Sing} S) > 0$ *or* $\overline{P}_6(S - \operatorname{Sing} S) > 0$. *In particular,* $\overline{P}_{12}(S - \operatorname{Sing} S) > 0$.
(3) *If S is smooth, then* $\overline{P}_2(S - \operatorname{Sing} S) > 0$.

Kuramoto [9] and Tsunoda [15] considered the problem finding the smallest positive integer m such that $\overline{P}_m(X) > 0$ for a smooth open algebraic surface X of $\overline{\kappa}(X) \geq 0$. The problem has not yet been solved completely when X is a rational surface of $\overline{\kappa}(X) \geq 1$ even in the case $\operatorname{char}(k) = 0$.

We infer from the results in [9] and [15] that, if $\operatorname{char}(k) = 0$ and if S is a normal affine surface with $\overline{\kappa}(S - \operatorname{Sing} S) = 0$ or 2 (resp. S is a smooth affine surface of $\overline{\kappa}(S) \geq 0$), then $\overline{P}_{12}(S - \operatorname{Sing} S) > 0$ (resp. $\overline{P}_{12}(S) > 0$). All the assertions (1)–(3) of Theorem 0.1 are not given in [9] and [15] even in the case of $\operatorname{char}(k) = 0$.

1. Preliminary results

A reduced effective divisor D is called an SNC-divisor if it has only simple normal crossings. We employ the following notations. For the definitions of $\overline{\kappa}$ and $\overline{P}_m$, see [5] (see also [6] for the definitions in the case where $\operatorname{char}(k) > 0$).

K_V: the canonical divisor on V.
$\overline{\kappa}(S)$: the logarithmic Kodaira dimension of S.
$\overline{P}_m(S)$: the logarithmic m-genus of S.
$g(B)$: the genus of a smooth projective curve B.
$p_a(F)$: the arithmetic genus of F.

$\rho(V)$: the Picard number of V.
$\#D$: the number of all irreducible components in $\mathrm{Supp}(D)$.
$\lfloor Q \rfloor$: the integral part of a $\mathbb{Q}$-divisor Q.
$\lceil Q \rceil := -\lfloor -Q \rfloor$: the roundup of a $\mathbb{Q}$-divisor Q.
$D_1 \sim D_2$: D_1 and D_2 are linearly equivalent.
$D_1 \equiv D_2$: D_1 and D_2 are numerically equivalent.

We recall some basic notions in the theory of peeling. For more details, see [12, Chapter 2] or [13, Chapter 1]. Let (V, D) be a pair of a smooth projective surface V and an SNC-divisor D on V. We call such a pair (V, D) an *SNC-pair.* A connected curve in D means a connected curve consisting of irreducible components of D. A connected curve T in D is called a *twig* if the dual graph of T is a linear chain and T meets $D - T$ in a single point at one of the end components of T, the other end of T is called the *tip* of T. A connected curve in D is called a *rod* (resp. *fork*) if it is a connected component of D and its dual graph is a linear chain (resp. its dual graph is the dual graph of the exceptional curves of a minimal resolution of a log terminal singular point and is not a linear chain). A connected curve E in D is said to be *rational* (resp. *admissible*) if it consists only of rational curves (resp. if there are no (-1)-curves in $\mathrm{Supp}(E)$ and the intersection matrix of E is negative definite). An admissible rational twig in D is said to be *maximal* if it is not extended to an admissible rational twig with more irreducible components of D. By a (-2)-rod (resp. (-2)-fork), we mean a rod (resp. fork) consisting only of (-2)-curves.

Let $\{T_\lambda\}$ (resp. $\{R_\mu\}$, $\{F_\nu\}$) be the set of all admissible rational maximal twigs (resp. all admissible rational rods, all admissible rational forks), where no irreducible components of T_λ's belong to R_μ's or F_ν's. Then there exists a unique decomposition of D as a sum of effective $\mathbb{Q}$-divisors $D = D^\# + \mathrm{Bk}(D)$ such that the following two conditions (i) and (ii) are satisfied:

(i) $\mathrm{Supp}(\mathrm{Bk}(D)) = (\cup_\lambda T_\lambda) \cup (\cup_\mu R_\mu) \cup (\cup_\nu F_\nu)$.
(ii) $(D^\# + K_V) \cdot Z = 0$ for every irreducible component Z of $\mathrm{Supp}(\mathrm{Bk}(D))$.

The $\mathbb{Q}$-divisor $\mathrm{Bk}(D)$ is called the *bark* of D.

Lemma 1.1. *Let (V, D) be an SNC-pair. Then every connected component of $D - \lceil D^\# \rceil$ is a (-2)-rod or a (-2)-fork.*

Proof. See [12, p. 94]. □

Definition 1.2. *An SNC-pair (V, D) is said to be* almost minimal *if, for* *every irreducible curve C on V, either $C\cdot(D^{\#}+K_V)\geq 0$ or $C\cdot(D^{\#}+K_V)<0$ and the intersection matrix of $C+\mathrm{Bk}(D)$ is not negative definite.*

Lemma 1.3. *Let (V,D) be an SNC-pair. Then there exists a birational morphism $\mu : V \to \tilde{V}$ onto a smooth projective surface $\tilde{V}$ such that the following conditions (1)–(4) are satisfied:*

1. *$\tilde{D} := \mu_*(D)$ is an SNC-divisor.*
2. *$\mu_*(\mathrm{Bk}(D)) \leq \mathrm{Bk}(\tilde{D})$ and $\mu_*(D^{\#}+K_V) \geq \tilde{D}^{\#}+K_{\tilde{V}}$.*
3. *$\overline{P}_n(V-D) = \overline{P}_n(\tilde{V}-\tilde{D})$ for every integer $n \geq 1$. In particular, $\overline{\kappa}(V-D) = \overline{\kappa}(\tilde{V}-\tilde{D})$.*
4. *The pair $(\tilde{V},\tilde{D})$ is almost minimal.*

Proof. See [12, Theorem 2.3.11.1 (p. 107)], which is the same as [13, Theorem 1.11]. □

We call the pair $(\tilde{V},\tilde{D})$ in Lemma 1.3 an *almost minimal model* of (V,D).

Lemma 1.4. *Let (V,D) be an almost minimal SNC-pair. Then the following assertions hold true:*

1. *$\overline{\kappa}(V-D)\geq 0$ if and only if $D^{\#}+K_V$ is nef.*
2. *If $\overline{\kappa}(V-D)\geq 0$, then $D^{\#}+K_V$ is semiample. Moreover, we have the following:*

   (2-1) $\overline{\kappa}(V-D)=0 \Longleftrightarrow D^{\#}+K_V\equiv 0$.
   (2-2) $\overline{\kappa}(V-D)=1 \Longleftrightarrow (D^{\#}+K_V)^2=0$ *and* $D^{\#}+K_V\not\equiv 0$.
   (2-3) $\overline{\kappa}(V-D)=2 \Longleftrightarrow (D^{\#}+K_V)^2>0$.

Proof. The assertion (1) follows from [12, Theorem 2.3.15.1 (p. 116)], which is the same as [13, Theorem 2.11]. To prove the assertion (2), we assume that $\overline{\kappa}(V-D)\geq 0$. Then $D^{\#}+K_V$ is nef by the assertion (1) and so $D+K_V\equiv(D^{\#}+K_V)+\mathrm{Bk}(D)$ gives rise to the Zariski decomposition of $D+K_V$, where $D^{\#}+K_V$ is the nef part. Then the assertion (2) follows from [2, Theorem (p. 685)]. □

In the subsequent argument, let (V,D) be an almost minimal SNC-pair of $\overline{\kappa}(V-D)\geq 0$. Then $D^{\#}+K_V$ is nef by (1) of Lemma 1.4.

Lemma 1.5. *Assume that there exists a (-1)-curve E such that $E\cdot(D^{\#}+K_V)=0$, $E\not\subset\mathrm{Supp}(\lfloor D^{\#}\rfloor)$ and the intersection matrix of $E+\mathrm{Bk}(D)$ is negative definite. Let $\sigma : V\to W$ be a composite of the contraction* *of E and the contractions of all subsequently contractible components of* $\mathrm{Supp}(\mathrm{Bk}(D))$. *Set* $B := \sigma_*(D)$. *Then the following assertions hold true.*

(1) *The divisor B is an SNC-divisor and each connected component of* $\sigma(\mathrm{Supp}(\mathrm{Bk}(D)))$ *is an admissible rational twig, an admissible rational rod or an admissible rational fork of B.*
(2) *For every integer $n \geq 1$, $\overline{P}_n(V - D) = \overline{P}_n(W - B)$.*
(3) $D^{\#} + K_V = \sigma^*(B^{\#} + K_W)$. *In particular, the pair (W, B) is an almost minimal SNC-pair with $\overline{\kappa}(W - B) = \overline{\kappa}(V - D)(\geq 0)$.*

Proof. All the assertions follow from [12, (4), (6) and (7) of Lemma 2.4.4.1 (p. 123)]. □

Let E be a (-1)-curve on V. Then E is called a *superfluous exceptional component* of D if $E \subset \mathrm{Supp}(\lfloor D^{\#} \rfloor)$, $E \cdot (D - E) = E \cdot (\lfloor D^{\#} \rfloor - E) = 2$ and E meets two adjacent components of $\lfloor D^{\#} \rfloor$ (i.e., E meets two irreducible components of $\lfloor D^{\#} \rfloor - E$). Assume that E is a superfluous exceptional component of D. Let $\mu : V \to W$ be the contraction of E and set $B := \mu_*(D)$. It is then clear that (W, B) is an SNC-pair and $D^{\#} + K_V \equiv \mu^*(B^{\#} + K_W)$. Further, $\overline{P}_n(V - D) = \overline{P}_n(W - B)$ for every integer $n \geq 1$. So, when we construct an almost minimal model, we may assume that there exist no superfluous exceptional components.

By using the argument as above and Lemmas 1.3 and 1.5, we have the following result.

Lemma 1.6. *Let (V, D) be an SNC-pair with $\overline{\kappa}(V - D) \geq 0$. Then there exists a birational morphism $f : V \to \tilde{V}$ onto a smooth projective surface $\tilde{V}$ such that the following conditions are satisfied:*

(1) *Set $\tilde{D} := f_*(D)$. Then $(\tilde{V}, \tilde{D})$ is an almost minimal SNC-pair with $\overline{P}_n(\tilde{V} - \tilde{D}) = \overline{P}_n(V - D)$ for every $n \geq 1$. In particular, $\overline{\kappa}(\tilde{V} - \tilde{D}) = \overline{\kappa}(V - D)$.*
(2) *There exist no superfluous exceptional components of $\tilde{D}$.*
(3) *There exist no (-1)-curves E such that $E \cdot (\tilde{D}^{\#} + K_{\tilde{V}}) = 0$, $E \not\subset \mathrm{Supp}\lfloor \tilde{D}^{\#} \rfloor$ and the intersection matrix of $E + \mathrm{Bk}(\tilde{D})$ is negative definite.*

In Lemma 1.6, we call the pair $(\tilde{V}, \tilde{D})$ a *strongly minimal model* of a given SNC-pair (V, D). An SNC-pair (V, D) with $\overline{\kappa}(V - D) \geq 0$ is said to be *strongly minimal* if (V, D) becomes a strongly minimal model of itself.

Let S be a smooth open algebraic surface of $\overline{\kappa}(S) \geq 0$. Then there exists an SNC-pair (V, D) such that $S \cong V - D$. To study S, it is convenient to study an almost minimal model (V', D') and a strongly minimal model (V'', D'') of (V, D). We often call the pair (V', D') (resp. (V'', D'')) an almost minimal model of S (resp. a strongly minimal model of S).

2. Structure of open algebraic surfaces of $\overline{\kappa} = 1$

In this section, we give a structure theorem for open algebraic surfaces of $\overline{\kappa} = 1$ in arbitrary characteristic. The main result of this section is Theorem 2.1. It is well-known that Theorem 2.1 was proved by Kawamata [7, Theorem 2.3] in the case where $\mathrm{char}(k) = 0$. A detailed proof of [7, Theorem 2.3] is given in [10, Chapter 2, Sections 2–4]. In fact, almost all the arguments in the proof of [7, Theorem 2.3] remain true also in the case where $\mathrm{char}(k) > 0$. By virtue of Lemma 1.4, we can give a shorter proof than that of [7, Theorem 2.3] in [10, Chapter 2, Sections 2–4]. Outline of our proof of Theorem 2.1 below is almost the same as in the proofs of [12, Theorems 2.6.1.5 (p. 175), 2.6.3.1 (p. 179) and 2.6.3.2 (p. 180)].

Theorem 2.1. *Let (V, D) be a strongly minimal SNC-pair of $\overline{\kappa}(V - D) = 1$. Let $h : V \to W$ be a successive contraction of all (-1)-curves E such that the intersection of E and the image of $D^{\#} + K_V$ equals zero. Set $C := h_*(D^{\#})$. Then, for a sufficiently large integer n, the complete linear system $|n(D^{\#} + K_V)|$ defines a fibration $\rho : V \to B$ from V onto a smooth projective curve B such that ρ is an elliptic fibration, a quasi-elliptic fibration or a $\mathbb{P}^1$-fibration, that $\pi := \rho \circ h^{-1} : W \to B$ is a relatively minimal model of ρ and that $F \cdot C = F \cdot \lfloor C \rfloor = 2$ (resp. $F \cdot C = F \cdot \lfloor C \rfloor = 0$) if $p_a(F) = 0$ (resp. $p_a(F) = 1$), where F is a general fiber of π. Moreover, the following assertions hold true.*

(I) *Assume that $p_a(F) = 1$ (i.e., π is an elliptic or quasi-elliptic fibration). Then we have:*

  (1) *$C = \sum_i d_i F_i$, where $0 < d_i \le 1$ and $m_i F_i$ is a scheme-theoretic fiber for some integer (multiplicity) $m_i \ge 1$.*

  (2) *Write $R^1 \pi_* \mathcal{O}_W = \mathcal{L} \oplus \mathcal{T}$, where $\mathcal{L}$ is a locally free $\mathcal{O}_B$-module and $\mathcal{T}$ is a torsion $\mathcal{O}_B$-module. Then the divisor $C + K_W$ can be expressed as follows:*

$$C + K_W = \pi^*(K_B + \delta) + \sum_s a_s E_s + \sum_i d_i F_i,$$

where δ is a divisor on B with $t := \deg \delta = \chi(\mathcal{O}_W) + \mathrm{length}\, \mathcal{T}$, $a_s E_s$ ranges over all multiple fibers of π with multiplicity m_s,

$0 \le a_s < m_s$, and $a_s = m_s - 1$ if $m_s E_s$ is not a wild fiber of π.

(II) *Assume that $p_a(F) = 0$ (i.e., π is a $\mathbb{P}^1$-fibration). Then we have:*

(1) *We set as $C = H + \sum_i d_i F_i$, where H is the sum of the horizontal components of C and the F_i's are fibers of π. Then H is an SNC-divisor and consists of either two sections or an irreducible 2-section of π.*

(2) *The divisor $C + K_W$ can be expressed as follows:*

$$C + K_W = \pi^*(K_B + \delta) + \sum_i d_i F_i,$$

where δ is a divisor on B such that $t := \deg \delta$ equals $H_1 \cdot H_2$ (resp. one half of the number of the branch points of $\pi|_H$, $1 - g(B)$) if $H = H_1 + H_2$ with sections H_1 and H_2 (resp. H is irreducible and $\pi|_H$ is separable, H is irreducible and $\pi|_H$ is not separable) and

$$d_i = \begin{cases} \frac{1}{2}\left(1 - \frac{1}{m_i}\right) & \text{if } \#(F_i \cap H) = 1, \\ 1 - \frac{1}{m_i} & \text{if } \#(F_i \cap H) = 2, \end{cases}$$

where m_i is a positive integer or $+\infty$.

In what follows, we prove Theorem 2.1. We retain the notations and assumptions as in Theorem 2.1.

Since $\overline{\kappa}(V - D) = 1$ and $\overline{P}_n(V - D) = h^0(V, n(D + K_V)) = h^0(V, \lfloor n(D^\# + K_V) \rfloor)$ for every integer $n \ge 1$, there exist positive integers n such that $n(D^\# + K_V)$ is a Cartier divisor and $\dim |n(D^\# + K_V)| \ge 1$. We take such an integer n. Lemma 1.4 implies that $D^\# + K_V$ is semiample. So we may assume that $|n(D^\# + K_V)|$ is base point free. Since $(D^\# + K_V)^2 = 0$ by Lemma 1.4, $|n(D^\# + K_V)|$ is composed of an irreducible pencil Λ without base points.

Let $\rho : V \to B$ be the fibration associated with Λ and let F' be a general fiber of ρ. Then $F' \cdot (D^\# + K_V) = 0$. We have two cases to treat separately.

Case A: $F' \cdot D^\# = 0$. Then $F'^2 = F' \cdot K_V = 0$ and so ρ is an elliptic or quasi-elliptic fibration. Moreover, every irreducible component of D is a fiber component of ρ.

Case B: $F' \cdot D^\# > 0$. Then $F' \cdot K_V < 0$ and so ρ is a $\mathbb{P}^1$-fibration. Every irreducible component D_i of $\mathrm{Supp}(\mathrm{Bk}(D))$ is a fiber component of ρ because $D_i \cdot (D^\# + K_V) = 0$. Therefore, $D^\# \cdot F' = \lfloor D^\# \rfloor \cdot F' = -F' \cdot K_V = 2$.

Since the fibration ρ is defined by $|n(D^{\#} + K_V)|$, the construction of h implies that $\pi = \rho \circ h^{-1}$ is a relatively minimal model of ρ. Thus, all the assertions before the assertion (I) are verified.

We prove the assertion (I) of Theorem 2.1. We consider the case where π is an elliptic or quasi-elliptic fibration and prove the assertions (1) and (2) of (I). Let F be a general fiber of π. Then $F \cdot C = F \cdot \lfloor C \rfloor = 0$. So every component of C is a fiber component of π. The canonical bundle formula (cf. [1, Theorem 2]) implies that

$$K_W = \pi^*(K_B + \delta) + \sum_s a_s E_s,$$

where δ and the $m_s E_s$'s are specified in the statement.

Let Z be a connected component of C. Then $0 \le Z \cdot (C + K_W) = Z^2 \le 0$ because $Z \cdot K_W = 0$. Hence $Z^2 = 0 = Z \cdot (C + K_W)$. This implies that $Z = d_i F_i$, where $0 < d_i \le 1$ and $m_i F_i$ is a (scheme-theoretic) fiber of π with multiplicity $m_i (\ge 1)$. This proves the assertions (1) and (2).

From now on, we prove the assertion (II) of Theorem 2.1. We consider the case where π is a $\mathbb{P}^1$-fibration. We recall the assumption of Theorem 2.1: The pair (V, D) is strongly minimal, namely, there exist no superfluous exceptional components in D and no (-1)-curves E such that $E \cdot (D^{\#} + K_V) = 0$, $E \not\subset \mathrm{Supp}(\lfloor D^{\#} \rfloor)$ and the intersection matrix of $E + \mathrm{Bk}(D)$ is negative definite.

Let F be a general fiber of π. Then $F \cdot (C + K_W) = 0$, $F \cdot K_W = -2$ and $F \cdot C = 2$. We notice that every component D_i of $\mathrm{Supp}(\mathrm{Bk}(D))$ is a fiber component of ρ since $D_i \cdot (D^{\#} + K_V) = 0$ (see the argument of Case B before Proof of the assertion (I)). By the construction of h, we know that every component of C meeting a general fiber of π (i.e., a horizontal component of C) has coefficient one in C. So we can write

$$C = H + \sum_i d_i F_i,$$

where H is either a sum of two sections of π or a 2-section of π and the F_i's are fibers of π.

We show that H is an SNC-divisor and calculate the values d_i's. Set $H' := h'(H)$ ($H'_i := h'(H_i)$ $(i = 1, 2)$ if H consists of two sections H_1 and H_2). The following lemma corrects [12, Lemma 2.6.3.3 (p. 181)].

Lemma 2.2. *With the same notations and assumptions as above, the following assertions hold true.*

(1) *Let A be a fiber component of ρ with $A^2 \leq -2$ and $A \not\subset \mathrm{Supp}(D)$. Then we have:*

(1-1) *A is a (-2)-curve.*
(1-2) *The divisor $A + D$ is an SNC-divisor.*
(1-3) *The connected component of $A+D$ containing A is a (-2)-rod or a (-2)-fork.*

(2) *Let E be a (-1)-curve contained in a fiber of ρ. Then we have:*

(2-1) *If E is not a component of D, then E meets H' in at most one point and transversally.*
(2-2) *If E is a component of D, then H' is irreducible and $E \cdot H' = 1$. Furthermore, the fiber G of ρ containing E has the dual graph in Figure 1, where $G = 2E + D_1 + D_2$.*

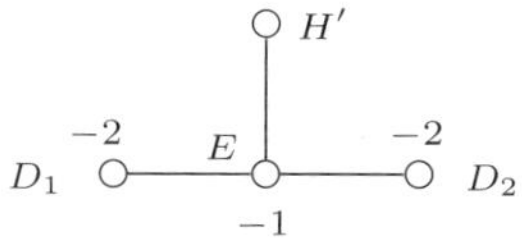

Figure 1

Proof. (1) By the hypothesis, we have $(D^{\#} + K_V) \cdot A = 0$, $A \cdot D^{\#} \geq 0$ and $A \cdot K_V \geq 0$. Then A is a (-2)-curve and $D^{\#} \cdot A = 0$. It follows from $D^{\#} \cdot A = 0$ and Lemma 1.1 that every connected component of D meeting A is a (-2)-rod or a (-2)-fork. Since every irreducible component of $\mathrm{Supp}(\mathrm{Bk}(D))$ is a fiber component of ρ, the intersection matrix of $A + \mathrm{Bk}(D)$ is negative definite. Therefore, we obtain the assertions (1-2) and (1-3).

(2) Let E be a (-1)-curve contained in a fiber, say G, of ρ. We consider the following two cases separately.

Case 1: $H' = H'_1 + H'_2$ is reducible. Suppose that E is a component of D. Then the coefficient of E in $D^{\#}$ equals one. Since $0 = (D^{\#} + K_V) \cdot E = H' \cdot E + (D^{\#} - H' - E) \cdot E - 2$, we have $H' \cdot E \leq 2$. We consider the following three subcases separately.

Subcase 1-1: $H' \cdot E = 2$. Then $(D^{\#} - H' - E) \cdot E = 0$. So every irreducible component of $D - H' - E$ meeting E is that of a (-2)-rod or a (-2)-fork in D. Then $E \cdot (D - \lceil D^{\#} \rceil) = 0$ because $E \subset \mathrm{Supp}(D)$. So, $(D - H' - E) \cdot E = (D^{\#} - H' - E) \cdot E = 0$. Since $H' \cdot E = H' \cdot G = 2$, E has coefficient one in G. By virtue of the assertion (1), we know that the dual graph of G looks like that in Figure 2, where $a \geq 0$ and $G = E + R_1 + \cdots + R_a + E'$.

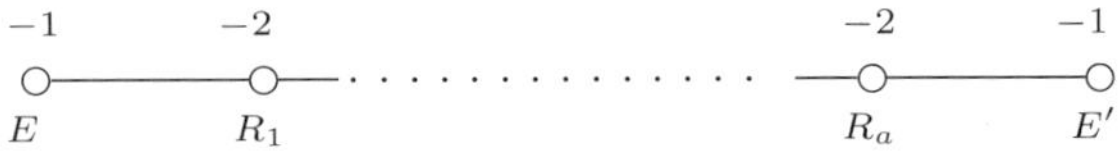

Figure 2

If E' is a component of D, then $E' \cdot (D - E') \leq 1$ and so $(D^\# + K_V) \cdot E' < 0$, which is a contradiction because $D^\# + K_V$ is nef. So, E' is not a component of D. Suppose that $a > 0$. Since R_a is either a component of a (-2)-rod in D or $R_a \not\subset \mathrm{Supp}(D)$, we conclude that $(D^\# + K_V) \cdot E' = -1$, which is a contradiction. Suppose that $a = 0$. Then we know that $E' \cdot (D^\# + K_V) = E' \cdot K_V + E' \cdot E = 0$ and the intersection matrix of $E' + \mathrm{Bk}(D)$ is negative definite. This contradicts the assumption that (V, D) is strongly minimal. Therefore, this subcase does not take place.

Subcase 1-2: $H' \cdot E = 1$. Since E meets a section of ρ, the coefficient of E in G equals one. Let $D_1, \ldots, D_r$ be all the components of $\mathrm{Supp}(D^\# - H' - E)$ meeting E. Here we note that E meets none of the components in $\mathrm{Supp}(D - \lceil D^\# \rceil)$ because $E \subset \mathrm{Supp}(D)$.

Suppose that $D_i \subset \mathrm{Supp}(\mathrm{Bk}(D))$ for every i, $1 \leq i \leq r$. Since $(D^\# - H' - E) \cdot E = 1$ and D_i $(1 \leq i \leq r)$ has coefficient < 1 in $D^\#$, we know that $r \geq 2$ and $D_1, \ldots, D_r$ are contained in $\mathrm{Supp}(G)$. It follows that E meets at least two irreducible components of $\mathrm{Supp}(\mathrm{Bk}(D))$. However, this is impossible because the contraction of E will produce two irreducible fiber components in the same fiber meeting a section.

Suppose that $D_i \not\subset \mathrm{Supp}(\mathrm{Bk}(D))$ for some i, $1 \leq i \leq r$. We may assume $D_1 \not\subset \mathrm{Supp}(\mathrm{Bk}(D))$. Then the coefficient of D_1 in $D^\#$ equals one. Since $(D^\# - H' - E) \cdot E = 1$, we know that $r = 1$ and D_1 is a component of G. So, E is a superfluous exceptional component in D, which contradicts the assumption.

Therefore, this subcase does not take place.

Subcase 1-3: $H' \cdot E = 0$. Then $(D^\# - H' - E) \cdot E = 2$. Note that E meets none of the components in $\mathrm{Supp}(D - \lceil D^\# \rceil)$ because $E \subset \mathrm{Supp}(D)$. Let $D_1, \ldots, D_r$ be all the components of $\mathrm{Supp}(D^\# - H' - E)$ meeting E and let α_i $(1 \leq i \leq r)$ be the coefficient of D_i in $D^\#$. Then D_i $(1 \leq i \leq r)$ is a component of G and $D_i \cdot E = 1$, here we note that every fiber of a $\mathbb{P}^1$-fibration contains no cycles of curves. Since $\alpha_1 + \cdots + \alpha_r = 2$, $r \geq 2$. On the other hand, since E is a (-1)-curve, we see that $r \leq 2$. Hence, $r = 2$. Then $\alpha_1 = \alpha_2 = 1$ and $E \cdot D_1 = E \cdot D_2 = 1$. So, E is a superfluous exceptional component in D, a contradiction.

As seen from Subcases 1-1–1-3, we know that E is not a component of D. Since $(D^{\#}+K_V)\cdot E=0$, we have $E\cdot D^{\#}=1$. Then $0\leq E\cdot H'\leq E\cdot D^{\#}=1$. Hence E meets H' in at most one point and transversally.

Case 2: H' is irreducible. In this case, H' is a 2-section of ρ. If E is not a component of D, then we see that E meets H' in at most one point and transversally by using the same argument as in the preceding paragraph. So we may assume that E is a component of D. We recall that G is the fiber of ρ containing E. Note that $0\leq H'\cdot E\leq 2$ because H' is a 2-section of ρ. We consider the following three subcases separately.

Subcase 2-1: $H'\cdot E=2$. Then the coefficient of E in G equals one. So we can derive a contradiction by using the same argument as in Subcase 1-1.

Subcase 2-2: $H'\cdot E=1$. In this subcase, $E\cdot(D^{\#}-H'-E)=1$ and the coefficient of E in G equals one or two. Note that E meets none of the components in $\mathrm{Supp}(D-\lceil D^{\#}\rceil)$ because $E\subset\mathrm{Supp}(D)$. Let $D_1,\ldots,D_r$ be all the components of $\mathrm{Supp}(D^{\#}-H'-E)$ meeting E and let α_i $(1\leq i\leq r)$ be the coefficient of D_i in $D^{\#}$. Suppose that $\alpha_i=1$ for some i, $1\leq i\leq r$. Then $r=1$ and E is a superfluous exceptional component of D, which is a contradiction. So, $\alpha_i<1$ for every i, $1\leq i\leq r$. Since $(D^{\#}-H'-E)\cdot E=\sum_{i=1}^{r}\alpha_i=1$, we know that $r\geq 2$. Moreover, since D_i $(1\leq i\leq r)$ is a component of G, we know that $r=2$ and the coefficient of E in G equals two. Since the coefficient of E in $D^{\#}$ equals one and $0\leq\alpha_1,\alpha_2<1$, D_i $(i=1,2)$ is a component of an admissible rational maximal twig, say T_i, in D and D_i is the end component of T_i, which is not a tip if $\#T_i\geq 2$. By the theory of peeling (see [12, Lemma 2.3.3.1 (p. 88)]), we know that $\alpha_i\geq\frac{1}{2}$ for $i=1,2$ and the equality holds if and only if $T_i=D_i$ is a (-2)-curve. Since $E\cdot(D^{\#}-H'-E)=\alpha_1+\alpha_2=1$, we conclude that $\alpha_1=\alpha_2=\frac{1}{2}$. Hence $F=2E+D_1+D_2$ and the dual graph of $H'+E+D_1+D_2$ is given in Figure 1.

Subcase 2-3: $H'\cdot E=0$. In this subcase, we can derive a contradiction by using the same argument as in Subcase 1-3.

The proof of Lemma 2.2 is thus completed. □

Lemma 2.3. *With the same notations and assumptions as above, the divisor H is an SNC-divisor.*

Proof. We use the same notations before Lemma 2.2. Let E be a (-1)-curve contained in a fiber G of ρ. Then, as seen from the proof of Lemma 2.2, we know that $H' \cdot E \leq 1$. We consider the following two cases separately.

Case 1: $H' \cdot E = 1$. Then the coefficient of E in G is either one or two. Since $(D^{\#} + K_V) \cdot E = 0$, we have $(D^{\#} - H') \cdot E = 0$. If E is a component of D, then it follows from (2) of Lemma 2.2 that G_{red} has the dual graph in Figure 1. If E is not a component of D, then it follows from Lemma 1.1 that every connected component of $\mathrm{Supp}(\mathrm{Bk}(D))$ meeting E is a (-2)-rod or a (-2)-fork in D. Since the pair (V, D) is strongly minimal, $E \not\subset \mathrm{Supp}(\lfloor D^{\#} \rfloor)$ and $E \cdot (D^{\#} + K_V) = 0$, it follows that the intersection matrix of $E + \mathrm{Bk}(D)$ is not negative definite. So we know that every component of $G_{\mathrm{red}} - E$ is a (-2)-curve. Suppose that the coefficient of E in G equals one. Then there exists another (-1)-curve E' in $\mathrm{Supp}(G)$. It follows that the intersection matrix of $E + \mathrm{Bk}(D)$ is negative definite, which is a contradiction. Hence the coefficient of E in G equals two. Since $H' \cdot E = 1$, H' is irreducible (i.e., H' is a 2-section of ρ).

We claim that E is the unique (-1)-curve in $\mathrm{Supp}(G)$. Indeed, if $E \subset \mathrm{Supp}(D)$, then the claim follows from (2) of Lemma 2.2. If $E \not\subset \mathrm{Supp}(D)$, then the claim follows beause $G_{\mathrm{red}} - E$ consists only of (-2)-curves. We know that the fiber G together with H' has one of the dual graphs in Figure 3. Then we can easily see that $h_*(G)$ is a smooth fiber touching the 2-section $H = h_*(H')$ in one point. In particular, H is smooth at $H \cap h_*(G)$.

(Case (a): $E \subset \mathrm{Supp}(D)$)

(a)

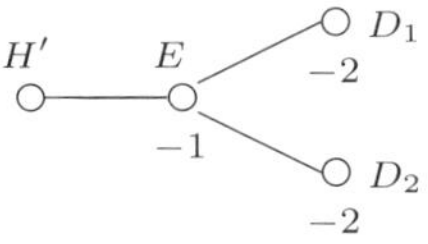

(Case (b): $E \not\subset \mathrm{Supp}(D)$)

(b)

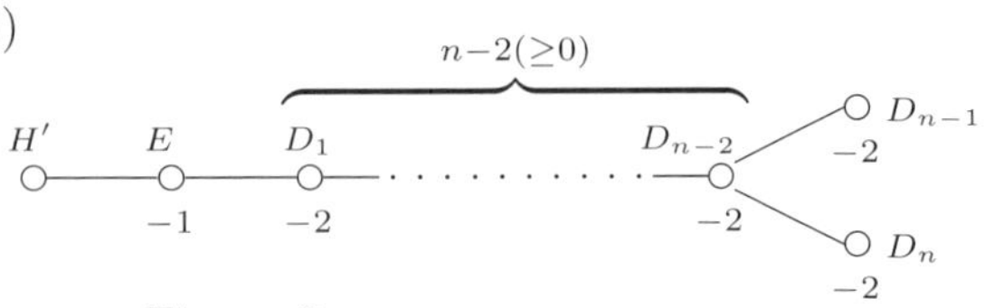

Figure 3

Case 2: $H' \cdot E = 0$. It follows from (2) of Lemma 2.2 that E is not a component of D. Since $E \cdot (D^{\#} + K_V) = 0$, we have $E \cdot (D^{\#} - H') = 1$. Let $D_1, \ldots, D_r$ be all the irreducible components of $\mathrm{Supp}(D)$ meeting E and let α_i $(1 \leq i \leq r)$ be the coefficient of D_i in $D^{\#}$, where we assume $\alpha_1 \geq \cdots \geq \alpha_r$. Then $\sum_{i=1}^{r} \alpha_i = E \cdot D^{\#} = 1$. It follows from $E \cdot H' = 0$ that $\mathrm{Supp}(E + D_1 + \cdots + D_r) \subset \mathrm{Supp}(G)$. So, $1 \leq r \leq 2$. Suppose that $\alpha_1 = 1$ (i.e., $D_1 \not\subset \mathrm{Supp}(\mathrm{Bk}(D))$). Then $E \cdot (D^{\#} + K_V) = 0$, $E \not\subset \mathrm{Supp}(\lfloor D^{\#} \rfloor)$ and the intersection matrix of $E + \mathrm{Bk}(D)$ is negative definite. This is a contradiction because (V, D) is strongly minimal. Hence $r = 2$ and $0 < \alpha_2 \leq \alpha_1 < 1$. Thus we see that E meets two connected components, say $D^{(1)}$ and $D^{(2)}$, of $\mathrm{Supp}(\mathrm{Bk}(D))$. Since $E \cdot (G_{\mathrm{red}} - E) = 2$, we know that E meets none of components of $\mathrm{Supp}(D - \lceil D^{\#} \rceil)$.

We note that $D^{(1)}$ and $D^{(2)}$ are contained in the fiber G. Since (V, D) is strongly minimal, we know that the intersection matrix of $E + D^{(1)} + D^{(2)}$ is not negative definite. Hence $\mathrm{Supp}(E + D^{(1)} + D^{(2)}) = \mathrm{Supp}(G)$. Since G has a unique (-1)-curve, there exist exactly two components of G whose coefficients in G equal one. Since $H' \cdot E = 0$ and $H' \cdot G = 2$, we may assume that $H' \cdot D^{(1)} \geq 1$. Then $D^{(1)}$ is an admissible rational maximal twig of D and so $H' \cdot D^{(1)} = 1$ and H' meets one of the terminal components of $D^{(1)}$. Let $D_1^{(1)}$ be the component of $D^{(1)}$ meeting H'.

Suppose that $D^{(2)} \cdot H' \geq 1$. Then $D^{(2)}$ is also an admissible rational maximal twig of D and H' meets one of the terminal components, say $D_1^{(2)}$, of $D^{(2)}$. Since $H' \cdot G = 2$, the coefficients of $D_1^{(1)}$ and $D_1^{(2)}$ in G equal one. Then h factors through the birational morphism $\mu : V \to W'$ that is a successive contraction of the (-1)-curve E and consecutively (smoothly contractible) curves in $\mathrm{Supp}(G)$ such that $\mu_*(F) = \mu(D_1^{(1)}) + \mu(D_1^{(2)})$. The birational morphism μ does not affect H'. Then we can easily see that $H = h(H')$ is smooth at $H \cap h_*(G)$.

Suppose that $D^{(2)} \cdot H' = 0$. Then $D_1^{(1)}$ is the unique component of G meeting H'. So H' is irreducible. Since $D_1^{(1)} \cdot H' = 1$, the coefficient of $D_1^{(1)}$ in G equals two. We note that $D^{(2)}$ is a connected component of D because $D^{(2)} \cdot H' = 0$. Namely, $D^{(2)}$ is an admissible rational rod or an admissible rational twig. We know that h factors through the birational morphism $\mu : V \to W'$ that is a successive contraction of the (-1)-curve E and consecutively (smoothly contractible) curves in $\mathrm{Supp}(G)$ such that $\mu_*(D_1^{(1)})$ is a (-1)-curve. Then $\mu_*(D_1^{(1)})$ is the unique (-1)-curves in $\mathrm{Supp}(h_*(G))$ and the coefficient of $\mu(D_1^{(1)})$ in $\mu_*(G)$ equals two. Further, the curve H' is not affected by μ. The weighted dual graph of $\mu_*(G) + \mu(H')$ looks like that in Figure 4. Then we can easily see that $H = h(H')$ is smooth at the

point $H \cap h_*(G)$. □

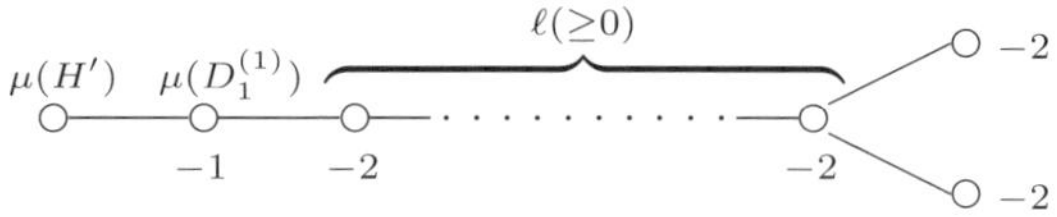

Figure 4

We note that $\pi : W \to B$ is a $\mathbb{P}^1$-bundle over B since π is relatively minimal. Let C_0 be a minimal section of the ruling π on W. Then $K_W \sim -2C_0 + \pi^*(K_B + C_0|_{C_0})$. Since H is a 2-section or a sum of two sections of π, there exists a divisor δ on B such that $H + K_W = \pi^*(K_B + \delta)$. Hence we have

$$C + K_W = \pi^*(K_B + \delta) + \sum_i d_i F_i.$$

We calculate $t := \deg \delta$.

Lemma 2.4. *With the same notations as above, the number $t := \deg \delta$ equals $H_1 \cdot H_2$ (resp. one half of the number of the branch points of $\pi|_H : H \to B$, $1 - g(B)$) if $H = H_1 + H_2$ with sections H_1 and H_2 (resp. if H is irreducible and $\pi|_H : H \to B$ is separable, if H is irreducible and $\pi|_H : H \to B$ is not separable).*

Proof. Assume that $H = H_1 + H_2$ with sections H_1 and H_2. Then we have

$$(H + K_W)|_{H_1} = K_{H_1} + H_2|_{H_1} = \pi^*(K_B + \delta)|_{H_1}.$$

So, $t = \deg H_2|_{H_1} = H_1 \cdot H_2$.

We consider the case where H is irreducible. Then $H' := h'(H)$ is a smooth curve and a 2-section of ρ. We infer from Lemma 2.3 that the curve H is smooth. So $(H + K_W)|_H = K_H$. If $\rho|_{H'} : H' \to B$ is separable, then we infer from the Riemann–Hurwitz formula that

$$t = \tfrac{1}{2}\{\text{ the number of the branch points of } \pi|_H\}.$$

Assume that $\rho|_{H'} : H' \to B$ is not separable. Then, $\rho|_{H'}$ is a purely inseparable covering of degree two and so is $\pi|_H : H \to B$. It then follows that $g(B) = g(H)$ (see [3, Proposition IV.2.5 (p. 302)]). Since H is a 2-section of π, we have $H \sim 2C_0 + \pi^*(\gamma)$ for some divisor γ on B. Then

$$2g(B) - 2 = (H + K_W) \cdot H = \pi^*(K_B + \delta) \cdot (2C_0 + \pi^*(\gamma)) = 2(2g(B) - 2 + t).$$

Therefore, $t = 1 - g(B)$. □

We compute the values d_i's in $C = H + \sum_i d_i F_i$. Although Lemma 2.5 can be proved by the argument as in the proof of [10, Lemma II.4.2], we reproduce the proof because the argument of the proof will be used in the proof of Theorem 0.1.

Lemma 2.5. *With the same notations as above, if $\#(F_i \cap H) = 2$ (resp. $\#(F_i \cap H) = 1$), then we have*

$$d_i = 1 - \frac{1}{m_i} \quad \left(\text{resp. } \frac{1}{2}\left(1 - \frac{1}{m_i}\right)\right),$$

where m_i is a positive integer or $+\infty$.

Proof. We consider the following two cases separately.

Case 1: $\#(F_i \cap H) = 2$. Set $G_i := h^*(F_i)$. If G_i is irreducible, then $G_i \subset \mathrm{Supp}(D)$ and the coefficient of G_i in $D^{\#}$ equals one. So $d_i = 1 = 1 - \frac{1}{+\infty}$. Assume that G_i is reducible. Here we note that the coefficient of $h'(F_i)$ in G_i equals one. So there exists a (-1)-curve $E_i \neq h'(F_i)$ contained in $\mathrm{Supp}(G_i)$. We claim that:

Claim 1.

(1) $E_i \not\subset \mathrm{Supp}(D)$.
(2) E_i is the unique (-1)-curve in $\mathrm{Supp}(G_i)$.
(3) $(G_i)_{\mathrm{red}} = B_1 + E_i + B_2$, where B_1 and B_2 are distinct admissible rational maximal twigs in D with $B_1 \cdot H' = B_2 \cdot H' = 1$.

Proof. (1) Suppose to the contrary that $E_i \subset \mathrm{Supp}(D)$. Then $E_i \cdot (D - E_i) \leq 2$. So, we know that E_i is a superfluous exceptional component of D, here we note that $E_i \neq h'(F_i)$. This contradicts the assumption that (V, D) is strongly minimal. So $E_i \not\subset \mathrm{Supp}(D)$.

(2) If there exists another (-1)-curve $E' \subset \mathrm{Supp}(G_i)$ other than E_i, then $E_i \cdot (D^{\#} + K_V) = 0$, $E_i \not\subset \mathrm{Supp}(\lfloor D^{\#} \rfloor)$ and the intersection matrix of $E_i + \mathrm{Bk}(D)$ is negative definite. This is a contradiction because (V, D) is strongly minimal. This proves the assertion (2).

(3) By using the same argument as in the proof of Lemma 2.3, we obtain the assertion. □

Moreover, by using the same argument as in the proof of Lemma 2.3, we know that:

Claim 2. Let B_{i1} $(i = 1, 2)$ be the component of B_i meeting H'. Then B_{11} and B_{12} are the two components of $\mathrm{Supp}(G_i)$ whose coefficients in G_i equal one. Moreover, $F_i = h_*(B_{i1})$ for $i = 1$ or 2.

Assume that $F_i = h_*(B_{11})$. Since $C = h_*(D^{\#})$, we know that the coefficient of B_{11} in $D^{\#}$ equals d_i. We infer from [12, p. 89] that $d_i = 1 - \frac{1}{m_i}$, where m_i is a positive integer.

Case 2: $\#(F_i \cap H) = 1$. We set as $F := F_i$ and $d := d_i$. Let $P := H \cap F$. We consider the case where $H = H_1 + H_2$ with two sections H_1 and H_2. Then $P = H_1 \cap H_2 \cap F$. Let $\sigma_1 : W_1 \to W$ be the blowing-up at P and let E be the proper transform of $\sigma_1^{-1}(P)$ on V. Since D is an SNC-divisor, $h : V \to W$ must be decomposed by σ_1. Then we have

$$\begin{aligned} h^*(K_W + H + dF) &= K_V + h'(H) + (1+d)E + (\text{other components}) \\ &\leq D + K_V, \end{aligned}$$

here we note that $h^*(C + K_W) = D^{\#} + K_V$. Hence we know that $d = 0 = \frac{1}{2}(1 - \frac{1}{1})$.

We consider the case where H is irreducible. Set $G := h^*(F)$ and $H' := h'(H)$. Since D is an SNC-divisor and $h'(F) \cup H' \subset \mathrm{Supp}(D)$, G is a reducible fiber of ρ. Let E be a (-1)-curve contained in $\mathrm{Supp}(G)$. If $E \subset \mathrm{Supp}(D)$, then we infer from (2) of Lemma 2.2 that the fiber G together with H' has the dual graph (a) in Figure 3 (which is the same as that in Figure 1), where $E = E_i$ and $G = 2E + D_1 + D_2$. It is then clear that the coefficient of D_i $(i = 1, 2)$ in $D^{\#}$ equals $\frac{1}{2}$. Since $F = h_*(D_1)$ or $h_*(D_2)$, we conclude that $d = \frac{1}{2} = \frac{1}{2}(1 - \frac{1}{+\infty})$ if $E \subset \mathrm{Supp}(D)$.

We assume further that $E \not\subset \mathrm{Supp}(D)$. As seen from the proof of Lemma 2.2, we know that $H' \cdot E \leq 1$. Suppose that $H' \cdot E = 1$. We then infer from the proof of Lemma 2.3 that the fiber G together with H' has the dual graph (b) in Figure 3 and that $G = 2(E + D_1 + \cdots + D_{n-2}) + D_{n-1} + D_n$. Since every connected component of $D_1 + \cdots + D_n$ is a (-2)-rod or a (-2)-fork in D, D_i $(1 \leq i \leq n)$ has coefficient zero in $D^{\#}$. Since $F = h_*(D_1)$ or $h_*(D_2)$, we conclude that $d = 0 = \frac{1}{2}(1 - \frac{1}{1})$ if $H' \cdot E = 1$.

Suppose further that $H' \cdot E = 0$. We use the same notations as in Case 2 of the proof of Lemma 2.3. Since H' is irreducible and $\#(H' \cap G) = 1$, we know that $D_1^{(2)} \cdot H' = 0$. Let $\mu : V \to W'$ be the same birational morphism as in Case 2 of the proof of Lemma 2.3. Since $D^{(i)}$ $(i = 1, 2)$ is a connected component of $\mathrm{Supp}(\mathrm{Bk}(D))$, we know that the weighted dual graph of $\mu_*(G) + \mu_*(H')$ looks like that in Figure 4. Since $D^{(1)}$ is an admissible maximal rational twig in D and μ has a unique fundamental point, we know that the dual graph of G is given as in Figure 5, where $(D_1^{(1)})^2 = -2 - n$. In particular, if $\ell \geq 1$, then we see that $D^{(2)}$ is either an admissible rational fork or an admissible rational rod with $\#D^{(2)} = 3$, whose terminal components are (-2)-curves. We can easily see that F is the image of D_1 or D_2. If $\ell = 0$, then the coefficient of D_i $(i = 1, 2)$ in $D^\#$ equals

$$\frac{n+1}{2n+3} = \frac{1}{2}\left(1 - \frac{1}{2n+3}\right),$$

where $n \geq 0$. If $\ell \geq 1$, then, by using [12, 2.3.5 (p. 93)], we know that the coefficient of D_i $(i = 1, 2)$ in $D^\#$ equals

$$\frac{1}{2}\left(1 - \frac{1}{m}\right),$$

where m is an integer ≥ 2. □

(Case: $\ell = 0$)

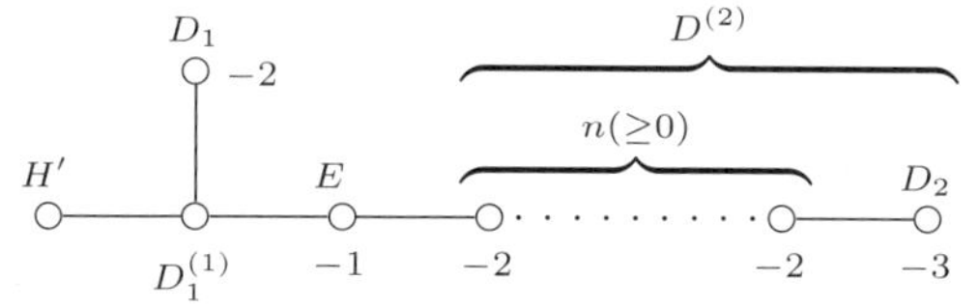

(Case: $\ell \geq 1$)

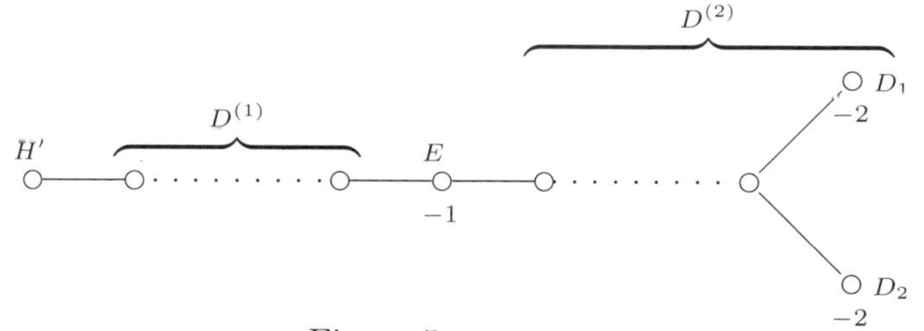

Figure 5

The proof of Theorem 2.1 is thus completed.

Remark 2.6. Let the notations and assumptions be the same as in Lemma 2.5. We note the following which will be used in the proof of Theorem 0.1.

(1) Suppose that $\#(F_i \cap H) = 2$ and $0 < d_i < 1$. Then, as seen from Case 1 of the proof of Lemma 2.5, we know that $\mathrm{Supp}(D) \cap \mathrm{Supp}(F_i)$ is not connected.

(2) Suppose that $\#(F_i \cap H) = 1$ and $0 < d_i < \frac{1}{2}$. Then, as seen from Case 2 of the proof of Lemma 2.5, we know that $\mathrm{Supp}(D) \cap \mathrm{Supp}(F_i)$ is not connected. In particular, $d_i = \frac{1}{2}$ if D is connected.

Remark 2.7. We use the same notations and assumptions as in Theorem 2.1. By using Theorem 2.1 and the Riemann–Roch theorem on the curve B, we obtain the followings:

(1) With the same assumptions as in Theorem 2.1 (I), we have

$$\overline{P}_n(V - D) = n(2g(B) - 2 + t) + \sum_s \lfloor \frac{na_s}{m_s} \rfloor + \sum_i \lfloor \frac{nd_i}{m_i} \rfloor + 1 - g(B)$$

for a sufficiently large integer n.

(2) With the same assumptions as in Theorem 2.1 (II), we have

$$\overline{P}_n(V - D) = n(2g(B) - 2 + t) + \sum_i \lfloor nd_i \rfloor + 1 - g(B).$$

for a sufficiently large integer n.

3. Logarithmic plurigenera of normal affine surfaces of $\overline{\kappa} = 1$

In this section, we give some results on logarithmic plurigenera of open algebraic surfaces of $\overline{\kappa} = 1$ in arbitrary characteristic and prove Theorem 0.1.

In Lemmas 3.1–3.3, we study logarithmic plurigenera of open algebraic surfaces of $\overline{\kappa} = 1$ by using Theorem 2.1. With the same notations as in Theorem 2.1 (II), we set $\delta_n := n(K_B + \delta) + \sum_i \lfloor nd_i \rfloor \pi(F_i)$ for a positive integer n. Here we note that $\deg(K_B + \delta) + \sum_i d_i > 0$ because $\overline{\kappa}(V - D) = 1$.

Lemma 3.1. *With the same notations and assumptions as in Theorem* 2.1 (II), *assume that* $g(B) \geq 2$. *Then* $\overline{P}_n(V - D) > 0$ *for every integer* $n \geq 2$. *Moreover, if* $\mathrm{char}(k) \neq 2$, *then* $\overline{P}_n(V - D) > 0$ *for every positive integer* n.

Proof. By the Riemann–Roch theorem, we have

$$\begin{aligned}\overline{P}_n(V - D) = h^0(B, \delta_n) &\geq \deg \delta_n + 1 - g(B) \\ &= (2n - 1)(g(B) - 1) + n \deg \delta + \sum_i \lfloor nd_i \rfloor.\end{aligned}$$

If $\deg \delta \geq 0$, then $h^0(B, \delta_n) > 0$ for every positive integer n because $g(B) \geq 2$ and $\sum_i \lfloor nd_i \rfloor \geq 0$. We consider the case $\deg \delta < 0$. It then follows from

(2) of Theorem 2.1 (II) that $\mathrm{char}(k) = 2$ and $\deg\delta = 1 - g(B)$. So, we have

$$h^0(B, \delta_n) \geq (n-1)(g(B)-1) + \sum_i \lfloor nd_i \rfloor.$$

Since $g(B) \geq 2$ and $\sum_i \lfloor nd_i \rfloor \geq 0$, we conclude that $\overline{P}_n(V-D) > 0$ for every integer $n \geq 2$. □

Lemma 3.2. *With the same notations as in Theorem* 2.1 (II), *assume that* $g(B) = 1$. *Then* $\overline{P}_n(V-D) > 0$ *for every integer* $n \geq 4$. *Moreover, if* $\mathrm{char}(k) \neq 2$, *then* $\overline{P}_n(V-D) > 0$ *for every integer* $n \geq 2$.

Proof. By the Riemann–Roch theorem, we have

$$\overline{P}_n(V-D) = h^0(B, \delta_n) \geq \deg\delta_n = n\deg\delta + \sum_i \lfloor nd_i \rfloor.$$

If $\deg\delta > 0$, then $\overline{P}_n(V-D) > 0$ for every integer $n > 0$. Assume that $\deg\delta \leq 0$. Since $\deg\delta \geq 0$ by (2) of Theorem 2.1 (II), $\deg\delta = 0$. We consider the following three cases separately.

Case 1: $H = H_1 + H_2$ with two sections H_1 and H_2. By (2) of Theorem 2.1 (II), we have $0 = \deg\delta = H_1 \cdot H_2$. Since $\deg\delta + \sum_i d_i = \sum_i d_i > 0$ by $\overline{\kappa}(V-D) = 1$, we can set as $\sum_i d_i = \sum_{i=1}^s d_i$, where $s \geq 1$ and $0 < d_i \leq 1$ for $i = 1, \ldots, s$. It follows from (2) of Theorem 2.1 (II) that $d_i = 1 - \frac{1}{m_i}$, where $m_i \geq 2$ or $m_i = +\infty$. Hence $d_i \geq \frac{1}{2}$ for $i = 1, \ldots, s$ and so $\sum_{i=1}^s \lfloor nd_i \rfloor > 0$ for every integer $n \geq 2$. Therefore, $\overline{P}_n(V-D) > 0$ for every integer $n \geq 2$.

Case 2: H is irreducible and $\pi|_H : H \to B$ is separable. Since $\deg\delta = 0$, it follows from (2) of Theorem 2.1 (II) that H is smooth and $\pi|_H : H \to B$ is étale. By using the same argument as in Case 1, we know that $\overline{P}_n(V-D) > 0$ for every integer $n \geq 2$.

Case 3: H is irreducible and $\pi|_H : H \to B$ is not separable. In this case, $\mathrm{char}(k) = 2$ and H is smooth (see Lemma 2.3). Since $\deg\delta + \sum_i d_i = \sum_i d_i > 0$ by $\overline{\kappa}(V-D) = 1$, we can set as $\sum_i d_i = \sum_{i=1}^s d_i$, where $s \geq 1$ and $0 < d_i \leq 1$ for $i = 1, \ldots, s$. Since $\pi|_H$ is a purely inseparable double covering, $\#(F \cap H) = 1$ for every fiber F of π. It then follows from (2) of Theorem 2.1 (II) that $d_i = \frac{1}{2}\left(1 - \frac{1}{m_i}\right)$, where m_i is an integer ≥ 2 or $m_i = +\infty$, for $i = 1, \ldots, s$. In particular, $\frac{1}{4} \leq d_i \leq \frac{1}{2}$ for $i = 1, \ldots, s$. For every integer $n \geq 4$, we have $\sum_{i=1}^s \lfloor nd_i \rfloor > 0$. Hence, $\overline{P}_n(V-D) > 0$ for every integer $n \geq 4$. □

Lemma 3.3. *With the same notations and assumptions as in Theorem 2.1 (II), assume that* $g(B) = 0$. *Then* $\overline{P}_n(V - D) > 0$ *for* $n = 8$ *or* 12. *Moreover, if* $\mathrm{char}(k) \neq 2$, *then* $\overline{P}_n(V - D) > 0$ *for* $n = 4$ *or* 6.

Proof. Since $g(B) = 0$, W is a rational surface. If $h^0(W, H + K_W) > 0$, then $\overline{P}_n(V - D) > 0$ for every integer $n > 0$. So, as in the subsequent argument, we assume further that $h^0(W, H + K_W) = 0$. Then H is an SNC-divisor and each irreducible component of H is a smooth rational curve (see [10, Lemma I.2.1.3 (p. 7)]). By the Riemann–Roch theorem, we have

$$\overline{P}_n(V - D) = h^0(B, \delta_n) \geq \deg \delta_n + 1$$
$$= -2n + n \deg \delta + \sum_i \lfloor n d_i \rfloor + 1.$$

Since $\sum_i \lfloor n d_i \rfloor \geq 0$, we know that if $\deg \delta \geq 2$ then $\overline{P}_n(V - D) > 0$ for every integer $n > 0$. We assume further that $\deg \delta \leq 1$. Since $\overline{\kappa}(V - D) = 1$, we know that

$$\deg \delta + \sum_i d_i > 2. \tag{3.1}$$

We set as $\sum_i d_i = \sum_{i=1}^s d_i$, where $0 < d_i \leq 1$ for $i = 1, \ldots, s$ and $s > 0$, and $A_n := \sum_{i=1}^s \lfloor n d_i \rfloor$. We consider the following two cases separately.

Case 1: $\deg \delta \leq 0$. By (2) of Theorem 2.1 (II), we know that $\deg \delta = 0$. Then $\sum_{i=1}^s d_i > 2$ by (3.1). In particular, $s \geq 3$.

Subcase 1-1: $H = H_1 + H_2$ with two sections H_1 and H_2. Since $0 = \deg \delta = H_1 \cdot H_2$ by (2) of Theorem 2.1 (II), we have $H_1 \cap H_2 = \emptyset$. So, it follows from (2) of Theorem 2.1 (II) that $d_i = 1 - \frac{1}{m_i}$, where $m_i \geq 2$ or $m_i = +\infty$, for $i = 1, \ldots, s$. If $s \geq 4$, then $A_n \geq 4\lfloor \frac{n}{2} \rfloor$ for every integer $n \geq 1$. In particular, if n is even, then $A_n \geq 2n$ and so $\overline{P}_n(V - D) \geq -2n + 1 + A_n \geq 1$. (In particular, $\overline{P}_2(V - D) > 0$.) Suppose that $s = 3$. We may assume that $m_1 \leq m_2 \leq m_3$. Since $d_1 + d_2 + d_3 > 2$ by (3.1), we have

$$1 > \frac{1}{m_1} + \frac{1}{m_2} + \frac{1}{m_3}.$$

So, if $m_1 = 2$, then $m_2 \geq 3$. Moreover, if $m_1 = 2$ and $m_2 = 3$ (resp. $m_1 = 2$ and $m_2 = 4$, $m_1 = m_2 = 3$), then $m_3 \geq 6$ (resp. $m_3 \geq 5$, $m_3 \geq 4$). So, A_n

satisfies one of the following inequalities:

$$A_n \geq \lfloor\frac{n}{2}\rfloor + \lfloor\frac{2n}{3}\rfloor + \lfloor\frac{5n}{6}\rfloor,$$
$$A_n \geq \lfloor\frac{n}{2}\rfloor + \lfloor\frac{3n}{4}\rfloor + \lfloor\frac{4n}{5}\rfloor,$$
$$A_n \geq 2\lfloor\frac{2n}{3}\rfloor + \lfloor\frac{3n}{4}\rfloor.$$

As seen from the inequalities as above, we see that $A_n \geq 2n$ for $n = 4$ or 6. Hence, $\overline{P}_n(V - D) \geq -2n + A_n + 1 > 0$ for $n = 4$ or 6.

Subcase 1-2: H is irreducible. By (2) of Theorem 2.1 (II), we know that $\pi|_H : H \to B(\cong \mathbb{P}^1)$ is a separable double cover. Here we note that H is smooth by Lemma 2.3. Moreover, since $\deg \delta = 0$, it follows from (2) of Theorem 2.1 (II) that $\pi|_H$ is unramified. However, this is not the case because $g(B) = g(H) = 0$. So this subcase does not take place.

Case 2: $\deg \delta = 1$. With the same notations as above, it suffices to find an integer n satisfying $A_n \geq n$ because

$$\overline{P}_n(V - D) = h^0(B, \delta_n) \geq 1 - n + A_n$$

for every integer $n \geq 1$. By (3.1), we have $\sum_{i=1}^{s} d_i > 1$. In particular, $s \geq 2$.

Subcase 2-1: $H = H_1 + H_2$ with two sections H_1 and H_2. Then $1 = \deg \delta = H_1 \cdot H_2$ by (2) of Theorem 2.1 (II). So, $\#(H_1 \cap H_2) = 1$. If $F_i \cap (H_1 \cap H_2) \neq \emptyset$ for some i, $1 \leq i \leq s$, then, as seen from the proof of Lemma 2.5, we know that $d_i = 0$. This is a contradiction. By using the same argument as in Subcase 1-1, we see that $d_i = 1 - \frac{1}{m_i}$, where $m_i \geq 2$ or $m_i = +\infty$, for $i = 1, \ldots, s$. If $s \geq 3$, then

$$A_n \geq \lfloor nd_1 \rfloor + \lfloor nd_2 \rfloor + \lfloor nd_3 \rfloor \geq 3\lfloor\frac{n}{2}\rfloor.$$

So $A_n > n$ if $n \geq 2$. Assume that $s = 2$. We may assume that $m_1 \leq m_2$. Then, since $d_1 + d_2 = 2 - (\frac{1}{m_1} + \frac{1}{m_2}) > 1$, we know that $m_2 \geq 3$. So,

$$A_n \geq \lfloor\frac{n}{2}\rfloor + \lfloor\frac{2n}{3}\rfloor.$$

Hence $A_n \geq n$ for every integer $n \geq 2$. Therefore, $\overline{P}_n(V - D) > 0$ for every integer $n \geq 2$.

Subcase 2-2: H is irreducible and $\pi|_H : H \to B$ is separable. Since H is a smooth rational curve by the assumption $h^0(W, H + K_W) = 0$ and $\deg \delta = 1$, we know that the number of the branch points of $\pi|_H$ equals two.

Let $P_1, P_2 \in B$ be the two branch points of $\pi|_H$. It follows from Lemma 2.5 that if $F_i = \pi^{-1}(P_1)$ or $\pi^{-1}(P_2)$ $(1 \le i \le s)$, then $d_i = \frac{1}{2}(1 - \frac{1}{m_i})$, where $m_i \ge 2$ or $m_i = +\infty$. In particular, $\frac{1}{4} \le d_i \le \frac{1}{2}$ if $F_i = \pi^{-1}(P_1)$ or $\pi^{-1}(P_2)$. Set $r := \#\{i \mid 1 \le i \le s, d_i < \frac{1}{2}\}$. Then $r \le 2$.

Suppose that $r = 0$. Then, by using the same argument as in Subcase 2-1, we know that $\overline{P}_n(V - D) > 0$ for every integer $n \ge 2$. Suppose that $r = 1$. We may assume that $d_1 < \frac{1}{2}$. If $s \ge 3$, then

$$A_n \ge \lfloor nd_1 \rfloor + \lfloor nd_2 \rfloor + \lfloor nd_3 \rfloor \ge \lfloor \frac{n}{4} \rfloor + 2\lfloor \frac{n}{2} \rfloor.$$

Hence $A_n \ge n$ if n is even. Assume that $s = 2$. Then

$$d_1 + d_2 = \frac{1}{2}\left(1 - \frac{1}{m_1}\right) + \left(1 - \frac{1}{m_2}\right),$$

where $m_1 \ge 2$ and $m_2 \ge 2$ or $m_2 = +\infty$. Since $d_1 + d_2 > 1$ by (3.1), $\frac{1}{2} > \frac{1}{2m_1} + \frac{1}{m_2}$. Hence $m_2 \ge 3$. Moreover, if $m_1 = 2$ (resp. $m_1 = 3$), then $m_2 \ge 5$ (resp. $m_2 \ge 4$). So, A_n satisfies one of the following inequalities:

$$A_n \ge \lfloor \frac{n}{4} \rfloor + \lfloor \frac{4n}{5} \rfloor =: A_{n,1},$$
$$A_n \ge \lfloor \frac{n}{3} \rfloor + \lfloor \frac{3n}{4} \rfloor =: A_{n,2},$$
$$A_n \ge \lfloor \frac{3n}{8} \rfloor + \lfloor \frac{2n}{3} \rfloor =: A_{n,3}.$$

Since $A_{3,2} = A_{3,3} = 3$ and $A_{4,1} = A_{4,2} = 4$, we know that $A_n \ge n$ for $n = 3$ or 4.

Suppose that $r = 2$. We may assume that $d_1, d_2 < \frac{1}{2}$. Then $d_i = \frac{1}{2}(1 - \frac{1}{m_i})$ for $i = 1, 2$, where m_i is an integer ≥ 2. Here we note that $s \ge 3$ because $\sum_{i=1}^{s} d_i > 1$ and $d_1, d_2 < \frac{1}{2}$. If $s \ge 4$, then, since $d_1, d_2 \ge \frac{1}{4}$ and $d_i \ge \frac{1}{2}$ for $i \ge 3$, we have

$$A_n \ge 2\lfloor \frac{n}{4} \rfloor + 2\lfloor \frac{n}{2} \rfloor.$$

Hence $A_n \ge n$ if n is even. If $s = 3$, then we have

$$A_n \ge 2\lfloor \frac{n}{4} \rfloor + \lfloor \frac{n}{2} \rfloor.$$

So $A_n \ge n$ if $4|n$. Therefore, we conclude that $\overline{P}_n(V - D) > 0$ for $n = 3$ or 4.

Subcase 2-3: H is irreducible and $\pi|_H : H \to B$ is not separable. (In this subcase, $\mathrm{char}(k) = 2$, $\pi|_H$ is purely inseparable and $\deg \delta = 1 - g(B) = 1$.) For every fiber F of π, $\#(F \cap H) = 1$. Then (2) of Theorem 2.1 (II) implies that $d_i = \frac{1}{2}(1 - \frac{1}{m_i})$, where $m_i \geq 2$ or $m_i = +\infty$, for $i = 1, \ldots, s$. Since $\sum_{i=1}^{s} d_i > 1$, $s \geq 3$. If $s \geq 4$, then

$$A_n \geq \lfloor nd_1 \rfloor + \lfloor nd_2 \rfloor + \lfloor nd_3 \rfloor + \lfloor nd_4 \rfloor \geq 4\lfloor \frac{n}{4} \rfloor.$$

So, $A_n \geq n$ if $4|n$. Suppose that $s = 3$. We may assume that $m_1 \leq m_2 \leq m_3$. Since $\sum_{i=1}^{3} d_i > 1$, we have

$$1 > \frac{1}{m_1} + \frac{1}{m_2} + \frac{1}{m_3}.$$

So, if $m_1 = 2$ then $m_2 \geq 3$. Moreover, if $m_1 = 2$ and $m_2 = 3$ (resp. $m_1 = 2$ and $m_2 = 4$, $m_1 = m_2 = 3$), then $m_3 \geq 7$ (resp. $m_3 \geq 5$, $m_3 \geq 4$). So, A_n satisfies one of the following inequalities:

$$A_n \geq \lfloor \frac{n}{4} \rfloor + \lfloor \frac{n}{3} \rfloor + \lfloor \frac{3n}{7} \rfloor =: A_{n,1},$$
$$A_n \geq \lfloor \frac{n}{4} \rfloor + \lfloor \frac{3n}{8} \rfloor + \lfloor \frac{2n}{5} \rfloor =: A_{n,2},$$
$$A_n \geq 2\lfloor \frac{n}{3} \rfloor + \lfloor \frac{3n}{8} \rfloor =: A_{n,3}.$$

Since $A_{12,1} = 12$, $A_{8,2} = 8$ and $A_{6,3} = 6$, we have $A_n \geq n$ for $n = 6, 8$ or 12. Therefore, we conclude that $\overline{P}_n(V - D) > 0$ for $n = 6, 8$ or 12. In particular, $\overline{P}_{24}(V - D) > 0$. □

Now we prove Theorem 0.1.

Proof of Theorem 0.1. Let $\varphi : \tilde{S} \to S$ be a good resolution (i.e., $\varphi^{-1}(\operatorname{Sing} S)$ is an SNC-divisor) and Δ the reduced exceptional divisor with respect to φ. Let (X, B) be an SNC-pair such that $X - B \cong \tilde{S}$. The pair $(X, B + \Delta)$ is then an SNC-pair with $\overline{\kappa}(X - (B + \Delta)) = \overline{\kappa}(S - \operatorname{Sing} S) = 1$. Let (V, D) be a strongly minimal model of $(X, B + \Delta)$ (cf. Section 1). There exists a birational morphism $\mu : X \to V$ such that $\mu_*(B + \Delta) = D$. Lemma 1.6 (1) implies that $\overline{P}_n(V - D) = \overline{P}_n(S - \operatorname{Sing} S)$ for every positive integer n. Since S is affine, $V - D$ contains no complete curves. Hence (V, D) satisfies the conditions in Theorem 2.1 (II). The assertions (1) and (2) then follow from Lemmas 3.1–3.3.

We prove the assertion (3). We use the same notations as in Theorem 2.1 (II). If $g(B) \geq 2$, then $\overline{P}_2(S) > 0$ by Lemma 3.1. Suppose next that $g(B) = 1$. As seen from the proof of Lemma 3.2, we know that $\overline{P}_2(S) > 0$ except for the case where H is irreducible and $\pi|_H : H \to B$ is a purely inseparable morphism of degree two. We consider this case. Since D is connected because S is smooth and affine, it follows from (2) of Remark

2.6 that $d_i \geq \frac{1}{2}$ for $i = 1, \dots, s$. So, as seen from Cases 1 and 3 of the proof of Lemma 3.2, we know that $\overline{P}_2(S) > 0$.

Suppose finally that $g(B) = 0$. We use the same notations as in the proof of Lemma 3.3. If $\deg \delta \geq 2$, then $\overline{P}_1(S) > 0$ by the proof of Lemma 3.3. Suppose that $\deg \delta \leq 0$. Then $\sum_{i=1}^{s} d_i > 2$, $H = H_1 + H_2$ with two sections H_1 and H_2 and $H_1 \cap H_2 = \emptyset$ by Case 1 of the proof of Lemma 3.3. So $d_i = 1 - \frac{1}{m_i}$, where $m_i \geq 2$ or $m_i = +\infty$. Since D is connected, we infer from (1) of Remark 2.6 that $d_i = 1$ for some i, $1 \leq i \leq s$. We may assume that $d_1 = 1$. Since $s \geq 3$ and $d_i \geq \frac{1}{2}$ for $i = 2, \dots, s$, we have

$$\overline{P}_2(S) \geq -2 \times 2 + \sum_{i=1}^{s} \lfloor 2d_i \rfloor + 1 = -1 + \sum_{i=2}^{s} \lfloor 2d_i \rfloor \geq 1.$$

Suppose that $\deg \delta = 1$. Then $\sum_{i=1}^{s} d_i > 1$ and so $s \geq 2$ by the proof of Lemma 3.3. Since D is connected, we infer from Remark 2.6 that $d_i \geq \frac{1}{2}$ for $i = 1, \dots, s$. So we have

$$\overline{P}_2(S) \geq -2 + \sum_{i=1}^{s} \lfloor 2d_i \rfloor + 1 = -1 + s \geq 1. \qquad \square$$

Acknowledgements

The author would like to express his gratitude to the referees for pointing out errors and for giving useful comments and suggestions which improved the paper. The author is supported by Grant-in-Aid for Scientific Research (No. 23740008) from JSPS.

References

1. E. Bombieri and D. Mumford, Enriques' classification of surfaces in char. p, II, Complex analysis and algebraic geometry, pp. 23–42, Iwanami Shoten, Tokyo, 1977.
2. T. Fujita, Fractionally logarithmic canonical rings of algebraic surfaces, J. Fac. Sci. Univ. Tokyo **30** (1984), 685–696.
3. R. Hartshorne, Algebraic Geometry, Graduate Texts in Mathematics 52, Springer-Verlag, New York, Heiderberg, Berlin, 1977.
4. S. Iitaka, On logarithmic Kodaira dimension of algebraic varieties, Complex analysis and algebraic geometry, pp. 175–189, Iwanami Shoten, Tokyo, 1977.
5. S. Iitaka, Algebraic Geometry, Graduate Texts in Mathematics 76, Springer-Verlag, New York, Heiderberg, Berlin, 1981.
6. T. Kambayashi, On Fujita's strong cancellation theorem for the affine plane, J. Fac. Sci. Univ. Tokyo **27** (1980), 535–548.
7. Y. Kawamata, On the classification of non-complete algebraic surfaces, Algebraic geometry (Proc. Summer Meeting, Univ. Copenhagen, Copenhagen,

1978), pp. 215–232, Lecture Notes in Mathematics 732, Springer, Berlin, 1979.
8. H. Kojima, Open surfaces of logarithmic Kodaira dimension zero in arbitrary characteristic, J. Math. Soc. Japan **53** (2001), 933–955.
9. Y. Kuramoto, On the logarithmic plurigenera of algebraic surfaces, Compos. Math. **43** (1981), 343–364.
10. M. Miyanishi, Non-complete algebraic surfaces, Lecture Notes in Mathematics 857, Springer-Verlag, Berlin-New York, 1981.
11. M. Miyanishi, On affine-ruled irrational surfaces, Invent. Math. **70** (1982), 27–43.
12. M. Miyanishi, Open algebraic surfaces, CRM Monograph Series 12, American Mathematical Society, Providence, RI, 2001.
13. M. Miyanishi and S. Tsunoda, Non-complete algebraic surfaces with logarithmic Kodaira dimension $-\infty$ and with non-connected boundaries at infinity, Japan. J. Math. **10** (1984), 195–242.
14. P. Russell, On affine-ruled rational surfaces, Math. Ann. **255** (1981), 287–302.
15. S. Tsunoda, Structure of open algebraic surfaces, I, J. Math. Kyoto Univ. **23** (1983), 95–125.

Some properties of $\mathbb{C}^*$ in $\mathbb{C}^2$

M. Koras

Institute of Mathematics, University of Warsaw, Warsaw, 02-097, Poland
E-mail: koras@mimuw.edu.pl

P. Russell

Department of Mathematics, McGill University, Montreal, H3A-0B9, Canada
E-mail: russell@math.mcgill.ca

We consider closed plane curves isomorphic to $\mathbb{C}^*$. We prove that with one exception the branches at infinity can be separated by an automorphism of $\mathbb{C}^2$. We also give a bound for the selfintersection number of the resolution curve.

Keywords: Embedding of $\mathbb{C}^*$, Cremona transformation, Kodaira dimension.

0. Introduction

0.1. Let U be a closed algebraic curve in $\mathbb{C}^2$ isomorphic to $\mathbb{C}^*$. Let $\overline{U}$ be the closure of U in $\mathbb{P}^2$. By L_∞ we denote the line at infinity in $\mathbb{P}^2$. Let $\Phi\colon \overline{S}' \to \mathbb{P}^2$ be the resolution of $\overline{U}$. By this we mean that Φ^{-1} is the minimal sequence of blow ups such that the reduced inverse image of the divisor $\overline{U} + L_\infty$ is an SNC-divisor. Let E' be the proper transform of $\overline{U}$ and let $D' = \Phi^{-1}(L_\infty)_{red}$. Let L'_∞ denote the proper transform of the line L_∞ in $\overline{S}'$. Let $\Psi\colon \overline{S}' \to \overline{S}$ be the *NC-minimalization of the divisor* D' *with respect to* E', i.e., Ψ is the successive contraction of possibly L'_∞ and then more D'-components such that $\Psi(D' + E')$ is an SNC-divisor and each (-1)-component of $\Psi(D')$ is a branching component of $\Psi(D'+E')$. We put $D = \Psi(D')$, $E = \Psi(E')$. Let $S = \overline{S} \setminus D$. Of course $S \simeq \mathbb{C}^2$. We note that $E' \cdot D' = E \cdot D = 2$. Since D has connected support, $D + E$ is not a chain.

Embeddings of $\mathbb{C}^*$ into $\mathbb{C}^2$ can be divided into two classes. The first class consists of embeddings which admit *a good asymptote*, see 0.2, the second class consists of those without any good asymptote. The embeddings from the first class are completely classified in [1].

Definition 0.2. We say that a rational curve $L \in \mathbb{P}^2$ is a *good asymptote* of U if $L \cap \mathbb{C}^2 \simeq \mathbb{C}^1$, and L meets $\overline{U}$ at most once at finite distance, i.e., $L \cdot U \leq 1$.

Notice that this definition differs slightly from the definition in [1], but the two definitions are equivalent up to an isomorphism of $\mathbb{C}^2$.

The main results of this article are Corollary 2.5 and Theorem 4.16. Corollary 2.5 gives a bound for E^2 and for $(K_{\overline{S}} + D + E)^2$ in the case where U does not admit a good asymptote. Theorem 4.16 says that in the case of no good asymptote the branches of $\overline{U}$ at infinity can be separated by an automorphism of $\mathbb{C}^2$. It follows from the classification given in [1] that with one exception this is also true in case where U admits a good asymptote. Hence throughout the paper we assume that U does not have a good asymptote.

Another remarkable property is proved in [8].

Theorem 0.3. $\overline{\kappa}(\overline{S} \setminus E) = -\infty$.

Theorem 0.3 and a theorem of Coolidge imply that U can be transformed into a line in $\mathbb{C}^2$ by a birational automorphism of $\mathbb{C}^2$, see [11].

The results proved in the paper are crucial for completing a classification of embeddings of $\mathbb{C}^*$ into a plane $\mathbb{C}^2$. This will be done in a forthcoming article.

1. Preliminaries

In the article we use several notions and results from the theory of open algebraic surfaces. We refer the reader to [13] for any undefined terms here. We will also use some results from T. Fujita's paper [2], particularly §3.

1.1. Let $\overline{M}$ be a complete, non-singular surface and $T = \sum_{i=1}^{n} m_i T_i$ a divisor on $\overline{M}$ with $T_1, \ldots, T_n$ distinct, irreducible curves.

(i) We write $\sim$ for linear equivalence of integral divisors. We write $\equiv$ for numerical equivalence of divisors, both over $\mathbb{Z}$ and over $\mathbb{Q}$.

(ii) We call T a simple normal crossing divisor (an SNC-divisor) if T is reduced, all its components are smooth and at most two of them meet at any point, and if so, transversally.

(iii) A (b)-curve on $\overline{M}$ is a curve $L \simeq \mathbb{P}^1$ with $L^2 = b$.

(iv) An SNC-divisor T is NC-minimal if every (-1)-component of T is a branching component.

(v) We call $Q(T) = (T_i \cdot T_j)_{1 \leq i,j \leq n}$ the intersection matrix of a reduced T and put $d(T) = det(-Q(T))$. We put $d(T) = 1$ if $\text{Supp}(T) = \emptyset$. For elementary results used in the computation of $d(Q)$ we refer to [2, §3], [10, 2.2.1] and [13, §2, section 3]

(vi) A divisor R is called *contractible* if it is the minimal resolution divisor of a quotient singular point. Hence R is a chain composed of smooth rational curves R_i such that $R_i^2 \leq -2$ or R is a fork of smooth rational curves with branches of type $(2, 2, n), (2, 3, 3), (2, 3, 4)$ or $(2, 3, 5)$ and the branching component of self-intersection number ≤ -2.

1.2. For the definition of *twig, tip, bark* of a divisor and their properties we refer to [2, §3] and [13, §2, section 3]. We recall only the definition of a *capacity* of a rational chain. Let

$$R = R_1 + \cdots + R_s$$

be a chain of smooth rational curves with dual graph

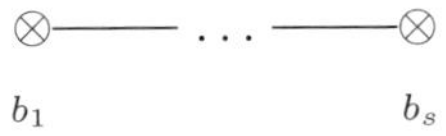

Suppose that R is admissible, i.e., that $b_i = R_i^2 \leq -2$, $i = 1, \cdots, s$. We recall that $Q(R)$ is negative definite and $d(R) \geq 2$. We put $e(R) = \frac{d(R_2 + \cdots + R_s)}{d(R)}$.

1.2.1 *If* $R_1^2 = -k$, *then* $e(R) \geq \frac{1}{k}$.

We recall

1.3. *Let T be a connected NC-minimal divisor consisting of smooth rational curves. Assume T is not a contractible divisor and let $T_1, \ldots, T_s$ be all the maximal twigs of T. Then* $\text{Bk}(T)^2 = -\sum e(T_i)$.

Lemma 1.4. *There is no curve $C \subset \overline{S}$ such that $C \cap S \simeq \mathbb{C}^1$ and $C \cdot E \leq 1$.*

Proof. Let L be the proper transform of C in $\mathbb{P}^2$. Then L is a good asymptote of $\overline{U}$; a contradiction. □

Corollary 1.5. *If $E^2 \neq -1$, then the pair $(\overline{S}, D+E)$ is relatively minimal (see [13, §2, section 3]) i.e. $K_{\overline{S}} + D + E \equiv P + Bk(D+E)$ is the Zariski decomposition, where $P = (K_{\overline{S}} + D + E)^+$.*

Lemma 1.6.

(i) $\overline{\kappa}(\overline{S} \setminus (D+E)) \geq 0$.
(ii) $(K_{\overline{S}} + D + E)^2 < 3$.

Proof. (i) If $\overline{\kappa}(\overline{S} \setminus (D+E)) = -\infty$, then $|K_{\overline{S}} + D + E| = \emptyset$, which implies $E \cdot D \leq 1$, but $E \cdot D = 2$. (ii) By [14], $(K_{\overline{S}} + D + E)^2 \leq 3(\chi(S \setminus E) + \frac{1}{12}N^2)$, where $N = (K_{\overline{S}} + D + E)^-$ is the negative part in the Zariski decomposition of the divisor $K_{\overline{S}} + D + E$. If $N \neq 0$ we are done since $\chi(S \setminus E) = 1$. Suppose that $N = 0$. It follows that the divisor $D+E$ has no twigs (see [2, §3], [13, §2, section 3]). Hence D is a chain and E meets the tips of D. If D has only one component, then $D^2 = 1$. Hence in all cases no component of D is a (-1)-curve since D is NC-minimal w.r.t E, see 0.1. Clearly D is not an admissible chain, see 1.2. Therefore there exists a component D_1 of D such that $D_1^2 \geq 0$. By some blowing up and down within D we can transform D into a chain Δ with a tip Δ_1 such that $\Delta_1^2 = 0$ and $E \cdot \Delta_1 = 1$. The linear system $|\Delta_1|$ induces a $\mathbb{C}$-ruling of S with E as a 1-section. The proper transform in $\mathbb{P}^2$ of a general member of the system is a good asymptote of U, contrary to our assumption. □

1.7. Write

$$(K_{\overline{S}} + D + E)^2 = 2 - \varepsilon$$

where $\varepsilon \geq 0$. We have $K_{\overline{S}} \cdot (K_{\overline{S}} + D + E) = 2 - \varepsilon$. We put $\gamma = -E^2$.

Lemma 1.8. $\gamma > 0$.

Proof. Suppose that $\gamma \leq 0$ After blowing up over one of the points in $E \cap D$ we may assume that $E^2 = 0$. Therefore U is a fiber of a $\mathbb{C}^*$-ruling of $\mathbb{C}^2$. There is a singular fiber with an irreducible component isomorphic to $\mathbb{C}$. This is a good asymptote of U and we reach a contradiction. □

1.9. Let $R_1, \cdots, R_s$ be all maximal twigs of $D+E$. Let $e_i = e(R_i)$.

Lemma 1.10. $-(Bk(D+E))^2 = \sum e_i \leq 1 + \varepsilon$.

Proof. Suppose that $\gamma \neq 1$. Then, by 1.5, the pair $(\overline{S}, D+E)$ is minimal. Let $K_{\overline{S}} + D + E = P + \mathrm{Bk}(D+E)$ be the Zariski decomposition. By Langer's version [12] of the Kobayashi inequality, see 1.13, $0 \leq P^2 \leq 3\chi(S \setminus E) = 3$. We have $P^2 = (K_{\overline{S}} + D + E)^2 - (\mathrm{Bk}(D+E))^2 = 2 - \varepsilon + \sum e_i$, so we are done. Suppose that $E^2 = -1$. We pass to a minimal model of the pair $(\overline{S}, D+E)$. In view of 1.4, we possibly contract E and further components of $D+E$, but we do not touch any of the maximal twigs of $D+E$. To the resulting divisor we apply the Kobayashi inequality and get the result. □

Lemma 1.11. $\varepsilon \leq 3$.

Proof. We claim that $\overline{S}$ is not isomorphic to a relatively minimal rational surface. By 1.8, $E^2 < 0$. Hence $\overline{S}$ is not isomorphic to $\mathbb{P}^2$, and if it is isomorphic to a Hirzebruch surface, then E is the only negative curve in $\overline{S}$. Also $D = D_1 + D_2$ has two irreducible components since the irreducible components of D generate $\mathrm{Pic}(\overline{S})$ freely. Now $D_1^2 \geq 0, D_2^2 \geq 0$, and since $E \cdot (D_1 + D_2) = 2$ we may assume $E \cdot D_1 \leq 1$, say. After some blowing up we may assume that $D_1^2 = 0$ and then the proper transform of a general member of the system $|D_1|$ is a good asymptote of U, in contradiction to our assumption.
Suppose that $\varepsilon \geq 4$. We have $(K_{\overline{S}} + D) \cdot (K_{\overline{S}} + E) = 2 - \varepsilon - K_{\overline{S}} \cdot E + (K_{\overline{S}} + D) \cdot E = 4 - \varepsilon \leq 0$. Since $E \cdot D = 2$, $|E + K_{\overline{S}} + D| \neq \emptyset$. By Fujita's Theorem [3] there exists m such that $|E + m(K_{\overline{S}} + D)| \neq \emptyset$, but $|E + (m+1)(K_{\overline{S}} + D)| = \emptyset$. We write

$$E + m(K_{\overline{S}} + D) = \sum A_i$$

with each A_i reduced and irreducible. We have $|A_i + D + K_{\overline{S}}| = \emptyset$ for every i. By a standard argument, [16, 2.1, 2.2] for example, A_i is a smooth rational curve and $A_i \cdot D \leq 1$. This implies $A_i \neq E$. Since $\overline{S}$ is not a relatively minimal surface we may assume that $A_i^2 < 0$ for every i. (If $A_i^2 \geq 0$ we replace A_i by a suitable singular member of the linear system $|A_i|$). We obtain $-2 \geq E \cdot (K_{\overline{S}} + E) + m(K_{\overline{S}} + D) \cdot (K_{\overline{S}} + E) = \sum A_i \cdot (K_{\overline{S}} + E)$. Hence there exists A_1 such that $A_1 \cdot (K_{\overline{S}} + E) < 0$. It follows that $A_1 \cdot K_{\overline{S}} < 0$. Hence $A_1^2 = -1$ and $A_1 \cdot E = 0$. Since $A_1 \cdot D \leq 1$, A_1 is not a branching component of D. Since $A_1 \cdot E = 0$ it follows from the NC-minimality of D w.r.t. E (see 0.1) that A_1 is not a component of D. Now the proper transform of A_1 in $\mathbb{P}^2$ is a good asymptote of U, a contradiction. □

Lemma 1.12. *Let $\overline{M}$ be a smooth projective surface. Let r be the rank the Neron-Severi group $NS(\overline{M})$. Then for any set $C_1, \ldots, C_r$ of distinct irreducible curves in $\overline{M}$ the matrix $[C_i \cdot C_j]$ is not negative definite.*

Proof. Suppose it is. Then in particular $C_1, \ldots, C_r$ are independent in $NS(\overline{M})\otimes\mathbb{Q}$ and hence form a basis of $NS(\overline{M})\otimes\mathbb{Q}$. We reach a contradiction with the Hodge Index Theorem. □

We will use an inequality of Bogomolov-Miyaoka-Yau type (simply the BMY-inequality) proved by R. Kobayashi, S. Nakamura and F. Sakai, [4, Lemma 8 and Corollary 9] (see also [14, Chapter 2, Theorem 6.6.2]). We state it as follows.

Lemma 1.13. *Let $\overline{X}$ be a smooth projective surface and let D be an SNC-divisor on $\overline{X}$. Let $D_1, \ldots, D_k$ be the connected components of D which are contractible divisors, see 1.1(vi). Let $G_i, 1 \leq i \leq k$, be the local fundamental group at the singular point obtained by the contraction of G_i to point. Assume that the pair (X, D) is almost minimal. Suppose that $\overline{\kappa}(X \setminus D) \geq 0$. Then*

$$((K_{\overline{X}} + D)^+)^2 \leq 3(\chi(X \setminus D) + \sum_{i=1}^{k} \frac{1}{|G_i|}).$$

The original BMY-inequality was proved in case $\overline{\kappa}(X\setminus D) = 2$. A. Langer [12] has extended it to the case $\overline{\kappa}(X \setminus D) = 0, 1$.

2. Basic inequality

2.1 Let $\psi\colon \overline{S} \to \overline{N}$ be a 2-reduction of the divisor D with respect to E, i.e., ψ is a sequence of successive contractions of (-1)-curves in D meeting E (and its successive images) once and such that the divisors $T = \psi(D)$ and $E_0 = \psi(E)$ satisfy the following:

(i) T is an NC-divisor.

(ii) for any (-1)-component T_i of T, $T_i \cdot E_0 \geq 2$ or T_i is a branching component of T.

Note that only curves meeting E once are contracted. In particular, E_0 is smooth and hence $E_0 \simeq \mathbb{P}^1$. A good way to think about the 2-reduction is that we stop the resolution process Φ of section 0 as soon as the proper transform of $\overline{U}$ has become non-singular. This, however, is not entirely correct if the NC-minimalization Ψ is non-trivial.

2.2. Let t denote the number of sprouting contractions in ψ. A subdivisional blowing down does not change the quantities $K \cdot (K + D)$ and

$E \cdot (K + D)$. Under a sprouting blowing down $K \cdot (K + D)$ increases by 1 and $E \cdot (K + D)$ decreases by 1. Here, by abuse of notation, K denotes the canonical divisor of the image of $\overline{S}$ at some stage of the contraction process ψ, and the images of E and D are denoted by the same letters. Hence

(a) $(E_0 + K_{\overline{N}}) \cdot (K_{\overline{N}} + T) = (E + K_{\overline{S}}) \cdot (K_{\overline{S}} + D)$.

(b) $(E_0 + 2K_{\overline{N}}) \cdot (K_{\overline{N}} + T) = (E + 2K_{\overline{S}}) \cdot (K_{\overline{S}} + D) + t = 6 - 2\varepsilon - K_{\overline{S}} \cdot E + t = 8 - 2\varepsilon - \gamma + t$.

We note the following for future reference.

2.2.1 A contribution (of 1) to t arises when there is a (-1)-curve in D that is non-branching in D, meets E once and has attached to it a maximal twig T of $D + E$ consisting of (-2)-curves. Note that if τ is the number of (-2)-components of T, then T contributes $\frac{\tau}{\tau+1}$ to $\sum e_i$ in 1.10.

Since $\overline{U}$ has two branches at infinity, E meets at most two components of D and hence

2.2.2 $t \leq 2$.

In the language 3.1, if the last characteristic pair of a branch is of the form $\binom{\tau}{1}$, it will in general produce a maximal twig as above and a contribution to t. For the precise analysis, however, the interaction of the branches and the NC-reduction will have to be considered.

Proposition 2.3. *Suppose that $(E_0 + 2K_{\overline{N}})(K_{\overline{N}} + T) \leq 0$ and that $\overline{N}$ is not a Hirzebruch surface or $\mathbb{P}^2$. Then there exists a (−1)-curve A in $\overline{N}$ such that $A \cdot E_0 \leq 1$.*

Proof. Suppose that such a curve does not exist. Let C_1, C_2 be the components of D which meet E. It may happen that $C_1 = C_2$.

Sub-Lemma 2.3.1. *There is no curve B in $\overline{N}$ such that $(E_0 + 2K_{\overline{N}}) \cdot B < 0$.*

Proof. Suppose B exists. Suppose first that $|B + K_{\overline{N}} + T| \neq \emptyset$. Let $F_m = B + m(K_{\overline{N}} + T)$. Arguing as in the proof of Lemma 1.11, we find m such that $|F_m| \neq \emptyset$ and $|F_{m+1}| = \emptyset$ and we have $B + m(K_{\overline{N}} + T) = \sum B_i$. Then $|B_i + K_{\overline{N}} + T| = \emptyset$ for every i. By the assumption in Proposition 2.3 we have $0 > B \cdot (E_0 + 2K_{\overline{N}}) \geq \sum B_i \cdot (E_0 + 2K_{\overline{N}})$. Hence there exists B_i such that $B_i \cdot (E_0 + 2K_{\overline{N}}) < 0$.

Free to replace B by B_i, we may assume that $|B + K_{\overline{N}} + T| = \emptyset$. Then B is a smooth rational curve and $B \cdot T \leq 1$. In particular $B \neq E_0$ since $E_0 \cdot T \geq 2$. So $B \cdot E_0 \geq 0$ and $K_{\overline{N}} \cdot B < 0$, i.e., $B^2 \geq -1$. Suppose that $B^2 \geq 0$. Since $\overline{N}$

is not a minimal rational surface there exists a singular member $\sum B_j$ of $|B|$ such that $B_j^2 < 0$ for every j. There exists B_j such that $B_j \cdot (E_0 + 2K_{\overline{N}}) < 0$. It follows that $B_j \cdot K_{\overline{N}} < 0$ hence $B_j^2 = -1$, and of course $|B_j + K_{\overline{N}} + T| = \emptyset$. We may replace B by B_j. Then $K_{\overline{N}} \cdot B = -1$, which implies $B \cdot E_0 \leq 1$, and B gives a good asymptote for U, a contradiction. The sub-lemma is proved. □

By Theorem 0.3 we have $\kappa(K_{\overline{N}} + E_0) = -\infty$. We argue as in [11, theorem 2.1].

(i) Suppose that $K_{\overline{N}} \cdot (K_{\overline{N}} + E_0) \leq 0$. Let L be a (-1)-curve in $\overline{N}$. Since $L \cdot E_0 \geq 2$, $|L + K_{\overline{N}} + E_0| \neq \emptyset$. As above we find $m \geq 1$ such that we have

$$F = L + m(K_{\overline{N}} + E_0) = \sum A_i$$

with, for each i, $A_i \simeq \mathbb{P}^1$, $A_i \cdot E_0 \leq 1$. If $A_i^2 \geq 0$ then, since $\overline{N}$ is not a minimal rational surface, we may replace A_i by a singular member of $|A_i|$ supported on negative curves. Since $F \cdot K_{\overline{N}} < 0$, there exists A_j such that $A_j \cdot K_{\overline{N}} < 0$. Hence we may assume that $A_i^2 < 0$ for every i. Hence $A_j^2 = -1$, so $A_j \cdot (E_0 + 2K_{\overline{N}}) < 0$, and we get contradiction with Lemma 2.3.1.

(ii) Suppose that $K_{\overline{N}} \cdot (K_{\overline{N}} + E_0) \geq 1$. Then $-K_{\overline{N}} - E_0 \geq 0$ by the Riemann-Roch theorem and, in fact, $-K_{\overline{N}} - E_0 > 0$ since $E_0 \cdot (-K_{\overline{N}} - E_0) = 2$. Let again L be a (-1)-curve in $\overline{N}$. Write $L = L + K_{\overline{N}} + E_0 + (-K_{\overline{N}} - E_0)$. Since $h^0(L) = 1$ and $L + K_{\overline{N}} + E_0 \geq 0$, $L + K_{\overline{N}} + E_0 = 0$. There exists a component $T_i = \psi(D_i)$ of T such that $T_i \cdot L > 0$. Then $T_i \cdot (K_{\overline{N}} + E_0) = T_i \cdot (-L) < 0$. It follows that $K_{\overline{N}} \cdot T_i < 0$ We obtain $T_i \cdot (E_0 + 2K_{\overline{N}}) = T_i \cdot (E_0 + K_{\overline{N}}) + T_i \cdot K_{\overline{N}} < 0$ in contradiction to lemma 2.3.1.

Proposition 2.4. $(E_0 + 2K_{\overline{N}})(K_{\overline{N}} + T) > 0$

Proof. We keep notation of 2.1 and 2.2. We have $\kappa(K_{\overline{N}} + E_0) = -\infty$. Suppose that $(E_0 + 2K_{\overline{N}}) \cdot (K_{\overline{N}} + T) \leq 0$. Suppose first that $\overline{N}$ is not isomorphic to a Hirzebruch surface or $\mathbb{P}^2$. Let A be a curve as in Proposition 2.3. Suppose that $|A + K_{\overline{N}} + T| \neq \emptyset$. We again find m such that $|A + m(K_{\overline{N}} + T)| \neq \emptyset$ and $|A + n(K_{\overline{N}} + T)| = \emptyset$ for $n > m$ and we write

$$F = A + m(K_{\overline{N}} + T) = \sum A_i$$

with, for each i, $A_i \simeq \mathbb{P}^1$, $A_i \cdot T \leq 1$. As above we may assume that $A_i^2 < 0$. We have $0 > (E_0 + 2K_{\overline{N}}) \cdot (A + m(K_{\overline{N}} + T)) = \sum (E_0 + 2K_{\overline{N}}) \cdot A_i$. Thus there exists A_j such that $A_j \cdot (E_0 + 2K_{\overline{N}}) < 0$. Since $A_j \cdot T \leq 1$, $A_j \neq E_0$. Thus

$A_j \cdot K_{\overline{N}} < 0$. Since $A_j^2 < 0$ we obtain $A_j^2 = -1$. It follows that $E_0 \cdot A_j \leq 1$. Therefore we may assume that $|A + K_{\overline{N}} + T| = \emptyset$. Then $A \cdot T \leq 1$, which implies that A is not a branching component of T. Also $A \cdot E_0 \leq 1$. By the properties of T it follows that A is not a component of $T + E_0$. The proper transform of A in $\mathbb{P}^2$ is a good asymptote of U and we reach a contradiction.

We have already seen that $\overline{N}$ cannot be isomorphic to $\mathbb{P}^2$. Suppose then that $\overline{N}$ is isomorphic to a Hirzebruch surface. Since the irreducible components of T generate $\mathrm{Pic}(\overline{N})$ freely, T has exactly two components. Write $T = T_1 + T_2$. Computing the determinant of T we get $-1 = T_1^2 T_2^2 - 1$. We may assume therefore that $T_1^2 = 0$. Let $T_2^2 = -n$. Let $a = T_1 \cdot E_0, b = T_2 \cdot E_0$. Then

$$E_0 \sim (an + b)T_1 + aT_2.$$

Now $p_a(E_0) = 0$ implies $-2 = (an + b)(2a - 2) + a(n - 2 - an)$ and consequently $(a - 1)(an + 2b - 2) = 0$. Thus

(i) $a \leq 1$

or

(ii) $a \geq 2$ and $2 = an + 2b$.

In case (i) the proper transform in $\mathbb{P}^2$ of a general member of the system $|T_1|$ is a good asymptote of U, so (i) cannot occur.
Consider (ii). We have $E_0^2 = a^2 n + 2ab = a(an + 2b) = 2a$. Suppose that $b = 0$. Then $2 = an$, so $a = 2$ and $n = 1$. But then $T_2^2 = -1$ and $T = T_1 + T_2$ is not 2-reduced w.r.t. E. Hence $b > 0$. Suppose that $b = 1$. Then $an = 0$, hence $n = 0$, but then the proper transform in $\mathbb{P}^2$ of a general member of the system $|T_2|$ is a good asymptote of U. Thus $b \geq 2$. Let $q = T_1 \cap T_2$.

Suppose that $q \notin E_0$. Then E_0 intersects T_1 and T_2 in points p_1 and p_2 respectively. The inverse of ψ involves blowing up over p_1 a times and blowing up over p_2 b times. We have $t = 2$ (i.e, ψ involves two sprouting contractions w.r.t D) and $K_{\overline{N}} \cdot (K_{\overline{N}} + T) = 2 + K_{\overline{S}} \cdot (K_{\overline{S}} + D) = 4 - \varepsilon - K_{\overline{S}} \cdot E$ by 2.2 (b). ψ involves $a+b$ contractions on E, hence $K_{\overline{N}} \cdot E_0 = K_{\overline{S}} \cdot E - (a+b)$. We obtain that $K_{\overline{N}} \cdot (K_{\overline{N}} + T) = 4 - \varepsilon - K_{\overline{N}} \cdot E_0 - a - b = 6 - \varepsilon + a - b$. On the other hand $K_{\overline{N}} \cdot (K_{\overline{N}} + T) = 8 + K_{\overline{N}} \cdot T_1 + K_{\overline{N}} \cdot T_2 = 4 + n$. Hence

$$(*) \qquad n = 2 - \varepsilon + a - b.$$

By our assumption

$$(**) \qquad (E_0 + 2K_{\overline{N}}) \cdot (K_{\overline{N}} + T) = 6 - a + b + 2n = 8 - \varepsilon + n \leq 0.$$

From (ii), since $b = 2 - \varepsilon + a - n$, we get

$$(***) \qquad (a-2)(n+2) = 2\varepsilon - 6.$$

($\star$1) By 1.11, $0 \leq \varepsilon \leq 3$.

Suppose that $\varepsilon = 0$. $D+E$ has two (-2)-twigs (maximal twigs with each component a (-2) curve) with determinants a and b. By 1.10, $a = b = 2$. From $(***)$ we get $0 = 2\varepsilon - 6$ i.e. $\varepsilon = 3$; a contradiction.

The following four results follow formally from $(*)$, $(**)$ and $(***)$ (without reference to q).

($\star$2) $n + 2 < 0$ if $\varepsilon \leq 2$.

($\star$3) Suppose that $\varepsilon = 1$. We have $n + 2 = -1$ or -2 or -4, so $n = -3$ or -4 or -6. But $(**)$ gives $n \leq -7$.

($\star$4) Suppose that $\varepsilon = 2$. From $(***)$ we get $n + 2 = -1$ or -2. So $n = -3$ or -4. But $(**)$ gives $n \leq -6$.

($\star$5) Suppose that $\varepsilon = 3$. From $(**)$ we obtain that $n \leq -5$. From $(***)$, $(a-2)(n+2) = 0$. It follows that $a = 2$.

Let T_3 be the member of the system $|T_1|$ passing through p_2. Since E_0 is tangent to T_2 at p_2, E_0 is transversal to T_3 at p_2. Hence E_0 meets T_3 transversally at a point $p_3 \neq p_2$. After the first blowing up E_0 meets the proper transform of T_2. It follows that the proper transform of T_3 in $\mathbb{P}^2$ is a good asymptote of U.

Now assume that $q \in E_0$.

Assume that E_0 meets T_2 also in a point $p_2 \neq q$. Then $E_0 \cap T_1 = \{q\}$. Hence E_0 is tangent to T_1 at q. It follows that E_0 is transversal to T_2 at q. Hence the local intersection of E_0 with T_2 at p_2 equals $b - 1$. Suppose that $b = 2$. Then $na = -2$, hence $a = 2$ and $n = -1$, i.e., $T_2^2 = 1$. After the first blowing up at q, E_0 leaves T_2. T_2 at this stage becomes a 0-curve T_2' which meets E_0 once. The proper transform of T_2' in $\mathbb{P}^2$ is a good asymptote of U, a contradiction. So $b \geq 3$. It follows that E_0 is tangent to T_2 at p_2. It follows that ψ involves one sprouting contraction w.r.t. D. On the other hand, ψ now involves $a + b - 1$ contractions on E. Computing $K_{\overline{N}} \cdot (K_{\overline{N}} + T)$ as above we get again have $(*)$, hence also $(**)$, $(***)$ and ($\star$1) to ($\star$5) .

Assume that $\varepsilon = 0$. From $(***)$ we get that $n+2$ divides -6. From $(**)$, $n \leq -8$. It follows that $n = -8$, $a = 3$. It follows further that $b = 13$. The proper transform of T_1 in $\overline{S}$ is a tip of $D + E$ and it is a (-3)-curve. $D + E$ also has a twig consisting of 11 (-2)-curves. The twig is created by blowing up over p_2. Hence $\sum e_i = \frac{1}{3} + \frac{11}{12} > 1$, a contradiction in view of 1.10. The cases $\varepsilon = 1, 2, 3$ we eliminate as above.

Assume that E_0 meets T_1 in a point $p_1 \neq q$. Then $E_0 \cap T_2 = \{q\}$. Since $b \geq 2$, E_0 is tangent to T_2 at q. Hence E_0 is transversal to T_1 at q. Suppose that $a = 2$. Then the proper transform of T_1 in $\overline{S}$ is a (-1)-curve and it meets E once. Thus $D + E$ is not NC-minimal w.r.t. E, a contradiction. Hence $a \geq 3$, i.e. E_0 is tangent to T_1 at p_1. As in the previous case ψ involves one sprouting contraction and $a + b - 1$ contractions on E. Again the $(*)$- and $(\star)$-results hold.
Suppose that $\varepsilon = 0$. As above we get $n = -8$, $a = 3$ and $b = 13$. Also $E_0^2 = 2a = 6$. Let C_1, C_2 be the two (-1)-components of D which meet E. $D - (C_1 + C_2)$ has three connected components, two single curves D_1, D_2 (they are tips of $D + E$) and one chain R. D_1 is the proper transform of the curve produced by the first blowing up over p_1 and D_2 is the proper transform of T_2. R is a chain which has the proper transform of T_1 as a tip. It is a (-3)-curve. The rest of R consists of 12 (-2)-curves. We have $D_1^2 = -2$, $D_2^2 = -5$, $E^2 = -9$.
2.4.1 Let $Q = D_1 + D_2 + R + E$. Consider the surface

$$Y = \overline{S} \setminus Q.$$

We claim that $\overline{\kappa}(Y) \geq 0$. We have $K_{\overline{S}} \cdot (K_{\overline{S}} + Q) = 2 + K_{\overline{S}} \cdot (K_{\overline{S}} + D + E) = 4 - \varepsilon = 4$. Since Q has 4 components that are rational trees we find $(2K_{\overline{S}} + Q) \cdot (K_{\overline{S}} + Q) = -\varepsilon = 0$. By the Riemann-Roch Theorem, $h^0(2K_{\overline{S}} + Q) + h^0(-K_{\overline{S}} - Q) > 0$. Suppose that $\overline{\kappa}(Y) = -\infty$. Suppose $|2K_{\overline{S}} + Q| = \emptyset$. Then $-K_{\overline{S}} - Q > 0$.
2.4.2 In view of 1.6, we have $h^0(-K_{\overline{S}} - D - E) = 0$ or $-K_{\overline{S}} - D - E = 0$. Hence by the Riemann-Roch theorem and 1.7

$$h^0(2K_{\overline{S}} + D + E) \geq 1 + K_{\overline{S}} \cdot (K_{\overline{S}} + D + E) = 3 - \epsilon$$

or

$$K_{\overline{S}} = -D - E \quad \text{and} \quad h^0(2K_{\overline{S}} + D + E) \geq 2 - \epsilon.$$

Hence $2K_{\overline{S}} + D + E \geq 0$. We obtain that $K_{\overline{S}} + C_1 + C_2 = 2K_{\overline{S}} + D + E + (-K_{\overline{S}} - Q) \geq 0$. This gives $K_{\overline{S}} \geq 0$, a contradiction. Hence $2K_{\overline{S}} + Q > 0$ and $\overline{\kappa}(Y) \geq 0$.

2.4.3 We claim that the pair $(\overline{S}, Q)$ is almost minimal. If it is not then there exists a (-1)-curve L such that $L \subset \operatorname{Supp}(K_{\overline{S}} + Q)^-$ and L is not a component of Q. But the intersection matrix of Q is negative definite and all irreducible components of Q are components of $(K_{\overline{S}} + Q)^-$. Since the rank of $\operatorname{Pic}(\overline{S})$ equals the number of irreducible components of Q plus 1 we reach contradiction with Lemma 1.12.

Since $\chi(Y) = -1$, the BMY-inequality (Langer's version, see 1.13) gives

$$\frac{1}{d(D_1)} + \frac{1}{d(D_2)} + \frac{1}{d(R)} + \frac{1}{d(E)} \geq 1.$$

This is a contradiction since $d(D_1) = 2, d(D_2) = 5, d(R) = 27, d(E) = 9$.

The cases $\varepsilon = 1, 2$ we eliminate as above. If $\varepsilon = 3$ we get, as above, that $a = 2$, but we already proved that $a \geq 3$.

Assume that $E_0 \cap T = \{q\}$. Then E_0 is singular, which we have seen is not the case. □

Corollary 2.5. *Let t denote the number of sprouting contractions in ψ, see 2.2. Then*

$$7 + t \geq 2\varepsilon + \gamma.$$

Proof. This follows from 2.2(b) and 2.4. □

3. Separation of branches I: The branches are tangent at infinity

3.1. Let $\lambda, \tilde{\lambda}$ be the branches of $\overline{U}$ at L_∞. The resolution process Φ, see 0, can be described in terms of Hamburger-Noether (HN-) pairs. For the definition in our context and basic properties of HN-pairs we refer to [1, 1.12]; see also [9, Appendix] or [15]. We remark also that to each HN-pair there is tacitly associated an $a \in \mathbb{C}$, a parameter that determines the location of the branch on the last exceptional curve produced by the blowups prescribed by the pair. Let $\binom{c_1}{p_1}, \ldots, \binom{c_h}{p_h}$ (resp. $\binom{\tilde{c}_1}{\tilde{p}_1}, \ldots, \binom{\tilde{c}_{\tilde{h}}}{\tilde{p}_{\tilde{h}}}$) be the sequence of HN-pairs of λ (resp. $\tilde{\lambda}$). We recall that, by definition, $c_1 = \lambda \cdot L_\infty$, $\tilde{c}_1 = \tilde{\lambda} \cdot L_\infty$, $c_{i+1} = GCD(c_i, p_i)$ and $c_i \geq p_i$. Let $\mu_1, \mu_2, \cdots$ (resp. $\tilde{\mu}_1, \tilde{\mu}_2, \cdots$) be the sequence of multiplicities of all singular points of λ infinitely near $\lambda \cap L_\infty$ (resp. of $\tilde{\lambda}$ infinitely near $\tilde{\lambda} \cap L_\infty$).

3.1.1 Then

(i) $$\sum_{i\geq 1} \mu_i = c_1 + p_1 + p_2 + \cdots + p_h - 1.$$

(ii) $$\sum_{i\geq 1} \mu_i^2 = c_1 p_1 + c_2 p_2 + \cdots + c_h p_h.$$

(iii) $$\sum_{i\geq 1} \tilde{\mu}_i = \tilde{c}_1 + \tilde{p}_1 + \tilde{p}_2 + \cdots + \tilde{p}_{\tilde{h}} - 1.$$

(iv) $$\sum_{i\geq 1} \tilde{\mu}_i^2 = \tilde{c}_1 \tilde{p}_1 + \tilde{c}_2 \tilde{p}_2 + \cdots + \tilde{c}_{\tilde{h}} \tilde{p}_{\tilde{h}}.$$

Throughout this section we assume that $\lambda \cap L_\infty = \tilde{\lambda} \cap L_\infty = q$ and that the branches cannot be separated by an automorphism of $\mathbb{C}^2$. At the end we will come to a contradiction. We will also assume that the resolution tree D' has the smallest possible number of irreducible components, i.e., if $\sigma\colon \mathbb{C}^2 \to \mathbb{C}^2$ is an automorphism, then the number of components of the resolution tree of $\overline{\sigma(U)}$ is not less than the number of components of D'.

3.1.2 Let s denote the number of *common pairs* of λ and $\tilde{\lambda}$. By this we mean that

$$\frac{c_i}{p_i} = \frac{\tilde{c}_i}{\tilde{p}_i} \quad \text{and} \quad a_i = \tilde{a}_i \quad \text{for} \quad i = 1, \ldots, s,$$

but one of these conditions is violated for $i = s + 1$. Then the branches separate somewhere along the chains created by the pairs $\binom{c_{s+1}}{p_{s+1}}, \binom{\tilde{c}_{s+1}}{\tilde{p}_{s+1}}$. Let $m_1, m_2 \cdots$ be the sequence of multiplicities of all singular points of $\overline{U}$ infinitely near to the point q.

3.1.3 Note that if, say, $GCD(c_i, p_i) = 1$, then λ is non-singular after the blow up according to $\binom{c_i}{p_i}$. If $\lambda, \tilde{\lambda}$ have separated at this stage, we have $h = i$ by definition. So in general there will be only one pair with $GCD(c_i, p_i) = 1$. If $\lambda, \tilde{\lambda}$ have not yet separated, i.e., if the pairs $\binom{c_i}{p_i}, \binom{\tilde{c}_i}{\tilde{p}_i}$ are common, an additional pair $\binom{c_{i+1}}{p_{i+1}} = \binom{1}{1}$ is recorded.

3.1.4 We have the following formulas, see [9], Appendix.

(*i*) $$\sum m_i = c_1 + \sum p_i - 1 + \tilde{c}_1 + \sum \tilde{p}_i - 1.$$

(*ii*)
$$\sum m_i^2 = \sum_{i=1}^{s}(p_i+\tilde{p}_i)(c_i+\tilde{c}_i)+\sum_{i>s} p_i c_i+\sum_{i>s} \tilde{p}_i\tilde{c}_i+2\min(\tilde{p}_{s+1}c_{s+1}, p_{s+1}\tilde{c}_{s+1}).$$

Lemma 3.2. *Let* $\gamma' = -E'^2$, *see 0.1. We obtain the following formulas.*

(*a*) $$\gamma' + 2d = \sum p_i + \sum \tilde{p}_i.$$

(*b*)
$$\gamma' + d^2 = \sum_{i=1}^{s}(p_i+\tilde{p}_i)(c_i+\tilde{c}_i)+\sum_{i>s} p_i c_i+\sum_{i>s} \tilde{p}_i\tilde{c}_i+2\min(\tilde{p}_{s+1}c_{s+1}, p_{s+1}\tilde{c}_{s+1})$$

where $d = c_1 + \tilde{c}_1$.

Proof. (a) We have $K_{\mathbb{P}^2} \cdot \overline{U} = -3d$, $K_{\overline{S}'} \cdot E' = -2 + \gamma'$. Blowing up a point of multiplicity m of a curve X increases the quantity $K \cdot X$ by m. Hence $K_{\overline{S}'} \cdot E' - K_{\mathbb{P}^2} \cdot \overline{U} = -2 + \gamma' + 3d = \sum m_i$. The statement follows from (i) above.

(b) We have $\overline{U}^2 - E'^2 = d^2 + \gamma' = \sum m_i^2$. The statement follows now from (ii). □

3.2.1 We put

$$c_1 - p_1 = \alpha c_2, \quad \tilde{c}_1 - \tilde{p}_1 = \tilde{\alpha}\tilde{c}_2 \quad \text{and} \quad \alpha_0 = \min(\alpha, \tilde{\alpha}).$$

Since $2\min(\tilde{p}_{s+1}c_{s+1}, p_{s+1}\tilde{c}_{s+1}) \leq \tilde{p}_{s+1}c_{s+1} + p_{s+1}\tilde{c}_{s+1}$, 3.2(b) gives

$$\gamma' + d^2 \leq \sum_{i=1}^{s}(p_i + \tilde{p}_i)(c_i + \tilde{c}_i) + \sum_{i>s} p_i c_i + \sum_{i>s} \tilde{p}_i\tilde{c}_i + \tilde{p}_{s+1}c_{s+1} + p_{s+1}\tilde{c}_{s+1}$$

and

$$\gamma' + d^2 \leq (p_1 + \tilde{p}_1)d + (c_2 + \tilde{c}_2)(P + \tilde{P}),$$

where $P = \sum_{i\geq 2} p_i$, $\tilde{P} = \sum_{i\geq 2} \tilde{p}_i$.

From this

(*i*) $$d(d - p_1 - \tilde{p}_1) + \gamma' \leq (c_2 + \tilde{c}_2)(P + \tilde{P}).$$

Since $d - p_1 - \tilde{p}_1 = c_1 - p_1 + \tilde{c}_1 - \tilde{p}_1 \geq \alpha_0(c_2 + \tilde{c}_2)$ and since $\gamma' \geq 1$ we get

$$(ii) \qquad \alpha_0 d < P + \tilde{P}.$$

Lemma 3.3. *Let h_Φ (resp. h_Ψ) be the number of sprouting contractions in Φ (resp. Ψ). Then $h_\Phi = 6 - K_{\overline{S}'} \cdot (K_{\overline{S}'} + D') = 2 + \varepsilon + \gamma + h_\Psi$. If E' is not touched by the contractions in Ψ, then $\gamma = \gamma'$. If, moreover, L'_∞, the proper transform of L_∞ in $\overline{S}'$, is not a (-1)-curve, then $h_\Psi = 0$.*

Proof. Under a subdivisional blowing up of a point on a divisor T the quantity $K \cdot (K + T)$ doesn't change. Under a sprouting blowing up the quantity decreases by 1. Hence $h_\Phi = K_{\mathbb{P}^2} \cdot (K_{\mathbb{P}^2} + L_\infty) - K_{\overline{S}'} \cdot (K_{\overline{S}'} + D')$ and $h_\Psi = K_{\overline{S}} \cdot (K_{\overline{S}} + D) - K_{\overline{S}'} \cdot (K_{\overline{S}'} + D')$. Since $K_{\overline{S}} \cdot E = -2 + \gamma$ the result follows from 1.7. □

Lemma 3.4. *We have $s = 0$.*

Proof. Suppose that $s \geq 1$. Note that then λ and $\tilde{\lambda}$ are both tangent to L_∞. (Otherwise both are not tangent to L_∞ and q is a point of multiplicity $deg(\overline{U})$ on $\overline{U}$.) Hence $c_1 > p_1, \tilde{c}_1 > \tilde{p}_1$. Also, both branches have more than one characteristic pair, i.e., $h > 1$ and $\tilde{h} > 1$. We put

$$\frac{c_1}{c_2} = k = \frac{\tilde{c}_1}{\tilde{c}_2}, \quad \text{and} \quad \frac{p_1}{c_2} = l = \frac{\tilde{p}_1}{\tilde{c}_2}.$$

We have $\alpha = \tilde{\alpha} = k - l \geq 1$.

Suppose that $\alpha = 1$, i.e., $k = l + 1$. The blowing up over q according to the pair $\binom{l+1}{l}$ produces a chain $L + C + M$, where L has l components with L_∞ as a (-1)-tip, C is the last exceptional curve and M is a $(-l-1)$-curve. The branches $\lambda, \tilde{\lambda}$ have common center q' on $C \setminus (L \cup M)$. In Φ^{-1} we now blow up q'. Let A be the resulting exceptional curve. Let us perform $l - 1$ successive additional sprouting blowups (they will not be part of Φ^{-1}), starting with a point on A that is *not* the center of λ or $\tilde{\lambda}$, creating a chain $A + B$ attached to C, with B of length $l - 1$. Let $L^\dagger_\infty$ be the last exceptional curve. As it is well known, we can now blow down, beginning with L_∞, the curves in L, then C, then $A + (B - L^\dagger_\infty)$, then M, producing a new completion of $\mathbb{C}^2$ with $L^\dagger_\infty$ as new line at infinity and a new completion $\overline{U}^\dagger$ of U.

3.4.1 Let us note for further reference that we have performed an *elementary automorphism*, or *DeJoncquières transformation*, of $\mathbb{C}^2$ determined by

$q \in L_\infty$, the pair $\binom{l+1}{l}$, the choice of $q' \in C$ and the choice of further fundamental points in creating the chain B.

The task of producing the NC-resolution for $\overline{U} + L_\infty$ is accomplished by further blowups over A. Let $A^\sharp$ be the resulting configuration of curves and put

$$\Gamma = L + M + C + A^\sharp + B.$$

Then, as a set, $D' = L + C + M + A^\sharp \subset \Gamma$. In constructing the minimal NC-resolution of $\overline{U}^\dagger + L_\infty^\dagger$ we have to reconstruct A, hence also B, and then $A^\sharp$. Hence also $D^\dagger \subset \Gamma$. There are three possibilities.
(i) The centers of λ and $\tilde{\lambda}$ on A are not on C and $l = 1$. Then $D^\dagger = A^\sharp$. (We have $L_\infty^\dagger = A$.)
(ii) The centers of λ and $\tilde{\lambda}$ on A are not on C and $l > 1$. Then $D^\dagger = B + A^\sharp + M$.
(iii) The center of λ or $\tilde{\lambda}$ on A is on C. Then $D^\dagger = B + A^\sharp + M + C$.
In each case $D^\dagger$ has fewer components than D', contrary to our assumption.

Hence $\alpha \geq 2$. Then either $L_\infty'^2 \leq -2$ or D' has a twig with an initial chain $L'_\infty + L_0$ with $L_\infty'^2 = -1$, L_0 a (-2)-chain attached to a (≤ -3)-curve in D' and $E' \cdot (L'_\infty + L_0) = 0$. This implies that E' is not touched by the contractions in $\Psi\colon \overline{S}' \to \overline{S}$. Hence $\gamma' = \gamma$.

By 2.5, $\gamma \leq 9$. Suppose that $\alpha \geq 3$. We have $3d < P + \tilde{P}$ by 3.2.1(ii). From 3.2(a) we obtain $P + \tilde{P} + p_1 + \tilde{p}_1 - \gamma = 2d$. We get that $d + p_1 + \tilde{p}_1 < \gamma \leq 9$. Thus $d < 9 - p_1 - \tilde{p}_1 \leq 7$. But $d = c_1 + \tilde{c}_1 \geq 2(\alpha + 1) \geq 8$, a contradiction.

Hence $\alpha = 2$, i.e., $k = l + 2$. Since $GCD(k, l) = 1$, l and k are odd. We have

$$(\dagger) \qquad c_1 = p_1 + 2c_2, \tilde{c_1} = \tilde{c_1} = \tilde{p_1}\tilde{c_2}, d - p_1 - \tilde{p_1} = 2(c_2 + \tilde{c}_2).$$

Substitute this into 3.2.1(i). We obtain

$$2d(c_2 + \tilde{c}_2) + \gamma \leq (c_2 + \tilde{c}_2)(P + \tilde{P})$$

and

$$(*) \qquad (c_2 + \tilde{c}_2)(2d - P - \tilde{P}) + \gamma \leq 0.$$

By 3.2(a), $2d - P - \tilde{P} = p_1 + \tilde{p}_1 - \gamma$. Hence

$$(c_2 + \tilde{c}_2)(l(c_2 + \tilde{c}_2) - \gamma) + \gamma \leq 0.$$

From this $l(c_2+\tilde{c}_2)<\gamma$. Thus $l(c_2+\tilde{c}_2)\le 8$, which implies $l\le 4$. Hence $l\le 3$.
Suppose that $l=3$. Then $c_2+\tilde{c}_2=2$ and we have $2(6-\gamma)+\gamma\le 0$ which gives $\gamma\ge 12$, a contradiction.
Therefore $l=1$ and $k=3$. $(*)$ takes the form $(c_2+\tilde{c}_2)(c_2+\tilde{c}_2-\gamma)+\gamma\le 0$. By simple algebra we get $(c_2+\tilde{c}_2-1)(c_2+\tilde{c}_2+1-\gamma)\le -1$. We find that

$$(**)\qquad p_1=c_2, \tilde{p}_1=\tilde{c}_2, c_2+\tilde{c}_2=p_1+\tilde{p}_1\le\gamma-2.$$

The proper transform of L_∞ in $\overline{S}$ is a (-2)-curve. Hence $D'=D$. Blowing up on L_∞ according to the pair $\binom{c_1}{p_1}$ produces a chain $L'_\infty+C+M$, where M consists of two (-2)-curves and C is branching in D. L'_∞ and M are maximal twigs in $D+E$ and contribute $\frac{1}{2}+\frac{2}{3}>1$ to $\sum e_i$. In view of 1.10 and 2.5,

$$(***)\qquad \varepsilon\ge 1 \quad\text{and}\quad \gamma\le 7+t-2\epsilon\le 5+t.$$

Hence $\gamma\le 7$. If $\gamma=7$, then $\epsilon=1$ and $t=2$, so $D+E$ has at least two maximal twigs with (-2)-tips and not contained in L, those produced by the pairs $\binom{c_h}{p_h}$ and $\binom{\tilde{c}_{\tilde{h}}}{\tilde{p}_{\tilde{h}}}$. In view of 1.2.1 we get that $\sum e_i>2$, in contradiction to 1.10.

Hence we have $\gamma\le 6$, so $c_2+\tilde{c}_2\le 4$.

Suppose that $c_2=\tilde{c}_2=2$. Let a be the number of common pairs of type $\binom{2}{2}$. Thus $s\ge a+1$. If $s\ge a+2$ then the next common pair is of type $\binom{2}{1}$, followed by a number, possibly zero, of common pairs of type $\binom{1}{1}$. The branches then both meet D transversally at different points of the last (-1)-curve . It follows that $t=0$ and $(***)$ gives $\gamma\le 5$. We reach a contradiction with $(**)$.

Hence $s=a+1$. Then either $\binom{c_{s+1}}{p_{s+1}}=\binom{\tilde{c}_{s+1}}{\tilde{p}_{s+1}}$ which is either $\binom{2}{2}$ or $\binom{2}{1}$, or, say, $\binom{c_{s+1}}{p_{s+1}}=\binom{2}{1}, \binom{\tilde{c}_{s+1}}{\tilde{p}_{s+1}}=\binom{2}{2}$. Hence $m:=\min(c_{s+1}\tilde{p}_{s+1},\tilde{c}_{s+1}p_{s+1})=2$ or 4. We have also $d=c_1+\tilde{c}_1=3(c_2+\tilde{c}_2)=12$. Note that $\binom{c_h}{p_h}=\binom{\tilde{c}_{\tilde{h}}}{\tilde{p}_{\tilde{h}}}=\binom{2}{1}$. Hence in 3.2(b) we have $c_hp_h+\tilde{c}_{\tilde{h}}\tilde{p}_{\tilde{h}}=2+2$ and all other individual terms on the RHS and the term d^2 on the LHS are divisible by 4. Hence γ is divisible by 4, so $\gamma=4$, and we have a contradiction with $(**)$.

Suppose now that $c_2=1$. Then $\tilde{c}_2\le 3$. Let a be the number of common pairs of type $\binom{\tilde{c}_2}{\tilde{c}_2}$. Then $s=1+a$. We have $h=s+1$ with $p_h=c_h=1$ and $\tilde{h}=1+a+b+1$, where $a+b$ is the total number of pairs equal to $\binom{\tilde{c}_2}{\tilde{c}_2}$. We have $m=\min(c_{s+1}\tilde{p}_{s+1},\tilde{c}_{s+1}p_{s+1})=\tilde{p}_{s+1}$. The formulas 3.2 take the form

$$(1)\qquad \gamma+2(3+3\tilde{c}_2)=s+1+(s+b)\tilde{c}_2+\tilde{p}_{\tilde{h}}.$$

and

(2) $\gamma+(3+3\tilde{c}_2)^2 = (3+3\tilde{c}_2)(1+\tilde{c}_2)+(s-1)(1+\tilde{c}_2)^2+2\tilde{p}_{s+1}+1+b\tilde{c}_2^2+\tilde{c}_2\tilde{p}_{\tilde{h}}$.

Suppose that $\tilde{c}_2 = 3$. Then $\gamma = 6$ by $(**)$. By $(***)$, $t \geq 1$. This implies that $\tilde{c}_{s+1} = \tilde{p}_{s+1} = 3$ and $\tilde{p}_{\tilde{h}} = 1$. (1) and (2) now give $6+24 = s+1+3(s+b)+1$, i.e., $28 = 4s+3b$ and $6+144 = 48+(s-1)16+9b+10$, i.e., $108 = 16s+9b$. The system of equations has no integer solutions.

Before proceeding with the analysis of cases we note the following. Since $\kappa(K_{\overline{S}} + E) = -\infty$ by [8], we have $h^0(2K_{\overline{S}} + E) = 0$ and by the Riemann-Roch theorem

$$(\sharp) \qquad h^0(-K_{\overline{S}} - E) \geq K_{\overline{S}} \cdot (K_{\overline{S}} + E) = K_{\overline{S}}^2 - 2 + \gamma.$$

As argued in 2.4.2

$$(\sharp\sharp) \quad h^0(2K_{\overline{S}}+D+E) \geq 1+K_{\overline{S}}\cdot(K_{\overline{S}}+D+E) = 3-\epsilon \text{ or } K_{\overline{S}} = -D-E.$$

Now suppose that $\tilde{c}_2 = 2$. By $(\dagger)$ and $(**)$ we have

$$c_1 = 3, p_1 = 1, \tilde{c}_1 = 6, \tilde{p}_1 = 2, d = 9.$$

(1) and (2) give $\gamma + 18 = s + 1 + (s + b)2 + 1$, i.e., $\gamma + 16 = 3s + 2b$ and $\gamma + 81 = 27 + (s-1)9 + 4b + 3 + 2\tilde{p}_{s+1}$, i.e., $\gamma + 60 = 9s + 4b + 2\tilde{p}_{s+1}$. We have $5 \leq \gamma \leq 6$ and $\tilde{p}_{s+1} \leq 2$. We find two solutions:

(i) $\gamma = 5, s = 7, b = 0, \tilde{p}_{s+1} = 1$
(ii) $\gamma = 6, s = 6, b = 2, \tilde{p}_{s+1} = 2$.

With the characteristic pairs fully determined it is now elementary to compute the terms on the right hand side of $\sharp$ and $\sharp\sharp$. We recall in particular that $K_{\overline{S}}^2 = 10-\#D$. Note that L'_∞ ia a (-2)-tip of D and we have $L'_\infty\cdot K_{\overline{S}} = 0$, $L'_\infty \cdot (-D-E) = 1$. Hence $K_{\overline{S}} \neq -D-E$.

In case (i) we find $\#D = 12$, hence $K_{\overline{S}}^2 = -2$, and $\varepsilon = 1$. By $(\sharp)$ and $(\sharp\sharp)$, $-K_{\overline{S}} - E \geq 0$ and $2K_{\overline{S}} + D + E \geq 0$. We obtain $K_{\overline{S}} + D = 2K_{\overline{S}} + D + E + (-K_{\overline{S}} - E) \geq 0$, a contradiction. In case (ii) we have $\#D = 13$, so $K_{\overline{S}}^2 = -3$, and $\varepsilon = 1$. We come to a contradiction by the same argument.

Suppose that $\tilde{c}_2 = 1$. we have

$$c_1 = 3, p_1 = 1, \tilde{c}_1 = 3, \tilde{p}_1 = 1, b = 0, d = 6.$$

The formulas give $\gamma + 10 = 2s + b$ and $\gamma + 24 = 4s + b$. We get the solution

(iii) $\gamma = 4, s = 7$.

We find $b_2(\overline{S}) = 11$, $K^2_{\overline{S}} = -1$, $\varepsilon = 2$. We come to a contradiction by the same argument.

□

3.5. We have shown that $s = 0$. Suppose that the branches stay together after the first blowing up, i.e., they both are tangent to L_∞. Let, as in 3.2.1,

$$c_1 - p_1 = \alpha c_2, \quad \tilde{c}_1 - \tilde{p}_1 = \tilde{\alpha}\tilde{c}_2.$$

We will show that, possibly at the cost of increasing the number of components of D', this case can be reduced to the case $\alpha = \tilde{\alpha} = 1$ and hence $\frac{c_1}{p_1} = \frac{\tilde{c}_1}{\tilde{p}_1} = \frac{l+1}{l}$ for some l. This case will be dealt with in 3.6.

Suppose that $\tilde{\alpha} = 1$. We show that we can then pass to a situation with $\alpha = \tilde{\alpha} = 1$.

Let $\tilde{p}_1 = l\tilde{c}_2$. Then $\tilde{c}_1 = (l+1)\tilde{c}_2$. We use the notation of 3.4.1. After blowing up according to $\binom{l+1}{l}$, the center $\tilde{p}$ of $\tilde{\lambda}$ is on $C \setminus (L \cup M)$. We now have three possibilities.
(i) The center p of λ is on M. Equivalently, $p_1 > l(c_1 - p_1)$.
(ii)The center p of λ is on $C \setminus (L \cup M)$. Equivalently, $p_1 = l(c_1 - p_1)$, or $\alpha = 1$. Moreover, $p \neq \tilde{p}$.
(iii) The center p of λ is on L. Equivalently, $p_1 < l(c_1 - p_1)$.

Suppose we have (i) or (ii). We then perform a DeJoncquières transformation exactly as in 3.4.1 with $q' = \tilde{p}$. The argument in the proof of 3.4, slightly modified, shows that we obtain a completion with smaller D'.

3.5.1 We remark that if the HN-sequence for $\tilde{\lambda}$ has at least l pairs $\binom{\tilde{c}_2}{\tilde{c}_2}$ following $\binom{\tilde{c}_1}{\tilde{p}_1}$, or if $\tilde{c}_2 = 1$, we can construct the above DeJoncquières transformation with blowups following $\tilde{\lambda}$, and it will then separates the branches. Hence this is not the case.

Suppose we have (iii). We now perform an elementary transformation as above, but with $q' \neq \tilde{p}$. Then we are in situation (ii) w.r.t. the new coordinate system, i.e., we have $\alpha = \tilde{\alpha} = 1$.

Assume now that $\alpha \geq 2$, $\tilde{\alpha} \geq 2$. We show that this is not possible. We write 3.2(b) as

$$\gamma + \alpha c_1 c_2 + \tilde{\alpha}\tilde{c}_1\tilde{c}_2 + 2c_1\tilde{c}_1 = \sum_{i\geq 2} p_i c_i + \sum_{i\geq 2} \tilde{p}_i\tilde{c}_i + 2\min(c_1\tilde{p}_1, \tilde{c}_1 p_1).$$

We have $2c_1\tilde{c}_1 = \tilde{c}_1(p_1+\alpha c_2)+c_1(\tilde{p}_1+\tilde{\alpha}\tilde{c}_2) = \tilde{c}_1p_1+c_1\tilde{p}_1+\alpha c_2\tilde{c}_1+\tilde{\alpha}c_1\tilde{c}_2$. Therefore

$$\begin{aligned}&\gamma+\tilde{c}_1p_1+c_1\tilde{p}_1+\alpha c_2\tilde{c}_1+\tilde{\alpha}c_1\tilde{c}_2+\alpha c_1c_2+\tilde{\alpha}\tilde{c}_1\tilde{c}_2\\ &=\sum_{i\geq 2}p_ic_i+\sum_{i\geq 2}\tilde{p}_i\tilde{c}_i+2\min(c_1\tilde{p}_1,\tilde{c}_1p_1)\end{aligned}$$

and

$$\gamma+\tilde{c}_1p_1+c_1\tilde{p}_1-2\min(c_1\tilde{p}_1,\tilde{c}_1p_1)+(c_1+\tilde{c}_1)(\alpha c_2+\tilde{\alpha}\tilde{c}_2)=\sum_{i\geq 2}p_ic_i+\sum_{i\geq 2}\tilde{p}_i\tilde{c}_i.$$

Let $\beta=\tilde{c}_1p_1+c_1\tilde{p}_1-2\min(c_1\tilde{p}_1,\tilde{c}_1p_1)\geq 0$. We get

$$(*)\qquad \gamma+\beta+(c_1+\tilde{c}_1)(\alpha c_2+\tilde{\alpha}\tilde{c}_2)=\sum_{i\geq 2}p_ic_i+\sum_{i\geq 2}\tilde{p}_i\tilde{c}_i.$$

From 3.2(a) we get

$$(**)\qquad \gamma+c_1+\tilde{c}_1+\alpha c_2+\tilde{\alpha}\tilde{c}_2=\sum_{i\geq 2}p_i+\sum_{i\geq 2}\tilde{p}_i.$$

We may assume by symmetry that $c_2\geq\tilde{c}_2$. Multiply (**) by c_2 and subtract (*). We obtain

$$\gamma c_2+(c_1+\tilde{c}_1)c_2+c_2(\alpha c_2+\tilde{\alpha}\tilde{c}_2)\geq\gamma+\beta+(c_1+\tilde{c}_1)(\alpha c_2+\tilde{\alpha}\tilde{c}_2).$$

So

$$\gamma(c_2-1)+(c_1+\tilde{c}_1)c_2\geq\beta+(\alpha c_2+\tilde{\alpha}\tilde{c}_2)(c_1+\tilde{c}_1-c_2).$$

Since $\alpha\geq 2,\tilde{\alpha}\geq 2,\beta\geq 0$ and $\tilde{c}_2\geq 1$ we have

$$\gamma(c_2-1)+(c_1+\tilde{c}_1)c_2\geq(2c_2+2)(c_1+\tilde{c}_1-c_2).$$

From this

$$\gamma(c_2-1)\geq(2c_2+2)(c_1+\tilde{c}_1)-(2c_2+2)c_2-(c_1+\tilde{c}_1)c_2.$$

So

$$\gamma(c_2-1)\geq(c_2+2)(c_1+\tilde{c}_1)-2(c_2+1)c_2.$$

We have $\gamma\leq 9$ by 2.5 and $\tilde{c}_1\geq 3$ since $\tilde{\alpha}\geq 2$. Also $c_1\geq 3c_2$ since $\alpha\geq 2$. We obtain

$$9c_2-9\geq(c_2+2)(3c_2+3)-2(c_2+1)c_2.$$

We get

$$0\geq c_2^2-2c_2+15.$$

This is a contradiction.

3.6. In this section we temporarily drop the assumption that D' has the smallest possible number of components. We consider here the case $s = 0$, $c_1 = (l+1)c_2, p_1 = lc_2$, $\tilde{c}_1 = (l+1)\tilde{c}_2, \tilde{p}_1 = l\tilde{c}_2$. We will prove that this case does not occur. Suppose the opposite. Let H' denote the (-1)-curve produced by the pair $\binom{c_1}{p_1}$ and let $H = \Psi(H')$. The branches meet H' in two different points. Ψ involves l successive contractions beginning with L'_∞. H' is not contracted by Ψ. Let F (resp. $\tilde{F}$) denote the part of D' produced by the pairs $\binom{c_2}{p_2}, \ldots, \binom{c_h}{p_h}$ (resp. $\binom{\tilde{c}_2}{\tilde{p}_2}, \ldots, \binom{\tilde{c}_{\tilde{h}}}{\tilde{p}_{\tilde{h}}}$). Let C (resp. $\tilde{C}$) be the unique (-1)-curve in F (resp. $\tilde{F}$)

Let r (resp. $\tilde{r}$) denote the number of pairs equal to $\binom{c_2}{c_2}$ (resp. $\binom{\tilde{c}_2}{\tilde{c}_2}$). Hence $p_{r+2} < c_{r+2} = c_2$ and $c_i \leq \frac{1}{2}c_2$ for $i > r+2$. We put $P' = \sum_{i \geq r+2} p_i$. In similar way we define $\tilde{P}'$. Notice that $c_2 > 1, \tilde{c}_2 > 1$ by the argument in 3.5.1. Therefore $h > r+1$, $\tilde{h} > \tilde{r}+1$, i.e., $P' \geq 1$ and $\tilde{P}' \geq 1$. Again by 3.5.1 we have $r \leq l-1, \tilde{r} \leq l-1$.

3.6.1 We note that $D+E$ has at least 3 maximal twigs, the $-(l+1)$-curve M (see 3.4.1), and one each in F an $\tilde{F}$ with a $(\geq -c_2)$- and a $(\geq -\tilde{c}_2)$-curve as tip respectively. By 1.2.1 they contribute at least $e = \frac{1}{l+1} + \frac{1}{c_2} + \frac{1}{\tilde{c}_2}$ to $\sum e_i$ in 1.10. In particular, $\varepsilon > 0$ if $e > 1$.

From 3.2(b) we get

$$\gamma + d^2 \leq (c_1 + \tilde{c}_1)(p_1 + \tilde{p}_1) + \sum_{i \geq 2} p_i c_i + \sum_{i \geq 2} \tilde{p}_i \tilde{c}_i$$
$$= d(p_1 + \tilde{p}_1) + rc_2^2 + p_{r+2}c_2 + \tilde{r}\tilde{c}_2^2 + \tilde{p}_{\tilde{r}+2}\tilde{c}_2 + \sum_{i \geq r+3} p_i c_i + \sum_{i \geq \tilde{r}+3} \tilde{p}_i \tilde{c}_i.$$

From this

$$\gamma + d(c_2 + \tilde{c}_2) \leq rc_2^2 + \tilde{r}\tilde{c}_2^2 + c_2 p_{r+2} + \frac{1}{2}c_2(P' - p_{r+2}) + \tilde{c}_2\tilde{p}_{r+2} + \frac{1}{2}\tilde{c}_2(\tilde{P}' - \tilde{p}_{\tilde{r}+2}).$$

From 3.2(a) we get

$$d = \frac{p_1 + \tilde{p}_1 + rc_2 + \tilde{r}\tilde{c}_2 + P' + \tilde{P}' - \gamma}{2}.$$

Hence

$$\gamma + \frac{1}{2}(c_2 + \tilde{c}_2)(p_1 + \tilde{p}_1 + rc_2 + \tilde{r}\tilde{c}_2 + P' + \tilde{P}') - \frac{1}{2}\gamma(c_2 + \tilde{c}_2)$$
$$\leq rc_2^2 + \tilde{r}\tilde{c}_2^2 + \frac{1}{2}c_2 p_{r+2} + \frac{1}{2}c_2 P' + \frac{1}{2}\tilde{c}_2\tilde{p}_{\tilde{r}+2} + \frac{1}{2}\tilde{c}_2\tilde{P}'.$$

From this

$$(*)\qquad (c_2+\tilde{c}_2)(p_1+\tilde{p}_1+rc_2+\tilde{r}\tilde{c}_2)+c_2\tilde{P}'+\tilde{c}_2P'-\gamma(c_2+\tilde{c}_2) < 2rc_2^2+2\tilde{r}\tilde{c}_2^2+c_2p_{r+2}+\tilde{c}_2\tilde{p}_{\tilde{r}+2}.$$

Since $p_1=lc_2, \tilde{p}_1=l\tilde{c}_2$ and since $P'\geq 1$, $\tilde{P}'\geq 1$ we have

$$(c_2+\tilde{c}_2)(l(c_2+\tilde{c}_2)+rc_2+\tilde{r}\tilde{c}_2) < 2rc_2^2+2\tilde{r}\tilde{c}_2^2+c_2p_{r+2}+\tilde{c}_2\tilde{p}_{r+2}+(\gamma-1)(c_2+\tilde{c}_2),$$

$$l(c_2+\tilde{c}_2)^2+(r+\tilde{r})c_2\tilde{c}_2 < rc_2^2+\tilde{r}\tilde{c}_2^2+c_2p_{r+2}+\tilde{c}_2\tilde{p}_{r+2}+(\gamma-1)(c_2+\tilde{c}_2).$$

Since $r,\tilde{r}\leq l-1$, $p_{r+2}\leq c_2-1$ and $\tilde{p}_{r+2}\leq \tilde{c}_2-1$, we have

$$l(c_2+\tilde{c}_2)^2+(r+\tilde{r})c_2\tilde{c}_2 < l(c_2^2+\tilde{c}_2^2)+(\gamma-2)(c_2+\tilde{c}_2).$$

Finally

$$(**)\qquad c_2\tilde{c}_2(2l+r+\tilde{r}) < (\gamma-2)(c_2+\tilde{c}_2).$$

Suppose that $l\geq 3$. Then $6c_2\tilde{c}_2 < 7(c_2+\tilde{c}_2)$ since $\gamma\leq 9$ by 2.5. This implies $c_2=\tilde{c}_2=2$. But then $\varepsilon>0$ by 3.6.1. This implies $\gamma\leq 7$ by 2.5. Now $(**)$ gives $24<20$, a contradiction.

Suppose that $l=2$. Notice that $\gamma<9$. Otherwise $\varepsilon=0$ and $t=2$, and there are two (-2)-tips in D. This gives a contradiction by 1.10 as before. $(**)$ gives $4c_2\tilde{c}_2 < 6(c_2+\tilde{c}_2)$. Let $c_2\leq\tilde{c}_2$.
Suppose that $c_2=2$. We obtain $8\tilde{c}_2<6(2+\tilde{c}_2)$, i.e., $\tilde{c}_2<6$. It follows by 3.6.1 that $\varepsilon>0$, so $\gamma\leq 7$. Now $(**)$ gives $8\tilde{c}_2<5(2+\tilde{c}_2)$, i.e., $\tilde{c}_2\leq 3$. If $\tilde{c}_2=2$, then γ is even by 3.2(b) since then $c_i=2$ and $\tilde{c}_i=2$ for every $i>1$. Hence $\gamma\leq 6$ and $(**)$ gives a contradiction. So $\tilde{c}_2=3$. From $(**)$ we obtain $r=\tilde{r}=0$. We have $P'=1, \tilde{P}'=\tilde{p}_2$, $p_1=lc_2=4$, $\tilde{p}_1=l\tilde{c}_2=6$. Now $(*)$ gives $51-5\gamma<\tilde{P}'$, a contradiction since $\tilde{P}'=1$ or 2 and $\gamma<9$.
Suppose that $c_2\geq 3$. Since $\gamma\leq 8$, $(**)$ gives $c_2(4\tilde{c}_2-6)<6\tilde{c}_2$ and $3(4\tilde{c}_2-6)<6\tilde{c}_2$. We get $\tilde{c}_2<3$, a contradiction.

Suppose $l=1$. Then by 3.5.1 $r=\tilde{r}=0$, i.e $c_2>c_3$ and $\tilde{c}_2>\tilde{c}_3$. We have $d=2c_2+2\tilde{c}_2$ and the formulas 3.2 take the form

$$(1)\qquad \gamma+3c_2+3\tilde{c}_2=\sum_{i\geq 2}p_i+\sum_{i\geq 2}\tilde{p}_i,$$

$$(2)\qquad \gamma+2c_2^2+2\tilde{c}_2^2+4c_2\tilde{c}_2=\sum_{i\geq 2}p_ic_i+\sum_{i\geq 2}\tilde{p}_i\tilde{c}_i.$$

We may assume that $c_2 \geq \tilde{c}_2$. The branches meet the (-1)-curve T_1 created by $\binom{c_1}{p_1}$ in distinct points. Hence, see 3.3, $h_\Phi = 1+(h-1)+\tilde{h}-1) = h+\tilde{h}-1$ since each of the HN-pairs gives exactly one sprouting contraction and the first pairs $\binom{c_1}{p_1}$ and $\binom{\tilde{c}_1}{\tilde{p}_1}$ give the common one. Also, T_1 is branching in D' and $(L'_\infty)^2 = -1$. Hence Ψ contracts only L'_∞, and it is a sprouting contraction, that is $h_\Psi = 1$. By 3.3, $h + \tilde{h} = 4 + \varepsilon + \gamma$. It follows from 2.5 that $\varepsilon + \gamma \leq 8$ ($\gamma = 9$ is ruled out as above). Hence $h + \tilde{h} \leq 12$. Since $c_2 > 1$, $h \geq 2$. Similarly $\tilde{h} \geq 2$. Hence $h, \tilde{h} \leq 10$.

We write $c_2 - p_2 = \mu c_3$, $\tilde{c}_2 - \tilde{p}_2 = \tilde{\mu}\tilde{c}_3$, $c_2 = kc_3$, $\tilde{c}_2 = \tilde{k}\tilde{c}_3$. Note that $\mu, \tilde{\mu} \geq 1$ and $k, \tilde{k} \geq 2$ since $r, \tilde{r} = 0$. We rewrite (2) in the form

$$(3) \qquad \gamma + c_2^2 + \tilde{c}_2^2 + 4c_2\tilde{c}_2 = -\mu c_2 c_3 + \sum_{i\geq 3} p_i c_i - \tilde{\mu}\tilde{c}_2\tilde{c}_3 + \sum_{i\geq 3} \tilde{p}_i\tilde{c}_i.$$

We get

$$\gamma + 4c_2\tilde{c}_2 \leq c_3^2(h - 2 - \mu k - k^2) + \tilde{c}_3^2(\tilde{h} - 2 - \tilde{\mu}\tilde{k} - \tilde{k}^2),$$

and, since $c_2 \geq \tilde{c}_2$,

$$(4) \qquad \gamma \leq c_3^2(h - 2 - \mu k - k^2) + \tilde{c}_3^2(\tilde{h} - 2 - \tilde{\mu}\tilde{k} - 5\tilde{k}^2).$$

We find $\tilde{h} - 2 - \tilde{\mu}\tilde{k} - 5\tilde{k}^2 \leq \tilde{h} - 24 < 0$ since $\tilde{h} \leq 10$. It follows from (4) that

$$(5) \qquad h - 2 - \mu k - k^2 > 0.$$

Since $h \leq 10$ we get $7 \geq k(\mu + k) \geq (\mu + 1)(2\mu + 1)$. We obtain $\mu = 1$ and $k = 2$ and $\binom{c_2}{p_2} = c_3\binom{2}{1}$. Hence
$(***)$ $D + E$ has at least three tips, two of them (-2)-tips. Hence $\varepsilon > 0$.

($\binom{c_1}{p_1}$ and $\binom{c_2}{p_2}$ produce (-2)-tips, $\binom{\tilde{c}_2}{\tilde{p}_2}$ a third tip.)

Claim. $\gamma + \varepsilon \leq 7$.

Proof. Suppose otherwise. Then $\varepsilon \geq 2$ is ruled out by 2.5, $\varepsilon > 0$ by $(***)$ and 1.10. Hence $\gamma = 7, \varepsilon = 1$. By 2.5, $t = 2$. Suppose that $h > 2$. Then $D + E$ has at least four (-2)-tips i.e. two produced by the pairs $\binom{c_1}{p_1}, \binom{c_2}{p_2}$ and two others which exist because $t = 2$. It follows from 1.10 that there are four tips, and they are maximal twigs of $D + E$. Hence $c_h = \tilde{c}_{\tilde{h}} = 2$. But now it follows from (2) that γ is even, a contradiction. Hence $h = 2$. This implies that $c_3 = 1$, so $c_2 = 2$. Since $c_2 \geq \tilde{c}_2 > 1$ we have $\tilde{c}_2 = 2$. We again reach contradiction with (2). $\square$

Since $\gamma+\varepsilon \le 7$, $h+\tilde{h} \le 11$. So $h \le 9$. (5) gives $h > 8$. Hence $h = 9$ and $\tilde{h} = 2$. Also $\gamma+\varepsilon = 7$. From (2) we get

$$\gamma + 6c_3^2 + 2\tilde{c}_2^2 + 4c_2\tilde{c}_2 = \sum_{i=3}^{9} p_i c_i + \tilde{p}_2\tilde{c}_2 \le 6c_3^2 + p_9c_9 + \tilde{c}_2^2.$$

It follows that $p_9 > 1$ since $c_9 < 4c_2\tilde{c}_2$. We have $\varepsilon > 0$ by $(***)$. Since $\gamma \ge 1$ by 1.8, $\varepsilon \ge 3$ is ruled out by 2.5. If $\varepsilon = 2$, then $\gamma = 5$, so $t = 2$ by 2.5, but $p_h = p_9 > 1$ implies $t \le 1$ since the last pair $\binom{c_9}{p_9}$ does not produce a (-2)-twig if $p_9 > 1$. Hence $\varepsilon = 1$ and $\gamma = 6$. By 2.5 $t \ge 1$. Since $p_9 > 1$, $\tilde{p}_{\tilde{h}} = \tilde{p}_2 = 1$. We rewrite (1) and (2) as follows.

$$(6) \qquad 5 + 6c_3 + 3\tilde{c}_2 = \sum_{i=2}^{9} p_i.$$

$$(7) \qquad 6 + 8c_3^2 + 2\tilde{c}_2^2 + 8c_3\tilde{c}_2 = \sum_{i=2}^{9} p_i c_i + \tilde{c}_2.$$

From this

$$(8) \qquad 6 + 6c_3^2 + 2\tilde{c}_2^2 + 8c_3\tilde{c}_2 = \sum_{i=3}^{9} p_i c_i + \tilde{c}_2$$

since $c_2 = 2c_3$. Suppose that there exists $4 \le j \le 8$ such that $c_j < c_3$. Then $c_j p_j \le \frac{c_3^2}{4}$ and $\sum_{i=3}^{9} p_i c_i \le (j-3)c_3^2 + (10-j)\frac{c_3^2}{4} \le 6c_3^2$. Now (8) gives a contradiction. Hence $c_i = c_3$ for $i < 8$. Suppose that $c_8 > p_8$. We write $c_8 - p_8 = \nu c_9$. Then $\sum_{i=3}^{9} p_i c_i \le 5c_3^2 + p_8c_8 + p_9c_9 = 6c_3^2 - \nu c_3c_9 + p_9c_9 \le 6c_3^2$ and again we reach contradiction with (8). Hence $p_i = c_i$ for $i \le 8$ and $c_9 = c_3$. From (6) we get

$$5 + 3\tilde{c}_2 = c_3 + p_9.$$

From (8) we get

$$(9) \qquad 6 + 2\tilde{c}_2^2 + 8c_3\tilde{c}_2 = p_9c_3 + \tilde{c}_2.$$

Now $p_9 = 5 + 3\tilde{c}_2 - c_3$ and (9) gives $6 + 2\tilde{c}_2^2 + 8c_3\tilde{c}_2 \le (5 + 3\tilde{c}_2 - c_3)c_3 + \tilde{c}_2$. Hence $6 + 2\tilde{c}_2^2 + 8c_3\tilde{c}_2 + c_3^2 \le 5c_3 + 3c_3\tilde{c}_2 + \tilde{c}_2$, i.e.,

$$6 + 2\tilde{c}_2^2 + 5c_3\tilde{c}_2 + c_3^2 \le 5c_3 + \tilde{c}_2.$$

It follows that $6 + c_3^2 < 5c_3$. This gives $2 < c_3 < 3$, a contradiction.

4. Separation of branches II: The branches separate on the first blowing up

In this section we rule out the last case in the proof of theorem 4.16, that of the branches separating on the first blowing up. We assume that the branch λ is tangent to L_∞ and $\tilde{\lambda}$ is not.

4.1. Let $\tilde{r}+1$ denote the number of pairs of $\tilde{\lambda}$ of the form $\binom{\tilde{c}_1}{\tilde{c}_1}$. So $\tilde{r} \geq 0$. We change slightly our usual labeling. The pairs of $\tilde{\lambda}$ we now label:

$$\binom{\tilde{c}_1}{\tilde{c}_1}, \dots, \binom{\tilde{c}_1}{\tilde{c}_1}, \binom{\tilde{c}_1}{\tilde{p}_1}, \dots, \binom{\tilde{c}_{\tilde{h}}}{\tilde{p}_{\tilde{h}}}$$

with either $\tilde{c}_1 = 1$ and $\tilde{r} = \tilde{h} = 0$, in which case we put $\tilde{p}_1 = 1$, or $\tilde{c}_1 > \tilde{p}_1$. Let $c_1 - p_1 = \alpha c_2$. We have $\alpha \geq 2$ since otherwise we may, as in 3.6, pass to an embedding with smaller resolution tree.

Let T_1 (resp. $\tilde{T}_1$) be the proper transform in $\overline{S}$ of the (-1)-curve produced by the pair $\binom{c_1}{p_1}$ (resp. $\binom{\tilde{c}_1}{\tilde{p}_1}$). Let C (resp. $\tilde{C}$) be the (-1)-curve produced by the last pair in the HN-sequence for λ (resp. $\tilde{\lambda}$). Since $\alpha \geq 2$ it is clear that T_1, $\tilde{T}_1$ and C, $\tilde{C}$, E' are not touched by Ψ, so have the same self-intersection in $\overline{S}'$ and $\overline{S}$. In particular $\gamma' = -E'^2 = -E^2 = \gamma$.

Let $S^\ddagger$ be the surface obtained by the first blowup. Let $H^\ddagger, L^\ddagger, E^\ddagger$ be the proper transforms in $S^\ddagger$ of the tangent line H to $\tilde{\lambda}, L_\infty, \overline{U}$. Then $H^\ddagger, L^\ddagger$ are fibers of a $\mathbb{P}^1$-ruling of $S^\ddagger$. We have $E^\ddagger \cdot L^\ddagger = \lambda \cdot L^\ddagger = c_1 - p_1$ and $E^\ddagger \cdot H^\ddagger = \tilde{\lambda} \cdot H^\ddagger + f$, where f is the intersection of H and $\overline{U}$ at finite distance. We have $f \geq 2$ since otherwise H is a good asymptote. If $\tilde{r} = 0$, then $\tilde{\lambda} \cdot H^\ddagger = \tilde{p}_1$. If $\tilde{r} > 0$, then $\tilde{\lambda} \cdot H^\ddagger \geq \tilde{c}_1$. Hence we have the following.

Lemma 4.2. *(a)* $c_1 - p_1 \geq \tilde{p}_1 + 2 \geq 3$.

(b) If $\tilde{r} > 0$, *then* $c_1 - p_1 \geq \tilde{c}_1 + 2 \geq 4$.

4.3. The formulas 3.2 take form

$$\gamma + 2c_1 + \tilde{c}_1 = \sum_{i\geq 1} p_i + \tilde{r}\tilde{c}_1 + \sum_{i\geq 1} \tilde{p}_i \tag{1}$$

and

$$\gamma + c_1^2 + 2c_1\tilde{c}_1 = \sum_{i\geq 1} p_i c_i + \tilde{r}\tilde{c}_1^2 + \sum_{i\geq 1} \tilde{p}_i\tilde{c}_i + 2p_1\tilde{c}_1. \tag{2}$$

We multiply (1) by $\tilde{c}_1$ and subtract (2). We obtain

$$\gamma(\tilde{c}_1 - 1) = (c_1 + \tilde{c}_1)(c_1 - \tilde{c}_1 - p_1) + \sum_{i\geq 2} p_i(\tilde{c}_1 - c_i) + \sum_{i\geq 2} \tilde{p}_i(\tilde{c}_1 - \tilde{c}_i). \tag{3}$$

Lemma 4.4. $\gamma \leq 8$.

Proof. Suppose that $\gamma = 9$. By 2.5, $\varepsilon = 0$ and $t = 2$. Hence for both λ and $\tilde{\lambda}$ we have the situation described in 2.2.1, i.e., we have two maximal twigs of $D + E$ composed of (-2)-curves. If either of these has more than one component, or if $D + E$ has a third maximal twig, we reach a contradiction with 1.10. Hence $D + E$ has precisely two maximal twigs, and they are (-2)-tips. It follows that $h = \tilde{h} = 1$. Let

$$L_\infty - -T - -T_1$$

be the upper chain created by the pair $\binom{c_1}{p_1}$, i.e., the chain having L_∞ and T_1 as tips. Then the chain $L_\infty - -T$ contracts to a (-2)-curve. So either
(i) $L_\infty^2 = -2$ and $T = \emptyset$ or
(ii) $L_\infty^2 = -1$ and T has the form $(-2) - - \cdots - -(-2) - -(-3)$ with a number $l \geq 0$ of (-2)-curves.
We find $p_1 = 1, c_1 = 3$ in the first case and $p_1 = 2l + 3, c_1 = 2l + 5$ in the second and we reach contradiction with 4.2(a). $\square$

Lemma 4.5. *$\tilde{c}_1 > 1$, i.e., $\tilde{\lambda}$ is not smooth. In particular, $\tilde{h} \geq 1$ and $\tilde{c}_{\tilde{h}} > \tilde{p}_{\tilde{h}}$.*

Proof. Suppose that $\tilde{c}_1 = 1$. The formulas 4(1) and (2) take the form

$$\gamma + 2c_1 + 1 = p_1 + \sum_{i \geq 2} p_i \tag{1}$$

and

$$\gamma + c_1^2 + 2c_1 = p_1 c_1 + 2p_1 + \sum_{i \geq 2} p_i c_i. \tag{2}$$

We write them in the following form.

$$\gamma + 1 + c_1 + \alpha c_2 = \sum_{i \geq 2} p_i \tag{3}$$

and

$$\gamma + \alpha c_1 c_2 + 2\alpha c_2 = \sum_{i \geq 2} p_i c_i. \tag{4}$$

We multiply (3) by c_2 and subtract (4). We get

$$c_2(1 + \gamma) + c_2 c_1 + \alpha c_2^2 \geq \gamma + \alpha c_1 c_2 + 2\alpha c_2. \tag{5}$$

From this

$$1+\gamma+c_1+\alpha c_2 > \alpha c_1 + 2\alpha.$$

Let $c_1 = kc_2, p_1 = lc_2$. Then $\alpha = k - l$. We get

$$\gamma - 2\alpha \geq c_2(k\alpha - \alpha - k). \tag{6}$$

Suppose that $\alpha \geq 3$ Then $k = \alpha + l \geq 4$. We obtain $\gamma - 6 \geq c_2(2k-3) \geq 5c_2$, a contradiction since $\gamma \leq 8$. Thus $\alpha = 2$. From (5) we get

$$c_2^2(k-2) + c_2(3-\gamma) + \gamma \leq 0.$$

Therefore $\Delta = (3-\gamma)^2 - 4\gamma(k-2) \geq 0$. Since $k \geq \alpha + 1 = 3$ we have $(3-\gamma)^2 - 4\gamma \geq 0$ and finally $\gamma^2 - 10\gamma + 9 \geq 0$. From this, since $\gamma > 2\alpha = 4$ by (6), we obtain $\gamma \geq 9$, a contradiction in view of 4.4. □

Lemma 4.6. $\tilde{c}_1 > c_i$ for $i \geq 2$.

Proof. It is enough to show that $c_2 \geq \tilde{c}_1$ is not possible. Multiply 4(1) by c_2 and subtract 4(2). We obtain

$$\begin{aligned}\gamma(c_2-1) = {} & -2c_1c_2 - \tilde{c}_1c_2 + c_1^2 + 2c_1\tilde{c}_1 + p_1c_2 - p_1c_1 - 2p_1\tilde{c}_1 \\ & + \sum_{i\geq 2} p_i(c_2 - c_i) + \sum_{i\geq 2} \tilde{p}_i(c_2 - \tilde{c}_i) + \tilde{r}\tilde{c}_1(c_2 - \tilde{c}_1) + \tilde{p}_1(c_2 - \tilde{c}_1).\end{aligned}$$

Let $c_1 = kc_2$, $p_1 = lc_2$. Then $\alpha = k - l$, hence $k \geq l + 2$.

If $c_2 \geq \tilde{c}_1$ we get

$$\gamma(c_2 - 1) > (-2k + k^2 + l - kl)c_2^2 + c_2(-\tilde{c}_1 + 2k\tilde{c}_1 - 2l\tilde{c}_1).$$

From this

$$\gamma > (-2k + k^2 + l - kl)c_2 - \tilde{c}_1 + 2k\tilde{c}_1 - 2l\tilde{c}_1.$$

Now $-2k+k^2+l-kl = (k-l)(k-2)-l \geq 2(k-2)-l = k+k-l-4 \geq k-2 \geq 1$. Since $c_2 > \tilde{c}_1$ we obtain

$$\gamma > \tilde{c}_1(k^2 - kl - l - 1) = \tilde{c}_1(k+1)(k-l-1) \geq 4\tilde{c}_1.$$

Now $\gamma \leq 8$ by 4.4, hence $\tilde{c}_1 < 2$, a contradiction in view of 4.5 □

Lemma 4.7. *Let* $\beta = c_1 - p_1 - \tilde{c}_1$. *If* $\tilde{r} > 0$ *then* $2 \leq \beta \leq 3$.

Proof. $\tilde{r} > 0$ implies $\beta \geq 2$ by 4.2(b). By 4(3), 4.6 and 4.2 we find $\gamma(\tilde{c}_1 - 1) \geq \beta(c_1 + \tilde{c}_1) \geq \beta(2\tilde{c}_1 + 3)$. In view of 4.6 this gives $\beta < \frac{\gamma}{2} \leq 4$. □

4.8. We consider again the surface Y introduced in 2.4.1. Let Q_1 (resp. $\tilde{Q}_1$) denote the maximal twig of $D+E$ which meets C (resp. $\tilde{C}$). If $h>1$ then Q_1 is the lower subchain produced by the pair $\binom{c_h}{p_h}$. If $h=1$ then Q_1 is the image under Ψ of the maximal twig of $D'+E'$ which has L'_∞ as a tip. In any case $\tilde{Q}_1$ is the lower subchain produced by the pair $\binom{\tilde{c}_{\tilde{h}}}{\tilde{p}_{\tilde{h}}}$. We write $D = Q_1 + C + Q_0 + \tilde{C} + \tilde{Q}_1$ and put

4.8.1 $$Q = Q_1 + Q_0 + \tilde{Q}_1 + E \quad \text{and} \quad Y = \overline{S} \setminus Q.$$

We note

4.8.2 $$\chi(Y) = -1.$$

Lemma 4.9. *If $\gamma \geq 6$ then $2K_{\overline{S}} + Q \geq 0$. In particular $\overline{\kappa}(Y) \geq 0$.*

Proof. If $\gamma \geq 6$ then $\varepsilon = 0$ or 1 by 2.5. As in 2.4.1 we have $K_{\overline{S}} \cdot (K_{\overline{S}} + Q) = K_{\overline{S}} \cdot (K_{\overline{S}} + D + E) - K_{\overline{S}} \cdot C - K_{\overline{S}} \cdot \tilde{C} = 4 - \varepsilon$. If $\varepsilon = 0$ we obtain the result as in 2.4.1 and 2.4.2.
Suppose that $\varepsilon = 1$. We have $K_{\overline{S}} \cdot (K_{\overline{S}} + Q) = 3$. By 2.5 we have $t \geq 1$. Hence Q_1 or $\tilde{Q}_1$, say $\tilde{Q}_1$, consists of (-2)-curves. Then the Riemann-Roch Theorem gives $h^0(-K_{\overline{S}} - Q_0 - Q_1 - E) + h^0(2K_{\overline{S}} + Q_0 + Q_1 + E) > 0$. By 2.4.2 we have $2K_{\overline{S}} + D + E \geq 0$. If $-K_{\overline{S}} - Q_0 - Q_1 - E \geq 0$ then $K_{\overline{S}} + C + \tilde{C} + \tilde{Q}_1 = 2K_{\overline{S}} + D + E + (-K_{\overline{S}} - Q_0 - Q_1 - E) \geq 0$. This implies that $K_{\overline{S}} \geq 0$, a contradiction. Thus $2K_{\overline{S}} + Q_0 + Q_1 + E \geq 0$ and hence $2K_{\overline{S}} + Q \geq 0$. □

Lemma 4.10. *If $\gamma \geq 6$ then the pair $(\overline{S}, Q)$ is almost minimal.*

Proof. Suppose that Q_0 is contractible (to a quotient singular point), i.e., has negative definite intersection matrix and is a chain or a contractible fork. Then Q has negative definite intersection matrix and the result follows as in 2.4.3.
Suppose that Q_0 is not contractible and that $(\overline{S}, Q)$ is not almost minimal. We need the following.

Sublemma *There is no (-1)-curve L in $\overline{S}$ such that $L \cdot Q_0 = 0$, L meets two connected components of $Q_1 + E + \tilde{Q}_1$ and together with these components contracts to a smooth point.*

Proof. Suppose that such an L exists. Let $\pi\colon \overline{S} \to \overline{X}$ be the contraction of L and the precisely two connected components of $Q_1 + E + \tilde{Q}_1$ it meets to a smooth point q_1. Let Q_2 be the third connected component. The surface

$\overline{S} \setminus Q_0$ is simply connected since it contains $\mathbb{C}^2$. Therefore $X = \overline{X} \setminus Q_0$ is simply connected. Let $\overline{X} \to X'$ be the contraction of Q_2 to a cyclic singular point q_2. Then $X' = \overline{X}' \setminus Q_0$ is simply connected. It is also easy to compute that $b_2(X') = 0$. Hence X' is contractible. Moreover $\overline{\kappa}(X') = \overline{\kappa}(X) = \overline{\kappa}(\overline{S} \setminus Q_0) = -\infty$. By [10] the logarithmic Kodaira dimension of the smooth locus of X' is negative. Since q_1 is smooth, $\overline{\kappa}(X' \setminus \{q_1, q_2\} = \overline{\kappa}(\overline{S} \setminus (Q \cup L)) = -\infty$. It follows that $\overline{\kappa}(Y) = \overline{\kappa}(S \setminus (Q_1 + E + \tilde{Q}_1)) = -\infty$, a contradiction in view of 4.9. □

Let $(\overline{Y'}, T')$ be an almost minimal model of $(\overline{S}, Q)$. $\overline{Y'}$ is obtained from $\overline{S}$ by a sequence of birational morphisms $p_i : \overline{Y}_i \to \overline{Y}_{i+1}, \overline{S} = \overline{Y}_0 \to \overline{Y}_1 \to \cdots \to \overline{Y}_\ell = \overline{Y}'$. Let $T_i = (p_{i-1})_*(T_{i-1}), T_0 = Q, T' = T_\ell$. Let $Y_i = \overline{Y}_i \setminus T_i$. For every i there exists a (-1)-curve $C_i \not\subseteq T_i$ such that $p_i : \overline{Y}_i \to \overline{Y}_{i+1}$ is the *NC-minimalization* of $C_i + T_i$. Finally, for the almost minimal model $(\overline{Y}', T')$, the negative part $(K_{\overline{Y}'} + T')^-$ coincides with the bark $\mathrm{Bk}(T')$. The contractions in this process involve only curves (or their images) contained in the support of $(K_{\overline{Y}} + T_0)^-$. We put

$$e(\overline{Y_i}, T_i) = \chi(\overline{Y_i} \setminus T_i) + \#\{\text{connected components of } T_i\}.$$

We find by an elementary calculation that $e(\overline{Y}_{i+1}, T_{i+1}) = e(\overline{Y}_i, T_i) - 1$. Hence $e(\overline{Y}', T') = e(\overline{S}, Q) - \ell = 3 - \ell$.

Let k denote the number of connected components of T' which contract to quotient singularities, with local fundamental groups G_j. Let u denote the number of connected components of T'. By 1.13 we have

$$\chi(Y') + \frac{k}{2} \geq \chi(Y') + \sum_{i=1}^{k} \frac{1}{|G_j|} \geq 0.$$

We have $\chi(Y') = e(\overline{Y}', T') - u = 3 - \ell - u$. We obtain

$$3 - \ell - u + \frac{k}{2} \geq 0.$$

Since Q_0 is not contractible, $k \leq u - 1$. Also $\ell \geq 1$ since $(\overline{S}, Q)$ is not almost minimal. We obtain $u \leq 3$ and $k \leq 2$. By the Sublemma above, $\chi(Y') \leq \chi(Y) = -1$ (in the minimalization process $\chi(Y_{i+1}) > \chi(Y_i)$ if and only if C_i meets two connected components of T_i and contracts to a smooth point together with these connected components). From 1.13 we get that $k > 1$. Hence $k = 2$, $\ell = 1$, $u = 3$. Also $\chi(Y') = -1 = \chi(Y)$. Again by 1.13 $(*)$ the two contractible connected components of T' are (-2)-curves.

We claim that C_0 meets Q_0.
Suppose otherwise. Suppose that C_0 meets only one connected component Q_2 of $Q_1+E+\tilde{Q}_1$. N ow $\chi(Y')=\chi(Y)$ implies that C_0+Q_2 must contract to a smooth point. It follows that $Q_2 \neq E$ since E is not a (-2)-curve (we have $\gamma \geq 6$) and hence $E+C_0$ cannot contract to a smooth point. Therefore E is untouched under p_0, so $E^2 \neq -2$ in T', and we have a contradiction to $(*)$. Thus C_0 meets two connected components of $Q_1+E+\tilde{Q}_1$ and together with these components contracts to a (-2)-curve. Let $\overline{X}$ be the image of $\overline{Y}'$ under the contraction of the two connected components of T' that are (-2)-curves to singular points. Put $X=\overline{X}\setminus Q_0$. We have $\overline{\kappa}(X)=-\infty$, X is simply-connected and has trivial Betti numbers. Hence X is contractible. By [10] the smooth locus of X has negative Kodaira dimension. It follows that $\overline{\kappa}(Y)=-\infty$, in contradiction to 4.9.

Hence C_0 meets Q_0 and, since $\chi(Y')=\chi(Y)$, one of connected components of $Q_1+E+\tilde{Q}_1$. The other two connected components are (-2)-curves. It follows that C_0 meets E and that $Q_1, \tilde{Q}_1$ are (-2)-curves.

Suppose that $h>1$. Then $d(Q_1)=c_h=2$, $d(\tilde{Q}_1)=\tilde{c}_{\tilde{h}}=2$. By 4.3(2), 4 divides γ. Thus $\gamma=8$, $\varepsilon=0$. Now we get contradiction with 1.10 since $D+E$ has two (-2)-tips and at least one other maximal twig which meets T_1. Hence $h=1$ and Q_1 is a tip of $D+E$ which meets T_1. We reach a contradiction as in the proof of 4.4.

4.11. Put $\omega=h_\Psi$. We have $h_\Phi=1+\tilde{r}+\tilde{h}+h-1$. By 3.3 we obtain

$$r+h+\tilde{h}=2+\varepsilon+\gamma+\omega.$$

Lemma 4.12. *If* $\gamma \geq 5$ *then* Q_0 *is not a chain.*

Proof.

We put $P=\sum_{i\geq 2} p_i$, $\tilde{P}=\sum_{i\geq 2}\tilde{p}_i$. With β as in 4.7 we get from 4

$$(*) \qquad \gamma(\tilde{c}_1-1)=\beta(c_1+\tilde{c}_1)+\sum_{i\geq 2} p_i(\tilde{c}_1-c_i)+\sum_{i\geq 2}\tilde{p}_i(\tilde{c}_1-\tilde{c}_i).$$

Suppose that Q_0 is a chain. We then have four cases:

(a) $h=1$, $\tilde{h}=1$

or

(b) $h=1$, $\tilde{h}=2$, $\tilde{p}_2=1$

or

(c) $h = 2$, $p_2 = 1$, $\tilde{h} = 1$

or

(d) $h = 2$, $p_2 = 1$, $\tilde{h} = 2$, $\tilde{p}_2 = 1$.

We note the following.
(i) If $h = 2$, then λ produces two tips in $D + E$, one of them a (-2)-tip.
(ii) If $\tilde{p}_{\tilde{h}} = 1$, in particular if $\tilde{h} = 2$, then $\tilde{\lambda}$ produces a (-2)-tip in $D + E$.

We observe that $h + \tilde{h} \leq 4$. From 4.11 we get $\tilde{r} \geq 3$. By 4.7 we have $2 \leq \beta \leq 3$. Notice that $P + \tilde{P} = h + \tilde{h} - 2$. From 4.11 we get $P + \tilde{P} \geq \gamma - \tilde{r}$. We have $c_1 - p_1 = \tilde{c}_1 + \beta$. From 4(1) we get

$$\gamma + c_1 + \beta + 2\tilde{c}_1 \geq \tilde{p}_1 + \tilde{r}(\tilde{c}_1 - 1) + \gamma.$$

So

$$c_1 + \tilde{c}_1 \geq \tilde{p}_1 - \beta + \tilde{r}(\tilde{c}_1 - 1) - \tilde{c}_1.$$

From $(*)$ we obtain

$$(**)\quad \gamma\tilde{c}_1 - \gamma \geq \beta(\tilde{p}_1 - \beta + \tilde{r}(\tilde{c}_1 - 1) - \tilde{c}_1) + \sum_{i\geq 2} p_i(\tilde{c}_1 - c_i) + \sum_{i\geq 2} \tilde{p}_i(\tilde{c}_1 - \tilde{c}_i).$$

(1) Suppose that $\beta = 3$. From 4.11 we have $\tilde{r} \geq \gamma - 2$ since $h + \tilde{h} \leq 4$. Using this we get

$$2\gamma + 3 \geq \tilde{c}_1(2\gamma - 9) + 3\tilde{p}_1.$$

Since $\beta = 3$, $\gamma \geq 7$ since $(*)$ and $c_1 = p_1 + \tilde{c}_1 + \beta$ gives $\gamma \geq 3(\frac{c_1+\tilde{c}_1}{\tilde{c}_1-1}) \geq 3(2 + \frac{6}{\tilde{c}_1-1}) > 6$. We obtain $17 \geq 5\tilde{c}_1 + 3\tilde{p}_1$. This implies $\tilde{c}_1 = 2$, $\tilde{p}_1 = 1$, $\tilde{h} = 1$. From $(**)$ we now obtain $\tilde{r} \leq \gamma - 1$. By 4.11, $1 + h + \gamma - 1 \geq 2 + \varepsilon + \gamma$ hence $h \geq 2 + \varepsilon$. This gives $\varepsilon = 0$, $h = 2$. In view of (i) and (ii) we reach contradiction with 1.10.

(2) Suppose that $\beta = 2$.
(2.1) Suppose also that $\tilde{r} \geq \gamma - 1$. $(**)$ gives

$$(***)\qquad \gamma + 2 \geq 2\tilde{p}_1 + (\gamma - 4)\tilde{c}_1.$$

Since $\gamma \geq 5$ we get $7 \geq 2\tilde{p}_1 + \tilde{c}_1$. This implies $\tilde{p}_1 = 1$ or $\tilde{p}_1 = 2$ and $\tilde{c}_1 = 3$. In both cases $\tilde{h} = 1$.
(2.1.1) Suppose also $h = 2$ and $\tilde{p}_1 = 1$. Then $\varepsilon = 1$, otherwise we reach contradiction with 1.10 as above. So $\tilde{r} \geq \gamma$ by 4.11 and $(**)$ gives $\gamma + 4 \geq (\gamma - 2)\tilde{c}_1 + 2\tilde{p}_1 + p_2(\tilde{c}_1 - c_2)$. For $\gamma \geq 6$ we get $10 \geq 2\tilde{p}_1 + 4\tilde{c}_1 + p_2(\tilde{c}_1 - c_2)$, so $5 > \tilde{p}_1 + 2\tilde{c}_1$ in view of 4.6 and we have a contradiction since $\tilde{c}_1 \geq 2$. For

$\gamma = 5$ we get $\tilde{c}_1 = 2$, $\tilde{c}_1 - c_2 = 1$ and hence $c_2 = 1$. But then $h = 1$.
(2.1.2) Suppose also $h = 2$ and $\tilde{p}_1 = 2$. Then $\gamma = 5$ by $(***)$. Now $(**)$ gives $15 \geq 4\tilde{r}$, but $\tilde{r} \geq \gamma - 1 = 4$, a contradiction.
(2.1.3) Suppose also $h = 1$. If $\varepsilon + \omega \geq 1$ then 4.11 gives $\tilde{r} \geq \gamma + 1$ and $(**)$ gives $\gamma + 6 \geq 2\tilde{p}_1 + \gamma\tilde{c}_1$ and further $11 \geq 2\tilde{p}_1 + 5\tilde{c}_1$; a contradiction. Hence $\varepsilon = \omega = 0$ and $\tilde{r} = \gamma$. From 4(1) and (3) we get

$$\gamma + 2c_1 + \tilde{c}_1 = p_1 + \tilde{p}_1 + \gamma\tilde{c}_1$$

and

$$\gamma(\tilde{c}_1 - 1) = 2(c_1 + \tilde{c}_1).$$

From the second equality we have $\gamma\tilde{c}_1 = \gamma + 2c_1 + 2\tilde{c}_1$. We substitute it to the first equality and get

$$\gamma = p_1 + \tilde{p}_1 + \tilde{c}_1 + \gamma,$$

a contradiction.
(2.2) Suppose also that $\tilde{r} \leq \gamma - 2$. From 4.11 we obtain $\gamma - 2 + h + \tilde{h} \geq 2 + \varepsilon + \gamma + \varepsilon$, i.e., $h + \tilde{h} \geq 4 + \varepsilon + \omega$. It gives $h = \tilde{h} = 2$ and $\varepsilon = 0$. We reach contradiction with 1.10 as before. □

Lemma 4.13. *If $\gamma \geq 6$ then Q_0 is not a contractible fork.*

Proof. Let H' the exceptional curve produced by the first blowing up in Φ^{-1}. Let H denote the proper transform of H' in $\overline{S}$. In view of 4.5 and $c_1 > p_1$ we have to blow up at least twice on H'. Hence $H^2 \leq -3$.

Suppose that Q_0 is a fork. Then either T_1 or $\tilde{T}_1$ is a branching component in Q_0.
Suppose T_1 is branching. Then the branches are: R_1, containing $\Psi(L'_\infty)$; R_2, containing H and $\tilde{T}_1$; R_3, meeting C. R_1 and R_2 are maximal twigs of $D + E$.
Suppose $\tilde{T}_1$ is branching. Then the branches are: R_1, containing $\Psi(L'_\infty), T_1, H$; R_2, the lower part of the chain produced by $\binom{\tilde{c}_1}{\tilde{p}_1}$; R_3, meeting $\tilde{C}$. R_1 and R_3 are maximal twigs of $D + E$.

Suppose that Q_0 is a contractible fork, but not of type $(2, 2, n)$. Suppose that T_1 is a branching component in Q_0.

If $h > 2$, then R_3 has a (≤ -3)-component and hence at most two components. It follows that $h \leq 3$. Also $\tilde{h} = 1$ or $\tilde{h} = 2$ and $\tilde{p}_2 = 1$. In any case $h + \tilde{h} \leq 5$. By 4.11, $\tilde{r} \geq 3$. But now the twig R_2 has at least 4 components and one of them, H, is a $(\leq -3)-$curve. This is impossible.

Suppose that $\tilde{T}_1$ is a branching component. Again $h+\tilde{h} \leq 5$, so $\tilde{r} \geq 3$. It follows that R_1 contains at least 4 components and we reach a contradiction as above.

Suppose that Q_0 is contractible of type $(2,2,n)$.
Suppose that T_1 is a branching component in Q_0. Since $H^2 \leq -3$, R_2 is the "long" n-twig of Q_0 and R_1, R_3 are single (-2)-curves. We have $d(Q_0) = 4(n(b-1) - \tilde{n})$ where $\tilde{n}$ denotes the determinant of the twig R_2 with the tip of R_2 meeting T_1 removed and $b = -T_1^2$. We have $h = 2$ and $p_2 = 2$ since R_3 is a single (-2)-curve. So $c_2 > p_2$, which implies in particular that $b \geq 3$. Since R_2 does not consist of (-2)-curves we have $n - \tilde{n} > 1$. We obtain $d(Q_0) \geq 4(2n - \tilde{n}) = 4(n + n - \tilde{n}) \geq 4(3+2) = 20$. We have $d(Q_1) = c_2 \geq 3$. From 1.13 we get

$$(*) \qquad 1 \leq \frac{1}{d(\tilde{Q}_1)} + \frac{1}{d(Q_1)} + \frac{1}{\gamma} + \frac{1}{d(Q_0)} \leq \frac{1}{d(\tilde{Q}_1)} + \frac{1}{3} + \frac{1}{6} + \frac{1}{20}.$$

This implies $d(\tilde{Q}_1) = 2$. It follows that $D + E$ has two (-2)-tips. Since it has at least three tips, $\varepsilon = 1$ in view of 1.10. Thus $\gamma = 6$ or 7 by 2.5. $(*)$ gives $1 \leq \frac{1}{2} + \frac{1}{d(Q_1)} + \frac{1}{6} + \frac{1}{20}$, which implies $d(Q_1) \leq 3$. Since $d(Q_1) = c_2 \geq 3$ we get $c_2 = 3$. Moreover, since $\tilde{c}_1 > c_2$ by 4.6, $\tilde{h} = 2$.

Suppose that $\omega = 0$. Then $R_1 = \Psi(L'_\infty)$ and $\frac{c_1}{c_2} = 3$, $\frac{p_1}{c_2} = 1$. Hence $c_1 = 9$, $p_1 = 3$. From 4.11 we obtain $\tilde{r} = \gamma - 1$. Now 4(1) gives $\gamma + 12 = \tilde{p}_1 + (\gamma - 2)\tilde{c}_1$. Since $\tilde{h} = 2$, $\tilde{c}_2 \geq 4$ and $\tilde{p}_1 \geq 2$. We obtain $\gamma + 12 \geq 2 + 4(\gamma - 2)$. This implies $\gamma \leq 6$, so $\gamma = 6$. Also $\tilde{c}_1 = 4$ and $\tilde{p}_1 = 2$, $\tilde{r} = 5$. From 4(3) we obtain $6(4-1) = 13 \cdot 2 + 2(\tilde{c}_1 - 3) + \tilde{c}_1 - 2$, a contradiction.

Thus $\omega \geq 1$. From 4.11 we now get $\tilde{r} \geq \gamma$. We have $c_1 - p_1 = \tilde{c}_1 + \beta$, so 4(1) gives $\gamma + c_1 + \beta + 2\tilde{c}_1 \geq \tilde{p}_1 + \gamma\tilde{c}_1 + 2 + 1$, i.e., $c_1 \geq \tilde{p}_1 + (\gamma - 2)\tilde{c}_1 + 3 - \gamma - \beta$. Now 4(3) gives

$$\gamma\tilde{c}_1 - \gamma \geq (\tilde{p}_1 + (\gamma - 1)\tilde{c}_1 + 3 - \gamma - \beta)\beta + \tilde{c}_1 - \tilde{c}_2 + 2(\tilde{c}_1 - 3).$$

If $\beta = 3$ then $2\gamma + 6 \geq 3\tilde{p}_1 + (2\gamma - 1)\tilde{c}_1 + \tilde{c}_1 - \tilde{c}_2$. Since $\tilde{p}_1 \geq 2$ we have $2\gamma > (2\gamma - 1)\tilde{c}_1$. It is a contradiction since $\gamma = 6$ or 7 and $\tilde{c}_1 \geq 4$. If $\beta = 2$, then $\gamma \geq 2\tilde{p}_1 + \gamma\tilde{c}_1 - 4 + \tilde{c}_1 - \tilde{c}_2 > \gamma\tilde{c}_1$; a contradiction.

Assume that $\tilde{T}_1$ is a branching in Q_0. Now R_2 and R_3 are single (-2)-curves. Hence $\tilde{h} = 2$, $\frac{\tilde{c}_1}{\tilde{c}_2} = 2$ and $\tilde{p}_2 = 2$. We again have $d(Q_0) \geq 20$ and, by 1.13, we get $d(Q_1) = 2$ and $d(\tilde{Q}_1) = \tilde{c}_2 = 3$. Hence $\tilde{c}_1 = 6, \tilde{p}_1 = 3$. From 4(3) we get

$$5\gamma = (c_1 + 6)\beta + \tilde{p}_2(\tilde{c}_1 - \tilde{c}_2) + p_2(\tilde{c}_1 - c_2).$$

Suppose that $h = 1$. Then $5\gamma = (c_1 + 6)\beta + 6$. As above, $\gamma = 6$ or 7. Since $\beta = 2$ or 3, β divides γ. Hence $\gamma = 6$. Now $30 \geq 2c_1 + 18$, which gives $c_1 \leq 6 = \tilde{c}_1$. But $c_1 = p_1 + \tilde{c}_1 + \beta > \tilde{c}_1$. We reach a contradiction.
Suppose that $h = 2$. Then $c_2 = d(Q_1) = 2$ and $p_2 = 1$. We get $5\gamma = (c_1 + 6)\beta + 2(\tilde{c}_1 - \tilde{c}_2) + \tilde{c}_1 - 2 = (c_1 + 6)\beta + 10$. Since $\gamma \leq 7$, we have $35 \geq 2c_1 + 12 + 10$ and again $c_1 \leq 6$, a contradiction. □

Proposition 4.14. $\gamma \leq 5$.

Proof. Suppose that $\gamma \geq 6$. By 4.12 and 4.13, Q_0 is not contractible. By 1.13 we have $\frac{1}{d(Q_1)} + \frac{1}{d(\tilde{Q}_1)} + \frac{1}{\gamma} \geq 1$. Since $\gamma \geq 6$ we have

(i) $d(Q_1) = d(\tilde{Q}_1) = 2$
 or

(ii) $\{d(Q_1), d(\tilde{Q}_1)\} = \{2, 3\}$, $\gamma = 6$. In this case $\overline{\kappa}(Y) = 0$ or 1 since otherwise $((K_{\overline{X}} + D)^+)^2 > 0$ in 1.13 by 4.9. We record that $(BkE)^2 = -\frac{4}{\gamma}$. Put $B_0 = (\mathrm{Bk}\, Q_1)^2 + (\mathrm{Bk}\, \tilde{Q}_1)^2$. Then $B_0 = -4$ if Q, $\tilde{Q}$ consist of (-2)-curves. Otherwise they are single curves and $B_0 = -\frac{4}{3} - 2 = -\frac{10}{3}$.

Consider (i). Then $\tilde{c}_{\tilde{h}} = d(\tilde{Q}_1) = 2$. Suppose that $h > 1$. Then $c_h = d(Q_1) = 2$. By 4(2), 4 divides γ. Hence $\gamma = 8$. So $\varepsilon = 0$ by 2.5 and we reach contradiction with 1.10. Suppose that $h = 1$. Then $C = T_1$ and Q_1 contains $\Psi(L'_\infty)$. We come to contradiction as in the proof of 4.4.

Consider (ii). By 4.9, $2K_{\overline{S}} + Q \geq 0$. Let $K_{\overline{S}} + Q = P + \mathrm{Bk}\, Q$ be the Zariski decomposition. We have $P \cdot (K_{\overline{S}} + Q) = P^2 = 0$ since $\kappa(Y) = 0$ or 1. Recall that P is *nef*. We get

$$0 = P \cdot (2K_{\overline{S}} + 2Q) = P \cdot (2K_{\overline{S}} + Q) + P \cdot Q \geq P \cdot Q.$$

Hence $P \cdot Q = P \cdot Q_0 = 0$. Fujita [2] classifies connected components Q_0 of a boundary divisor of an almost minimal surface such that $P \cdot Q_0 = 0$. In our case Q_0 is one of the following:

(a) a chain,

(b) a tree with exactly two branching components and four maximal twigs being (-2)-tips,

(c) a fork of type (d_1, d_2, d_3) where $\frac{1}{d_1} + \frac{1}{d_2} + \frac{1}{d_3} = 1$.

Case (a) is ruled out by 4.12.
Consider (b). Then $(\mathrm{Bk}\, Q_0)^2 = -2$. Now $-4 - \varepsilon = (K_{\overline{S}} + Q)^2 = (\mathrm{Bk}\, Q)^2 =$

$-2-\frac{4}{\gamma}+B_0$ and $-\frac{4}{\gamma}+B_0$ is an integer. Since $\gamma \geq 6$ this implies $B_0=-\frac{10}{3}$ and $\gamma=6$. We get $-4-\varepsilon=(\operatorname{Bk} Q)^2=-2-4=-6$, which gives $\varepsilon=2$, a contradiction by 2.5.

Consider (c). We have

$$(*) \qquad -4-\varepsilon=(K_{\overline{S}}+Q)^2=(\operatorname{Bk} Q_0)^2+B_0-\frac{4}{6}.$$

Q_0 is of the type (3,3,3), (2,4,4) or (2,3,6). We find that $(\operatorname{Bk} Q_0)^2 \leq -1$. Since $\varepsilon \leq 1$ it follows from $(*)$ that $\varepsilon=1$. It follows next from $(*)$ that $B_0=-\frac{10}{3}$, i.e, that Q_1 and $\tilde{Q}_1$ are single curves, and that $(\operatorname{Bk} Q_0)^2=-1$. By examining all possibilities we see that every twig of Q_0 is a tip, i.e., $\#Q_0=4$. Hence $\#Q=7$, $b_2(\overline{S})=8$, $K_{\overline{S}}^2=2$. From $K_{\overline{S}}\cdot(K_{\overline{S}}+Q)=5$ we get $K_{\overline{S}}\cdot Q=-7$. Let B be the branching component of Q_0. Examining all possibilities we find that $B^2>0$. But $B=T_1$ or $B=\tilde{T}_1$ and both T_1 and $\tilde{T}_1$ are untouched under Ψ and hence are negative curves. We reach a contradiction. □

Lemma 4.15. $\gamma-\tilde{c}_1-p_1-\tilde{p}_1 \leq 0$.

Proof. Suppose the opposite. By 4.14, $\gamma \leq 5$, so $4 \geq \tilde{c}_1+p_1+\tilde{p}_1$. In view of 4.5 we get $\tilde{c}_1=2, \tilde{p}_1=1$ and $p_1=1$. It follows that $\tilde{h}=1$ and $c_2=1$. Hence $h=1$. By 4.11, $\tilde{r}=\gamma+\varepsilon+\omega$. Hence $\tilde{r}>0$, otherwise $\gamma=0$, which is impossible by 1.8. By 4.2, $2 \leq \beta \leq 3$. 4(3) gives $5 \geq \gamma=(c_1+2)\beta$. It follows that $c_1=0$, a contradiction. □

Theorem 4.16. *If U has no good asymptote then the branches of $\overline{U}$ at infinity can be separated by an automorphism of $\mathbb{C}^2$.*

Proof. Suppose the opposite. By results of section 3 we may assume that things are as in 4.1. By 4.14 we have $\gamma \leq 5$.

From 4(2) we get

$$(1) \qquad \gamma+\alpha c_2 c_1+2\alpha c_2 \tilde{c}_1=\sum_{i\geq 2} p_i c_i+\sum_{i\geq 2}\tilde{p}_i\tilde{c}_i+\tilde{r}\tilde{c}_1^2+\tilde{p}_1\tilde{c}_1.$$

Since $\tilde{p}_1 \leq \alpha c_2-2$ by 4.2(a)

$$(2) \qquad \gamma+\alpha c_2 c_1+\alpha c_2\tilde{c}_1+2\tilde{c}_1 \leq \sum_{i\geq 2} p_i c_i+\sum_{i\geq 2}\tilde{p}_i\tilde{c}_i+\tilde{r}\tilde{c}_1^2.$$

4(1) takes the form

$$(3) \qquad \gamma+2c_1+\tilde{c}_1-p_1-\tilde{p}_1=\sum_{i\geq 2} p_i+\sum_{i\geq 2}\tilde{p}_i+\tilde{r}\tilde{c}_1$$

and

$$2(c_1+\tilde{c}_1)+\gamma-\tilde{c}_1-p_1-\tilde{p}_1=\sum_{i\geq 2}p_i+\sum_{i\geq 2}\tilde{p}_i+\tilde{r}\tilde{c}_1.$$

By 4.15, $\gamma-\tilde{c}_1-p_1-\tilde{p}_1\leq 0$. Hence $c_1+\tilde{c}_1\geq\frac{1}{2}(\sum_{i\geq 2}p_i+\sum_{i\geq 2}\tilde{p}_i)+\frac{1}{2}\tilde{r}\tilde{c}_1$.

From (2) we get

$$\gamma+\frac{\alpha c_2}{2}(\sum_{i\geq 2}p_i+\sum_{i\geq 2}\tilde{p}_i+\tilde{r}\tilde{c}_1)+2\tilde{c}_1\leq\sum_{i\geq 2}p_ic_i+\sum_{i\geq 2}\tilde{p}_i\tilde{c}_i+\tilde{r}\tilde{c}_1^2.$$

Suppose that $\tilde{r}=0$. Then

$$\gamma+\frac{\alpha c_2}{2}(\sum_{i\geq 2}p_i+\sum_{i\geq 2}\tilde{p}_i)+2\tilde{c}_1\leq\sum_{i\geq 2}p_ic_i+\sum_{i\geq 2}\tilde{p}_i\tilde{c}_i.$$

Since $\alpha\geq 2$ this implies that $\frac{\alpha c_2}{2}<\tilde{c}_2$ and further $c_2<\tilde{c}_2$. Indeed, if $\tilde{c}_2\leq\frac{\alpha c_2}{2}$, then the above inequality gives

$$\gamma+\sum_{i\geq 2}p_ic_2+\sum_{i\geq 2}\tilde{p}_i\tilde{c}_2+2\tilde{c}_1\leq\sum_{i\geq 2}p_ic_i+\sum_{i\geq 2}\tilde{p}_i\tilde{c}_i$$

which is a contradiction. It follows that $\tilde{c}_2=\tilde{p}_1$, otherwise $\tilde{c}_2\leq\frac{\tilde{p}_1}{2}\leq\frac{\alpha c_2-2}{2}$. We rewrite (3) as

$$\gamma+c_1+\alpha c_2+\tilde{c}_1-\tilde{c}_2=\sum_{i\geq 2}p_i+\sum_{i\geq 2}\tilde{p}_i$$

and (1) as

$$\gamma+2\alpha c_2\tilde{c}_1+\alpha c_1c_2-\tilde{c}_2\tilde{c}_1=\sum_{i\geq 2}p_ic_i+\sum_{i\geq 2}\tilde{p}_i\tilde{c}_i.$$

Multiply the first equality by $\tilde{c}_2$ and subtract the second one. We obtain

$$\gamma(\tilde{c}_2-1)=(\alpha c_2-\tilde{c}_2)(c_1+2\tilde{c}_1-\tilde{c}_2)+\sum_{i\geq 2}p_i(\tilde{c}_2-c_i)+\sum_{i\geq 2}\tilde{p}_i(\tilde{c}_2-\tilde{c}_i).$$

Since $\tilde{c}_2>\frac{\alpha c_2}{2}\geq c_2\geq c_i$ for $i\geq 2$

$$\gamma(\tilde{c}_2-1)\geq(\alpha c_2-\tilde{c}_2)(c_1+2\tilde{c}_1-\tilde{c}_2).$$

We have $\tilde{c}_1\geq 2\tilde{c}_2$. Since $c_1>\tilde{p}_1=\tilde{c}_2$ and $\alpha c_2-\tilde{c}_2=c_1-p_1-\tilde{p}_1\geq 2$ by 4.2(a) we obtain $\gamma(\tilde{c}_2-1)\geq 2\cdot 4\tilde{c}_2$. It follows that $\gamma=9$, a contradiction.

Thus $\tilde{r}>0$. By 4.7, $2\leq\beta\leq 3$.

Suppose that $\tilde{c}_1\geq 2c_2$. Then $\tilde{c}_1-c_i\geq\frac{\tilde{c}_1}{2}$ for every $i\geq 2$. Also $\tilde{c}_1-\tilde{c}_i\geq\frac{\tilde{c}_1}{2}$ for every $i\geq 2$.

From 4(3) we obtain

$$\gamma(\tilde{c}_1 - 1) = (c_1 + \tilde{c}_1)\beta + \sum_{i\geq 2} p_i(\tilde{c}_1 - c_i) + \sum_{i\geq 2} \tilde{p}_i(\tilde{c}_1 - \tilde{c}_i) \tag{4}$$
$$\geq (c_1 + \tilde{c}_1)\beta + \frac{\tilde{c}_1}{2}(\sum_{i\geq 2} p_i + \sum_{i\geq 2} \tilde{p}_i).$$

It follows that $\gamma \geq 5$, i.e., $\gamma = 5$, and further $5 > 4 + \frac{1}{2}(\sum_{i\geq 2} p_i + \sum_{i\geq 2} \tilde{p}_i)$. It gives $\sum_{i\geq 2} p_i + \sum_{i\geq 2} \tilde{p}_i \leq 1$. It follows that $h = 1$ or $h = 2$ and $p_2 = 1$, and similarly for $\tilde{h}$, $\tilde{p}_2$. It follows that Q_0 is a chain in contradiction to 4.12.

Hence $\tilde{c}_1 < 2c_2$. 4.3(3) and 4.6 give $5\tilde{c}_1 > 2(c_1 + \tilde{c}_1)$ i.e. $c_1 < \frac{3}{2}\tilde{c}_1$. But $c_1 \geq 3c_2 > \frac{3}{2}\tilde{c}_1$ since $\alpha \geq 2$, a contradiction. □

References

1. P. Cassou-Nogues, M. Koras, P. Russell, Closed embeddings of $\mathbb{C}^*$ in $\mathbb{C}^2$, part I, J. Algebra **322**(2009).
2. T. Fujita, *On the topology of non-complete surfaces*, J. Fac. Sci. Univ. Tokyo, **29**(1982), 503-566.
3. T. Fujita, *On Zariski problem*, Proc. Japan. Acad. **55**(1979), 106-110.
4. R. V. Gurjar and M. Miyanishi, Affine lines on logarithmic $\mathbb{Q}$-homology planes, Math. Ann. **294**(1992), 463-482.
5. S. Iitaka, *On logarithmic Kodaira dimension of algebraic varieties*, Complex Analysis and Algebraic Geometry, Iwanami Shoten, Tokyo, 1977, 175-189.
6. Y. Kawamata, *On the classification of non-compact algebraic surfaces*, Lecture Notes in Mathematics **732**, Springer (1979).
7. R. Kobayashi, *Uniformization of complex surfaces*, Adv. Stud. Pure Math., **18**(1990), 313-394.
8. M. Koras, *$\mathbb{C}^*$ in $\mathbb{C}^2$ is birationally equivalent to a line*, Affine Algebraic Geometry: The Russell Festschrift, CRM Proceedings & Lecture Notes, 2011.
9. M. Koras, P. Russell, *$\mathbb{C}^*$-actions on $\mathbb{C}^3$: The smooth locus is not of hyperbolic type*, J. Algebraic Geometry, **8**(1999), 603-694.
10. M. Koras, P. Russell, *Contractible affine surfaces with quotient sigularities*, Transformation Groups, **12**, 2007, 293-340.
11. M. Kumar and P. Murthy: *Curves with negative self intersection on rational surfaces*, J. Math. Kyoto Univ., 22-4(1983), 767-777.
12. A. Langer, Logarithmic orbifold Euler numbers of surfaces with applications, Proc. London Math. Soc. (3) **2003**, no.2, 358-396.
13. M. Miyanishi, *Open Algebraic Surfaces*, CRM monograph series, Amer. Math. Soc., 2001.
14. Y. Miyaoka, *The maximal number of quotient singularities on surfaces with given numerical invariants*, Math. Ann., **268**(1984), 159-171., American Math. Society, Providence, Rhode Island, 2001.

15. P. Russell, *Hamburger-Noether expansions and approximate roots of polynomials*, Manuscripta Math., **31**(1980), 25-95.
16. P. Russell, *On affine-ruled rational surfaces*, Math. Ann., **255**(1981), 287-302.

Acknowledgements

The first author was supported by Polish Grant N N201 608640. The second author was supported by a grant from NSERC, Canada. The authors thank the referee for very careful reading, pointing out several errors and many comments.

Abhyankar-Sathaye Embedding Conjecture for a geometric case

Tomoaki Ohta

Department of Mathematics, Fukuoka University of Education, Akama-Bunkyomachi, Munakata, Fukuoka, 811-4192, Japan
E-mail: tomohta@fukuoka-edu.ac.jp

Dedicated to Professor Masayoshi Miyanishi on the occasion of his 70th birthday

We obtain the affirmative answer for a special case of the Abhyankar-Sathaye Embedding Conjecture for closed embeddings of $\mathbb{A}^{n-1}$ into $\mathbb{A}^n$ with $n \geq 3$. Indeed, let S be an irreducible hypersurface in $\mathbb{A}^n$ which is isomorphic to $\mathbb{A}^{n-1}$. Then we determine the structure of the canonical compactification of $S \cong \mathbb{A}^{n-1}$ in $\mathbb{P}^n$ under a geometric condition. For such a hypersurface S, we show that there exists a tame automorphism of $\mathbb{A}^n$ which transforms S onto a coordinate hyperplane in $\mathbb{A}^n$. Moreover, we obtain the standard equation of S up to affine automorphisms of $\mathbb{A}^n$.

Keywords: Affine space, automorphism, compactification.

1. Introduction

Let k be an algebraically closed field of characteristic $\mathrm{char}(k) \geq 0$, which is the ground field, and let n be a natural number. Let $k[x_1, \ldots, x_n]$ be the polynomial ring of n variables $x_1, \ldots, x_n$ over k. Let $\mathrm{Aut}(\mathbb{A}^n)$ be the group of polynomial automorphisms of the affine n-space $\mathbb{A}^n$ over k. Let us consider the *Abhyankar-Sathaye Embedding Conjecture*, which is closely related to the structure of $\mathrm{Aut}(\mathbb{A}^n)$, as follows:

Conjecture 1.1. *Let $S = \{f = 0\}$ be an irreducible hypersurface in $\mathbb{A}^n$ with defining polynomial $f \in k[x_1, \ldots, x_n]$ such that S is isomorphic to $\mathbb{A}^{n-1}$. Then there exists an automorphism of $\mathbb{A}^n$ which transforms S onto a coordinate hyperplane in $\mathbb{A}^n$.*

2010 *Mathematics Subject Classification*: Primary 14R10; Secondary 32J05.

This conjecture is usually considered under the assumption $\mathrm{char}(k) = 0$ since there exist some counterexamples to it when $\mathrm{char}(k) > 0$ (cf. [**Mo**], [**Sa**]). Now, for simplicity, we assume that k is the complex number field $\mathbb{C}$. For the case $n = 1$, the conjecture is true obviously. For the case $n = 2$, Abhyankar-Moh [**AM**] and Suzuki [**Su**] showed that the conjecture is true. For the case $n \geq 3$, the conjecture is still unsolved. In this paragraph, we summarize some partial affirmative answers for the case $n = 3$. Sathaye [**Sa**], Russell [**Ru**] and Miyanishi [**Mi1**] proved that the conjecture is true when $f = gx_3 + h$ with some $g, h \in \mathbb{C}[x_1, x_2]$. Wright [**Wr**] proved that the conjecture is true when $f = gx_3^m + h$ with some $g, h \in \mathbb{C}[x_1, x_2]$ and $m \geq 2$. Let $d := \deg f$ be the degree of f. For the case $d \leq 2$, we see that the conjecture is true easily. For the case $d = 3$, in Ohta [**Oh1**], we determined all the standard forms of f, which consists of nine different types. For the case where $d = 4$ and the closure of S in the projective 3-space $\mathbb{P}^3$ is normal, in Ohta [**Oh2**] and [**Oh3**], we determined all the standard forms of f, which consists of twenty-one different types. In both cases, we showed that the conjecture is true by constructing explicit automorphisms of $\mathbb{A}^3$ which transform S onto a coordinate hyperplane. On the other hand, Kishimoto [**Ki**] proved that the conjecture is true under a geometric condition for the closure of S in $\mathbb{P}^3$. Our main result in this paper is a generalization of his result. Before stating our result and his result, we give some notations and assumptions needed later.

From now on to the end of this paper, we assume that k is an algebraically closed field of $\mathrm{char}(k) \geq 0$ and $n \geq 3$. Let $S = \{f = 0\}$ be an irreducible hypersurface in $\mathbb{A}^n$ with defining polynomial $f \in k[x_1, \ldots, x_n]$ of degree $d := \deg f \geq 2$ such that S is isomorphic to $\mathbb{A}^{n-1}$. Here we embed $\mathbb{A}^n$ into $\mathbb{P}^n$ canonically as the complement of the hyperplane $H_0 := \{x_0 = 0\}$, where $(x_0 : x_1 : \cdots : x_n)$ is the homogeneous coordinate of $\mathbb{P}^n$. Let X be the closure of S in $\mathbb{P}^n$. Moreover, we assume the following:

Assumption 1.1. *There exists a linear $(n-2)$-space L contained in $X \cap H_0$ such that X has the multiplicity $d-1$ along L, that is,* $\mathrm{mult}_L X = d - 1$.

Then our main result in this paper is the following:

Theorem 1.1. *Let k be an algebraically closed field of* $\mathrm{char}(k) \geq 0$. *Let $S = \{f = 0\}$ be an irreducible hypersurface in $\mathbb{A}^n$ satisfying $n \geq 3$, $d \geq 2$, $S \cong \mathbb{A}^{n-1}$ and Assumption* 1.1. *Then there exists a tame automorphism of $\mathbb{A}^n$ which transforms S onto a coordinate hyperplane in $\mathbb{A}^n$. Moreover, up to affine automorphisms of $\mathbb{A}^n$, the standard equation of S is given by*

$$x_1^{d-1}(\nu_1 x_1 + \nu_2 x_2) + x_3 + x_1 (g_1 + g_2 x_2 + \cdots + g_n x_n) = 0$$

with $(\nu_1, \nu_2) = (1,0), (0,1)$ *and* $g_1, \ldots, g_n \in k[x_1]$ *satisfying the condition*

$$(\nu_2 x_1^{d-1} + x_1 g_2, 1 + x_1 g_3, x_1 g_4, \ldots, x_1 g_n) = k[x_1].$$

Remark 1.1. (1) For the closure X of S in $\mathbb{P}^n$, we note that the intersection $X \cap H_0$ coincides with L or $L \cup L'$, where L' is a linear $(n-2)$-space contained in $X \cap H_0$ with $L' \neq L$. In the standard equation of S, we note that $X \cap H_0 = L$ if and only if $(\nu_1, \nu_2) = (1,0)$, and that $X \cap H_0 = L \cup L'$ if and only if $(\nu_1, \nu_2) = (0,1)$. See §2 and §3 for the more detailed geometric structure of X.
(2) Kishimoto [**Ki**] proved Theorem 1.1 for the case where $k = \mathbb{C}$, $n = 3$ and $X \cap H_0 = L$. His proof was based on the result of Kaliman-Zaidenberg [**KZ**] concerning the generic fiber of an $\mathbb{A}^2$-fibration. We generalize his result to the case where k is any algebraically closed field and any $n \geq 3$. Moreover, our proof of Theorem 1.1 is very elementary since we shall use only fundamental facts of algebraic geometry.
(3) An analogue of Theorem 1.1 holds when $n = 2$. Indeed, let $S = \{f = 0\} \cong \mathbb{A}^1$ be an irreducible curve in $\mathbb{A}^2$ with $d := \deg f \geq 2$ satisfying Assumption 1.1 for the case $n = 2$. Then we note that L is a point of $H_0 \subset \mathbb{P}^2$. Since $\mathrm{mult}_L X = d - 1$, we may assume that $S = \{g_1(x_1) + g_2(x_1)x_2 = 0\}$ for some $g_1, g_2 \in k[x_1]$ by applying a suitable affine automorphism of $\mathbb{A}^2$. By Proposition 3.1 in §3, we have $(g_2(x_1)) = k[x_1]$, that is, g_2 is a non-zero constant in $k[x_1]$. Thus the curve S is transformed onto a coordinate line in $\mathbb{A}^2$ by a tame automorphism of $\mathbb{A}^2$: $x_1' = x_1$, $x_2' = g_1 + g_2 x_2$. In particular, the standard equation of S is given by $x_2 + g(x_1) = 0$ for some $g \in k[x_1]$ with $\deg g = d$, up to affine automorphisms of $\mathbb{A}^2$.

In the last part of this section, we give the definition of tame automorphisms of $\mathbb{A}^n$ precisely, although we have already used it. First we introduce some special subgroups of $\mathrm{Aut}(\mathbb{A}^n)$. Let $(x_1, \ldots, x_n)$ and $(x_1', \ldots, x_n')$ be two affine coordinates of $\mathbb{A}^n$. For an element (a_{ij}) of the general linear group $GL(n,k)$ over k and $b_1, \ldots, b_n \in k$, there exists an automorphism of $\mathbb{A}^n$ such that $x_i' = \sum_{j=1}^n a_{ij} x_j + b_i$ $(i = 1, \ldots, n)$. This type of automorphism is called an *affine automorphism* of $\mathbb{A}^n$. The set $A(n,k)$ of all affine automorphisms of $\mathbb{A}^n$ is a subgroup of $\mathrm{Aut}(\mathbb{A}^n)$. For $c_1, \ldots, c_n \in k^\times := k \setminus \{0\}$, $p_i \in k[x_{i+1}, \ldots, x_n]$ $(i = 1, \ldots, n-1)$ and $p_n \in k$, there exists an automorphism of $\mathbb{A}^n$ such that $x_i' = c_i x_i + p_i$ $(i = 1, \ldots, n)$. This type of automorphism is called a *de Jonquières automorphism* of $\mathbb{A}^n$. The set $J(n,k)$ of all de Jonquières automorphisms of $\mathbb{A}^n$ is a subgroup of $\mathrm{Aut}(\mathbb{A}^n)$. Let us denote by $T(n,k)$ the subgroup of $\mathrm{Aut}(\mathbb{A}^n)$ generated by $A(n,k)$ and $J(n,k)$. An element of $T(n,k)$ is called a *tame automorphism* of $\mathbb{A}^n$.

2. Preliminaries

In this section, we shall describe geometric properties of a hypersurface X of $\mathbb{P}^n$ satisfying $S \cong \mathbb{A}^{n-1}$ and Assumption 1.1 in §1. We use the same notations as those in §1. We note that $n \geq 3$ and $d \geq 2$. We also note that, since $\text{mult}_L X = d-1$, there exists a hyperplane L' in $H_0 \cong \mathbb{P}^{n-1}$ such that the restricted Cartier divisor $X|_{H_0} = (d-1)L + L'$, where the case $L' = L$ is allowed. Since we consider a blowing-up of $\mathbb{P}^n$ mainly, we often use the general facts of blowing-ups in Griffiths-Harris [**GH**] and Hartshorne [**Ha**] without comments.

Let $\sigma : V \to \mathbb{P}^n$ be a blowing-up of $\mathbb{P}^n$ along $L \cong \mathbb{P}^{n-2}$ with exceptional divisor $\overline{E} \cong \mathbb{P}^{n-2} \times \mathbb{P}^1$. Let $\psi : \mathbb{P}^n \cdots \to \mathbb{P}^1$ be a projection of $\mathbb{P}^n$ from L and $\overline{\psi} : V \to \mathbb{P}^1$ the resolution of indeterminacy of ψ. We note that $\overline{\psi} : V \to \mathbb{P}^1$ is a $\mathbb{P}^{n-1}$-bundle over $\mathbb{P}^1$ and the Picard group $\text{Pic}(V)$ of V is isomorphic to $\text{Pic}(\mathbb{P}^1) \oplus \mathbb{Z} \cong \mathbb{Z} \oplus \mathbb{Z}$ (cf. [**Ha**]). Let $\overline{T}$ be the proper transform of a closed algebraic subset T of $\mathbb{P}^n$ by σ, where T is not contained in L. Then we obtain a canonical divisor $K_V \sim \sigma^* K_{\mathbb{P}^n} + \overline{E} \sim -(n+1)\overline{H_0} - n\overline{E}$ of V. We also obtain a linear equivalence $\overline{X} \sim d\overline{H_0} + \overline{E}$ on V since $\text{mult}_L X = d - 1$.

Now we summarize fundamental properties of $\overline{E}$. First of all, we note that the restriction $\sigma|_{\overline{E}} : \overline{E} \to L$ is a trivial $\mathbb{P}^1$-bundle over $L \cong \mathbb{P}^{n-2}$ and the restriction $\overline{\psi}|_{\overline{E}} : \overline{E} \to \mathbb{P}^1$ is a trivial $\mathbb{P}^{n-2}$-bundle over $\mathbb{P}^1$. Thus we have an isomorphism $(\sigma|_{\overline{E}}, \overline{\psi}|_{\overline{E}}) : \overline{E} \cong L \times \mathbb{P}^1 \cong \mathbb{P}^{n-2} \times \mathbb{P}^1$. Let M_0 be a fiber of $\overline{\psi}|_{\overline{E}}$. Let F_0 be the inverse image of a hyperplane in $L \cong \mathbb{P}^{n-2}$ by $\sigma|_{\overline{E}}$. Let f_0 be a fiber of $\sigma|_{\overline{E}}$, which is isomorphic to $\mathbb{P}^1$. Then we have two isomorphisms $\sigma|_{M_0} : M_0 \cong L \cong \mathbb{P}^{n-2}$ and $(\sigma|_{F_0}, \overline{\psi}|_{F_0}) : F_0 \cong \mathbb{P}^{n-3} \times \mathbb{P}^1$. By noting that $\overline{E} \cong \mathbb{P}^{n-2} \times \mathbb{P}^1$, we obtain a canonical divisor $K_{\overline{E}} \sim (\sigma|_{\overline{E}})^* K_{\mathbb{P}^{n-2}} + (\psi|_{\overline{E}})^* K_{\mathbb{P}^1} \sim -(n-1)F_0 - 2M_0$ of $\overline{E}$. By the adjunction formula on $\overline{E}$, we obtain $\overline{E}|_{\overline{E}} \sim -M_0 + F_0$. From this, we obtain $\overline{X}|_{\overline{E}} \sim (d-1)M_0 + F_0$.

From now on, we investigate the $\mathbb{P}^{n-1}$-bundle $\overline{\psi} : V \to \mathbb{P}^1$ over $\mathbb{P}^1$ in detail. Let T be a closed subvariety of $\mathbb{P}^n$ with $\dim_k T \geq 1$ which is not contained in any hyperplane of $\mathbb{P}^n$ containing L. We note that X is one of such T. Then the morphism $\overline{\psi}|_{\overline{T}} : \overline{T} \to \mathbb{P}^1$ is surjective. Every fiber of the morphism $\overline{\psi}$, $\overline{\psi}|_{\overline{E}}$, $\overline{\psi}|_{\overline{T}}$ is the Cartier divisor $\overline{H}\,(\cong \mathbb{P}^{n-1})$, the restricted Cartier divisor $\overline{H}|_{\overline{E}}\,(\cong \mathbb{P}^{n-2})$, $\overline{H}|_{\overline{T}}$ respectively, where H is a hyperplane in $\mathbb{P}^n$ containing L. Any two fibers of the morphism $\overline{\psi}$, $\overline{\psi}|_{\overline{E}}$, $\overline{\psi}|_{\overline{T}}$ are linearly equivalent on V, $\overline{E}$, $\overline{T}$ respectively since any two points of $\mathbb{P}^1$ are linearly equivalent. Then we obtain the following lemmas:

Lemma 2.1. (cf. Proposition 3.2.1 in Beltrametti-Sommese [**BS**]) *Let Y and Z be smooth projective varieties over an algebraically closed field k of*

$\mathrm{char}(k) \geq 0$ with $\dim_k Z \geq 1$ and $\delta := \dim_k Y - \dim_k Z \geq 1$. Let $\rho : Y \to Z$ be a surjective morphism and $\mathcal{A}$ an ample invertible sheaf on Y. Assume that $F \cong \mathbb{P}^\delta$ and $\mathcal{A}|_F \cong \mathcal{O}_{\mathbb{P}^\delta}(1)$ for any closed fiber F of ρ. Then the push-forward $\mathcal{B} := \rho_*\mathcal{A}$ is an ample locally free sheaf on Z of $\mathrm{rank}\,\mathcal{B} = \delta + 1$ and the morphism $\rho : Y \to Z$ is isomorphic to a $\mathbb{P}^\delta$-bundle over Z with $(Y, \mathcal{A}) \cong (\mathbb{P}(\mathcal{B}), \mathcal{O}_{\mathbb{P}(\mathcal{B})}(1))$. In particular, $\mathrm{Pic}(Y) \cong \mathrm{Pic}(Z) \oplus \mathbb{Z}$.

Lemma 2.2. *One obtains the following:*

(i) *$\mathcal{L} := \mathcal{O}_V(\overline{X}) = \mathcal{O}_V(d\overline{H_0} + \overline{E})$ is an ample invertible sheaf on V.*
(ii) *$\mathcal{L}|_{\overline{H}} \sim \overline{X}|_{\overline{H}} \sim \mathcal{O}_{\mathbb{P}^{n-1}}(1)$ for every fiber $\overline{H} \cong \mathbb{P}^{n-1}$ of $\overline{\psi}$, where H is a hyperplane in $\mathbb{P}^n$ containing L. This means that $\overline{X}$ intersects $\overline{H}$ transversally in V, and $\overline{X} \cap \overline{H}$ is a hyperplane in $\overline{H} \cong \mathbb{P}^{n-1}$. In particular, $\overline{X}$ is smooth.*
(iii) *$\overline{\psi} : V \to \mathbb{P}^1$ is a $\mathbb{P}^{n-1}$-bundle over $\mathbb{P}^1$ with $(V, \mathcal{L}) \cong (\mathbb{P}(\mathcal{E}), \mathcal{O}_{\mathbb{P}(\mathcal{E})}(1))$, where $\mathcal{E} := \overline{\psi}_*\mathcal{L}$.*
(iv) *$\overline{\psi}|_{\overline{X}} : \overline{X} \to \mathbb{P}^1$ is a $\mathbb{P}^{n-2}$-bundle over $\mathbb{P}^1$ with $(\overline{X}, \mathcal{L}_1) \cong (\mathbb{P}(\mathcal{E}_1), \mathcal{O}_{\mathbb{P}(\mathcal{E}_1)}(1))$, where $\mathcal{L}_1 := \mathcal{L}|_{\overline{X}}$ and $\mathcal{E}_1 := (\overline{\psi}|_{\overline{X}})_*\mathcal{L}_1$. In particular, $\mathrm{Pic}(\overline{X}) \cong \mathbb{Z} \oplus \mathbb{Z}$.*
(v) *$\overline{X} \cap (\overline{H_0} \cup \overline{E})$ is a connected divisor of $\overline{X}$ which has exactly two irreducible components.*

Proof. (i) Let $\overline{\mathrm{NE}}(V)$ be the Mori cone of V, that is, the closure of the cone of effective 1-cycles on V modulo numerical equivalence. Let l_0 be a line in $\overline{H_0} \cong \mathbb{P}^{n-1}$. Then we note that $\overline{\mathrm{NE}}(V) = \mathbb{R}_+[l_0] + \mathbb{R}_+[f_0]$, where $\mathbb{R}_+$ is the set of non-negative real numbers, and the intersection numbers $(\overline{H_0} \cdot l_0)_V = 0$, $(\overline{H_0} \cdot f_0)_V = 1$, $(\overline{E} \cdot l_0)_V = 1$ and $(\overline{E} \cdot f_0)_V = -1$. Hence we have $(\mathcal{L} \cdot l_0)_V = 1 > 0$ and $(\mathcal{L} \cdot f_0)_V = d - 1 > 0$. By Kleiman's ampleness criterion (cf. [**KM**]), we obtain (i).
(ii) Since $\mathcal{L} = \mathcal{O}_V(\overline{X}) = \mathcal{O}_V(d\overline{H_0} + \overline{E})$, we have $\mathcal{L}|_{\overline{H}} \sim \overline{X}|_{\overline{H}} \sim \overline{E}|_{\overline{H}} \sim \mathcal{O}_{\mathbb{P}^{n-1}}(1)$ for every hyperplane H in $\mathbb{P}^n$ containing L. Thus we obtain (ii).
(iii) By (i), (ii) and Lemma 2.1, we obtain (iii).
(iv) We note that $\mathcal{L}_1 := \mathcal{L}|_{\overline{X}}$ is an ample invertible sheaf on $\overline{X}$ and

$$\mathcal{L}_1|_{\overline{X} \cap \overline{H}} \sim (d\overline{H_0} + \overline{E})|_{\overline{H}}|_{\overline{X} \cap \overline{H}} \sim \overline{E}|_{\overline{H}}|_{\overline{X} \cap \overline{H}} \sim \mathcal{O}_{\mathbb{P}^{n-2}}(1)$$

for every fiber $\overline{X} \cap \overline{H} \cong \mathbb{P}^{n-2}$ of $\overline{\psi}|_{\overline{X}}$. By Lemma 2.1, we obtain (iv).
(v) Since the pair $(\overline{X}, \overline{X} \cap (\overline{H_0} \cup \overline{E}))$ is a completion of the affine variety S, the boundary $\overline{X} \cap (\overline{H_0} \cup \overline{E})$ is a connected divisor of $\overline{X}$ (cf. [**Mi2**]). Since $\mathrm{Pic}(\overline{X}) \cong \mathbb{Z} \oplus \mathbb{Z}$ and the affine coordinate ring of $\overline{X} \setminus (\overline{H_0} \cup \overline{E}) \cong S$ is UFD, the boundary $\overline{X} \cap (\overline{H_0} \cup \overline{E})$ has exactly two irreducible components

and $\mathrm{Pic}(\overline{X})$ is generated freely by the two components (cf. [**Ha**]). Thus we obtain (v). □

Here we put $\overline{L_0} := \overline{X} \cap \overline{H_0}\ (\cong \mathbb{P}^{n-2})$ on V although we do not define L_0 on $\mathbb{P}^n$. By Lemma 2.2(ii), we have $\overline{X}|_{\overline{H_0}} = \overline{L_0}$. As mentioned in the beginning of this section, there exists a hyperplane L' in $H_0 \cong \mathbb{P}^{n-1}$ such that $X|_{H_0} = (d-1)L + L'$, where the case $L' = L$ is allowed. We note that $\sigma(\overline{L_0}) = L'$. If $L' = L$, then we obtain $\overline{L_0} = \overline{E} \cap \overline{H_0} \subset \overline{E}$ and $\overline{L_0} \sim M_0$ on $\overline{E}$. If $L' \neq L$, then we obtain $\overline{L_0} = \overline{L'} \not\subset \overline{E}$. By Lemma 2.2(v), we note that $\overline{X} \cap (\overline{H_0} \cup \overline{E})$ has exactly two irreducible components. Since $\overline{L_0}$ is the one component obviously, we denote by $\overline{C}$ the other component. Thus we have $\overline{X} \cap (\overline{H_0} \cup \overline{E}) = \overline{C} \cup \overline{L_0}$. If $L' = L$, then we obtain $\overline{X} \cap \overline{E} = \overline{C} \cup \overline{L_0}$ since $\overline{L_0} \subset \overline{E}$. If $L' \neq L$, then we obtain $\overline{X} \cap \overline{E} = \overline{C}$ since $\overline{L_0} \not\subset \overline{E}$. Then we obtain the following lemma:

Lemma 2.3. *One obtains the following:*

(i) $\overline{X}|_{\overline{E}} = \overline{C} + (d-1-\lambda)\overline{L_0}$, *where* $0 \leq \lambda \leq d-1$ *and* $\overline{C} \sim \lambda M_0 + F_0$. *If* $L' = L$, *then* $0 \leq \lambda \leq d-2$. *If* $L' \neq L$, *then* $\lambda = d-1$.

(ii) *If* $1 \leq \lambda \leq d-1$, *then* $\sigma|_{\overline{C}} : \overline{C} \to L \cong \mathbb{P}^{n-2}$ *is a surjective generically finite morphism of degree* λ. *If* $\lambda = 0$, *then* $\overline{C}$ *is the inverse image of a hyperplane in* $L \cong \mathbb{P}^{n-2}$ *by* $\sigma|_{\overline{E}}$.

(iii) $\overline{C}|_{\overline{X} \cap \overline{H}} \sim \mathcal{O}_{\mathbb{P}^{n-2}}(1)$ *for every fiber* $\overline{X} \cap \overline{H} \cong \mathbb{P}^{n-2}$ *of* $\overline{\psi}|_{\overline{X}}$, *where* H *is a hyperplane in* $\mathbb{P}^n$ *containing* L. *This means that* $\overline{C}$ *intersects* $\overline{X} \cap \overline{H}$ *transversally in* $\overline{X}$, *and* $\overline{C} \cap (\overline{X} \cap \overline{H}) = \overline{C} \cap \overline{H}$ *is a hyperplane in* $\overline{X} \cap \overline{H} \cong \mathbb{P}^{n-2}$. *In particular,* $\overline{C}$ *intersects* $\overline{L_0}$ *transversally in* $\overline{X}$, *and* C *is smooth.*

(iv) *For* $n = 3$, $\overline{\psi}|_{\overline{C}} : \overline{C} \cong \mathbb{P}^1$. *For* $n \geq 4$, $\overline{\psi}|_{\overline{C}} : \overline{C} \to \mathbb{P}^1$ *is a* $\mathbb{P}^{n-3}$-*bundle over* $\mathbb{P}^1$ *with* $(\overline{C}, \mathcal{L}_2) \cong (\mathbb{P}(\mathcal{E}_2), \mathcal{O}_{\mathbb{P}(\mathcal{E}_2)}(1))$, *where* $\mathcal{L}_2 := \mathcal{L}|_{\overline{C}}$ *and* $\mathcal{E}_2 := (\overline{\psi}|_{\overline{C}})_*\mathcal{L}_2$. *In particular,* $\mathrm{Pic}(\overline{C}) \cong \mathbb{Z} \oplus \mathbb{Z}$.

Proof. (i) Assume that $L' = L$. Since $\overline{X} \cap \overline{E} = \overline{C} \cup \overline{L_0}$, we have $\overline{X}|_{\overline{E}} = \alpha\overline{C} + \beta\overline{L_0} \sim (d-1)M_0 + F_0$ on $\overline{E}$, where $\alpha, \beta \geq 1$ and $\overline{L_0} \sim M_0$ on $\overline{E}$. For any fiber $M_0 \cong \mathbb{P}^{n-2}$ of $\overline{\psi}|_{\overline{E}}$, we have $\alpha\overline{C}|_{M_0} \sim F_0|_{M_0} \sim \mathcal{O}_{\mathbb{P}^{n-2}}(1)$ and hence $\alpha = 1$. For any fiber $f_0 \cong \mathbb{P}^1$ of $\sigma|_{\overline{E}}$, we obtain $\lambda := (\overline{C} \cdot f_0)_{\overline{E}} = d-1-\beta \leq d-2$. Then we note that $\beta = d-1-\lambda$. Since $\overline{C} \subsetneq \overline{E}$, there exists a fiber $f_0 \cong \mathbb{P}^1$ of $\sigma|_{\overline{E}}$ such that $f_0 \not\subset \overline{C}$. Thus we also obtain $\lambda = (\overline{C} \cdot f_0)_{\overline{E}} \geq 0$. Assume that $L' \neq L$. Since $\overline{X} \cap \overline{E} = \overline{C}$, we have $\overline{X}|_{\overline{E}} = \alpha\overline{C} \sim (d-1)M_0 + F_0$ on $\overline{E}$, where $\alpha \geq 1$. Similarly to the case $L' = L$, we obtain $\alpha = 1$ and $\lambda := (\overline{C} \cdot f_0)_{\overline{E}} = d-1$. Thus we obtain (i).

(ii) By noting that $\lambda = (\overline{C} \cdot f_0)_{\overline{E}}$ for any fiber $f_0 \cong \mathbb{P}^1$ of $\sigma|_{\overline{E}}$, we obtain (ii).
(iii) By (i), we also have $\overline{E}|_{\overline{X}} = \overline{C} + (d-1-\lambda)\overline{L_0}$. From this, we have

$$\overline{C}|_{\overline{X}\cap\overline{H}} \sim \overline{E}|_{\overline{X}}|_{\overline{X}\cap\overline{H}} \sim \overline{E}|_{\overline{H}}|_{\overline{X}\cap\overline{H}} \sim \mathcal{O}_{\mathbb{P}^{n-2}}(1)$$

for every fiber $\overline{X} \cap \overline{H} \cong \mathbb{P}^{n-2}$ of $\overline{\psi}|_{\overline{X}}$, where H is a hyperplane in $\mathbb{P}^n$ containing L. Thus we obtain (iii).
(iv) Assume that $n = 3$. By (i), we have $(\overline{C} \cdot M_0)_{\overline{E}} = (\overline{X}|_{\overline{E}} \cdot M_0)_{\overline{E}} = ((d-1)M_0 + F_0 \cdot M_0)_{\overline{E}} = 1$ for any fiber $M_0 \cong \mathbb{P}^1$ of $\overline{\psi}|_{\overline{E}} : \overline{E} \cong \mathbb{P}^1 \times \mathbb{P}^1 \to \mathbb{P}^1$. Thus we obtain the assertion for $n = 3$. Assume that $n \geq 4$. Then we note that $\mathcal{L}_2 := \mathcal{L}|_{\overline{C}}$ is an ample invertible sheaf on $\overline{C}$ and

$$\mathcal{L}_2|_{\overline{C}\cap\overline{H}} \sim (d\overline{H_0} + \overline{E})|_{\overline{H}}|_{\overline{C}\cap\overline{H}} \sim \overline{E}|_{\overline{H}}|_{\overline{C}\cap\overline{H}} \sim \mathcal{O}_{\mathbb{P}^{n-3}}(1)$$

for every fiber $\overline{C} \cap \overline{H} \cong \mathbb{P}^{n-3}$ of $\overline{\psi}|_{\overline{C}}$. By Lemma 2.1, we obtain the assertion for $n \geq 4$. Thus we obtain (iv). □

Proposition 2.1. *One obtains the following:*

(i) *For every hyperplane H in $\mathbb{P}^n$ containing L with $H \neq H_0$, there exists a unique hyperplane L_H in $H \cong \mathbb{P}^{n-1}$ with $L_H \neq L$ such that $X|_H = (d-1)L + L_H$ and $\overline{X}|_{\overline{H}} = \overline{L_H}$. One notes that $X|_{H_0} = (d-1)L + L'$ and $\overline{X}|_{\overline{H_0}} = \overline{L_0}$.*
(ii) *Assume that $L' \neq L$. Then there exists a hyperplane H in $\mathbb{P}^n$ containing L with $H \neq H_0$ such that $L_H \cap L \neq L' \cap L$, where L_H is defined in* (i).

Proof. (i) Let H be a hyperplane in $\mathbb{P}^n$ containing L with $H \neq H_0$. Then we have an isomorphism $\sigma|_{\overline{H}} : \overline{H} \cong H\,(\cong \mathbb{P}^{n-1})$. By Lemma 2.2(ii), we note that $\overline{X} \cap \overline{H}$ is a hyperplane in $\overline{H} \cong \mathbb{P}^{n-1}$. We also note that $\overline{E} \cap \overline{H}$ is a hyperplane in $\overline{H} \cong \mathbb{P}^{n-1}$ and $\sigma(\overline{E} \cap \overline{H}) = L$. Now we show that $\overline{X} \cap \overline{H} \neq \overline{E} \cap \overline{H}$. Assume to the contrary that $\overline{X} \cap \overline{H} = \overline{E} \cap \overline{H}$. Since $\overline{X} \cap \overline{H}$ is contained in $\overline{E}$, it is an irreducible component of $\overline{X} \cap (\overline{H_0} \cup \overline{E}) = \overline{C} \cup \overline{L_0}$. Since $\overline{L_0} = \overline{X} \cap \overline{H_0}$, we have $\overline{C} = \overline{X} \cap \overline{H}$. Thus, since $\overline{C}$ and $\overline{L_0}$ are two distinct fibers of $\overline{\psi}|_{\overline{X}}$, we have $\overline{C} \cap \overline{L_0} = \emptyset$. By Lemma 2.3(iii), this is a contradiction. Hence we have $\overline{X} \cap \overline{H} \neq \overline{E} \cap \overline{H}$. By the isomorphism $\sigma|_{\overline{H}}$, we obtain $\sigma(\overline{X} \cap \overline{H}) \neq L$. By putting $L_H := \sigma(\overline{X} \cap \overline{H})$, we obtain (i).
(ii) Assume that $L' \neq L$. We note that $\overline{X} \cap \overline{E} = \overline{C}$ and $\overline{L'} = \overline{L_0} \not\subset \overline{E}$. Let H be any hyperplane in $\mathbb{P}^n$ containing L. Then we have an isomorphism $\sigma|_{\overline{E}\cap\overline{H}} : \overline{E} \cap \overline{H} \cong L\,(\cong \mathbb{P}^{n-2})$. If $H \neq H_0$, then by (i) we obtain $\sigma(\overline{C} \cap \overline{H}) = \sigma((\overline{X} \cap \overline{H}) \cap (\overline{E} \cap \overline{H})) = \sigma(\overline{L_H} \cap (\overline{E} \cap \overline{H})) = L_H \cap L$, which is

a hyperplane in $L \cong \mathbb{P}^{n-2}$. On the other hand, we obtain $\sigma(\overline{C} \cap \overline{H_0}) = \sigma(\overline{L_0} \cap (\overline{E} \cap \overline{H_0})) = \sigma(\overline{L'} \cap (\overline{E} \cap \overline{H_0})) = L' \cap L$, which is also a hyperplane in $L \cong \mathbb{P}^{n-2}$. Here we note that $\sigma|_{\overline{C}} : \overline{C} \to L \cong \mathbb{P}^{n-2}$ is surjective by Lemma 2.3(i),(ii). If $\sigma(\overline{C} \cap \overline{H}) = \sigma(\overline{C} \cap \overline{H_0})$ for any hyperplane H in $\mathbb{P}^n$ containing L with $H \neq H_0$, then $\sigma|_{\overline{C}}$ is not surjective. This is a contradiction. Hence there exists a hyperplane H in $\mathbb{P}^n$ containing L with $H \neq H_0$ such that $\sigma(\overline{C} \cap \overline{H}) \neq \sigma(\overline{C} \cap \overline{H_0})$, that is, $L_H \cap L \neq L' \cap L$. Thus we obtain (ii). □

3. Proof of Theorem 1.1

In this section, we shall prove Theorem 1.1. We use the same notations as those in the previous sections. Let $(x_0 : x_1 : \cdots : x_n)$ be the homogeneous coordinate of $\mathbb{P}^n$ with $n \geq 3$. Let $\mathrm{Aut}(\mathbb{P}^n)$ be the group of automorphisms of $\mathbb{P}^n$. Let $F \in k[x_0, x_1, \ldots, x_n]$ be the homogenization of the defining polynomial $f \in k[x_1, \ldots, x_n]$ of S. We note that F is a defining polynomial of X with $\deg F = \deg f = d \geq 2$. We note that $H_0 = \{x_0 = 0\}$ and we may assume that $L = \{x_0 = x_1 = 0\}$ by applying a suitable automorphism of $\mathbb{P}^n$. Then we obtain the following lemma:

Lemma 3.1. *The following conditions are equivalent:*

(i) $\mathrm{mult}_L X = d - 1$.

(ii) *There exist homogeneous polynomials $F_1, \ldots, F_n \in k[x_0, x_1]$ such that*

$$F = F_1(x_0, x_1) + F_2(x_0, x_1)x_2 + \cdots + F_n(x_0, x_1)x_n$$

and at least one of $F_2, \ldots, F_n$ is not zero.

Under the condition (i) or (ii), the restriction $\overline{X}|_{\overline{E}}$ is expressed by the equation

$$F_2(y_0, y_1)x_2 + \cdots + F_n(y_0, y_1)x_n = 0,$$

where $((x_2 : \cdots : x_n), (y_0 : y_1))$ is a bihomogeneous coordinate of $\overline{E} \cong \mathbb{P}^{n-2} \times \mathbb{P}^1$.

Proof. First we express the defining equation of X as follows:

$$F = \sum_{0 \leq i_2 + \cdots + i_n \leq d} F_{i_2 \cdots i_n}(x_0, x_1) x_2^{i_2} \cdots x_n^{i_n} = 0,$$

where $i_2, \ldots, i_n$ are non-negative integers and $F_{i_2 \cdots i_n} \in k[x_0, x_1]$ are homogeneous polynomials of degree $d - i_2 - \cdots - i_n$.

Let us consider the blowing-up $\sigma : V \to \mathbb{P}^n$ of $\mathbb{P}^n$ along $L = \{x_0 = x_1 = 0\} \cong \mathbb{P}^{n-2}$ with the exceptional divisor $\overline{E} \cong \mathbb{P}^{n-2} \times \mathbb{P}^1$. Since we investigate the multiplicity of X along L, we would like to take an affine open subset of $\mathbb{P}^n$ containing a Zariski open subset of L. Indeed, we may take an affine open subset $U_2 := \{x_2 \neq 0\}$ of $\mathbb{P}^n$ with canonical isomorphism $U_2 \cong \mathbb{A}^n_{(x_0,x_1,x_3,\dots,x_n)}$. We note that $L \cap U_2 = \{x_0 = x_1 = 0\}$ in $U_2 \cong \mathbb{A}^n$. Here we put $V_{2,0} := \mathbb{A}^n_{(x_0,s,x_3,\dots,x_n)}$ and $V_{2,1} := \mathbb{A}^n_{(t,x_1,x_3,\dots,x_n)}$, and we patch $V_{2,0}$ and $V_{2,1}$ by the relations $x_0 = tx_1$, $x_1 = sx_0$, $s = 1/t$. Then we note that the union $V_{2,0} \cup V_{2,1}$ can be identified with a Zariski open subset of V and $\sigma^{-1}(U_2) = V_{2,0} \cup V_{2,1}$. We also note that the blowing-up morphism σ on $V_{2,0}$ or on $V_{2,1}$ is expressed by $(x_0, s, x_3, \dots, x_n) \mapsto (x_0, sx_0, x_3, \dots, x_n)$ or $(t, x_1, x_3, \dots, x_n) \mapsto (tx_1, x_1, x_3, \dots, x_n)$ respectively.

Now we show the equivalence of (i) and (ii). Since X is given by the equation

$$\sum_{0 \le i_2 + \cdots + i_n \le d} F_{i_2 \cdots i_n}(x_0, x_1) x_3^{i_3} \cdots x_n^{i_n} = 0 \quad \text{on} \quad U_2,$$

we obtain the total transform $\sigma^* X$ with two defining equations

$$\sum_{0 \le i_2 + \cdots + i_n \le d} x_0^{d - i_2 - \cdots - i_n} F_{i_2 \cdots i_n}(1, s) x_3^{i_3} \cdots x_n^{i_n} = 0 \quad \text{on} \quad V_{2,0},$$

$$\sum_{0 \le i_2 + \cdots + i_n \le d} x_1^{d - i_2 - \cdots - i_n} F_{i_2 \cdots i_n}(t, 1) x_3^{i_3} \cdots x_n^{i_n} = 0 \quad \text{on} \quad V_{2,1}.$$

By noting that the exceptional divisor $\overline{E}$ is given by the equation $x_0 = 0$ on $V_{2,0}$ or the equation $x_1 = 0$ on $V_{2,1}$, we see that the condition $\operatorname{mult}_L X = d - 1$ is equivalent to the condition that every $(i_2, \dots, i_n)$ with non-zero $F_{i_2 \cdots i_n}$ satisfies $d - i_2 - \cdots - i_n \ge d - 1$ and at least one of all the homogeneous polynomials $F_{i_2 \cdots i_n}$ with $d - i_2 - \cdots - i_n = d - 1$ is not zero. By putting $F_1 := F_{00\cdots 0}$, $F_2 := F_{10\cdots 0}$, $\dots$, $F_n := F_{0\cdots 01}$, we obtain the equivalence of (i) and (ii). Moreover, by continuing the above argument, we have the proper transform $\overline{X}$ with defining equation

$$x_0 F_1(1, s) + \{F_2(1, s) + F_3(1, s) x_3 + \cdots + F_n(1, s) x_n\} = 0$$

on $V_{2,0} \cong \mathbb{A}^n_{(x_0,s,x_3,\dots,x_n)}$. By putting $x_0 = 0$ and by taking a completion, we obtain the defining equation of $\overline{X}|_{\overline{E}}$ as required. Thus we complete the proof. □

By our assumption $\operatorname{mult}_L X = d - 1$ and Lemma 3.1, we see that X is given by the equation $F_1(x_0, x_1) + F_2(x_0, x_1) x_2 + \cdots + F_n(x_0, x_1) x_n = 0$ on

$\mathbb{P}^n$, where $F_1, \ldots, F_n \in k[x_0, x_1]$ are homogeneous polynomials and at least one of $F_2, \ldots, F_n$ is not zero. By putting $x_0 = 1$, we obtain the hypersurface S with defining equation $f = F_1(1, x_1) + F_2(1, x_1)x_2 + \cdots + F_n(1, x_1)x_n = 0$ on $\mathbb{A}^n$, where at least one of $F_2(1, x_1), \ldots, F_n(1, x_1)$ is not zero. Since S is irreducible, we note that $(F_1(1, x_1), F_2(1, x_1), \ldots, F_n(1, x_1)) = k[x_1]$. By our assumption $S \cong \mathbb{A}^{n-1}$ and the following proposition, we complete the proof of the former half of Theorem 1.1.

Here we give some notations needed in the following proposition. Let us denote by $\Gamma(\mathcal{O}_T)$ the coordinate ring of an affine variety T over k. Let us denote by $R^\times$ the unit group of a k-algebra R. For non-zero polynomials $h_1, \ldots, h_l \in k[x_1]$ with $l \geq 1$, we denote by $\gcd(h_1, \ldots, h_l)$ the unique monic polynomial $h \in k[x_1]$ such that $(h_1, \ldots, h_l) = (h)$ as equality of ideals. We note that the case $l = 1$ is allowed.

Proposition 3.1. *Let k be an algebraically closed field of* $\mathrm{char}(k) \geq 0$, *and let m be a natural number with $m \geq 2$. Let $(x_1, \ldots, x_m)$ be an affine coordinate of $\mathbb{A}^m$. Let T be an irreducible hypersurface in $\mathbb{A}^m$ over k with defining equation $a_1(x_1) + a_2(x_1)x_2 + \cdots + a_m(x_1)x_m = 0$, where $a_1, \ldots, a_m$ are polynomials of $k[x_1]$ satisfying $(a_2, \ldots, a_m) \neq (0)$ as ideals of $k[x_1]$. Then the following conditions are equivalent:*

(i) $\Gamma(\mathcal{O}_T)$ *is UFD and* $\Gamma(\mathcal{O}_T)^\times = k^\times$.
(ii) $T \cong \mathbb{A}^{m-1}$.
(iii) T *is transformed onto a coordinate hyperplane in $\mathbb{A}^m$ by a tame automorphism of $\mathbb{A}^m$ fixing the variable x_1.*
(iv) $(a_2, \ldots, a_m) = k[x_1]$.

Moreover, if $k = \mathbb{C}$, then the above conditions are equivalent to the following:

(v) T *is smooth and acyclic.*

Proof. Since T is irreducible, we note that $(a_1, \ldots, a_m) = k[x_1]$. We also note that a composite of tame automorphisms of $\mathbb{A}^m$ is tame.
(1) By a suitable permutation of $x_2, \ldots, x_m$, which is a tame automorphism of $\mathbb{A}^m$ fixing the variable x_1, and by the associated replacement of indices of $a_2, \ldots, a_m$, we may assume that $a_2, \ldots, a_{m_0}$ are non-zero polynomials and $(a_2, \ldots, a_m) = (a_2, \ldots, a_{m_0})$ as equality of ideals for some $2 \leq m_0 \leq m$. By putting $b := \gcd(a_2, \ldots, a_{m_0})$, we have non-zero polynomials $\alpha_2, \ldots, \alpha_{m_0} \in k[x_1]$ satisfying $a_i = b\,\alpha_i$ for $2 \leq i \leq m_0$. Then we note that $T = \{b(x_1)(\alpha_2(x_1)x_2 + \cdots + \alpha_{m_0}(x_1)x_{m_0}) + a_1(x_1) = 0\}$ with

$\gcd(\alpha_2, \ldots, \alpha_{m_0}) = 1$, $b \neq 0$ and $(a_1, b) = k[x_1]$. We also have polynomials q, $r \in k[x_1]$ satisfying $a_1 = qb + r$ and $\deg b > \deg r$.

First we show that T is transformed onto a hypersurface $b(x_1)x_2 + r(x_1) = 0$ in $\mathbb{A}^m$ by a tame automorphism of $\mathbb{A}^m$ fixing the variable x_1. Assume that $m_0 = 2$. Since α_2 is a non-zero constant, we have the assertion by applying the tame automorphism of $\mathbb{A}^m$: $x_2' = \alpha_2 x_2 + q$, $x_i' = x_i$ $(i \neq 2)$. From now on, we may assume that $m_0 \geq 3$. We take a notice of the coefficients α_2 and α_3. There exist non-zero polynomials $r^{(-1)}$, $r^{(0)} \in k[x_1]$ such that $\alpha_2 = \gcd(\alpha_2, \alpha_3) r^{(-1)}$ and $\alpha_3 = \gcd(\alpha_2, \alpha_3) r^{(0)}$. Then we note that $(r^{(-1)}, r^{(0)}) = k[x_1]$. By the Euclidean algorithm, there exist polynomials $q^{(1)}, r^{(1)}, \ldots, q^{(N)}, r^{(N)} \in k[x_1]$ with some $N \geq 1$ such that $r^{(i-2)} = q^{(i)} r^{(i-1)} + r^{(i)}$ and $\deg r^{(i-1)} > \deg r^{(i)}$ for $1 \leq i \leq N$, and $r^{(N-1)}$ is a non-zero constant and $r^{(N)} = 0$. Applying the tame automorphism of $\mathbb{A}^m$: $x_2' = x_3 + q^{(1)} x_2$, $x_3' = x_2$, $x_i' = x_i$ $(i \neq 2, 3)$, we have that T is transformed onto a hypersurface

$$b\left\{\gcd(\alpha_2, \alpha_3)\left(r^{(0)} x_2 + r^{(1)} x_3\right) + \alpha_4 x_4 + \cdots + \alpha_{m_0} x_{m_0}\right\} + a_1 = 0.$$

Applying the tame automorphism of $\mathbb{A}^m$: $x_2' = x_3 + q^{(2)} x_2$, $x_3' = x_2$, $x_i' = x_i$ $(i \neq 2, 3)$, we obtain a hypersurface

$$b\left\{\gcd(\alpha_2, \alpha_3)\left(r^{(1)} x_2 + r^{(2)} x_3\right) + \alpha_4 x_4 + \cdots + \alpha_{m_0} x_{m_0}\right\} + a_1 = 0.$$

Repeating the above arguments, we obtain a hypersurface

$$b\left\{\gcd(\alpha_2, \alpha_3)\left(r^{(N-1)} x_2 + r^{(N)} x_3\right) + \alpha_4 x_4 + \cdots + \alpha_{m_0} x_{m_0}\right\} + a_1 = 0.$$

Applying the tame automorphism of $\mathbb{A}^m$: $x_2' = r^{(N-1)} x_2$, $x_i' = x_i$ $(i \neq 2)$, we obtain a hypersurface

$$b\left\{\gcd(\alpha_2, \alpha_3) x_2 + \alpha_4 x_4 + \cdots + \alpha_{m_0} x_{m_0}\right\} + a_1 = 0.$$

Repeating the above arguments, we finally obtain a hypersurface

$$b \cdot \gcd(\alpha_2, \alpha_3, \ldots, \alpha_{m_0}) x_2 + a_1 = 0.$$

Then we note that $\gcd(\alpha_2, \alpha_3, \ldots, \alpha_{m_0}) = 1$. Moreover, applying the tame automorphism of $\mathbb{A}^m$: $x_2' = x_2 + q$, $x_i' = x_i$ $(i \neq 2)$, we obtain a hypersurface $b(x_1)x_2 + r(x_1) = 0$. Thus we see that T is transformed onto a hypersurface $b(x_1)x_2 + r(x_1) = 0$ in $\mathbb{A}^m$ by a tame automorphism of $\mathbb{A}^m$ fixing the variable x_1.

Now we show the equivalence of (i) $\sim$ (iv). We put $C := \{b(x_1)x_2 + r(x_1) = 0\}$ in $\mathbb{A}^2$, where $b \neq 0$, $(a_1, b) = k[x_1]$ and $\deg b > \deg r$. Then we note that $T \cong C \times \mathbb{A}^{m-2}$ and $\Gamma(\mathcal{O}_T) \cong \Gamma(\mathcal{O}_C)[Y_1, \ldots, Y_{m-2}]$, where

$Y_1, \ldots, Y_{m-2}$ are variables. First we show (iv)$\Rightarrow$(iii). Assume (iv), which means that $b = 1$. Since $\deg b > \deg r$, we have $r = 0$. Hence the hypersurface $b(x_1)x_2 + r(x_1) = 0$ in $\mathbb{A}^m$ is a coordinate hyperplane in $\mathbb{A}^m$. Thus we obtain (iii). Since (iii)$\Rightarrow$(ii)$\Rightarrow$(i) is obvious, it suffices to show (i)$\Rightarrow$(iv). Assume (i). Since $\Gamma(\mathcal{O}_T) \cong \Gamma(\mathcal{O}_C)[Y_1, \ldots, Y_{m-2}]$, we see that $\Gamma(\mathcal{O}_C)$ is UFD and $\Gamma(\mathcal{O}_C)^\times = k^\times$. Then it is well-known that $C \cong \mathbb{A}^1$ (cf. Lemma 2.8 in Freudenburg [**Fr**]). Here we embed the affine plane $\mathbb{A}^2$ into the projective plane $\mathbb{P}^2$ canonically as the complement of the line $L_0 = \{x_0 = 0\}$, where $(x_0 : x_1 : x_2)$ is the homogeneous coordinate of $\mathbb{P}^2$. Since $C \cong \mathbb{A}^1$, the curve C in $\mathbb{A}^2$ has one place at infinity, that is, the closure of C in $\mathbb{P}^2$ meets L_0 at only one point. If $\deg b \geq 1$, then the closure of C in $\mathbb{P}^2$ meets L_0 at two points $(0:0:1)$ and $(0:1:0)$ since $\deg b > \deg r$. This is a contradiction. Hence we have $b = 1$, which means that $(a_2, \ldots, a_m) = (a_2, \ldots, a_{m_0}) = (b) = k[x_1]$. Thus we obtain (iv).
(2) Assume that $k = \mathbb{C}$. We show the equivalence of (ii) and (v). Since (ii)$\Rightarrow$(v) is obvious, it suffices to show (v)$\Rightarrow$(ii). Assume that T is smooth and acyclic. By the arguments in (1), we obtain $T \cong C \times \mathbb{A}^{m-2}$, where C is an irreducible affine curve. Thus we see that C is also smooth and acyclic. Then it is well-known that $C \cong \mathbb{A}^1$. Hence we obtain $T \cong \mathbb{A}^1 \times \mathbb{A}^{m-2} \cong \mathbb{A}^{m-1}$. Thus we obtain (ii). □

From now on, we shall prove the latter half of Theorem 1.1.

Proposition 3.2. *The standard equations of X and S are given by*

$$\begin{cases} X : x_1^{d-1}(\nu_1 x_1 + \nu_2 x_2) + x_0^{d-1}x_3 + x_0 x_1 (G_1 + G_2 x_2 + \cdots + G_n x_n) = 0 \\ S : x_1^{d-1}(\nu_1 x_1 + \nu_2 x_2) + x_3 + x_1 (g_1 + g_2 x_2 + \cdots + g_n x_n) = 0 \end{cases}$$

with $(\nu_1, \nu_2) = (1,0), (0,1)$, $G_i \in k[x_0, x_1]$, $g_i := G_i(1, x_1) \in k[x_1]$ and $(\nu_2 x_1^{d-1} + x_1 g_2, 1 + x_1 g_3, x_1 g_4, \ldots, x_1 g_n) = k[x_1]$ up to $\mathrm{Aut}(\mathbb{P}^n)$ *and* $A(n,k)$ *respectively, where $H_0 = \{x_0 = 0\}$, $L = \{x_0 = x_1 = 0\}$ and $L' = \{x_0 = \nu_1 x_1 + \nu_2 x_2 = 0\}$.*

Proof. Now we note that $H_0 = \{x_0 = 0\}$ and $L = \{x_0 = x_1 = 0\}$. By Lemma 3.1 and Proposition 3.1, we obtain the hypersurface X with defining equation $F_1(x_0, x_1) + F_2(x_0, x_1)x_2 + \cdots + F_n(x_0, x_1)x_n = 0$, where $F_1, \ldots, F_n \in k[x_0, x_1]$ are homogeneous polynomials with $(F_2(1, x_1), \ldots, F_n(1, x_1)) = k[x_1]$.
(1) Assume that $L' = L$. We take any hyperplane H in $\mathbb{P}^n$ containing L with $H \neq H_0$. By Proposition 2.1(i), there exists a hyperplane L_H in $H \cong \mathbb{P}^{n-1}$

with $L_H \neq L$ such that $X|_H = (d-1)L + L_H$. By applying an automorphism of $\mathbb{P}^n$ fixing H_0 and L, we may assume that $X|_{H_0} = \{x_0 = x_1^d = 0\}$, $X|_H = \{x_1 = x_0^{d-1}x_3 = 0\}$ and $H = \{x_1 = 0\} =: H_1$.
(2) Assume that $L' \neq L$. By Proposition 2.1(ii), there exists a hyperplane H in $\mathbb{P}^n$ containing L with $H \neq H_0$ such that $L_H \cap L \neq L' \cap L$. By applying an automorphism of $\mathbb{P}^n$ fixing H_0 and L, we may assume that $X|_{H_0} = \{x_0 = x_1^{d-1}x_2 = 0\}$, $X|_H = \{x_1 = x_0^{d-1}x_3 = 0\}$, $L_H \cap L = \{x_0 = x_1 = x_3 = 0\}$ and $H = \{x_1 = 0\} =: H_1$.
(3) Summarizing (1) and (2), we may assume that $X|_{H_0} = \{x_0 = x_1^{d-1}(\nu_1 x_1 + \nu_2 x_2) = 0\}$ and $X|_{H_1} = \{x_1 = x_0^{d-1}x_3 = 0\}$ with $(\nu_1, \nu_2) = (1,0), (0,1)$, where $H_0 = \{x_0 = 0\}$, $H_1 = \{x_1 = 0\}$ and $L = \{x_0 = x_1 = 0\}$. Hence we obtain the standard equation of X as required. We note that $G_1 \in k$ and $G_2 = \cdots = G_n = 0$ for the case $d = 2$. Moreover, putting $x_0 = 1$, we obtain the standard equation of S as required. □

Remark 3.1. Let us describe the relation of the standard equation of X and the integer λ in Lemma 2.3(i). Let μ be the maximal integer with $0 \leq \mu \leq d-1$ such that the polynomial $\nu_2 x_1^{d-1}x_2 + x_0^{d-1}x_3 + x_0x_1\{G_2(x_0,x_1)x_2 + \cdots + G_n(x_0,x_1)x_n\}$ is exactly divisible by x_0^{μ}. Then we have $1 \leq \mu \leq d-1$ if $L' = L$, and $\mu = 0$ if $L' \neq L$. Now we see that $\mu = d-1-\lambda$. Indeed, this is showed as follows. By Lemma 3.1, the restriction $\overline{X}|_{\overline{E}} = \overline{C} + (d-1-\lambda)\overline{L_0}$ is given by the equation

$$\nu_2 y_1^{d-1}x_2 + y_0^{d-1}x_3 + y_0y_1\{G_2(y_0,y_1)x_2 + \cdots + G_n(y_0,y_1)x_n\} = 0$$

where $((x_2 : \cdots : x_n), (y_0 : y_1))$ is the bihomogeneous coordinate of $\overline{E} \cong \mathbb{P}^{n-2} \times \mathbb{P}^1$. Assume that $L' = L$. By noting that $\overline{L_0} = \{y_0 = 0\}$ on $\overline{E}$, we obtain $\mu = d-1-\lambda$. Assume that $L' \neq L$. Since $\lambda = d-1$, we obtain $\mu = 0 = d-1-\lambda$. Thus we complete the proof.

In the last part of this paper, we shall investigate the singular locus $\operatorname{Sing} X$ of X and the multiplicities of singularities of X.

Lemma 3.2. *For the standard equation of X, one obtains the following:*

(i) *Assume that $d = 2$. If $L' = L$, then $\operatorname{Sing} X = \{x_0 = x_1 = x_3 = 0\} \subsetneq L$. If $L' \neq L$, then $\operatorname{Sing} X = \{x_0 = x_2 = x_3 + G_1x_1 = 0\} \subsetneq L'$ and $\operatorname{Sing} X \not\subset L$, where $G_1 \in k$. In both cases, $\operatorname{mult}_p X = 2$ for any $p \in \operatorname{Sing} X$.*
(ii) *Assume that $d \geq 3$. Let L_i be the set of all the points $p \in L$ satisfying $\operatorname{mult}_p X = i$ for each $i = d-1, d$. Then $\operatorname{Sing} X = L = L_{d-1} \cup L_d$*

and L_d is the subset of L of all points $p = (0:0:p_2:\cdots:p_n) \in L$ satisfying the conditions $\nu_2 p_2 = 0, p_3 = 0$ and

$$p_2G_2(y_0, y_1) + p_4G_4(y_0, y_1) + \cdots + p_nG_n(y_0, y_1) = 0 \text{ in } k[y_0, y_1].$$

In particular, L_{d-1} is a non-empty Zariski open subset of L and L_d is a Zariski closed subset of L. Moreover, X is a cone if and only if $L_d \neq \emptyset$.

Proof. Let us consider the standard equation of X in Proposition 3.2. Since $X \setminus H_0 = S \cong \mathbb{A}^{n-1}$, we have $\operatorname{Sing} X \subset X \cap H_0 = L \cup L'$. If $d = 2$, then we obtain (i) by the Jacobian criterion. Thus it suffices to show (ii). From here, we assume that $d \geq 3$. By the Jacobian criterion and Lemma 3.1, we can check that $\operatorname{Sing} X = L$ and $\operatorname{mult}_L X = d-1$ directly. Since $\operatorname{mult}_L X = d-1$, we note that $\operatorname{mult}_p X = d-1$ for general points $p \in L$ and that $d-1 \leq \operatorname{mult}_p X \leq d$ for any point $p \in L$. Moreover, we also note that $\operatorname{mult}_p X = d$ if and only if the fiber $(\sigma|_{\overline{X}})^{-1}(p) = \overline{X} \cap \overline{E} \cap \sigma^{-1}(p)$ is isomorphic to $\mathbb{P}^1$, and if and only if

$$\nu_2 p_2 y_1^{d-1} + p_3 y_0^{d-1} + y_0 y_1\{p_2G_2(y_0, y_1) + \cdots + p_nG_n(y_0, y_1)\} = 0 \text{ in } k[y_0, y_1]$$

for each point $p = (0:0:p_2:\cdots:p_n) \in L$. From this, we obtain (ii). □

Acknowledgments

The author would like to express his hearty gratitude to Professor Mikio Furushima for his invaluable advice and warm encouragement from the first meeting at Kyushu University until the accomplishment of this paper. The author also would like to express his sincere gratitude to Professor Takashi Kishimoto for his useful comments and helpful discussions from the first meeting at Osaka University until the accomplishment of this paper. Furthermore, the author is grateful to the referees for their suggestions and corrections.

References

[AM]. S.S. Abhyankar and T.T. Moh, *Embeddings of the line in the plane*, J. reine angew. Math. **276** (1975), 148–166.

[BS]. M.C. Beltrametti and A.J. Sommese, *The Adjunction Theory of Complex Projective Varieties*, de Gruyter, Berlin, 1995.

[Fr]. G. Freudenburg, *Algebraic Theory of Locally Nilpotent Derivations*, Springer, Berlin, 2006.

[GH]. P. Griffiths and J. Harris, *Principles of Algebraic Geometry*, John Wiley & Sons, New York, 1978.

[Ha]. R. Hartshorne, *Algebraic Geometry*, Springer, New York, 1977.

[KZ]. S. Kaliman and M. Zaidenberg, *Families of affine planes: the existence of a cylinder*, Michigan Math. J. **49** (2001), 353–367.

[Ki]. T. Kishimoto, *Abhyankar-Sathaye Embedding Problem in dimension three*, J. Math. Kyoto Univ. **42-4** (2002), 641–669.

[KM]. J. Kollár and S. Mori, *Birational Geometry of Algebraic Varieties*, Cambridge University Press, Cambridge, 1998.

[Mi1]. M. Miyanishi, *Simple birational extensions of a polynomial ring* $k[x, y]$, Osaka J. Math. **15** (1978), 663–677.

[Mi2]. M. Miyanishi, *Open algebraic surfaces*, CRM Monograph Series 12, AMS Providence, RI, 2001.

[Mo]. T.T. Moh, *On the classification problem of embedded lines in characteristic p*, Algebraic Geometry and Commutative Algebra, in Honor of Masayoshi Nagata, Vol. I (1988), 267–279.

[Oh1]. T. Ohta, *The structure of algebraic embeddings of* $\mathbb{C}^2$ *into* $\mathbb{C}^3$ (*the cubic hypersurface case*), Kyushu J. Math. **53** (1999), 67–106.

[Oh2]. T. Ohta, *The structure of algebraic embeddings of* $\mathbb{C}^2$ *into* $\mathbb{C}^3$ (*the normal quartic hypersurface case.* I), Osaka J. Math. **38** (2001), 507–532.

[Oh3]. T. Ohta, *The structure of algebraic embeddings of* $\mathbb{C}^2$ *into* $\mathbb{C}^3$ (*the normal quartic hypersurface case.* II), Osaka J. Math. **46** (2009), 563–597.

[Ru]. P. Russell, *Simple birational extensions of two dimensional affine rational domains*, Compositio Math. **33** (1976), 197–208.

[Sa]. A. Sathaye, *On linear planes*, Proc. Amer. Math. Soc. **56** (1976), 1–7.

[Su]. M. Suzuki, *Propriétés topologiques des polynômes de deux variables complexes, et automorphismes algébriques de l'espace* $\mathbb{C}^2$, J. Math. Soc. Japan **26** (1974), 241–257.

[Wr]. D. Wright, *Cancellation of variables of the form* $bT^n - a$, J. Algebra **52** (1978), 94–100.

Some subgroups of the Cremona groups

Vladimir L. Popov*

Steklov Mathematical Institute,
Russian Academy of Sciences,
Gubkina 8, Moscow 119991, Russia
and
National Research University Higher School of Economics,
20 Myasnitskaya Ulitsa, Moscow 101000, Russia
E-mail: popovvl@mi.ras.ru

To M. Miyanishi on his 70th birthday

We explore algebraic subgroups of the Cremona group $\mathcal{C}_n$ over an algebraically closed field of characteristic zero. First, we consider some class of algebraic subgroups of $\mathcal{C}_n$ that we call flattenable. It contains all tori. Linearizability of the natural rational actions of flattenable subgroups on $\mathbf{A}^n$ is intimately related to rationality of the invariant fields and, for tori, is equivalent to it. We prove stable linearizability of these actions and show the existence of nonlinearizable actions among them. This is applied to exploring maximal tori in $\mathcal{C}_n$ and to proving the existence of nonlinearizable, but stably linearizable elements of infinite order in $\mathcal{C}_n$ for $n \geqslant 5$. Then we consider some subgroups $\mathcal{J}(x_1, \ldots, x_n)$ of $\mathcal{C}_n$ that we call the rational de Jonquières subgroups. We prove that every affine algebraic subgroup of $\mathcal{J}(x_1, \ldots, x_n)$ is solvable and the group of its connected components is Abelian. We also prove that every reductive algebraic subgroup of $\mathcal{J}(x_1, \ldots, x_n)$ is diagonalizable. Further, we prove that the natural rational action on $\mathbf{A}^n$ of any unipotent algebraic subgroup of $\mathcal{J}(x_1, \ldots, x_n)$ admits a rational cross-section which is an affine subspace of $\mathbf{A}^n$. We show that in this statement "unipotent" cannot be replaced by "connected solvable". This is applied to proving a conjecture of A. Joseph on the existence of "rational slices" for the coadjoint representations of finite-dimensional algebraic Lie algebras $\mathfrak{g}$ under the assumption that the Levi decomposition of $\mathfrak{g}$ is a direct product. We then consider some overgroup $\widehat{\mathcal{J}}(x_1, \ldots, x_n)$ of $\mathcal{J}(x_1, \ldots, x_n)$ and prove that every torus in $\widehat{\mathcal{J}}(x_1, \ldots, x_n)$ is linearizable. Finally, we prove the existence of an element $g \in \mathcal{C}_3$ of order 2 such that $g \notin G$ for every connected affine algebraic subgroup G of $\mathcal{C}_\infty$; in particular, g is not stably linearizable.

*Supported by grants РФФИ 11-01-00185-а, НШ–5139.2012.1, and the program *Contemporary Problems of Theoretical Mathematics* of the Russian Academy of Sciences, Branch of Mathematics.

Keywords: Cremona group, linearizability, tori, invariant fields, the rational de Jonquières subgroups, cross-sections.

1. Introduction

The last three decades were marked by growing interest in problems related to the affine Cremona group $\mathcal{C}_n^{\text{aff}}$ (the group of biregular automorphisms of the affine n-dimensional space $\mathbf{A}^n$). Despite of a remarkable progress made during these years, some fundamental problems still remain unsolved. For instance, at the moment the linearization problem for algebraic tori is solved only for $n \leqslant 3$ and its difficult solution for $n = 3$ is one of the highlights of the theory.

Some of these problems may be formulated entirely in terms of group-theoretic structure of $\mathcal{C}_n^{\text{aff}}$. Thereby, they admit the birational counterparts related to the full Cremona group $\mathcal{C}_n$ (the group of birational automorphisms of $\mathbf{A}^n$). It is of interest to explore them. We have not seen publications purposefully developing this viewpoint. A step in this direction is made in this paper.

In Section 2 we first consider a class of algebraic subgroups of $\mathcal{C}_n$ that we call flattenable. Linearizability of their natural rational actions on $\mathbf{A}^n$ is intimately related to rationality of their invariant fields, the subject of classical Noether problem. All algebraic tori in $\mathcal{C}_n$ are contained in this class and, for them, these two properties, linearizability and rationality, are equivalent. We show that flattenable groups are special in the sense of Serre (see [26]) and that every rational locally free action on $\mathbf{A}^n$ of a special group is stably linearizable; in particular, this is so for tori. On the other hand, we show that there are stably linearizable, but nonlinearizable rational locally free actions on $\mathbf{A}^n$ of connected affine algebraic groups, in particular, that of tori. We then apply this to the problem of describing maximal tori in $\mathcal{C}_n$ and show that nowadays one can say more on it than in the time when Białynicki-Birula and Demazure wrote their papers [4], [10]. Namely, apart from n-dimensional maximal tori (that are all conjugate), $\mathcal{C}_n$ for $n \geqslant 5$ contains maximal tori of dimension $n-3$ (and does not contain maximal tori of dimensions $n-2$, $n-1$ and $> n$). This answers a question of Hirschowitz in [13, Sect. 3]. As another application, we prove the existence of nonlinearizable, but stably linearizable elements of infinite order in $\mathcal{C}_n$ for $n \geqslant 5$.

In Sections 3 and 4 we consider a natural counterpart of the classical de Jonquières subgroups of $\mathcal{C}_n^{\text{aff}}$ that we call the rational de Jonquières subgroups of $\mathcal{C}_n$. We prove that their affine algebraic subgroups are solvable and have Abelian groups of connected components. We also prove that reductive algebraic subgroups of the rational de Jonquières subgroups of $\mathcal{C}_n$ are diagonalizable. Then we prove that for the natural rational action on $\mathbf{A}^n$ of any unipotent algebraic subgroup of a rational de Jonquières subgroup of $\mathcal{C}_n$ there exists an affine subspace of $\mathbf{A}^n$ which is a rational cross-section for this action (recall that for rational actions of connected solvable affine algebraic subgroups of $\mathcal{C}_n$ on $\mathbf{A}^n$, the existence of *some* rational cross-sections, not necessarily affine subspaces of $\mathbf{A}^n$, is ensured by the general Rosenlicht's theorem, see [23, Theorem 10]). We also show that in this result "unipotent" cannot be replaced by "connected solvable". We then apply this result to a conjecture of A. Joseph ([14, Sect. 7.11]) on the existence of "rational slices" for the coadjoint representations of finite-dimensional algebraic Lie algebras $\mathfrak{g}$ and prove this conjecture under the assumption that the Levi decomposition of $\mathfrak{g}$ is a direct product. Further, we consider a certain natural class of overgroups of the rational de Jonquières subgroups and, using the results of Section 2, show that the natural action on $\mathbf{A}^n$ of any subtorus of such an overgroup is linearizable. Finally, we prove the existence of an element $g \in \mathcal{C}_3$ of order 2 such that $g \notin G$ for every connected affine algebraic subgroup G of the direct limit $\mathcal{C}_\infty$ of the tower of natural inclusions $\mathcal{C}_1 \hookrightarrow \mathcal{C}_2 \hookrightarrow \cdots$; in particular, g is not stably linearizable.

Conventions, notation and some generalities

Below "variety" means "algebraic variety". We assume given an algebraically closed field k of characteristic zero which serves as domain of definition for each of the varieties considered below. Each variety is identified with its set of points rational over k. Along with the standard notation and conventions of [5] we use the following ones.

- $\operatorname{Aut} X$ is the automorphism group of a variety X.
- $\operatorname{Bir} X$ is the group of birational automorphisms of an irreducible variety X.
- $X \approx Y$ means that X and Y are birationally isomorphic irreducible varieties.
- If f is a rational function on the product $X \times Y$ of varieties and $x \in X$ is a point such that $f|_{x \times Y}$ is well defined, then $f(x)$ is the element of $k(Y)$ such that $f(x, y) = f(x)(y)$ for every point $(x, y) \in X \times Y$ where f is defined.
- Given a dominant rational map $\varphi\colon X \dashrightarrow Y$ of varieties, φ^* is the embedding $k(Y) \hookrightarrow k(X)$, $f \mapsto f \circ \varphi$.

— Given an action

$$\alpha\colon G \times X \to X \tag{1}$$

of a group G on a set X and the elements $g \in G$, $x \in X$, then $\alpha(g,x) \in X$ is denoted by $g \cdot x$. If H is a subgroup of G, then $\alpha|_H$ is the restriction of α to $H \times X$.

— $\mathbf{A}^n \times \mathbf{A}^m$ is identified with $\mathbf{A}^{n+m}$ by means of the isomorphism

$$\mathbf{A}^n \times \mathbf{A}^m \to \mathbf{A}^{n+m},$$
$$((a_1,\ldots,a_n),(b_1,\ldots,b_m)) \mapsto (a_1,\ldots,a_n,b_1,\ldots,b_m).$$

— $K^\times$ is the multiplicative group of a field K.
— K^+ is the additive group of a field K.
— If K/F is a field extension, then K is called *pure* (resp. *stably pure*) over F if K is generated over F by a finite collection of algebraically independent elements (resp. if K is contained in a field that is pure over both K and F).
— G^0 is the identity component of an algebraic group G.
— "Torus" means "affine algebraic torus".

Let G be an algebraic group and let X be a variety.

If (1) is a morphism, then α is called a *regular action*. In this case, for every element $g \in G$, the map $X \to X$, $x \mapsto g \cdot x$, is an automorphism of X and the image of the homomorphism $G \to \operatorname{Aut} X$, $g \mapsto \{x \mapsto g \cdot x\}$ is called an *algebraic subgroup of* $\operatorname{Aut} X$. A regular action α is called *locally free* if there is a dense open subset U of X such that the G-stabilizer of every point of U is trivial.

From now on we assume that X is irreducible. The map

$$\operatorname{Bir} X \to \operatorname{Aut}_k k(X), \qquad \varphi \mapsto (\varphi^*)^{-1}, \tag{2}$$

is a group isomorphism. We always identify $\operatorname{Bir} X$ and $\operatorname{Aut}_k k(X)$ by means of (2) when we speak about action of a subgroup of $\operatorname{Bir} X$ by k-automorphisms of $k(X)$ and, conversely, action of a subgroup of $\operatorname{Aut}_k k(X)$ by birational automorphisms of X.

Let $\theta\colon G \to \operatorname{Bir} X$ be an abstract group homomorphism. It determines an action of G on X by birational isomorphisms. If the domain of definition of the partially defined map $G \times X \to X$, $(g,x) \mapsto \theta(g)(x)$, contains a dense open subset of $G \times X$ and coincides on it with a rational map $\varrho\colon G \times X \dashrightarrow X$, then this action (and ϱ) is called a *rational action* of G on X and $\theta(G)$ is called an *algebraic subgroup of* $\operatorname{Bir} X$.

There is a method for constructing algebraic subgroups of Bir X. Namely, let Y be another irreducible variety and let $\gamma\colon Y \dashrightarrow X$ be a birational isomorphism. Then $\operatorname{Bir} Y \to \operatorname{Bir} X$, $g \mapsto \gamma \circ g \circ \gamma^{-1}$, is a group isomorphism and the image of any algebraic subgroup of $\operatorname{Aut} Y$ under it is an algebraic subgroup of $\operatorname{Bir} X$. In fact, by [23, Theorem 1], this method is universal, i.e., every algebraic subgroup of $\operatorname{Bir} X$ is obtained in this manner for the appropriate Y and γ. In other words, for every rational action of G on X there is a regular action of G on an irreducible variety Y, the open subsets X_0 and Y_0 of resp. X and Y, and an isomorphism $Y_0 \to X_0$ such that the induced field isomorphism $k(X) = k(X_0) \to k(Y_0) = k(Y)$ is G-equivariant. If the action of G on Y is locally free, then the rational action ϱ is called *locally free.*

Let $\varrho\colon G \times X \dashrightarrow X$ be a rational action of G on X and let f be an element of $k(X)$. Then $\{g \cdot f \mid g \in G\}$ is an "algebraic family" of rational functions on X in the following sense: there is a rational function $\widehat{f} \in k(G \times X)$ such that $g \cdot f = \widehat{f}(g)$ for every $g \in G$. Indeed, $\varrho^*(f) \in k(G \times X)$ and $\varrho^*(f)(g, x) = (g^{-1} \cdot f)(x)$ for every point $(g, x) \in G \times X$ where $\varrho^*(f)$ is defined; whence the claim.

If X and Y are irreducible varieties endowed with rational actions of G such that there is a G-equivariant birational isomorphism $X \dashrightarrow Y$, then we write

$$X \overset{G}{\approx} Y.$$

In order to avoid a confusion, in some cases when several rational actions are simultaneously considered, we denote X endowed with a rational action ϱ of G by

$${}_{\varrho}X.$$

If Y is another variety, then $X \times Y$ endowed with the rational action of G via the first factor by means of ϱ is denoted by

$${}_{\varrho}X \times Y.$$

We denote by

$${}_{\lambda}G$$

the underlying variety of G endowed with the action of G by left translations.

If ϱ is a rational action of G on X, then

$$\pi_{G,X}\colon X \dashrightarrow X /\!\!/ G$$

is a rational quotient of ϱ, i.e., X/G and $\pi_{G,X}$ are resp. a variety and a dominant rational map such that $\pi_{G,X}^*(k(X/G)) = k(X)^G$ (see [21, Sect. 2.4]). Depending on the situation we choose X/G as a suitable variety within the class of birationally isomorphic ones. A *rational section* (resp., *cross-section*) for ϱ is a rational map $\sigma\colon X/G \dashrightarrow X$ such that $\pi_{G,X} \circ \sigma = \mathrm{id}$ (resp., a subvariety S of X such that $\pi_{G,X}|_S\colon S \dashrightarrow X/G$ is a birational isomorphism). The closure of the image of a rational section is a rational cross-section and, since $\operatorname{char} k = 0$, the closure of every cross-section is obtained in this manner.

The group

$$\mathcal{C}_n := \operatorname{Aut}_k k(\mathbf{A}^n)$$

is called the *Cremona group of rank n* (over k). It is endowed with a topology, the *Zariski topology of* $\mathcal{C}_n$, in which families of elements of $\mathcal{C}_n$ "algebraically parametrized" by algebraic varieties are closed, [27, Sect. 1.6]. For every algebraic subgroup G of $\mathcal{C}_n$ and its subset S, the closure of S in $\mathcal{C}_n$ coincides with the closure of S in G in the Zariski topology of G. In particular, G is closed in $\mathcal{C}_n$. Left and right translations of $\mathcal{C}_n$ are homeomorphisms.

We denote by $x_1, \ldots, x_n \in k[\mathbf{A}^n]$ the *standard coordinate functions* on $\mathbf{A}^n$:

$$x_i((a_1, \ldots, a_n)) = a_i. \tag{3}$$

They are algebraically independent over k and $k(\mathbf{A}^n) = k(x_1, \ldots, x_n)$. For every $n \geqslant 2$, we identify $\mathbf{A}^{n-1}$ with the image of the embedding $\mathbf{A}^{n-1} \hookrightarrow \mathbf{A}^n$, $(a_1, \ldots, a_{n-1}) \mapsto (a_1, \ldots, a_{n-1}, 0)$, and denote the restriction $x_i|_{\mathbf{A}^{n-1}}$ for $i = 1, \ldots, n-1$ still by x_i. Correspondingly, we have the embedding $\mathcal{C}_{n-1} \hookrightarrow \mathcal{C}_n$, $g \mapsto \widehat{g}$, where $\widehat{g}\cdot x_i := g\cdot x_i$ if $i = 1, \ldots, n-1$ and $\widehat{g}\cdot x_n := x_n$. The direct limit for the tower of these embeddings $\mathcal{C}_1 \hookrightarrow \mathcal{C}_2 \hookrightarrow \cdots \hookrightarrow \mathcal{C}_n \hookrightarrow \cdots$ is the *Cremona group* $\mathcal{C}_\infty$ *of infinite rank*. We identify every $\mathcal{C}_n$ with the subgroup of $\mathcal{C}_\infty$ by means of the natural embedding $\mathcal{C}_n \hookrightarrow \mathcal{C}_\infty$. A subgroup G of $\mathcal{C}_\infty$ is called *algebraic* if there exists an integer $n > 0$ such that G is an algebraic subgroup of $\mathcal{C}_n$.

We distinguish the following two algebraic subgroups of $\mathcal{C}_n$:

$$\begin{aligned}\mathrm{GL}_n &:= \{g \in \mathcal{C}_n \mid g \cdot x_i = \textstyle\sum_{j=1}^n \alpha_{ij} x_j, \ \ \alpha_{ij} \in k\},\\ D_n &:= \{g \in \mathcal{C}_n \mid g \cdot x_i = \alpha_i x_i, \ \ \alpha_i \in k\};\end{aligned}$$

D_n is the maximal torus in GL_n.

Let g be an element and let G be a subgroup of $\mathcal{C}_n$. If $g \in \mathrm{GL}_n$ (resp. $G \subseteq \mathrm{GL}_n$), then g (resp. G) is called a *linear element* (resp. a *linear subgroup*). If

g (resp. G) is conjugate to a linear element (resp. a linear subgroup), then it is called a *linearizable element* (resp. a *linearizable subgroup*). If g (resp. G) is a linearizable element (resp. a linearizable subgroup) of some $\mathcal{C}_m$ for $m \geqslant n$, then it is called a *stably linearizable element* (resp. a *stably linearizable subgroup*). A rational action ϱ of an algebraic group H on $\mathbf{A}^n$ is called resp. a *linear*, *linearizable* or *stably linearizable action* if the image of H in $\mathcal{C}_n$ corresponding to ϱ is resp. a linear, linearizable or stably linearizable subgroup of $\mathcal{C}_n$.

Acknowledgment. I am grateful to MING-CHANG KANG who drew my attention to paper [28].

2. Flattening, linearizability, tori

Definition 2.1. An affine algebraic group G is called *flattenable* if the underlying variety of G endowed with the action of G by left translations admits an equivariant open embedding into some $\mathbf{A}^n$ endowed with a rational linear action of G. The G-module $\mathbf{A}^n$ is then called a *flattening* of G.

Every flattenable group is connected.

Example 2.1. $\mathbf{A}^n$ endowed with the natural action of $\mathbf{G}_m^n$,

$$\operatorname{diag}(\varepsilon_1, \ldots, \varepsilon_n) \cdot (a_1, \ldots, a_n) := (\varepsilon_1 a_1, \ldots, \varepsilon_n a_n), \tag{4}$$

is a flattening of $\mathbf{G}_m^n$. Hence, every torus is flattenable.

Example 2.2. The underlying vector space of the algebra $\mathrm{Mat}_{n\times n}$ of all $(n \times n)$-matrices with entries in k endowed with the action of $\mathbf{GL}_n$ by left multiplications, $g \cdot a := ga$, $g \in \mathbf{GL}_n$, $a \in \mathrm{Mat}_{n\times n}$, is a flattening of $\mathbf{GL}_n$. Hence, GL_n is flattenable.

Example 2.3. Let $G_1, \ldots, G_s$ be affine algebraic groups and let $G := G_1 \times \cdots \times G_s$. If $\mathbf{A}_{n_i}$ endowed with an action of G_i is a flattening of G_i, then $\mathbf{A}_{n_1} \times \cdots \times \mathbf{A}_{n_s}$ endowed with the natural action of G is a flattening of G. Hence, G is flattenable if every G_i is.

Example 2.4. Consider a finite-dimensional associative (not necessarily commutative) k-algebra A with an identity element. The group of all invertible elements of A is then a connected affine algebraic group G whose underlying variety is an open subset of that of A. The action of G on A by left multiplications is linear and the identity map is an equivariant embedding of G into A. Thus, A is a flattening of G and G is flattenable. If A is the product of n copies of the k-algebra k, we obtain Example 2.1. Taking $A = \mathrm{Mat}_{n\times n}$, we obtain Example 2.2.

In general, flattening of G is not unique.

Example 2.5. $\mathrm{Mat}_{n\times n}$ endowed with the action of $\mathbf{GL}_n$ given by $g \cdot a := (g^t)^{-1}a$, $g \in \mathbf{GL}_n$, $a \in \mathrm{Mat}_{n\times n}$, where g^t is the transpose of g, is a flattening of $\mathbf{GL}_n$. It is not isomorphic to that from Example 2.1 (as the highest weights of these two flattenings are not equal).

Lemma 2.1. *If the underlying variety of a connected affine reductive algebraic group $G \neq \{e\}$ is isomorphic to an open subset U of $\mathbf{A}^n$, then $U \neq \mathbf{A}^n$ and the center of G is at least one-dimensional.*

Proof. As $G \neq \{e\}$ is reductive, it contains a torus T of positive dimension. For the action of T on G by left translations, the fixed point set is empty. But for any regular action of T on $\mathbf{A}^n$, the fixed point set is nonempty, see [4, Theorem 1]. Hence, $U \neq \mathbf{A}^n$.

Since $D := \mathbf{A}^n \setminus U \neq \varnothing$ and U is affine, the dimension of every irreducible component of D is $n-1$, see [19, Lemma 3]. Since $\operatorname{Pic} \mathbf{A}^n = 0$, this entails that D is the zero set of some regular function f on $\mathbf{A}^n$. Therefore, $f|_U$ is a nonconstant invertible regular function on U. By [24, Theorem 3], every such function is, up to a scalar multiple, a character of G. So there is a nontrivial character of G. On the other hand, as G is a connected reductive group, $G = G' \cdot C$ where G' is the derived group of G (it is semisimple), C is the connected component of the identity in the center of G (it is a torus), and $G' \cap C$ is finite, see [5, Sect. 14.2]. This entails that the character group of G is a free abelian group of rank $\dim C$. Hence, $\dim C \geqslant 1$. □

Corollary 2.1. *There are no nontrivial semisimple flattenable groups.*

Recall from [26] that an algebraic group G is called *special* if every principal homogeneous space under G over every field K containing k is trivial. By [26] special group is automatically connected and affine. Special groups are classified: a connected affine algebraic group G is special if and only if a maximal connected semisimple subgroup of G is isomorphic to

$$\mathbf{SL}_{n_1} \times \cdots \times \mathbf{SL}_{n_r} \times \mathbf{Sp}_{m_1} \times \cdots \times \mathbf{Sp}_{m_s} \tag{5}$$

for some integers $r \geqslant 0, s \geqslant 0, n_i, m_j$ (by [26] such groups are special, and by [11] only these are).

Lemma 2.2. *Every flattenable group G is special.*

Proof. Let $\mathbf{A}^n$ endowed with a rational linear action α of G be a flattening of G and let $\pi_{G,{}_\alpha\mathbf{A}^n}\colon {}_\alpha\mathbf{A}^n \dashrightarrow {}_\alpha\mathbf{A}^n/G$ be a rational quotient for this action. By Definition 2.1, α is locally free. Hence, by [20, Theorem 1.4.3], proving that G is special is equivalent to proving that $\pi_{G,{}_\alpha\mathbf{A}^n}$ admits a rational section. But the existence of such a rational section is clear because Definition 2.1 entails that ${}_\alpha\mathbf{A}^n/G$ is a single point. □

By Lemma 2.2 a maximal connected semisimple subgroup of every flattenable group is isomorphic to a group of type (5). Hence, every *reductive* flattenable group is a quotient $(T \times S)/C$ where T is a torus, S is a group of type (5) and C is a finite central subgroup.

Conjecture. *The following properties of a connected reductive algebraic group G are equivalent:*

(i) *G is flattenable;*
(ii) *G is isomorphic to $\mathbf{GL}_{n_1} \times \cdots \times \mathbf{GL}_{n_r}$.*

Theorem 2.1. *Let α be a locally free rational action of a flattenable group G on $\mathbf{A}^m$. If the invariant field $k({}_\alpha\mathbf{A}^m)^G$ is pure over k, then α is linearizable.*

Proof. Consider for α a rational quotient,

$$\pi_{G,{}_\alpha\mathbf{A}^m}\colon {}_\alpha\mathbf{A}^m \dashrightarrow {}_\alpha\mathbf{A}^m/G. \tag{6}$$

As explained in Introduction, there is a variety X endowed with a regular locally free action α' of G such that ${}_{\alpha'}X \overset{G}{\approx} {}_\alpha\mathbf{A}^m$. By [8, Theorem 2.13], shrinking X if necessary, we may assume that the geometric quotient

$${}_{\alpha'}X \to {}_{\alpha'}X/G$$

for α' exists and is a torsor over ${}_{\alpha'}X/G$. As G is special by Lemma 2.2, this torsor is locally trivial in Zariski topology. Hence, ${}_{\alpha'}X \overset{G}{\approx} {}_\lambda G \times ({}_{\alpha'}X/G)$ and therefore,

$${}_\alpha\mathbf{A}^m \overset{G}{\approx} {}_\lambda G \times ({}_\alpha\mathbf{A}^m/G). \tag{7}$$

Let ${}_\beta\mathbf{A}^n$ be a flattening of G. Definition 2.1 yields

$${}_\lambda G \overset{G}{\approx} {}_\beta\mathbf{A}^n. \tag{8}$$

From (7) and (8) we obtain

$${}_\alpha\mathbf{A}^m \overset{G}{\approx} {}_\beta\mathbf{A}^n \times ({}_\alpha\mathbf{A}^m/G). \tag{9}$$

The assumption of purity and (9) yield

$$ {}_{\alpha}\mathbf{A}^{m}/\!\!/G \approx \mathbf{A}^{m-n}. \tag{10} $$

Consider the action γ of G on $\mathbf{A}^m$ defined by

$$ {}_{\gamma}\mathbf{A}^{m} := {}_{\beta}\mathbf{A}^{n} \times \mathbf{A}^{m-n}. \tag{11} $$

From (9), (10), and (11) we deduce that

$$ {}_{\alpha}\mathbf{A}^{m} \overset{G}{\approx} {}_{\gamma}\mathbf{A}^{m}. \tag{12} $$

But γ is linear because β is. This and (12) complete the proof. □

Lemma 2.3. *For every affine algebraic group G and every integer r there exists a rational locally free linear action of G on $\mathbf{A}^s$ for some $s > r$.*

Proof. By [5, Prop. 1.10], we may assume that G is a closed subgroup of some GL_n. As there is a closed embedding of GL_n in GL_{n+1}, we may in addition assume that $n^2 > r$. By Example 2 there is a rational locally free linear action α of GL_n on $\mathbf{A}^{n^2}$. Hence, $\alpha|_G$ shares the requested properties. □

Theorem 2.2. *Every rational locally free action α of a special algebraic group G on $\mathbf{A}^m$ is stably linearizable.*

Proof. The same argument as in the proof of Theorem 2.1 shows that (7) holds. By Lemma 2.3 there is a rational locally free linear action γ of G on $\mathbf{A}^s$ for some $s \geqslant m$. Like for α, for γ we have

$$ {}_{\gamma}\mathbf{A}^{s} \overset{G}{\approx} {}_{\lambda}G \times ({}_{\gamma}\mathbf{A}^{s}/\!\!/G). \tag{13} $$

Let $d := \dim G$. Since by [7] the underlying variety of G is rational, we have

$$ G \approx \mathbf{A}^{d}. \tag{14} $$

From (7), (13), and (14) we then obtain

$$ \begin{aligned} \mathbf{A}^{m} &\approx \mathbf{A}^{d} \times ({}_{\alpha}\mathbf{A}^{m}/\!\!/G), \\ \mathbf{A}^{s} &\approx \mathbf{A}^{d} \times ({}_{\gamma}\mathbf{A}^{s}/\!\!/G). \end{aligned} \tag{15} $$

In turn, (7), (13), and (15) imply

$$ \begin{aligned} {}_{\alpha}\mathbf{A}^{m} \times \mathbf{A}^{d} &\overset{G}{\approx} {}_{\lambda}G \times ({}_{\alpha}\mathbf{A}^{m}/\!\!/G) \times \mathbf{A}^{d} \overset{G}{\approx} {}_{\lambda}G \times \mathbf{A}^{m}, \\ {}_{\gamma}\mathbf{A}^{s} \times \mathbf{A}^{d} &\overset{G}{\approx} {}_{\lambda}G \times ({}_{\gamma}\mathbf{A}^{s}/\!\!/G) \times \mathbf{A}^{d} \overset{G}{\approx} {}_{\lambda}G \times \mathbf{A}^{s}. \end{aligned} \tag{16} $$

Since $s \geqslant m$, we have $d + s - m \geqslant 0$ and from (16) we deduce

$$\begin{aligned} {}_\alpha\mathbf{A}^m \times \mathbf{A}^{d+s-m} &= {}_\alpha\mathbf{A}^m \times \mathbf{A}^d \times \mathbf{A}^{s-m} \\ &\overset{G}{\approx} {}_\lambda G \times \mathbf{A}^m \times \mathbf{A}^{s-m} \\ &= {}_\lambda G \times \mathbf{A}^s \\ &\overset{G}{\approx} {}_\gamma\mathbf{A}^s \times \mathbf{A}^d. \end{aligned} \tag{17}$$

Since the action of G on ${}_\gamma\mathbf{A}^s \times \mathbf{A}^d$ is linear, (17) completes the proof. □

The next theorem implies that "stably linearizable" in Theorem 2.2 cannot be replaced by "linearizable".

Theorem 2.3. *For every connected semisimple algebraic group $G \neq \{e\}$, there exists a rational nonlinearizable locally free action of G on $\mathbf{A}^d$ for $d = \dim G$.*

Proof. Since (14) holds, there exists a rational locally free action α of G on $\mathbf{A}^d$ such that ${}_\lambda G \overset{G}{\approx} {}_\alpha\mathbf{A}^d$. We claim that α is nonlinearizable. For, otherwise, we would get a rational locally free linear (hence, regular) action of G on $\mathbf{A}^d$. Since $d = \dim G$, one of its orbits is open in $\mathbf{A}^d$ and isomorphic to the underlying variety of G. Therefore, by Lemma 2.1 the center of G is at least one-dimensional — a contradiction because G is semisimple. □

For tori we can get an additional information.

Lemma 2.4. *Let X be an irreducible variety endowed with a rational faithful action α of a torus T. Then*

(i) *α is locally free;*
(ii) *$\dim T \leqslant \dim X$;*
(iii) $\operatorname{tr} \deg_k k(X)^T = \dim X - \dim T$.

Proof. By [29, Cor. 2 of Lemma 8] (see also [4, Cor. 1 of Prop. 1]) there is an irreducible affine variety Y endowed with a regular action of T such that $X \overset{T}{\approx} Y$. Hence, we may (and shall) assume that X is affine and α is regular. By [21, Theorem 1.5], we also may (and shall) assume that X is a closed T-stable subset of a finite-dimensional algebraic T-module V not contained in a proper T-submodule of V.

As α is faithful, the kernel of the action of T on V is trivial. As T is a torus, V is the direct sum of T-weight subspaces. Hence, if U is the complement in V to the union of these spaces, this kernel coincides with the T-stabilizer of every point of U. Thus, these stabilizers are trivial. But by the construction, $X \cap U$ is a nonempty open subset of X. This proves (i) that, in turn, entails (ii) and, by [21, Cor. in Sect. 2.3], also (iii). □

Corollary 2.2 ([10]). *The dimension of every torus in $\mathcal{C}_n$ is at most n.*

Corollary 2.3. *Every rational action of a torus on $\mathbf{A}^n$ is stably linearizable.*

Proof. Since tori are special groups, this follows from Lemma 2.4(i) and Theorem 2.2. □

Theorem 2.4. *The following properties of a rational action α of a torus T on $\mathbf{A}^n$ are equivalent:*

(i) *α is linearizable;*
(ii) *the invariant field $k({}_\alpha\mathbf{A}^n)^T$ is pure over k.*

Proof. Assume that (ii) holds. Let T_0 be the kernel of the action of T on X. By Lemma 2.4, the induced action of T/T_0 on $\mathbf{A}^n$ is locally free. Hence, replacing T by T/T_0, we may assume that the action of T on $\mathbf{A}^n$ is locally free. Since T is flattenable, in this case (ii)⇒(i) follows from Theorem 2.1.

(i)⇒(ii) is the corollary of the following more general statement.

Lemma 2.5. *For any rational linear action α of a diagonalizable affine algebraic group D on $\mathbf{A}^n$, the invariant field $k({}_\alpha\mathbf{A}^n)^D$ is pure over k.*

Proof of Lemma 2.5. By [5, Prop. 8.2(d)], the image of D under the homomorphism $D \to \mathrm{GL}_n$ determined by α is conjugate to a subgroup of D_n. Hence, we may (and shall) assume that D is a closed subgroup of D_n. Since $\mathbf{A}^n$ with the natural action of D_n is a flattening of D_n (see Example 2.1), we have ${}_\alpha\mathbf{A}^n \overset{D}{\approx} {}_\lambda D_n$. Therefore,

$$ {}_\alpha\mathbf{A}^n /\!\!/ D \approx D_n/D. \tag{18} $$

But D_n/D is a torus, see [5, Props. 8.4 and 8.5], hence, a rational variety. The claim now follows from (18). □

Corollary 2.4 ([4, Cor. 2 of Prop. 1]).

(a) *Every faithful rational action of a torus T on $\mathbf{A}^n$ is linearizable in either of the following cases:*

(i) $\dim T \geqslant n-2$;
(ii) $n \leqslant 3$.

(b) *Every d-dimensional torus in $\mathcal{C}_n$ for $d = n-2, n-1, n$ is conjugate to a subgroup of D_n. In particular, every n-dimensional torus in $\mathcal{C}_n$ is conjugate to D_n.*

Proof. (a) By Corollary 2.2, if $T \neq \{e\}$, then (ii)$\Rightarrow$(i). Assume that (i) holds. Then $\operatorname{tr deg}_k k(\mathbf{A}^n)^T \leqslant 2$ by Lemma 2.4(iii). As $k(\mathbf{A}^n)^T$ is unirational, it is then pure over k by the Lüroth and Castelnuovo theorems; whence the claim by Theorem 2.4.

Part (b) follows from (a). □

By Corollaries 2.2 and 2.4(b) all n-dimensional tori in $\mathcal{C}_n$ are maximal and conjugate and there are no maximal $(n-1)$- and $(n-2)$-dimensional tori in $\mathcal{C}_n$. In dimension $n-3$ the situation is different:

Theorem 2.5. *Let $n \geqslant 5$. Every $(n-3)$-dimensional connected affine algebraic group G can be realized as an algebraic subgroup of $\mathcal{C}_n$ such that*

(i) *$k(\mathbf{A}^n)^G$ is not pure, but stably pure over k;*
(ii) *the natural rational action of G on $\mathbf{A}^n$ is locally free.*

Proof. By [28] there exists a nonrational threefold X such that $\mathbf{A}^2 \times X \approx \mathbf{A}^5$. Then $\mathbf{A}^n \approx \mathbf{A}^{n-3} \times X$. This and (14) yield that there exists a rational locally free action γ of G on $\mathbf{A}^n$ such that ${}_\gamma\mathbf{A}^n \overset{G}{\approx} {}_\lambda G \times X$. Since $k({}_\lambda G)^G = k$, by [25, Lemma 3], we have ${}_\gamma\mathbf{A}^n/G \approx X$; whence the claim. □

Corollary 2.5. *Let $n \geqslant 5$. Then*

(a) *there is a rational locally free nonlinearizable action of an $(n-3)$-dimensional torus on $\mathbf{A}^n$;*
(b) *$\mathcal{C}_n$ contains an $(n-3)$-dimensional maximal torus.*

Proof. Use the notation of Theorem 2.5 and its proof and let G be a torus. Then γ is nonlinearizable by Theorem 2.4. This proves (a). As the torus G is not conjugate to a subgroup of D_n, Corollary 2.4(b) implies that it is maximal. This proves (b). □

Corollary 2.6. *Every $\mathcal{C}_n$ for $n \geqslant 5$ contains a nonlinearizable, but stably linearizable element of infinite order.*[a]

Proof. For any subset X of $\mathcal{C}_n$ denote by $\overline{X}$ the closure of X in the Zariski topology of $\mathcal{C}_n$ (see Section 1). By Corollary 2.5(b), $\mathcal{C}_n$ contains an $(n-3)$-dimensional maximal torus T. By [5, Sect. III.8.8], there exists an element $g \in T$ such that $T = \overline{S}$ for $S := \{g^d \mid d \in \mathbf{Z}\}$. Corollary 2.3 yields that g is stably linearizable. Assume that g is linearizable and let $h \in \mathcal{C}_n$ be an element such that $hgh^{-1} \in D_n$. Then $S \subset h^{-1}D_nh$. Since left and right translations of $\mathcal{C}_n$ are homeomorphisms and $\overline{D_n} = D_n$, we obtain

$$T = \overline{S} \subset \overline{h^{-1}D_nh} = h^{-1}\overline{D_n}h = h^{-1}D_nh.$$

This contradicts the maximality of T because $h^{-1}D_nh$ is an n-dimensional torus. □

The next statement yields a rectification of Corollaries 2.3 and 2.6.

Theorem 2.6. *Every torus T in $\mathcal{C}_m$ is conjugate in $\mathcal{C}_{m+\dim T}$ to a subgroup of $D_{m+\dim T}$.*

Proof. Let α be the natural rational action of T on $\mathbf{A}^m$ and let $d := \dim T$. By Lemma 2.4, α is locally free. By [25, Lemma 3], (7) and (14) we have

$$\begin{aligned} (\mathbf{A}^d \times {}_\alpha\mathbf{A}^m)/\!\!/T &\approx \mathbf{A}^d \times ({}_\alpha\mathbf{A}^m/\!\!/T), \\ {}_\alpha\mathbf{A}^m &\overset{T}{\approx} {}_\lambda T \times ({}_\alpha\mathbf{A}^m/\!\!/T), \\ T &\approx \mathbf{A}^d. \end{aligned} \tag{19}$$

From (19) we deduce that $k({}_\alpha\mathbf{A}^m \times \mathbf{A}^d)^T$ is pure over k. Since T is flattenable, Theorem 2.1 then entails that $\mathbf{A}^d \times {}_\alpha\mathbf{A}^m \overset{T}{\approx} {}_\gamma\mathbf{A}^{m+d}$ for a rational linear action γ; whence the claim. □

3. Subgroups of the rational de Jonquières groups

Let $t_1, \ldots, t_n$ be a system of generators of $k(\mathbf{A}^n)$ over k,

$$k(\mathbf{A}^n) = k(t_1, \ldots, t_n).$$

---

[a] In the preprint version of this paper (see `arXiv:1110.2410v1-v3`) the inequality $n \geqslant 5$ in Theorem 2.5 and Corollaries 2.5, 2.6, and 4.4 was replaced by $n \geqslant 6$ because I used the result of [3] in place of that of [28], about which I was unaware.

The elements $t_1, \dots, t_n$ are algebraically independent over k and determine the following flag of subfields of $k(\mathbf{A}^n)$:

$$K_n \subset K_{n-1} \subset \cdots \subset K_0, \text{where } K_i := \begin{cases} k(t_{i+1}, t_{i+2}, \dots, t_n) & \text{if } i \leqslant n-1, \\ k & \text{if } i = n. \end{cases} \tag{20}$$

For any elements $f_i \in K_i$ and $\mu_i \in k^\times$, $i = 1, \dots, n$, put

$$t'_i := \mu_i t_i + f_i \quad \text{and} \quad K'_i := \begin{cases} k(t'_{i+1}, t'_{i+2}, \dots, t'_n) & \text{if } i \leqslant n-1, \\ k & \text{if } i = n. \end{cases} \tag{21}$$

It follows from (21) that there are elements $f'_i \in K'_i$, $i = 1, \dots, n$, such that

$$t_i = \mu_i^{-1} t'_i + f'_i.$$

Hence, $K'_i = K_i$ for every i. In particular, $t'_1, \dots, t'_n$ is an algebraically independent system of generators of $k(\mathbf{A}^n)$ over k, so there is an element $g \in \mathcal{C}_n$ such that

$$g \cdot t_i = \mu_i t_i + f_i \qquad \text{for every} \quad i = 1, \dots, n. \tag{22}$$

The set $\mathcal{J}(t_1, \dots, t_n)$ of all such elements g is a subgroup of $\mathcal{C}_n$. It stabilizes the flag of subfields (20):

$$g \cdot K_i = K_i \qquad \text{for all } g \in \mathcal{J}(t_1, \dots, t_n) \text{ and } i = 0, \dots, n. \tag{23}$$

If $s_1, \dots, s_n$ is another system of generators of $k(\mathbf{A}^n)$ over k, then the subgroups $\mathcal{J}(t_1, \dots, t_n)$ and $\mathcal{J}(s_1, \dots, s_n)$ are conjugate in $\mathcal{C}_n$.

Given an analogy of the construction of $\mathcal{J}(t_1, \dots, t_n)$ with that of the de Jonquières subgroup of $\mathrm{Aut}_k k[t_1, \dots, t_n]$, cf. [31, p. 85], we call $\mathcal{J}(t_1, \dots, t_n)$ the *rational de Jonquières subgroup* of $\mathcal{C}_n$ with respect to $t_1, \dots, t_n$.

Example 3.1. By the Lie–Kolchin theorem every closed connected solvable subgroup G of GL_n is conjugate in GL_n to a subgroup of $\mathcal{J}(x_1, \dots, x_n)$. Hence, G lies in $\mathcal{J}(t_1, \dots, t_n)$ where $t_1, \dots, t_n$ are the homogeneous linear forms in $x_1, \dots, x_n$.

In the notation of (22), for every $i = 1, \dots, n$, we have the following maps:

$$\begin{aligned} \chi_i \colon \mathcal{J}(t_1, \dots, t_n) &\to k^\times, \quad g \mapsto \mu_i, \\ \varphi_i \colon \mathcal{J}(t_1, \dots, t_n) &\to K_i, \quad g \mapsto f_i. \end{aligned} \tag{24}$$

Lemma 3.1. *For every $i = 1, \dots, n$,*

(a) *χ_i is a homomorphism of groups;*

(b) *for all* $g_1, g_2 \in \mathcal{J}(t_1, \ldots, t_n)$,

$$\varphi_i(g_1 g_2) = \chi_i(g_2)\varphi_i(g_1) + g_1 \cdot (\varphi_i(g_2)); \tag{25}$$

(c) *if* G *is an algebraic subgroup of* $\mathcal{C}_n$ *contained in* $\mathcal{J}(t_1, \ldots, t_n)$, *then* $\chi_i|_G$ *is a regular function on* G *and there is a rational function* $F_i \in k(G \times \mathbf{A}^n)$ *such that* $F_i(g) = \varphi_i(g)$ *for all* $g \in G$;

(d) *the order of every element* $g \in \bigcap_{i=1}^n \ker \chi_i$, $g \neq e$, *is infinite.*

Proof. Let $g_1, g_2 \in \mathcal{J}(t_1, \ldots, t_n)$. Then (22) and (24) yield

$$\begin{aligned}\chi_i(g_1 g_2)t_i + \varphi_i(g_1 g_2) &= g_1 g_2 \cdot t_i = g_1 \cdot (g_2 \cdot t_i)\\ &= g_1 \cdot \big(\chi_i(g_2)t_i + \varphi_i(g_2)\big)\\ &= \chi_i(g_2)\big(\chi_i(g_1)t_i + \varphi_i(g_1)\big) + g_1 \cdot \big(\varphi_i(g_2)\big).\end{aligned} \tag{26}$$

As the image of φ_i lies in the $\mathcal{J}(t_1, \ldots, t_n)$-stable field K_i, (26) and algebraic independence of $t_1, \ldots, t_n$ over k yield that (25) and $\chi_i(g_1 g_2) = \chi_i(g_1)\chi_i(g_2)$ hold. This proves (a) and (b).

(c) Let $\alpha\colon G \times \mathbf{A}^n \dashrightarrow \mathbf{A}^n$ be the natural rational action of G on $\mathbf{A}^n$ and let $\beta\colon G \times \mathbf{A}^n \to G \times \mathbf{A}^n$, $(g, a) \mapsto (g^{-1}, a)$. Put $S_i := \beta^*(\alpha^*(t_i)) \in k(G \times \mathbf{A}^n)$. Then $S_i(g, a) = t_i(\alpha(\beta(g, a))) = t_i(\alpha(g^{-1}, a)) = t_i(g^{-1} \cdot a) = (g \cdot t_i)(a)$ for every (g, a) in the domain of definition. Hence, $S_i(g) = \chi_i(g)t_i + \varphi_i(g)$ for every $g \in G$. Given that $S_i \in k(G \times \mathbf{A}^n) = k(G)(t_1, \ldots, t_n)$ and $\varphi_i(g) \in K_i$, this implies (c).

(d) As $g \neq e$ and $\chi_i(g) = 1$ for every i, (22) and (24) entail that there is j such that $\varphi_j(g) \neq 0$. Let d be the largest j with this property. Then $g \cdot f = f$ for every $f \in K_d$. As $g \cdot t_d = t_d + \varphi_d(g)$ and $\varphi_d(g) \in K_d$, this yields

$$g^s \cdot t_d = t_d + s\varphi_d(g) \quad \text{for every } \ s \in \mathbf{Z}. \tag{27}$$

Since $\varphi_d(g) \neq 0$ and $\operatorname{char} k = 0$, (27) implies that $g^s \neq e$ for every $s \neq 0$. This proves (d). □

Theorem 3.1. *Let* G *be an affine algebraic subgroup of* $\mathcal{J}(t_1, \ldots, t_n)$. *Then* G *is solvable and* G/G^0 *is Abelian.*

Proof. First, consider the case where G is finite; we then have to prove that G is Abelian. Consider the homomorphism

$$\delta\colon \mathcal{J}(t_1, \ldots, t_n) \to D_n, \quad g \mapsto \operatorname{diag}(\chi_1(g), \ldots, \chi_n(g)).$$

Since $\ker \delta = \bigcap_{i=1}^n \ker \chi_i$ and G has no elements of infinite order, Lemma 3.1(d) implies that $G \cap \ker \delta = \{e\}$. Therefore, δ embeds G into the Abelian group D_n; whence, the claim.

Now consider the general case. By [6, Lemma 5.11], there is a finite subgroup H of G that intersects every connected component of G. Hence, the restriction to H of the canonical homomorphism $G \to G/G^0$ is a surjective homomorphism $H \to G/G^0$. According to what we have already proved, H is Abelian. This shows that G/G^0 is Abelian. By [12, Theorem 9.2.5], the problem is then reduced to proving that G^0 is solvable.

Since $\operatorname{char} k = 0$, there exists a Levi subgroup L in G^0, see [5, 11.22]. It is a connected reductive group and we have to show that L is a torus, i.e., that the derived subgroup L' of L is trivial. Arguing on the contrary, assume that $L' \neq \{e\}$. Then L' contains an element $g \neq e$ of finite order. Indeed, L' contains a torus $\neq \{e\}$ (see [5, Cor. 2 in Sect. IV.13.17 and Theorem 12.1(b)]), but every torus $\neq \{e\}$ has a nontrivial torsion (see [5, Prop. 8.9(d)]). On the other hand, $L' \subseteq \bigcap_{i=1}^{n} \ker \chi_i$ as every homomorphism $L \to k^{\times}$ contains L' in its kernel. By Lemma 3.1(d) this entails that the order of g is infinite. This contradiction completes the proof. □

Corollary 3.1. *Every finite subgroup of $\mathcal{J}(t_1, \ldots, t_n)$ is Abelian.*

Theorem 3.2. *Let G be a reductive algebraic subgroup of $\mathcal{J}(t_1, \ldots, t_n)$. Then G is a diagonalizable group.*

Proof. By Theorem 3.1 the reductive group G^0 is solvable. Hence, G^0 is a torus. Let H be the subgroup of G from the proof of Theorem 3.1. It acts on G^0 by conjugation because G^0 is normal in G. The fixed point set F of this action is a closed subgroup of G^0. Assume that $F \neq G^0$. Then, since the torsion subgroup of G^0 is dense in G^0 (see [5, Cor. III.8.9]), there exists an element $g \in G^0 \setminus F$ whose order is finite. Let S be the subgroup of G^0 generated by the set $\{hgh^{-1} \mid h \in H\}$. Since G^0 is Abelian and the orders of g and H are finite, S is finite as well. Since S is stable with respect to the action of H on G^0 by conjugation, this implies that the subgroup generated by S and H is finite, too. Corollary 3.1 then yields that this subgroup is Abelian. Hence, $g \in F$ — a contradiction. Therefore, $F = G^0$, i.e., G^0 and H commute. Since the Abelian groups H and G^0 generate G, this implies that G is Abelain. The claim then follows by [5, Prop. III.8.4(4) and Cor. III.4.4(1)]. □

4. Affine subspaces as cross-sections

By [23, Theorem 10], for every rational action of a connected solvable algebraic group there exists a rational cross-section. The next theorem refines this for some rational actions on $\mathbf{A}^n$ by showing that there exist cross-sections that are affine subspaces of $\mathbf{A}^n$.

Theorem 4.1. *Let $G \neq \{e\}$ be a unipotent affine algebraic subgroup of $\mathcal{J}(x_1, \ldots, x_n)$ and let α be the corresponding rational action of G on $\mathbf{A}^n$. Then there exist a sequence $1 \leqslant i_1 < \cdots < i_m \leqslant n$ of natural numbers and a sequence $\Theta_1, \ldots, \Theta_m$ of nonempty open subsets of k such that for every $(c_1, \ldots, c_m) \in \Theta_1 \times \cdots \times \Theta_m$ the affine subspace of $\mathbf{A}^n$ defined by the equations (see (3)):*

$$x_{i_1} = c_1, \ldots, x_{i_m} = c_m,$$

is a rational cross-section for α.

For the proof of Theorem 4.1 we need the following

Lemma 4.1. *Let K be a field of characteristic 0 and let $f(x)$ be a rational function in a variable x with the coefficients in K. Let K' be a subfield of K. If*

$$f(a_1 + a_2) = f(a_1) + f(a_2) \tag{28}$$

whenever f is defined at a_1, a_2 and $a_1 + a_2 \in K'$, then there is an element $c \in K$ such that $f(x) = cx$.

Proof of Lemma 4.1. We may (and shall) assume that $f \neq 0$. Let $\overline{K}$ be an algebraic closure of K. First, we claim that (28) holds whenever f is defined at a, b and $a + b \in \overline{K}$. Indeed, by (28) the rational function $F(x_1, x_2) := f(x_1) + f(x_2) - f(x_1 + x_2)$ (see (3)) vanishes at every point of $\mathbf{A}^2(K')$ where it is defined. Since $\mathbf{A}^2(K')$ is Zariski dense in $\mathbf{A}^2$, this yields $F = 0$; whence the claim.

Thus, f is a rational partially defined endomorphism of the algebraic group $\overline{K}^+$. But by [34] (cf. also [17, Sect. 11.1.1]) every rational partially defined homomorphism of algebraic groups is, in fact, an everywhere defined algebraic homomorphism. This entails that $f(x) \in K[x]$. Since f has only finitely many roots, $\ker f$ is finite. Therefore, $f(\overline{K}^+)$ is a one-dimensional closed subgroup of $\overline{K}^+$; whence $f(\overline{K}^+) = \overline{K}^+$. On the other hand, since $\operatorname{char} K = 0$, there are no nonzero elements of finite order in $\overline{K}^+$. Hence, $\ker f = \{0\}$. Thus, f is an isomorphism; whence the claim. □

Proof of Theorem 4.1. We shall use the notation of (20), (24) with

$$t_1 = x_1, \ldots, t_n = x_n.$$

Since $\operatorname{char} k = 0$, G is connected. As G is a nontrivial unipotent group, it contains a one-dimensional normal subgroup U isomorphic to k^+. We identify U with k^+ by an isomorphism $U \to k^+$. Since G is unipotent, there are no nontrivial algebraic homomorphisms $G \to k^\times$, therefore, by Lemma 3.1 there are rational functions $F_i \in k(G \times \mathbf{A}^n)$ such that

$$\begin{aligned} g \cdot x_i &= x_i + F_i(g), \\ F_i(g) &\in K_i, \end{aligned} \qquad \text{for every } g \in G \text{ and } i. \tag{29}$$

Since $U \neq \{e\}$, (29) entails that $F_j(u) \neq 0$ for some $u \in U$ and j. Let d be the largest j appearing in this fashion. Then (29) and (20) yield

$$K_d^U = K_d. \tag{30}$$

In turn, from (30) and (25) we infer that

$$F_d(u_1 + u_2) = F_d(u_1) + F_d(u_2) \qquad \text{for all } \ u_1, u_2 \in U.$$

By Lemma 4.1, this implies that there is a nonzero element $s \in K_d$ such that

$$F_d(u) = us \qquad \text{for every } \ u \in U. \tag{31}$$

Thus, by (29) and (31),

$$u \cdot x_d = x_d + us, \qquad \text{for every } \ u \in U. \tag{32}$$

By [23, Theorem 1], there exists a nonempty open subset $\mathbf{A}_0^n$ of $\mathbf{A}^n$ and its embedding in an irreducible variety Y,

$$Y \hookleftarrow \mathbf{A}_0^n \subseteq \mathbf{A}^n,$$

such that the rational action of U on Y determined by $\alpha|_U$ and by this embedding is regular. We identify $\mathbf{A}_0^n$ with the image of this embedding. By [23], Theorem 2, shrinking Y if necessary, we may (and shall) assume that there exists a geometric quotient of Y by this action of U,

$$\pi_{U,Y} : Y \to Y/U.$$

Then $\pi_{U,Y}|_{\mathbf{A}_0^n}$ is the restriction to $\mathbf{A}_0^n$ of a rational quotient for $\alpha|_U$,

$$\pi_{U,\mathbf{A}^n} : \mathbf{A}^n \dashrightarrow Y/U =: \mathbf{A}^n /\!\!\!/ U.$$

Let $H := G/U$. Then α induces a rational action β of H on Y/U. Consider a rational quotient for β,

$$\pi_{H,Y/U} : Y/U \dashrightarrow (Y/U) /\!\!\!/ H.$$

Then the composition

$$\pi_{G,\mathbf{A}^n} := \pi_{H,Y/U} \circ \pi_{U,\mathbf{A}^n}$$

is a rational quotient for α,

$$\pi_{G,\mathbf{A}^n} \colon \mathbf{A}^n \dashrightarrow (Y/U)/H =: \mathbf{A}^n/G.$$

Shrinking $\mathbf{A}^n_0$ and Y if necessary, we may (and shall) assume that

(i) $\pi_{H,Y/U}$ is a morphism;
(ii) $s|_{\mathbf{A}^n_0}$ is regular and vanishes nowhere.

To sum up, we have the following commutative diagram:

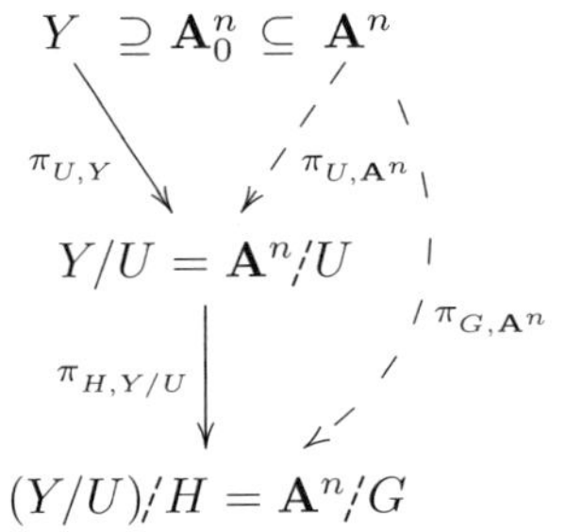

(33)

For every element $c \in k$, denote by L_c the hyperplane in $\mathbf{A}^n$ defined by the equation $x_d = c$. The set

$$\Omega := \{c \in k \mid L_c \cap \mathbf{A}^n_0 \neq \varnothing\}$$

is nonempty and open in k. Take an element $c \in \Omega$ and a point $a \in \mathbf{A}^n_0$. By property (ii) above, s is regular and does not vanish at a. Consider the U-orbit of a in Y. Formula (31) shows that there is a unique $u_0 \in U$ such that the value of $x_d \in k(Y)$ at $u_0 \cdot a$ is c, namely,

$$u_0 = \frac{x_d - c}{s}(a). \tag{34}$$

This means that every U-orbit in Y intersects $L_c \cap \mathbf{A}^n_0$ at most at one point, i.e., $\pi_{U,Y}|_{L_c \cap \mathbf{A}^n_0}$ is injective. Since $\dim L_c \cap \mathbf{A}^n_0 = \dim Y/U$ and $\operatorname{char} k = 0$, this implies that $\pi_{U,Y}|_{L_c \cap \mathbf{A}^n_0} \colon L_c \cap \mathbf{A}^n_0 \to Y/U$ is a birational isomorphism. Hence, L_c intersects the domain of definition of $\pi_{U,\mathbf{A}^n}$ and

$$\pi_{U,\mathbf{A}^n}|_{L_c} \colon L_c \dashrightarrow Y/U = \mathbf{A}^n/U, \tag{35}$$

is a birational isomorphism. This means that L_c is a rational cross-section for $\alpha|_U$. In particular, this implies that shrinking $\mathbf{A}^n_0$ if necessary, we may (and shall) assume that

(iii) for every point of $\mathbf{A}^n_0$, its U-orbit in Y intersects L_c.

Now we argue by induction on $\dim G$. If $\dim G = 1$, then $G = U$ and the claim is proved since every L_c for $c \in \Omega$ is a rational cross-section for α (so in this case $s = 1$, $i_1 = d$ and $\Theta_1 = \Omega$).

Now assume that $\dim G > 1$. The action β and the birational isomorphism (35) determine a rational action γ of H on L_c such that (35) becomes an H-equivariant birational isomorphism. From (33) we then deduce that

$$\pi_{G,\mathbf{A}^n}|_{L_c}\colon L_c \dashrightarrow \mathbf{A}^n/G$$

is a rational quotient for γ. We identify L_c with $\mathbf{A}^{n-1}$ by means of the isomorphism $(a_1, \ldots, a_{d-1}, c, a_{d+1}, \ldots, a_n) \mapsto (a_1, \ldots, a_{d-1}, a_{d+1}, \ldots, a_n)$ and, for every function $f \in k(\mathbf{A}^n)$ whose domain of definition intersects L_c, put

$$\overline{f} := f|_{L_c} \in k(L_c).$$

Then $\overline{x_1}, \ldots, \overline{x_{d-1}}, \overline{x_{d+1}}, \ldots, \overline{x_n}$ are the standard coordinate functions on L_c.

We claim that the image of H in $\mathrm{Aut}_k k(L_c) = \mathcal{C}_{n-1}$ determined by the action γ is contained in $\mathcal{J}(\overline{x_1}, \ldots, \overline{x_{d-1}}, \overline{x_{d+1}}, \ldots, \overline{x_n})$. If this is proved, then, by the inductive assumption, there exist a nonempty set of indices $i_1, \ldots, i_r$ and a nonempty open subsets $\Theta_1, \ldots, \Theta_r$ of k such that for every $(c_1, \ldots, c_r) \in \Theta_1 \times \cdots \times \Theta_r$ the affine subspace S of L_c defined by the equations

$$\overline{x_{i_1}} = c_1, \ldots, \overline{x_{i_r}} = c_r,$$

is a rational cross-section for γ, i.e.,

$$\pi_{G,\mathbf{A}^n}|_S\colon S \dashrightarrow \mathbf{A}^n/G$$

is a birational isomorphism. As S is an affine subspace in $\mathbf{A}^n$ defined by the equations $x_d = c, x_{i_1} = c_1, \ldots, x_{i_r} = c_r$, this will complete the proof.

It remains to prove the claim. To this end, consider in $k(\mathbf{A}^n)$ the subfield $k(\mathbf{A}^n)^U$ of U-invariants elements with respect to $\alpha|_U$. Since L_c is a rational cross-section of $\pi_{U,\mathbf{A}^n}$, the map

$$k(\mathbf{A}^n)^U \to k(L_c), \qquad f \mapsto \overline{f},$$

is a well-defined k-isomorphism of fields that is H-equivariant with respect to the actions of H on $k(\mathbf{A}^n)^U$ and $k(L_c)$ determined resp. by α and γ. Let

$$k(L_c) \to k(\mathbf{A}^n)^U, \qquad t \mapsto \widehat{t},$$

be the inverse isomorphism. Below we will consider β and γ as the actions of G with the kernel U.

Take a point $a \in \mathbf{A}_0^n$. By the above discussion and property (iii), the U-orbit of a in Y intersects L_c at a single point $u_0 \cdot a$ where u_0 is given by (34). As $\widehat{\overline{x_i}} \in k(\mathbf{A}^n)^U$, this yields

$$\widehat{\overline{x_i}}(a) = \widehat{\overline{x_i}}(u_0 \cdot a) = \overline{x_i}(u_0 \cdot a) = x_i(u_0 \cdot a) = ((-u_0) \cdot x_i)(a). \tag{36}$$

Let $z, y_1, \ldots, y_{n-1}$ be the variables over k. It follows from (29), (34) and (36) that there are $d-1$ rational functions

$$R_j(z, y_j, y_{j+1}, \ldots, y_{n-1}) \in k(z, y_j, y_{j+1}, \ldots, y_{n-1}), \qquad j = 1, \ldots, d-1,$$

such that

$$\widehat{\overline{x_i}} = \begin{cases} x_i + R_i\Big(\dfrac{c - x_d}{s}, x_{i+1}, \ldots, x_n\Big) & \text{if } i \leqslant d-1, \\ x_i & \text{if } i \geqslant d+1. \end{cases} \tag{37}$$

In turn, from (37), (29), (20) and (23) we infer that

$$g \cdot \widehat{\overline{x_i}} - x_i \in k(x_{i+1}, \ldots, x_{d-1}, x_d, x_{d+1}, \ldots, x_n) \quad \text{for all } g \in G \text{ and } i;$$

whence,

$$g \cdot \overline{x_i} - \overline{x_i} \in k(\overline{x_{i+1}}, \ldots, \overline{x_{d-1}}, \overline{x_d}, \overline{x_{d+1}}, \ldots, \overline{x_n}) \quad \text{for all } g \in G \text{ and } i. \tag{38}$$

The claim now follows from (38) because $\overline{x_d} = c \in k$. □

Corollary 4.1. *For every unipotent algebraic subgroup G of GL_n, there exists an affine subspace L of $\mathbf{A}^n$ such that L is a rational cross-section for the natural action of G on $\mathbf{A}^n$.*

Proof. There exists an element $g \in \mathrm{GL}_n$ such that $gGg^{-1} \subset \mathcal{J}(x_1, \ldots, x_n)$ (see Example 3.1). By Theorem 4.1 there exists an affine subspace S of $\mathbf{A}^n$ that is a rational cross-section for the natural action of gGg^{-1} on $\mathbf{A}^n$. Then the affine subspace $g^{-1}(S)$ is a rational cross-section for the natural action of G on $\mathbf{A}^n$. □

Here is the application of Corollary 4.1. Let G be a connected affine algebraic group and let $\mathfrak{g}$ be the Lie algebra of G. Joseph put forward the following

Conjecture A ([14, Sect. 7.11]). *For the coadjoint action of G on $\mathfrak{g}^*$, there exists an affine subspace L of $\mathfrak{g}^*$ such that $k(\mathfrak{g}^*)^G \to k(L)$, $f \mapsto f|_L$, is a well-defined isomorphism of fields.*

Joseph calls such L a *rational slice.*

According to the Levi decomposition, $\mathfrak{g}$ is a semidirect product of a reductive Lie algebra $\mathfrak{r}$ and the unipotent radical $\mathfrak{u}$,

$$\mathfrak{g} = \mathfrak{r} \ltimes \mathfrak{u}. \tag{39}$$

Corollary 4.2. *If* (39) *is a direct product,* $\mathfrak{g} = \mathfrak{r} \times \mathfrak{u}$, *then Conjecture A is true.*

Proof. Let R and U be the closed connected subgroups of G whose Lie algebras are resp. $\mathfrak{r}$ and $\mathfrak{u}$. Assume that $\mathfrak{g} = \mathfrak{r} \times \mathfrak{u}$. In this case, if $L_{\mathfrak{r}}$ and $L_{\mathfrak{u}}$ are the rational slices for the coadjoint actions of resp. R and U, then $L_{\mathfrak{r}} \times L_{\mathfrak{u}}$ is a rational slice for the coadjoint action of G. The existence of $L_{\mathfrak{r}}$ is proved in [16] and the existence of $L_{\mathfrak{u}}$ is ensured by Corollary 4.1. □

Remark 4.1. Another application is that Theorem 4.1 yields the results of [22, Part II, Chap. I, §7].

The rational de Jonquières subgroup $\mathcal{J}(t_1, \ldots, t_n)$ lies in another interesting subgroup of $\mathcal{C}_n$. Namely, as for $\mathcal{J}(t_1, \ldots, t_n)$, one checks that, for every $f_i \in K_i$ and $\mu_i \in K_i^{\times}$, there exists an element $g \in \mathcal{C}_n$ for which (22) holds and that the set $\widehat{\mathcal{J}}(t_1, \ldots, t_n)$ of all such elements g is a subgroup of $\mathcal{C}_n$. The flag of subfields (20) is stable with respect to $\widehat{\mathcal{J}}(t_1, \ldots, t_n)$:

$$g \cdot K_i = K_i \qquad \text{for all } g \in \widehat{\mathcal{J}}(t_1, \ldots, t_n) \text{ and } i. \tag{40}$$

If $s_1, \ldots, s_n$ is another system of generators of $k(\mathbf{A}^n)$ over k, then the subgroups $\widehat{\mathcal{J}}(t_1, \ldots, t_n)$ and $\widehat{\mathcal{J}}(s_1, \ldots, s_n)$ are conjugate in $\mathcal{C}_n$.

The following fact is known; it provides an information on tori in $\widehat{\mathcal{J}}(t_1, \ldots, t_n)$ (see Corollary 4.3 below).

Theorem 4.2. *For every (not necessarily algebraic) subgroup G of $\widehat{\mathcal{J}}(t_1, \ldots, t_n)$, the invariant field $k(\mathbf{A}^n)^G$ is pure over k.*

Proof. We shall sketch a proof since our argument provides a bit more information (equality (43)) than that of [18] and [15]. The key ingredient is the following Miyata's lemma:

Lemma 4.2 ([18, Lemma], cf. [15, Lemme 1.1], [1, Sect. 3]). *Let F be a field, let z be a variable over F, and let H be a group that acts on $F[z]$ by ring automorphisms leaving F stable*[b]. *Then the subfield of $F(z)$ gener-*

[b]It is not assumed that F is pointwise fixed.

ated by $F(z)^H$ over F is, in fact, generated by a single element $x \in F[z]^H$:

$$F(F(z)^H) = F(x). \qquad \square \tag{41}$$

Turning to the proof of Theorem 4.2, we first show that, in the notation of Lemma 4.2,

$$F(z)^H = F^H(x). \tag{42}$$

Indeed, $F(x)^H \subseteq F(z)^H$ since $F(x) \subseteq F(z)$. On the other hand, (41) entails that $F(z)^H \subseteq F(x)^H$. Hence, $F(z)^H = F(x)^H$. Therefore, (42) would be proved if the equality

$$F(x)^H = F^H(x) \tag{43}$$

is established. To prove (43), consider two cases: (a) $x \in F$, (b) $x \notin F$. If (a) holds, then $F(x) = F$, hence, $F(x)^H = F^H$. On the other hand, (a) and $x \in F[z]^H$ yield that $x \in F^H$, hence, $F^H(x) = F^H$. This proves (43) if (a) holds. Now assume that (b) holds. Then x is transcendental over F by [32, §73, Theorem]. Consider an element $f \in F(x)^H$. It can be written as $f = p/q$ where

$$p = \sum_{i=0}^{s} a_i x^i, \qquad q = \sum_{j=0}^{r} b_j x^j, \quad a_i, b_j \in F, \quad a_s b_r \neq 0, \tag{44}$$

and p and q are relatively prime polynomials in x with the coefficients in F. Since $F[x]$ is a factorial ring, the relative primeness of p and q and H-invariance of f imply that there is a map $\gamma\colon H \to F^*$ (in fact, a 1-cocycle) such that

$$h \cdot p = \gamma(h)p, \qquad h \cdot q = \gamma(h)q \quad \text{for every } h \in H. \tag{45}$$

Since x is H-invariant, (44) and (45) yield

$$h \cdot a_i = \gamma(h)a_i, \qquad h \cdot b_j = \gamma(h)b_j \quad \text{for all } h \in H \text{ and } i, j. \tag{46}$$

From (46) we infer that $f = a_s^{-1}p/a_s^{-1}q \in F^H(x)$. Thus, $F(x)^H \subseteq F^H(x)$. Since x is H-invariant, the inverse inclusion is clear. This proves (43). Thus, (42) holds and, moreover, either $x \in F^H$ or x is transcendental over F.

Now let G be a subgroup of $\widehat{\mathcal{J}}(t_1, \ldots, t_n)$. We have $K_{i-1} = K_i(t_i)$ and t_i is transcendental over K_i for every $i = 1, \ldots, n$. By (40) and the definition of $\widehat{\mathcal{J}}(t_1, \ldots, t_n)$ the action of G on K_{i-1} satisfies the conditions of Lemma 4.2 (with $F = K_i$, $z = t_i$, $H = G$). Hence, as is proved above, there is an element $z_i \in K_i[t_i]^G$ such that $K_{i-1}^G = K_i(t_i)^G = K_i^G(z_i)$ and either

$z_i \in K_i^G$ or z_i is transcendental over K_i. Respectively, either $K_{i-1}^G = K_i^G$ or K_{i-1}^G is pure over K_i^G of transcendental degree 1. Since

$$k = K_n^G \subseteq K_{n-1}^G \subseteq \cdots \subseteq K_1^G \subseteq K_0^G = k(\mathbf{A}^n)^G,$$

this completes the proof. □

Corollary 4.3. *Every torus in* $\widehat{\mathcal{J}}(t_1, \ldots, t_n)$ *is conjugate in* $\mathcal{C}_n$ *to a subgroup of* D_n.

Proof. This follows from Theorems 2.4 and 4.2. □

Corollary 4.4. *Let* $n \geqslant 5$. *Every* $(n-3)$*-dimensional connected affine algebraic group can be realized as an algebraic subgroup of* $\mathcal{C}_n$ *such that*

(i) G *is not conjugate to a subgroup of* $\widehat{\mathcal{J}}(t_1, \ldots, t_n)$;
(ii) *the natural rational action of* G *on* $\mathbf{A}^n$ *is locally free.*

Proof. This follows from Theorems 2.5 and 4.2. □

Remark 4.2. The assumption that k is algebraically closed is not used in the proof of Theorem 4.2.

Remark 4.3. In [18], Lemma 4.2 is used for proving that $k(\mathbf{A}^n)^G$ is pure over k if G is a subgroup of $\mathrm{GL}_n \cap \mathcal{J}(x_1, \ldots, x_n)$. Note that in this case, if G is finite, then purity of $k(\mathbf{A}^n)^G$ over k follows from Corollary 3.1 and Lemma 2.5.

Remark 4.4. A weakened version of Theorem 4.2 is the subject of [33]. In it, G is an affine algebraic group and $\widehat{\mathcal{J}}(x_1, \ldots, x_n)$ is replaced by $\mathcal{J}(x_1, \ldots, x_n)$. However, the argument in [33] does not amount to complete and accurate proof. Indeed, it is based on the claim, left unproven, that if G is reductive, then G is conjugate in $\mathcal{J}(x_1, \ldots, x_n)$ to a subgroup of D_n. Further, the claim that, for a one-dimensional unipotent algebraic group U, "every point is U-equivalent to a unique point of the subspace $S = \{x \in k^n : x_m = 0\}$" is false because $u \cdot s$ may be not defined for $u \in U$ and $s \in S$. Ditto for the claim that $F_i \in k(x_{i+1}, \ldots, x_n) \otimes k[t]$ (counterexample: $n = 3$ and the action is given by $t \cdot x_1 = x_1 - t/x_2(x_2 + t)$, $t \cdot x_2 = x_2 + t$, $t \cdot x_3 = x_3$), so the equality $F_m(x_{m+1}, \ldots, x_n; t) = tF_m(x_{m+1}, \ldots, x_n)$ remains unproven.

Remark 4.5. One cannot replace $\widehat{\mathcal{J}}(t_1, \ldots, t_n)$ in Theorem 4.2 by the $\mathcal{C}_n$-stabilizer of the flag of subfields (20). Indeed, by [30], for $k = \mathbf{C}$, $n = 3$, this stabilizer contains a subgroup G of order 2 such that $k(\mathbf{A}^n)^G$ is not pure over k.

Combining the construction from [30] with Corollary 2.3 and Lemma 2.5 we obtain the following

Theorem 4.3. *Let $k = \mathbf{C}$ and let $\mathcal{A}$ be the union of all connected affine algebraic subgroups of $\mathcal{C}_\infty$. There exists an element $g \in \mathcal{C}_3$ of order 2 such that $g \notin \mathcal{A}$. In particular, g is not stably linearizable.*

Proof. Let X be the three-dimensional counterexample of Artin and Mumford to the Lüroth problem ([2], see also [9]): X is a smooth projective unirational threefold such that

$$\mathrm{H}^3(X, \mathbf{Z})_{\mathrm{tors}} \neq 0. \tag{47}$$

Since the torsion subgroup of the third integral cohomology group of a smooth complex variety is a birational invariant and, in particular, is zero if the variety is rational, (47) implies that X is not rational.

In [30] is constructed a subgroup G of order 2 in $\mathcal{C}_3$ such that $k(\mathbf{A}^3)^G$ is k-isomorphic to $k(X)$. Let g be the generator of G. Arguing on the contrary, assume that g is contained in a connected affine algebraic subgroup H of $\mathcal{C}_\infty$. Since the order of g is finite, g is a semisimple element of H. Hence, g lies in a maximal torus T of H (see [5, Theorems III.10.6(6) and IV.11.10]). By Corollary 2.3 there exists a positive integer n_0 such that $T \subset \mathcal{C}_{n_0}$ and T is conjugate in $\mathcal{C}_{n_0}$ to a subtorus of D_{n_0}. Fix an integer $n \geqslant \max\{n_0, 3\}$. Then $G \subset \mathcal{C}_n$ and G is conjugate in $\mathcal{C}_n$ to a subgroup of D_n. This and Lemma 2.5 yield that for the natural action of G on $\mathbf{A}^n$ the field $k(\mathbf{A}^n)^G$ is pure over k. Since $G \subset \mathcal{C}_3$, by [25, Lemma 3], we have

$$\mathbf{A}^n/G \approx \mathbf{A}^3/G \times \mathbf{A}^{n-3} \approx X \times \mathbf{P}^{n-3}. \tag{48}$$

From (48) we infer that the smooth projective variety $X \times \mathbf{P}^{n-3}$ is rational and therefore $\mathrm{H}^3(X \times \mathbf{P}^{n-3})_{\mathrm{tors}} = 0$. On the other hand, the Künneth formula and (47) yield that $\mathrm{H}^3(X \times \mathbf{P}^{n-3})_{\mathrm{tors}} \neq 0$ — a contradiction. □

We conclude by an example which shows that in the formulation of Corollary 4.1 "unipotent" cannot be replaced by "connected solvable" (recall that if G is connected solvable, then the existence of *some* rational cross-section is ensured by [23, Theorem 10]).

Example 4.1. Fix a choice of two integers d_1 and d_2 such that

$$d_1 - d_2 \geqslant 2, \tag{49}$$
$$|d_1| \geqslant 2, \quad |d_2| \geqslant 2, \tag{50}$$
$$\gcd(d_1, d_2) = 1. \tag{51}$$

Consider the one-dimensional subtorus

$$T := \{\operatorname{diag}(t^{d_1}, t^{d_2}) \mid t \in k^*\} \tag{52}$$

of $\mathbf{G}_m^2$ and its rational linear action α on $\mathbf{A}^2$ defined by formula (4).

In view of (51), the T-stabilizer of every point $a \in \mathbf{A}^2$, $a \neq (0,0)$, is trivial.

Claim. *There is no affine subspace in $\mathbf{A}^2$ that is a rational cross-section for α.*

Proof. Assume that some affine subspace L of $\mathbf{A}^2$ is a rational cross-section for α. Since T-orbits in general position are one-dimensional, L is a line. Let

$$\mu_1 x_1 + \mu_2 x_2 + \nu = 0, \qquad \mu_1, \mu_2, \nu \in k \tag{53}$$

be its equation. Since L is a rational cross-section, there is a nonempty open subset U of $\mathbf{A}^2$ such that for every point $a = (a_1, a_2) \in U$, the T-orbit of a intersects L at a single point, i.e., by (52) and (53), the following equation in t

$$\mu_1 a_1 t^{d_1} + \mu_2 a_2 t^{d_2} + \nu = 0 \tag{54}$$

has a single nonzero solution. Shrinking U, we may assume that $a_1 a_2 \neq 0$ for every $a \in U$.

If $\mu_1\mu_2 = 0$, then (54) becomes an equation of the form $\mu t^d + \nu = 0$ where $\mu \in k$, $\mu \neq 0$, and $|d| \geqslant 2$ by (50). If $\nu = 0$, it does not have nonzero solutions; if $\nu \neq 0$, there are at least two such solutions. So this case is impossible.

If $\mu_1\mu_2 \neq 0$ and $\nu = 0$, then the solutions of (54) coincide with the roots of $\mu_1 a_1 t^{d_1 - d_2} + \mu_2 a_2$. In view of (49), there are at least two distinct roots, so this case is impossible as well.

Let $\mu_1\mu_2\nu \neq 0$ and $d_2 > 0$. Denote by f be the right-hand side of (54). Set

$$h := d_1 f - t\frac{df}{dt} = (d_1 - d_2)\mu_2 a_2 t^{d_2} + d_1 \nu. \tag{55}$$

By (54) and (55), for a fixed a_2, there are only finitely many a_1's such that the polynomials f and h have a common root. Since every multiple root of f is also a root of h, this means that there are points $a \in U$ such that f does not have multiple roots. From (49), (50) it then follows that for such a point a equation (54) has at least two nonzero solutions. Thus, this case is also impossible.

Finally, let $\mu_1\mu_2\nu \neq 0$ and $d_2 < 0$. Then the solutions of equation (54) coincide with the roots of the polynomial $q := \mu_1 a_1 t^{d_1-d_2} + \nu t^{-d_2} + \mu_2 a_2$. We have

$$p := (d_1 - d_2)q - t\frac{dq}{dt} = (d_1 - d_2)\mu_2 a_2 + d_1 \nu t^{-d_2}. \tag{56}$$

Then the same argument as above with f and h replaced resp. by q and p shows that this case is impossible as well.

This contradiction completes the proof. □

References

1. H. Ahmad, M. Hajja, M.-c. Kang, *Rationality of some projective linear actions*, J. Algebra **228** (2000), 643–658.
2. M. Artin, D. Mumford, *Some elementary examples of unirational varieties which are not rational*, Proc. London Math. Soc. (3) **25** (1972), 75–95.
3. A. Beauville, J.-L. Colliot-Thélène, J.-J. Sansuc, P. Swinnerton-Dyer, *Varietes stablement rationneles non rationneles*, Annals of Math. **121** (1985), no. 2, 283–318.
4. A. Białynicki-Birula, *Remarks on the action of an algebraic torus on k^n*, Bull. Acad. Polon. Sci. Sér. Sci. Math., Astr., Phys. **14** (1966), no. 4, 177–181.
5. A. Borel, *Linear Algebraic Groups*, Graduate Texts in Mathematics, Vol. 126, 2nd edn. (Springer-Verlag, New York, 1991).
6. A. Borel, J.-P. Serre, *Théorèmes de finitude en cohomogie galoisienne*, Comm. Math. Helv. **39** (1964), 111–164.
7. C. Chevalley, *On algebraic group varieties*, J. Math. Soc. Japan **6** (1954), nos. 3–4, 303–324.
8. J.-L. Colliot-Thélène, B. Kunyavskiĭ, V. L Popov, Z. Reichstein, *Is the function field of a reductive Lie algebra purely transcendental over the field of invariants for the adjoint action?*, Compositio Math. **147** (2011), no. 02, 428–466.
9. P. Deligne, *Variétés unirationnelles non rationnelles* (*d'après M. Artin et D. Mumford*), in *Séminaire Bourbaki*, Lecture Notes in Mathematics, Vol. 317, Exp. no. 402 (Springer-Verlag, Berlin, 1973), pp. 45–57.
10. M. Demazure, *Sous-groupes algébriques de rang maximum du groupe de Cremona*, Ann. Sci. de l'E.N.S., 4e sér. **3** (1970), fasc. 4, 507–588.

11. A. Grothendieck, *Torsion homologique et sections rationnelle*, in *Anneaux de Chow et Applications*, Séminaire Claude Chevalley, Vol. 3, Exp. no. 5 (Secrétariat mathématique, Paris, 1958), pp. 1–29.
12. M. Hall, Jr., *The Theory of Groups* (Macmillan, New York, 1959).
13. A. Hirschowitz, *Le groupe de Cremona d'après Demazure*, in *Séminaire N. Bourbaki*, Lecture Notes in Mathematics, Vol. 317, Exp. no. 413 (Springer-Verlag, Berlin, 1973), pp. 261–276.
14. A. Joseph, *An algebraic slice in the coadjoint space of the Borel and the Coxeter element*, Adv. Math. **227** (2011), no. 1, 522–585.
15. M. Kervaire, T. Vust, *Fractions rationnelles invariantes par un groupe fini: Quelques exemples*, in: *Algebraic Transformation Gropus and Invariant Theory*, DMV Seminar, Band 13, Birkhäuser, Basel, 1989, pp. 157–179.
16. B. Kostant, *Lie group representations on polynomial rings*, Amer. J. Math. **85** (1963), 327–404.
17. Y. I. Merzlyakov, *Rational Groups* (Nauka, Moscow, 1980) (in Russian).
18. T. Miyata, *Invariants of certain groups* I, Nagoya Math. J. **41** (1971), 69–73.
19. V. L. Popov, *On the stability of the action of an algebraic group on an algebraic variety*, Math. USSR Izv. **6** (1972), 367–379 (1973).
20. V. L. Popov, *Sections in invariant theory*, in *Proceedings of The Sophus Lie Memorial Conference, Oslo* 1992 (Scandinavian University Press, Oslo, 1994), pp. 315–362.
21. V. L. Popov, E. B. Vinberg, *Invariant theory*, in *Algebraic Geometry* IV, Encyclopaedia of Mathematical Sciences, Vol. 55 (Springer-Verlag, Berlin, 1994), pp. 123–284.
22. L. Pukanszky, *Leçons sur les Représentations des Groupes*, Monographies de la Société Mathématique de France, Vol. 2 (Dunod, Paris, 1967).
23. M. Rosenlicht, *Some basic theorems on algebraic groups*, Amer. J. Math. **78** (1956), 401–443.
24. M. Rosenlicht, *Toroidal algebraic groups*, Proc. Amer. Math. Soc. **12** (1961), 984–988.
25. M. Rosenlicht, *On quotient variaties and the affine embedding of certain homogeneous spaces*, Trans. Amer. Math. Soc. **101** (1961), 211–223.
26. J.-P. Serre, *Espaces fibrés algébriques*, in *Anneaux de Chow et Applications*, Séminaire Claude Chevalley, Vol. 3, Exp. no. 1 (Secrétariat mathématique, Paris, 1958), pp. 1–37.
27. J.-P. Serre, *Le groupe de Cremona et ses sous-groupes finis*, in *Séminaire N. Bourbaki*, Vol. 2008/2009, Astérisque Vol. 332, Exp. no. 1000 (Société Mathématique de France, 2010), pp. 75–100.
28. N.-I. Shepherd-Barron, *Stably rational irrational varieties*, in: *The Fano Conference*, Univ. Torino, Turin, 2004, pp. 693–700.
29. H. Sumihiro, *Equivariant completion*, J. Math. Kyoto Univ. **14** (1974), no. 1, 1–28.
30. D. D. Triantaphyllou, *Invariants of finite groups acting nonlinearly on rational function fields*, J. Pure Appl. Algebra **18** (1980), 315–331.
31. A. van den Essen, *Polynomial Automorphisms*, Progress in Mathematics, Vol. 190 (Birkhäuser, Basel, 2000).

32. B. L. van der Waerden, *Algebra* I (Springer-Verlag, Berlin, 1967).
33. E. B. Vinberg, *Rationality of the field of invariants of a triangular group*, Mosc. Univ. Math. Bull. **37** (1992), no. 2, 27–29.
34. A. Weil, *On algebraic groups of transformations*, Amer. J. Math. **77** (1955), no. 2, 355–391.

The gonality of singular plane curves II

Fumio Sakai

Department of Mathematics, Graduate School of Science and Engineering, Saitama University, Shimo–Okubo 255, Sakura–ku, Saitama 338–8570, Japan

E-mail: fsakai@rimath.saitama-u.ac.jp

Let C be an irreducible singular plane curve of type (d,ν). Let G denote the *gonality* of C. Criteria for the equality: $G = d - \nu$ have been discussed ([2], [3], [4], [9]). The purpose of this paper is to improve the criteria given in our previous paper [9]. In particular, in case $\nu = 3$, we obtain an effective criterion which seems to be optimal. However, for many singular plane curves, the equality is not the case. We therefore at the same time discuss the lower bounds of G. We generalize our previous methods employed in [9].

Keywords: Plane curves, singular points, gonality.

1. Introduction

Let C be an irreducible plane curve of degree d over $\mathbf{C}$. To a non-constant rational function φ on C, we can associate a holomorphic map $\varphi : \tilde{C} \to \mathbf{P}^1$, where the $\tilde{C}$ is the non-singular model of C. The *gonality* of C, denoted by Gon(C), is defined to be the minimum of the degrees of such holomorphic maps. For simplicity, we write $G = \mathrm{Gon}(C)$. Let ν denote the *maximal multiplicity* of the singular points on C. In this situation, we say that C is of *type* (d,ν). The projection from the point with multiplicity ν to a line, we obtain a rational function of degree $d-\nu$. So we have the upper bound: $\mathrm{Gon}(C) \le d - \nu$. Let g denote the genus of C. Let δ be the delta invariant of C such that $g = (d-1)(d-2)/2 - \delta$. The purpose of this paper is to improve the criteria for the equality: $G = d - \nu$ given in [9]. We also discuss the lower bounds of G. Namely, we prove criteria for the inequality: $G \ge d - \nu - q$ with $q \ge 0$.

We prepare some notations. Let $m_1, \ldots, m_n$ denote the *multiplicities* of all singular points of C. We here include *infinitely near singular points*. We have the formula: $\delta = \sum_{i=1}^{n} m_i(m_i - 1)/2$. We use the notation: $\mathrm{Data}(C) = [m_1, \ldots, m_n]$. We call it the *data* of C. Sometimes, we write

$[3_\alpha, 2_\beta]$ instead of $[\overbrace{3,\ldots,3}^{\alpha},\overbrace{2,\ldots,2}^{\beta}]$. Let ν_i denote the i-th largest multiplicity. By definition, we have $\nu_1 = \nu$ and $\nu_i = 1$ for $i > n$.

Definition 1. We define the following non-negative invariants:

(i) $V = \sum_{i=1}^{n} m_i(m_i - 2)/4$,
(ii) $\eta = \sum_{i=1}^{n} (m_i/\nu)^2$ (See [9]),
(iii) $\sigma = (\nu_2/\nu) + (\nu_3/\nu) + (\nu_4/\nu)$.

We have the equality: $\delta - V = (1/4)\sum_{i=1}^{n} m_i^2 = (1/4)\nu^2\eta$.

In order to discuss the lower bound of the gonality G, let us choose a non-negative integer q so that $1 < d - \nu - q \le d - \nu$.

Definition 2. Set

$$k_0 = \begin{cases} \max\{[d/\nu], 3\} & \text{if } d/\nu \ge \sigma - q/\nu, \\ \max\{[d/\nu], 2\} & \text{otherwise.} \end{cases}$$

We define a quadratic function: $Q(x) = x(d - x)$, which is a monotone increasing function for $x \le d/2$. We denote by $[x]$ the Gauss integer of x.

Theorem 1. *Let C be an irreducible singular plane curve of type (d, ν) with $d \ge 6$. Let q be as above. Then we have $G \ge d - \nu - q$, if the following two conditions are satisfied:*

(A) $\delta \le Q([d/2]) - (d - \nu - q)$,
(B) $\delta \le Q(k_0) - (d - \nu - q) + V$.

In particular, if $k_0 = [d/\nu]$ and $\nu \ge 6$, then Condition (A) is not necessary.

Remark 1.

(1) In case $q = 0$, we get the equality: $G = d - \nu$.
(2) For the case in which $\nu = 2$, we have $V = 0$. So Conditions (A) and (B) coincide. Letting $q = 0$, we recover Coppens-Kato's criterion [4].
(3) In [9], we considered the case in which $q = 0$ and used the condition: $\delta \le Q(k_0) - (d - \nu)$. Since $Q(k_0) \le Q([d/2])$, $k_0 \ge [d/\nu]$ and $V \ge 0$, Theorem 1 is a sharper criterion than that in [9], Proposition 2.

For the case in which $\nu = 3$, we can remove Condition (B) in Theorem 1. Letting $\mathrm{Data}(C) = [3_\alpha, 2_\beta]$, we have $\delta = 3\alpha + \beta$.

Theorem 2. *Let C be an irreducible singular plane curve of type $(d, 3)$ with $d \geq 6$. Let q be as above. Then we have $G \geq d-3-q$, if Condition (A) in Theorem 1 is satisfied.*

Remark 2. We shall see in Example 7 that unlike the cases in which $\nu = 2, 3$ (See Remark 1, (ii), Theorem 2), for $\nu \geq 5$, under Condition (A) alone, the assertion in Theorem 1 may not hold.

To state the lower bound version of another type of criterion proved in [9], we introduce some notations.

Definition 3. (Cf. [9]) Let q be a non-negative integer as above. We define the following functions of the invariants η, ν and q.

$$\begin{cases} h(\eta, \nu, q) = \dfrac{\eta}{2(1+q/\nu)} + \dfrac{1+q/\nu}{2}, \\ f_k(\eta, \nu, q) = \dfrac{k\sqrt{\eta} - (1+1/\nu+q/\nu)}{k-1}. \end{cases}$$

Theorem 3. *Let C be an irreducible singular plane curve of type (d, ν). Let q be as above. Then, we have the inequality:*

$$G \geq d - \nu - q,$$

if the following two conditions are satisfied:

(E) $\quad d/\nu > h(\eta, \nu, q)$,
(F) $\quad d/\nu > f_{k_0}(\eta, \nu, q)$.

In particular, if $d/\nu \geq (5/2)(1+q/\nu)$, then Condition (F) is not necessary.

Remark 3. In general, if d/ν is large, then the criterion in Theorem 1 is sharper than that in Theorem 3. The converse is true if d/ν is small. See Examples 1, 11 and the table in Example 12. Cf., Proposition 5 in [9].

In Sec. 2, we prepare the basic settings. In Sec. 3, we prove Theorem 1. We provide some extremal examples for the case in which $\nu = 2$. In Sec. 4, we prove Theorem 2. We discuss examples for the case in which $\nu = 3$. In Sec. 5, we prove Theorem 3. We also reformulate Theorem 3 in terms of the invariant δ. In Sec. 6, we discuss the lower bound of the genus for which the equality $G = d - \nu$ holds.

2. Preliminaries

Let C be an irreducible plane curve of type (d, ν). We consider a rational function φ on C. Write $r = \deg(\varphi)$. Suppose φ is induced by a rational function $\Phi = \Phi_1/\Phi_2$ on $\mathbf{P}^2$, where Φ_1 and Φ_2 are relatively prime homogeneous polynomials of the same degree, say k of three variables x, y, z.

We can resolve the base points of the rational map $\Phi : \mathbf{P}^2 \to \mathbf{P}^1$ as well as the singular points of C, by a sequence of successive blowing ups $\pi_i : X_i \to X_{i-1}$ at points $P_i \in X_{i-1}$, $i = 1, \ldots, s$ with $X_0 = \mathbf{P}^2$. Letting $X = X_s$, $\pi = \pi_1 \circ \cdots \circ \pi_s$, we obtain a morphism $\Phi \circ \pi : X \to \mathbf{P}^1$. By construction, the strict transform $\tilde{C}$ of C is nothing but the non-singular model of C. Let m_i denote the multiplicity of C at P_i. We have $\tilde{C} \sim \pi^* O(d) - \sum m_i E_i$, where the E_i is the total transform of the exceptional curve for π_i. We observed in [9] that there exist non-negative integers a_i such that a fiber of $\tilde{\Phi} \sim \pi^* O(k) - \sum a_i E_i$. As a consequence, we obtain

$$r = dk - \sum a_i m_i, \qquad k^2 = \sum a_i^2.$$

Lemma 1. *We have*

$$\delta - V \geq Q(k) - r.$$

Proof. See the proof of Lemma 1 in [9], where we neglected the term V. □

Proposition 1. *If $r < d - \nu - q$ for a non-negative integer q, then we have*

$$\delta - V > Q(k) - (d - \nu - q).$$

Proof. This follows from Lemma 1. □

We further discuss the estimation of the term $\sum a_i m_i$. Let b denote the number of a_i's with $a_i \neq 0$. Note that $b \leq k^2$. We denote by ν_i the i-th largest multiplicity of points on C (see Introduction).

Lemma 2. *Assume $r < d - \nu - q$ for a non-negative integer q. We have*

(i) $k \geq 2$,
(ii) $k \geq [d/\nu]$,
(iii) $k \geq 3$, *if* $d/\nu \geq \sigma - q/\nu$.

Proof. Cf., [9]. (i) If $k = 1$, then we have $b = 1$, hence $r = d - m_i$ for some i. Thus, we obtain $r \geq d - \nu$. So under the assumption $r < d - \nu - q$, we must have $k \geq 2$.

(ii) By using the Cauchy-Schwartz inequality:

$$\sum a_i m_i \leq k\sqrt{\sum_{a_i \neq 0} m_i^2} \leq \nu k^2,$$

we have $d(k-1) < \sum a_i m_i - \nu - q \leq \nu(k^2-1) - q$. It follows that $k > d/\nu - 1$, which implies that $k \geq [d/\nu]$.

(iii) If $k = 2$, then we have either $b = 1$ or $b = 4$. In case $b = 1$, we would have $r \geq 2d - 2\nu > d - \nu$. In case $b = 4$, there are four distinct numbers $i_1, \ldots, i_4$ such that $a_{i_1} = a_{i_2} = a_{i_3} = a_{i_4} = 1$ and $a_i = 0$ for $i \neq i_1, \ldots, i_4$. It follows that $r = 2d - m_{i_1} - m_{i_2} - m_{i_3} - m_{i_4} \geq 2d - \nu - \nu_2 - \nu_3 - \nu_4$. Since $r < d - \nu - q$, we have $d < \nu_2 + \nu_3 + \nu_4 - q$. Dividing both sides by ν, we obtain the inequality: $d/\nu < \sigma - q/\nu$. □

Lemma 3. *If $r < d - \nu - q$ for a non-negative integer q, then we have*

$$k \geq k_0.$$

Proof. In view of the definition of k_0 (see Definition 2), the assertion immediately follows from Lemma 2. □

Since the a_i's and the m_i's are non-negative integers, the values $\sum a_i m_i$ are more restrictive. We renumber the a_i's so that $a_1 \geq a_2 \geq \cdots \geq a_b$ ($a_i = 0$ for $i > b$).

Definition 4. Set

$$A(k) = \left\{ (a_1, \ldots, a_b) \;\middle|\; \begin{array}{l} a_i \text{ are positive integers,} \\ a_1 \geq a_2 \geq \cdots \geq a_b, \\ k^2 = \sum_{i=1}^{b} a_i^2 \end{array} \right\}.$$

We abbreviate $(2, 1, 1, 1, 1, 1) \in A(3)$ as $(2, 1_5)$. For instance, we have the list $A(3) = \{(3), (1_9), (2, 1_5), (2_2, 1)\}$.

For an element $\mathbf{a} = (a_1, \ldots, a_b) \in A(k)$, letting $S(\mathbf{a}) = \sum_{i=1}^{b} a_i \nu_i$, we define

$$S(k) = \max_{\mathbf{a} \in A(k)} S(\mathbf{a}).$$

Lemma 4. *For any element $\mathbf{a} = (a_1, \ldots, a_b) \in A(k)$, we have the inequality:*

$$\sum_{i=1}^{b} a_i m_i \leq S(k).$$

Proof. It suffices to note that $\sum_{i=1}^{b} a_i m_i \leq \sum_{i=1}^{b} a_i \nu_i$. $\square$

We now consider the case in which $\nu = 3$. We write $\mathrm{Data}(C) = [3_\alpha, 2_\beta]$.

Lemma 5. *If $\nu = 3$, then we have*

$$\text{(i)} \qquad S(1_{k^2}) = \begin{cases} k^2 + 2\alpha + \beta & \text{if } \alpha + \beta \leq k^2, \\ 2k^2 + \alpha & \text{if } \alpha \leq k^2 < \alpha + \beta, \\ 3k^2 & \text{if } k^2 < \alpha. \end{cases}$$

$$\text{(ii)} \qquad S(1_{k^2}) \leq 2k^2 + \alpha.$$

Proof. (i) If $\alpha+\beta \leq k^2$, then we have $S(1_{k^2}) = 3\alpha+2\beta+(k^2-\alpha-\beta) = k^2+2\alpha+\beta$. If $\alpha \leq k^2 < \alpha+\beta$, then we have $S(1_{k^2}) = 3\alpha+2(k^2-\alpha) = 2k^2+\alpha$.
(ii) The assertion is obvious. $\square$

Lemma 6. *If $\nu = 3$, then we have $S(k) = S(1_{k^2})$.*

Proof. Take an element $\mathbf{a} = (a_1, \ldots, a_b) \in A(k)$. Write $T = S(1_{k^2}) - S(\mathbf{a})$.

Case (1). $k^2 \geq \alpha + \beta$.

(1a). $b \geq \alpha + \beta$. We have

$$\begin{aligned} T &= \sum_{i=1}^{b} a_i^2 + 2\alpha + \beta - \left(3\sum_{i=1}^{\alpha} a_i + 2\sum_{i=\alpha+1}^{\alpha+\beta} a_i + \sum_{i=\alpha+\beta+1}^{b} a_i\right) \\ &= \sum_{i=1}^{\alpha}(a_i-1)(a_i-2) + \sum_{i=\alpha+1}^{\alpha+\beta}(a_i-1)^2 + \sum_{i=\alpha+\beta+1}^{b} a_i(a_i-1) \geq 0. \end{aligned}$$

(1b). $\alpha \leq b < \alpha + \beta$. We have

$$\begin{aligned} T &= \sum_{i=1}^{b} a_i^2 + 2\alpha + \beta - \left(3\sum_{i=1}^{\alpha} a_i + 2\sum_{i=\alpha+1}^{b} a_i\right) \\ &= \sum_{i=1}^{\alpha}(a_i-1)(a_i-2) + \sum_{i=\alpha+1}^{b}(a_i-1)^2 + (\alpha+\beta-b) \geq 0. \end{aligned}$$

(1c). $b < \alpha \leq \alpha + \beta$. We have

$$T = \sum_{i=1}^{b} a_i^2 + 2\alpha + \beta - \left(3\sum_{i=1}^{b} a_i\right) = \sum_{i=1}^{b}(a_i-1)(a_i-2) + (2\alpha+\beta-2b) \geq 0.$$

Case (2). $\alpha \le k^2 < \alpha + \beta$.

(2a). $\alpha \le b$. We have

$$T = 2\sum_{i=1}^{b} a_i^2 + \alpha - \left(3\sum_{i=1}^{\alpha} a_i + 2\sum_{i=\alpha+1}^{b} a_i\right)$$
$$= \sum_{i=1}^{\alpha}(2a_i - 1)(a_i - 1) + \sum_{i=\alpha+1}^{b} 2a_i(a_i - 1) \ge 0.$$

(2b). $b < \alpha$. We have

$$T = 2\sum_{i=1}^{b} a_i^2 + \alpha - \left(3\sum_{i=1}^{b} a_i\right) = \sum_{i=1}^{b}(2a_i - 1)(a_i - 1) + (\alpha - b) \ge 0.$$

Case (3). $k^2 < \alpha$. In this case, we have

$$T = 3\sum_{i=1}^{b} a_i^2 - \left(3\sum_{i=1}^{b} a_i\right) = \sum_{i=1}^{b} 3a_i(a_i - 1) \ge 0.$$

□

Remark 4. For $\nu \ge 4$, the case $S(\mathbf{a}) > S(1_{k^2})$ may occur for some $\mathbf{a} \in A(k)$. For $\mathrm{Data}(C) = [4_3, 2]$ and $\mathbf{a} = (2_4)$, we have $S(\mathbf{a}) = 28$, but $S(1_{16}) = 26$. In general, if $\mathrm{Data}(C) = [4_t, 3_\alpha, 2_\beta]$, then we can show the upper bound: $S(k) \le S(1_{k^2}) + t$.

3. Proof of Theorem 1

We first recall the following

Lemma 7. (Lemma 7 in [9]. See also [4], [5]) *Let C be a plane curve of degree d. Let φ be a rational function on C. If there is a positive integer $l < d$ satisfying the inequality:*

$$r + \delta < Q(l + 1),$$

where $r = \deg(\varphi)$, then there exists a rational function Φ on $\mathbf{P}^2$ with degree $k \le l$ which induces φ.

Remark 5. For the sake of completeness, we shall give a sketchy proof of Lemma 7 in Appendix.

Proposition 2. *Let φ be a rational function on a plane curve C of type (d, ν). Suppose $r = \deg(\varphi) < d - \nu - q$ for a non-negative integer q. Under Condition (A), there exists a rational function Φ on $\mathbf{P}^2$ of degree k with $k < [d/2]$ which induces φ on C.*

Proof. Under Condition (A), we obtain $r+\delta < Q([d/2])$. Therefore, it suffices to put $l=[d/2]-1$ in Lemma 7. □

Proof of Theorem 1. Assume to the contrary that $G < d-\nu-q$, which implies that there is a rational function φ on C with $r=\deg(\varphi) < d-\nu-q$. By Proposition 2, under Condition (A), we can find a rational function Φ of degree $k < [d/2]$ on $\mathbf{P}^2$, which induces φ.

By Lemma 3, we must have $k \geq k_0$. Hence, we obtain $Q(k) \geq Q(k_0)$, since $k < d/2$. Using Proposition 1, we have

$$\delta \geq Q(k)+V-r > Q(k_0)+V-(d-\nu-q),$$

which contradicts Condition (B).

Now we assume that $k_0=[d/\nu]$ and $\nu \geq 6$. In this case, we can show that Condition (B) implies Condition (A). The assertion follows from the first part. We first prove the relation: $\delta \leq 2(1-1/\nu)(\delta - V)$. For each i, we have $m_i(\nu-1) \geq \nu(m_i-1)$, hence we obtain $(1-1/\nu)\,m_i^2 \geq m_i(m_i-1)$. By summing up, we get the required inequality.

As a consequence, Condition (B) implies the inequality:

$$\delta \leq 2(1-1/\nu)\{Q([d/\nu])-(d-\nu-q)\}.$$

Let $P(d,\nu)$ denote the following difference:

$$2(1-1/\nu)\{Q(d/\nu)-(d-\nu-q)\}-\{Q((d-1)/2)-(d-\nu-q)\}.$$

Note that $Q([d/\nu]) \leq Q(d/\nu)$ and $Q((d-1)/2) \leq Q([d/2])$. So, if $P(d,\nu) < 0$, then Condition (A) is satisfied. By computation, we have

$$P(d,\nu) = -\frac{\nu(\nu-4)^2-8}{4\nu^3}\cdot d^2 + \frac{1}{4} - \frac{\nu-2}{\nu}\cdot(d-\nu-q).$$

Since $\nu \geq 6$ and $d > \nu$, we have

$$\begin{aligned}\frac{\nu(\nu-4)^2-8}{4\nu^3}\cdot d^2 - \frac{1}{4} &> \frac{\nu(\nu-4)^2-8}{4\nu} - \frac{1}{4}\\ &= \frac{(\nu-1)\{(\nu-1)(\nu-6)+2\}}{4\nu} > 0.\end{aligned}$$

Thus we see that $P(d,\nu) < 0$. □

Remark 6. In case $k_0=[d/\nu]$ and $\nu=5$, if $q=0$, then Condition (A) is not necessary for $d \leq 94$. Indeed, by computation, we have $P(d,5) < 0$ for $d \leq 94$.

Example 1. Let us consider the following curve of type $(36,6)$:

$$C : y^{30} = \prod_{i=1}^{2}(x-a_i)^6 \prod_{j=1}^{24}(x-b_j),$$

where the a_i's and the b_j's are mutually distinct complex numbers. We see that $\mathrm{Data}(C) = [6_{16}]$. By applying Theorem 1, we find that $G = 30$.

We now discuss the particular case in which $\nu = 2$. In this case, Condition (A) = Condition (B).

Corollary 1. *Let C be an irreducible plane curve of type $(d,2)$ with $d \geq 6$. If Condition (A) for $q = 0$ is not satisfied, then we have the lower bound:*

$$G \geq Q([d/2]) - \delta.$$

Proof. Apply Theorem 1 with $q = \delta - \{Q([d/2]) - (d-2)\}$. □

Example 2. In Coppens-Kato [4], Examples 4.1, 4.2, for all $d \geq 6$, they constructed irreducible nodal plane curves C of degree d such that $\delta = Q([d/2]) - (d-2) + 1$ and $G \leq d-3$. For those curves C, we infer from Corollary 1 that $G \geq d-3$, hence we conclude that $G = d-3$.

Example 3. We examine the following plane curves:

$$C_{2k+1} : \; y\prod_{i=1}^{k}(x-a_i)^2 - c\prod_{j=1}^{k}(y-b_j)^2 = 0,$$

$$C_{2k+2} : \; y(x-a_0)\prod_{i=1}^{k}(x-a_i)^2 - c\prod_{j=1}^{k+1}(y-b_j)^2 = 0,$$

where the a_i's and the b_j's are mutually distinct ($b_j \neq 0$) and the constant c is generally chosen. It turns out that C_{2k+1} (resp. C_{2k+2}) is an irreducible nodal plane curve with k^2 (resp. k^2+k) nodes at $P_{ij} = (a_i, b_j)$ for $i, j \geq 1$. The irreducibility follows from Eisenstein's criterion applied to the homogenization of the defining polynomial. We infer from Corollary 1 that $G \geq k$ for C_{2k+1} and $G \geq k+1$ for C_{2k+2}.

On the other hand, the rational function $\Phi = \prod_{i=1}^{k}(y-b_j)/\prod_{j=1}^{k}(x-a_i)$ induces a rational function φ on C_{2k+1} with $\deg\varphi = k$. Thus $G \leq k$. Hence, we conclude that $G = k$ for C_{2k+1}. Similarly, we can show that $G = k+1$ for C_{2k+2}. See also [9], Example 1.

Remark 7. For general irreducible nodal plane curves C of degree d, it was proved by Coppens [2] that if $\delta < (d^2 - 7d + 18)/2$ (or equivalently, if $g \geq 2d - 7$), then $G = d - 2$. In particular, for general nodal plane curves C of degree $2k + 1$ (resp. $2k + 2$) with k^2 (resp. $k(k + 1)$) nodes for $k \geq 4$ (resp. $k \geq 5$) , one has $G = 2k - 1$ (resp. $= 2k$). So we understand that the curves given in Example 3 are *special* nodal curves.

Example 4. We give other examples which attain the lowest bound in Corollary 1. The following curve of type $(2k + 1, 2)$:

$$C : y^{2k+1} = \prod_{i=1}^{k} (x - a_i)^2$$

attains the lowest bound $G = k$. The curve C has k singular points with the multiplicity sequence (2_k). Note that $\delta = k^2$ and $Q(k) - \delta = k$. It follows from Corollary 1 that $G \geq k$. Letting $Y = \prod_{i=1}^{k}(x - a_i)/y^k$, since $y = Y^2$, we find that $\mathbf{C}(\mathbf{C}) = \mathbf{C}(\mathbf{x}, \mathbf{Y})$. So C is birational to the curve:

$$C' : Y^{2k+1} = \prod_{i=1}^{k} (x - a_i).$$

Clearly, C' has a singular point $(1, 0, 0)$ with multiplicity $k+1$. Thus $G \leq k$. We conclude that $G = k$.

4. Proof of Theorem 2

Let C be a plane curve of type $(d, 3)$ with $d \geq 6$. Let q be a non-negative integer with $d - 3 - q \geq 2$. Assume that $G < d - 3 - q$. Let φ be a rational function on C with $r = \deg(\varphi) < d - 3 - q$. We assume that Condition (A) is satisfied. By Proposition 2, φ is induced by a rational function Φ on $\mathbf{P}^2$ of degree k with $k < [d/2]$. We have seen that $k \geq [d/3] \geq 2$ (Lemma 2).

Set Data$(C) = [3_\alpha, 2_\beta]$. Note that $r = dk - \sum a_i m_i$ and $\sum a_i^2 = k^2$. We have $d - 3 - q > r = dk - \sum a_i m_i$. Using Lemma 6, we obtain the inequality:

$$d(k - 1) < \sum a_i m_i - 3 - q \leq \sum_{i=2}^{k^2} \nu_i - q.$$

Since $\sum_{i=2}^{k^2} \nu_i \leq 2(k^2 - 1) + \alpha - 1$ (Lemma 5, (ii)), we obtain

$$k > \frac{d}{2} - 1 - \frac{\alpha - 1 - q}{2(k - 1)}.$$

We must have $\alpha - 1 - q > 0$, since $k < [d/2]$. Thus, we have the inequality:

$$k > \frac{d}{2} - 1 - \frac{\alpha - 1 - q}{2([d/3] - 1)}.$$

Definition 5. Define

$$\lambda = \frac{\alpha - 1 - q}{2([d/3] - 1)}, \quad e = \begin{cases} [\lambda] & (\text{ if } d \text{ is even}), \\ [\lambda + \frac{1}{2}] & (\text{ if } d \text{ is odd}). \end{cases}$$

We consider the four subcases: (Ia) d is even and $\lambda \in \mathbf{N}$, (Ib) d is even and $\lambda \notin \mathbf{N}$, (IIa) d is odd and $\lambda + \frac{1}{2} \in \mathbf{N}$, (IIb) d is odd and $\lambda + \frac{1}{2} \notin \mathbf{N}$.

Lemma 8. *We have $k \geq k_1$, where*

$$k_1 = [d/2] - e + \begin{cases} 0 \text{ for Cases (Ia), (IIb)}, \\ -1 \text{ for Case (Ib)}, \\ 1 \text{ for Case (IIa)}. \end{cases}$$

Proof. By definition, the assertion is obvious. □

Remark 8. Since $k < [d/2]$, we must have

$$e \geq \begin{cases} 1 \text{ for Cases (Ia), (IIb)}, \\ 0 \text{ for Case (Ib)}, \\ 2 \text{ for Case (IIa)}. \end{cases}$$

Remark 9. We have the following inequalities:

(i) $\alpha \geq 1 + q + 2e([d/3] - 1)$ (if d is even),
(ii) $\alpha \geq 1 + q + (2e - 1)([d/3] - 1)$ (if d is odd).

Note that the inequalities are strict for Cases (I_b), (II_b).

Lemma 9. *We also have*

(i) $\alpha \geq 1 + q + 2e(d - 5)/3$ *(if d is even),*
(ii) $\alpha \geq 1 + q + (2e - 1)(d - 5)/3$ *(if d is odd).*

Proof. It suffices to note that $[d/3] \geq (d - 2)/3$. □

Lemma 10. *We have*

$$\alpha \leq \begin{cases} \dfrac{(d-2)^2+8+4q}{12} & \text{if } d \text{ is even,} \\ \dfrac{(d-2)^2+7+4q}{12} & \text{if } d \text{ is odd.} \end{cases}$$

Proof. Since $\delta = 3\alpha + \beta$, the assertion follows from Condition (A). □

Lemma 11. *We have the bound:*

$$d \geq \begin{cases} 8e - 7/2 & \text{if } d \text{ is even,} \\ 8e - 7 & \text{if } d \text{ is odd,} \end{cases}$$

except if $e = 0$.

Proof. Suppose d is even. Combining the inequalities in Lemma 9 and in Lemma 10, we obtain

$$d \geq 8e - 1 - \frac{5-8q}{d-5}.$$

In case $q > 0$, we have $d \geq 8e - 1$. In case $q = 0$, we have $d \geq 8e - 7/2$ except if $d = 6$. If $d = 6$, then, by Remark 9, we have $\alpha \geq 1 + 2e$. On the other hand, by Lemma 10, we must have $\alpha \leq 2$. As a consequence, we have $e = 0$.

Suppose d is odd. By Lemma 9 and Lemma 10, we obtain

$$d \geq 8e - 5 - \frac{4-8q}{d-5}.$$

In case $q > 0$, we have $d \geq 8e - 5$. In case $q = 0$, we have $d \geq 8e - 7$. □

In order to prove Theorem 2, we have to show that $V \geq Q([d/2]) - Q(k_1)$. Indeed, in a similar manner to that as in the proof of Theorem 1, since $k \geq k_1$ (Lemma 8), by Proposition 1, we would have $\delta - V > Q(k_1) - (d-3-q)$. So if $V \geq Q([d/2]) - Q(k_1)$, then we would have $\delta > Q([d/2]) - (d-3-q)$, which contradicts Condition (A). We now set

$$J = V - \{Q([d/2]) - Q(k_1)\}.$$

Lemma 12. *We have*

$$J = 3\alpha/4 - \begin{cases} e^2 & \text{for Case (Ia),} \\ (e+1)^2 & \text{for Case (Ib),} \\ e(e-1) & \text{for Case (IIa),} \\ e(e+1) & \text{for Case (IIb).} \end{cases}$$

Proof. Note that $V = 3\alpha/4$. Letting $k_1 = [d/2] - t$, we have $Q([d/2]) - Q(k_1) = t(t + d - 2[d/2])$. The assertion then follows from the definition of k_1 in Lemma 8. □

Lemma 13. *If $e \geq 3$, then we have $J \geq 0$. In case $e = 2$, for Cases (Ia), (Ib), (IIa), we also have $J \geq 0$ except for Case (Ib) with $q = 0$.*

Proof. By Lemma 9 and Lemma 11, we have

$$J \geq \begin{cases} 3e(e - 17/12) + 3(q+1)/4 & \text{for Case (Ia)}, \\ 3e(e - 25/12) + (3q-1)/4 & \text{for Case (Ib)}, \\ 3e(e - 7/3) + 3(q+5)/4 & \text{for Case (IIa)}, \\ 3e(e-3) + 3(q+5)/4 & \text{for Case (IIb)}. \end{cases}$$

Now the assertions are clear except if $e = 2$ for Case (Ib). In this case, the right hand side is equal to $3(q-1)/4$, which is non-negative for $q \geq 1$. □

Proof of Theorem 2. We have to prove that $J \geq 0$. In view of Lemma 13 and Remark 8, It suffices to check the following cases: Case (Ia) $e = 1$, Case (Ib), $e = 0, 1, 2$ ($q = 0$ if $e = 2$) and Case (IIb) $e = 1, 2$. Assume to the contrary that $J < 0$.

Case (Ia). $e = 1$. We have $\alpha = 1$, which implies $e \leq 0$, a contradiction.

Case (Ib). (i) $e = 0$. We have $\alpha = 1, q = 0$, which is Case (Ia). (ii) $e = 1$. We have $\alpha \leq 5$. In view of Remark 9, we have either (1) $\alpha = 5$, $q \leq 1$, $d = 6, 8$, or (2) $\alpha = 4$, $q = 0$, $d = 6, 8$. But, by Lemma 10, if $d = 6$ and $q < 1$, then we have $\alpha \leq 2$ and if $d = 8$ and $q = 1$ (resp. $q = 0$), then $\alpha \leq 4$ (resp. $\alpha \leq 3$). Thus we arrived at a contradiction. (iii) $e = 2, q = 0$. We have $\alpha \leq 11$. In view of Remark 9, we have $2 \leq [d/3] < (\alpha + 3)/4$. Hence, we have either (1) $6 \leq \alpha \leq 9$, $d = 6, 8$, or (2) $\alpha = 10, 11$, $d \leq 10$. But, by Lemma 10, if $d \leq 10$ (resp. $d = 6, 8$), then we obtain $\alpha \leq 6$ (resp. $\alpha \leq 3$). We arrived at a contradiction.

Case (IIb). (i) $e = 1$. We have $\alpha \leq 2$. But, by Remark 9, we must have $\alpha > 2$. (ii) $e = 2$. We have $\alpha \leq 7$. By Remark 9, we must have $5 \leq \alpha \leq 7$, $q \leq 2$, $d = 7$. But, in this case, by Lemma 10, we have $\alpha \leq 3$, which is a contradiction. □

Corollary 2. *Let C be an irreducible plane curve of type $(d, 3)$ with $d \geq 6$. If Condition (A) for $q = 0$ is not satisfied, then we have the lower bound:*

$$G \geq Q([d/2]) - \delta.$$

Remark 10. Let C be a plane curve of type $(d,3)$. In case $d = 7$, by Theorem 2, if $\delta \le 8$, then we have $G = 4$. If $\delta = 9$, then Data(C) is one of the data $[3_3], [3_2, 2_3], [3, 2_6]$. By using Corollary 3, (iii) we can show that $G = 4$ for those curves with data $[3_3]$ or $[3_2, 2_3]$. In case $d = 8$, by Theorem 2, if $\delta \le 11$, then we have $G = 5$. For $\delta = 12$, we obtain $G \ge 4$. For the data $[3_4]$, there is an example with $G = 4$. See Example 7.

Example 5. We examine the septic curve

$$C : y^5 = (x-a)^3(x-b)^2\prod_{i=1}^{2}(x-c_i),$$

where a, b, c_i are mutually distinct. This curve C has the data $[3, 2_6]$. We have $g = 6$. By letting $Y = y^2/(x-a)(x-b)$, we see that C is birational to the curve $C' : Y^5(x-b) = (x-a)\prod_{i=1}^{2}(x-c_i)^2$, which has the data $[2_4]$, hence we infer from Theorem 1 that $G = 4$.

Example 6. For $k \ge 2$, we define the following plane curve of type $(2k+2, 3)$:

$$C_k : y^{2k-1}(x-a)(x-b) = \prod_{i=1}^{k+1}(x-c_i)^2,$$

where a, b, c_i are mutually distinct. The point $(0,1,0) \in C_k$ is an ordinary triple point. We have Data(C_k) $= [3, 2_{k^2-1}]$. Hence, we have $\delta = k^2 + 2$. Since $Q(k+1) - (2k-1) = \delta$, by Theorem 2, we conclude that $G = 2k - 1$. Note that $g = (k-1)(k+2)$.

Example 7. For $m \ge 3$, we define the following plane curve of type $(3m-1, m)$:

$$C_m : y\prod_{i=1}^{2}(x-a_i)^m\prod_{i=3}^{m}(x-a_i) - c\prod_{j=1}^{2}(y-b_j)^m\prod_{j=3}^{m}(y-b_j) = 0$$

where the a_i's (resp. the b_j's) are mutually distinct ($b_j \ne 0$) and the constant c is generally chosen. We find that Data(C) $= [m_4]$, hence $\delta = 2m(m-1)$. Letting $d = 3m-1$, we have $d - m = 2m - 1$. The rational function $\Phi = \prod_{j=1}^{2}(y-b_j)/\prod_{i=1}^{2}(x-a_i)$ induces a rational function φ on C_m with $\deg(\varphi) = 2(3m-1) - 4m = 2m-2$. So we obtain $G \le 2m - 2 < d - m$. On the other hand, we have

$$Q([d/2]) - (d-m) = \frac{(9m-5)(m-1)}{4} - \begin{cases} 0 & \text{if } m \text{ is odd,} \\ 1/4 & \text{if } m \text{ is even.} \end{cases}$$

We easily see that for $m \geq 5$, Condition (A) with $q = 0$ is satisfied. But, since $G < d - m$, Condition (A) alone is not sufficient for the equality: $G = d - m$. Note that $g = (m-1)(5m-6)/2$.

Finally, we remark that C_3 is an octic curve with data $[3_4]$. We have $G \leq 4$. By Theorem 2 with $q = 1$, we know that $G \geq 4$. So we conclude that $G = 4$. See Remark 10.

5. Proof of Theorem 3

We now pass to the proof of Theorem 3. Let $\pi : X \to \mathbf{P}^2$ be the minimal resolution of the singularities of C. We do not require that the inverse image $\pi^{-1}(C)$ has normal crossings. We defined the functions $h(\eta, \nu, q)$ and $f_k(\eta, \nu, q)$ in Definition 3.

Proposition 3. *Let C be an irreducible plane curve of type (d, ν). Let q be a non-negative integer. Assume that $d/\nu > h(\eta, \nu, q)$. Let φ be a rational function on C with $r = \deg \varphi < d - \nu - q$. Then we can find a rational function Φ on $\mathbf{P}^2$ which induces φ on C such that the map $\Phi \circ \pi : X \to \mathbf{P}^1$ is a morphism. Furthermore, the invariant η and the degree k of Φ satisfy the inequalities:*

$$1 + 1/\nu + q/\nu < \sqrt{\eta} < d/\nu \leq f_k(\eta, \nu, q).$$

Proof. We follow the arguments in [9], Lemma 9. By Theorem 3.1 in Serrano [12], there exists such a rational function Φ if $\tilde{C}^2 > (r+1)^2$. Let k denote the degree of Φ. Under the hypothesis, we have

$$\begin{aligned}
\tilde{C}^2 - (r+1)^2 &\geq d^2 - \sum m_i^2 - (d - \nu - q)^2 \\
&= (2d - \nu - q)(\nu + q) - \sum m_i^2 \\
&= \nu^2 \{2(1 + q/\nu)(d/\nu) - (\eta + (1 + q/\nu)^2)\} \\
&= 2\nu^2 (1 + q/\nu)(d/\nu - h(\eta, \nu, q)) > 0.
\end{aligned}$$

Now we see that $d/\nu - \sqrt{\eta} > 0$, because

$$d/\nu - \sqrt{\eta} = (1 + q/\nu)^{-1} \{\sqrt{\eta} - (1 + q/\nu)\}^2 / 2 + \{d/\nu - h(\eta, \nu, q)\} > 0.$$

On the other hand, by Lemma 6 in [9], we have $d - \nu - q - 1 \geq r \geq k\nu(d/\nu - \sqrt{\eta})$. We then obtain the inequality:

$$k \leq \frac{d/\nu - (1 + (1+q)/\nu)}{d/\nu - \sqrt{\eta}} = 1 + \frac{\sqrt{\eta} - (1 + 1/\nu + q/\nu)}{d/\nu - \sqrt{\eta}}.$$

It follows from the fact $k \geq 2$ (See Lemma 2) that $\sqrt{\eta}-(1+1/\nu+q/\nu) > 0$. We also have $k(d/\nu - \sqrt{\eta}) \leq d/\nu - 1 - 1/\nu - q/\nu$. We infer that

$$(k-1)d/\nu \leq k\sqrt{\eta} - (1 + 1/\nu + q/\nu),$$

which gives the required inequality: $d/\nu \leq f_k(\eta, \nu, q)$. □

Lemma 14. *Suppose* $t > h(\eta, \nu, q)$. *If* $t \geq (5/2)(1 + q/\nu)$, *then* $t > f_3(\eta, \nu, q)$.

Proof. We can write the inequality: $t > h(\eta, \nu, q)$ as $\eta < \tilde{\eta}$, where

$$\tilde{\eta} = 2(1 + q/\nu)(t - (1 + q/\nu)/2).$$

So we have $f_3(\eta, \nu, q) < f_3(\tilde{\eta}, \nu, q)$. We have

$$t - f_3(\tilde{\eta}, \nu, q) = \frac{2t + (1 + 1/\nu + q/\nu) - 3\sqrt{2(1 + q/\nu)(t - (1 + q/\nu)/2)}}{2}.$$

Now we consider the quadratic function:

$$U(t) = (2t + 1 + 1/\nu + q/\nu)^2 - 18(1 + q/\nu)\{t - (1 + q/\nu)/2\}.$$

Let z_+, z_- $(z_+ > z_-)$ be the two roots of the equation: $U = 0$. We see that if $t > z_+$, then $U(t) > 0$. By computation, we have

$$z_+ = (7/4)(1 + q/\nu) - (1/2\nu) + \frac{3\sqrt{(1 + q/\nu)^2 - 4(1 + q/\nu)/\nu}}{4}.$$

Clearly, we have $z_+ < (5/2)(1 + q/\nu)$. □

Proof of Theorem 3. Assume that there is a rational function φ on C with $r = \deg\varphi < d - \nu - q$. Under Condition (E), by Proposition 3, we can find a rational function Φ on $\mathbf{P}^2$ which induces φ on C such that the degree k of Φ satisfies the inequality: $d/\nu \leq f_k(\eta, \nu, q)$. Note that f_k is a decreasing function of k for $k > 1$, since $\sqrt{\eta} - (1 + 1/\nu + q/\nu) > 0$ (Proposition 3). Using the inequality: $k \geq k_0$ (Lemma 3), we must have $d/\nu \leq f_k(\eta, \nu, q) \leq f_{k_0}(\eta, \nu, q)$, which contradicts Condition (F).

We prove the latter half. Under Condition (E), in case $d/\nu \geq (5/2)(1 + q/\nu)$, by Lemma 14, we have the inequality: $d/\nu > f_3(\eta, \nu, q)$. Since we assumed $k_0 \geq 3$, we also have $f_{k_0}(\eta, \nu, q) \leq f_3(\eta, \nu, q)$, as we noted in the above proof. Thus we obtain Condition (F). □

In particular, by letting $q = 0, 1$, we obtain the following criteria.

Corollary 3. (Cf., Proposition 4 in [9]) *We have* $G = d - \nu$, *if*

(i) $\quad d/\nu > \dfrac{\eta+1}{2} \quad$ (*if* $k_0 \geq 3,\ d/\nu \geq 5/2$),

(ii)

$$d/\nu > \max\Big\{\frac{\eta+1}{2}, \frac{3\sqrt{\eta}-(1+1/\nu)}{2}\Big\} \quad (\textit{if } k_0 \geq 3,\ d/\nu < 5/2),$$

(iii) $\quad d/\nu > \max\Big\{\dfrac{\eta+1}{2}, 2\sqrt{\eta}-(1+1/\nu)\Big\} \quad$ (*if* $k_0 = 2$).

Corollary 4. *We have* $G \geq d-\nu-1$, *if*

(i) $\quad d/\nu > h(\eta,\nu,1) \quad$ (*if* $k_0 \geq 3,\ d/\nu \geq (5/2)(1+1/\nu)$),

(ii) $\quad d/\nu > \max\Big\{h(\eta,\nu,1), \dfrac{3\sqrt{\eta}-(1+2/\nu)}{2}\Big\}$

(*if* $k_0 \geq 3,\ d/\nu < (5/2)(1+1/\nu)$),

(iii) $\quad d/\nu > \max\Big\{h(\eta,\nu,1), 2\sqrt{\eta}-(1+2/\nu)\Big\} \quad$ (*if* $k_0 = 2$),

where $h(\eta,\nu,1) = \{\eta+(1+1/\nu)^2\}/\{2(1+1/\nu)\}$.

Example 8. Let C be the plane curve of degree 11 defined by the equation:

$$y^{11} = x^4(x-1)(x-a) \qquad (a \neq 0,1).$$

We easily see that C has two singular points $P=(0,0,1)$ and $Q=(1,0,0)$ with multiplicity sequences $(4_2,3)$ and (5_2), respectively. We have $g=10$. So $\nu=\nu_2=5$, $\nu_3=\nu_4=4$. Thus, we have $\eta=3.64$ and $\sigma=13/5$. Letting $q=2$, we have $d/\nu=11/5=2.2>h(\eta,\nu,2)=2$, $d/\nu>f_3(\eta,\nu,2)=2.06...$ We also have $d/\nu=\sigma-2/\nu=11/5$, hence $k_0=3$. It follows from Theorem 3 that $G\geq 4$.

On the other hand, we can show that C is birational to the following curve C' of type $(13,9)$:

$$C' : Y^{11} = X^9(X-1)^3(X-a).$$

It suffices to set $X=a(x-1)/(x-a), Y=by^9/x^3(x-a)^2$, where $b=\sqrt[11]{a^{10}(a-1)^4}$. Hence we obtain $G=\mathrm{Gon}(C')\leq 13-9=4$. As a consequence, we have $G=4$. We remark that the rational function $\Phi=y^9/x^3(x-1)(x-a)$ induces a rational function φ on C which computes the gonality.

Remark 11. Let us calculate the invariant η of a plane curve C having a singular point P, which is locally defined by the equation: $y^\nu - x^\mu = 0$ with $\nu \leq \mu$. Consider the Euclidean algorithm: $\mu=\rho_1 m_1+m_2$, $m_1=\rho_2 m_2+m_3$, $\ldots$, $m_{s-1}=\rho_s m_s$, where $\nu=m_1>m_2>\cdots>m_s=\mathrm{GCD}(\nu,\mu)$. Then we have $\nu\mu=\sum_{i=1}^{s}\rho_i m_i^2$. The singular point P has the multiplicity sequence: $\overbrace{m_1, \ldots, m_1}^{\rho_1}, \overbrace{m_2, \ldots, m_2}^{\rho_2}, \ldots, \overbrace{m_s, \ldots, m_s}^{\rho_s}$. Thus, if C has only this singular point P, then $\eta = \mu/\nu - m_{s-1}/\nu^2$ (if $\mathrm{GCD}(\nu, \mu) = 1$) and $\eta = \mu/\nu$ (if $\mathrm{GCD}(\nu, \mu) \geq 2$). Note that $\eta > 1$, unless $\mu = \nu + 1$.

Example 9. Consider the plane curve C:

$$y^n = \prod_{i=1}^{k} (x - a_i),$$

where the a_i's are mutually distinct. The degree d of C is equal to $\max\{n, k\}$. We have

$$G = \begin{cases} k & \text{if } n \geq k+1, \\ n-1 & \text{if } n = k, \\ n & \text{if } n < k. \end{cases}$$

If $n = 1$ or $k = 1$, then C is a rational curve, hence $G = 1$. If $n = k$ or $n = k + 1$, then C is a smooth curve. Hence we know that $G = n - 1$. In what follows, we exclude these cases.

1) $n \geq k + 2$. In this case, the point $(1, 0, 0)$ is the unique singular point defined locally by $z^{n-k} - y^n = 0$. Thus $\nu = n - k$. Since $k \geq 2$, we have $\eta > 1$. By Remark 11, we have $\eta \leq n/(n - k) = d/\nu$. Since $\eta-(\eta+1)/2 = (\eta-1)/2 > 0$ and $\eta-\{2\sqrt{\eta}-(1+1/\nu)\} = (\sqrt{\eta}-1)^2+1/\nu > 0$, we can apply Corollary 3, (iii) and we conclude that $G = k$.

2) $n < k$. In this case, C has the unique singular point $(0, 1, 0)$, which is locally defined by $z^{k-n} - x^k = 0$. Thus $\nu = k - n$. Since $n \geq 2$, we have $\eta > 1$. By Remark 11, we have $\eta \leq k/(k - n) = d/\nu$. We can also apply Corollary 3, (iii) and we obtain $G = n$.

Example 10. For $m \geq 2$, we consider the plane curve C_k of type $(km + 1, m)$:

$$y^{mk+1} = \prod_{i=1}^{k} (x - a_i)^m,$$

where the a_i's are mutually distinct. We easily see that C_k is birational to the curve: $Y^{mk+1} = \prod_{i=1}^{k} (x - a_i)$. So we infer from the computation in Example 9 that $G = k$. Cf. Example 4.

Example 11. We now give a special case so that $k_0 = 4$ and Condition (F) is effective. Let us take the following plane curve C of type $(17, 4)$:

$$y^{17} = \prod_{i=1}^{2} (x - a_i)^4 \prod_{i=3}^{5} (x - a_i)^2 \prod_{i=6}^{7} (x - a_i),$$

where the a_i's are mutually distinct. We find that C has the data $[4_8, 2_{24}]$, hence $\eta = 14$. Letting $q = 5$, we have $h(14, 4, 5) = 4.23\ldots$, $f_3(14, 4, 5) = 4.36\ldots$, $f_4(14, 4, 5) = 4.15\ldots$. Since $k_0 = 4$ and $17/4 = 4.25$, We can apply Theorem 3 and we obtain the inequality: $G \geq 8$.

As a corollary of Theorem 3 with $q = 0$, in terms of the invariant δ, we obtain the following criterion.

Theorem 4. *Let C be an irreducible singular plane curve of type (d, ν) with $\nu \geq 3$. Then, we have*

$$G = d - \nu,$$

if the following two conditions are satisfied

$$\text{(M)} \qquad \delta < \frac{(d-1)\nu}{2},$$

$$\text{(N)} \qquad \delta < \left(\frac{2d + \nu + 1}{6}\right)^2 + \frac{\nu(\nu - 2)}{4}.$$

In particular, in case $d/\nu \geq 5/2$, Condition (N) is not necessary.

Proof. Write Data$(C) = [m_1, \ldots, m_n]$. We prove the upper bound: $\eta \leq \eta^*$, where $\eta^* = (2/\nu)^2\delta + 2/\nu - 1$. Indeed, by definition, we have

$$\nu^2\eta = \sum m_i^2 = 2\sum m_i(m_i - 1) - \sum m_i(m_i - 2) \leq 4\delta - \nu(\nu - 2),$$

which gives the required upper bound. The equality holds if and only if C has the data $[\nu, 2_{n-1}]$.

Assume that Condition (M) is satisfied. Namely, we have $d/\nu > 2\delta/\nu^2 + 1/\nu$. We show that $d/\nu \geq \sigma$. It follows that $k_0 \geq 3$ (See Definition 2 for $q = 0$). Since $\delta \geq \nu(\nu - 1)/2 + \sum_{i=2}^4 \nu_i(\nu_i - 1)/2$, we have

$$2\delta/\nu^2 \geq 1 - 1/\nu + \sum_{i=2}^{4}(\nu_i/\nu)^2 - \sigma/\nu.$$

Using Schwarz's inequality $\sum_{i=2}^4(\nu_i/\nu)^2 \geq \sigma^2/3$, we obtain

$$d/\nu > \sigma^2/3 - \sigma/\nu + 1.$$

It suffices to prove that $\sigma^2/3 - \sigma/\nu + 1 \geq \sigma - (1/\nu - 1/\nu^2)$. Indeed, we then have $d > \nu\sigma - (1 - 1/\nu)$, which implies that $d \geq \nu\sigma$. Now we have

$$\begin{aligned} &\sigma^2/3 - \sigma/\nu + 1 - \{\sigma - (1/\nu - 1/\nu^2)\} \\ &\quad = \frac{1}{3}\left\{\sigma - \frac{3}{2}(1 + 1/\nu)\right\}^2 + \frac{(\nu - 4)(1 + 1/\nu)}{4\nu} + \frac{(\nu - 3)}{4\nu^2}. \end{aligned}$$

So we get the assertion unless $\nu = 3$. In case $\nu = 3$, the right hand side is equal to $\{(\sigma - 2)^2 - 1/3\}/3$, which is non-negative unless $\nu_2 + \nu_3 + \nu_4 \le 7$. If $d/3 < \sigma$, then we see that $d = 6$ and $(\nu_2, \nu_3, \nu_4) = (3,3,1)$ or $(3,2,2)$. But for these cases, Condition (M) is not satisfied.

Now we write $h(\eta, \nu) = h(\eta, \nu, 0) = (\eta+1)/2$ and $f_3(\eta, \nu) = f_3(\eta, \nu, 0) = \{3\sqrt{\eta} - (1 + 1/\nu)\}/2$. We can reformulate the inequality: $d/\nu > h(\eta^*, \nu)$ as Condition (M) and the inequality $d/\nu > f_3(\eta^*, \nu)$ as Condition (N). Clearly, the functions h and f_3 are increasing functions of the variable η. Since $\eta \le \eta^*$, we have $h(\eta, \nu) \le h(\eta^*, \nu)$ and $f_3(\eta, \nu) \le f_3(\eta^*, \nu)$. Thus Condition (E) in Theorem 3 is satisfied. Note that $f_3(\eta, \nu) \ge f_{k_0}(\eta, \nu)$, since $k_0 \ge 3$ and Condition (E) is satisfied. Consequently, if both Conditions (M) and (N) are satisfied, then the assumptions for Theorem 3 with $q = 0$ are satisfied. Therefore, by Theorem 3, we conclude that $G = d - \nu$.

Finally, in case $k_0 \ge 3$ and $d \ge 5/2$, Condition (F) is not necessary in Theorem 3. So we can omit Condition (N). We can also show this by the following direct computation. We have

$$\left(\frac{2d + \nu + 1}{6}\right)^2 + \frac{\nu(\nu - 2)}{4} - \frac{(d-1)\nu}{2}$$
$$= (1/18)\{(2(d - \nu) + 1)(d - (5\nu - 1)/2) + 9\nu/2\} > 0.$$

So in case $d/\nu \ge 5/2$, Condition (M) implies Condition (N). □

6. Gonality and the genus

We now discuss the relation between the gonality G and the genus g of an irreducible plane curve C. For $\nu = 2, 3$, by Theorem 1 and by Theorem 2, we obtain the following optimal results.

Proposition 4. *Let C be an irreducible singular plane curve of type $(d, 2)$ with $d \ge 6$. If*

$$g \ge \frac{(d+1)(d-3)}{4} - \begin{cases} 1/4 & \text{if } d \text{ is even,} \\ 0 & \text{if } d \text{ is odd,} \end{cases}$$

then the equality: $G = d - 2$ holds.

Proposition 5. *Let C be an irreducible singular plane curve of type $(d, 3)$ with $d \ge 6$. If*

$$g \ge \frac{(d+2)(d-4)}{4} + \begin{cases} 0 & \text{if } d \text{ is even,} \\ 1/4 & \text{if } d \text{ is odd,} \end{cases}$$

then the equality: $G = d - 3$ holds.

We recall the following facts.

Lemma 15. (Cf., Namba [8], Corollary 2.4.5) *Let C be an irreducible singular plane curve of type (d, ν).*

(i) Suppose $d - \nu = 3$. If $g \geq 3$, then $G = 3$.
(ii) Suppose $d - \nu = 5$. If $g \geq 13$, then $G = 5$.

Lemma 16 (Brill-Neother bound). *Let C be an irreducible singular plane curve of type (d, ν). If $g \leq 2(d - \nu - 2)$, then we have $G < d - \nu$.*

Proof. The assertion is a consequence of the Brill-Noether upper bound (see [1], [7]): $G \leq (g+3)/2$. □

Example 12. For $\nu \leq 5$, we collect the results which are obtained by our criteria for $G = d - \nu$. Let C be an irreducible singular plane curve of type (d, ν). We have the equality: $G = d - \nu$ if $g \geq B(d, \nu)$, where the values $B(d, \nu)$ are listed in the following table:

d	6	7	8	9	10	11	12	13	14	15	16	17	18	19	20
$B(d,2)$	5	8	11	15	19	24	29	35	41	48	55	63	71	80	89
$B(d,3)$	3	7	10	14	18	23	28	34	40	47	54	62	70	79	88
$B(d,4)$	$*$	3	8	13	19	26	34	43	53	64	67	79	92	106	110
$B(d,5)$	$*$	$*$	3	9	13	21	28	37	46	57	68	81	94	109	119

For $\nu = 2$ and $\nu = 3$, we obtain the values $B(d, \nu)$ by the formulas in Proposition 4 and in Proposition 5. For $\nu = 4, 5$, we choose the better bounds between those given in Theorem 1 and Theorem 4. Since $V \geq \nu(\nu-2)/4$, the inequality $\delta \leq Q([d/\nu]) - (d - \nu) + \nu(\nu - 2)/4$ implies Condition (B). So if

$$\delta \leq \min \{Q([d/2]) - (d - \nu),\ Q([d/\nu]) - (d - \nu) + \nu(\nu - 2)/4\},$$

then we can apply Theorem 1. Similarly, to apply Theorem 4, we compute the minimum of the right hand sides of the inequalities in Conditions (M) and (N). It turns out that, in terms of the genus g, for $\nu = 4$, if $d \leq 11$, then Theorem 4 gives the lower values and if $d \geq 16$, then Theorem 1 gives the lower values. Both Theorems give the same values for $12 \leq d \leq 15$. For $\nu = 5$, Theorem 1 gives the better results for $d \geq 20$. For the case in which $d-\nu = 3$ or 5, we can use Lemma 15. So we put $B(6,3) = B(7,4) = B(8,5) = 3$ and $B(9,4) = B(10,5) = 13$.

As we observed in Example 2, the bounds $B(d,2)$ are best possible. The curve C_3 given in Example 7 is of type $(8,3)$ with $g = 9$, $G = 4$. Also the curve C_4 (resp. C_5) is of type $(11,4)$ (resp. $(14,5)$) with $g = 21$, $G \leq 6$ (resp. $g = 38$, $G \leq 8$).

Acknowledgment

(1) The author was partially supported by Grants-in-Aid for Scientific Research (C) (No. 23540041), JSPS.
(2) The author would like to thank the referee for pointing out some mistakes in the first version of this paper.

Appendix A. Proof of Lemma 7

The purpose of this Appendix is to give a comprehensive proof of Lemma 7 in Sec. 3, which was stated in our previous paper [9]. We essentially follow the arguments in Coppens-Kato [4], [5].

Proof of Lemma 7. Assume to the contrary that there exist no rational functions on $\mathbf{P}^2$ with degree less than or equal to l which induces φ. Let $\pi : X \to \mathbf{P}^2$ be a composition of blowing ups so that the strict transform $\tilde{C}$ of C is smooth. On $\tilde{C}$, we write $\Delta = (\varphi)_\infty$. By definition, we have $\varphi \in L(\Delta)$.

Step 1. Let H be a line. Write $h = \pi^* H|_{\tilde{C}}$. We have

$$\dim H^0(\tilde{C}, \mathcal{O}_{\tilde{C}}(lh + \Delta)) \geq (l+1)(l+2).$$

Proof. Let $W_l = H^0(\mathbf{P}^2, \mathcal{O}(l))$ be the vector space of homogeneous polynomials of degree l. We know that $\dim W_l = (l+1)(l+2)/2$. Let H also denote the linear form defining the line H. We difine a homomorphism:

$$W_l^2 \ni (\Phi_1, \Phi_2) \xrightarrow{\tau} \pi^* \left(\frac{\Phi_1}{H^l}\right)|_{\tilde{C}} - \pi^* \left(\frac{\Phi_2}{H^l}\right)|_{\tilde{C}} \cdot \varphi \in L(lh + \Delta).$$

We see that $(\Phi_1, \Phi_2) \in \mathrm{Ker}(\tau)$ if and only if $\varphi = (\Phi_1/\Phi_2)|_{\tilde{C}}$. By the assumption, we infer that $\mathrm{Ker}(\tau) = \{0\}$. It follows that $W_l^2 \subset L(lh + \Delta)$. Hence, we obtain the desired inequality. □

Step 2. By Riemann-Roch Theorem and the inequality in Step 1, we have

$$\dim H^0(\tilde{C}, \mathcal{O}_{\tilde{C}}(K_{\tilde{C}} - (lh + \Delta))) \\ \geq (l+1)(l+2) + (d-1)(d-2)/2 - 1 - ld - (\delta + r).$$

Step 3. For any integer l, we have the inequality:

$$(l+1)(l+2) + (d-1)(d-2)/2 - 1 - ld \geq Q(l+1).$$

Proof. The left hand side is equal to $-Q(l+1)+(l+1)+d(d-1)/2$. Thus, we obtain

$$\begin{aligned} L.H.S - R.H.S &= 2(l+1)(l+1-d) + (l+1) + d(d-1)/2 \\ &= 2\Big(l+1-\frac{d}{2}\Big)\Big(l+1-\frac{d-1}{2}\Big) \geq 0 \end{aligned}$$

for integers l. Note that the equality holds if and only if $l = [d/2] - 1$. □

Step 4. Using the inequalities in Steps 2, 3 and the assumption, we have

$$\dim H^0(\tilde{C}, \mathcal{O}_{\tilde{C}}(K_{\tilde{C}} - (lh + \Delta)) \geq Q(l+1) - (\delta + r) > 0.$$

So, there exists an effective divisor A on $\tilde{C}$ such that $A \sim K_{\tilde{C}} - (lh + \Delta)$. In Step 1, it was possible to choose genral lines $H_1, \ldots, H_l$, so that the points $C \cap H_i = \{p_{i1}, \ldots, p_{id}\}$ are away from the centers of the blowing ups, for $i = 1, \ldots, l$. In this case, we have $h_i = \pi^* H_i|_{\tilde{C}} = \sum_j p_{ij}$, Hence, replacing lh with $\sum h_i$, we obtain $A + \sum p_{ij} + \Delta \sim K_{\tilde{C}}$.

Step 5. Since X is a rational surface, we have a surjection:

$$H^0(X, \mathcal{O}(K_X + \tilde{C})) \to H^0(\tilde{C}, \mathcal{O}_{\tilde{C}}(K_{\tilde{C}})).$$

Proof. The short exact sequence $0 \to \mathcal{O}(K_X) \to \mathcal{O}(K_X + \tilde{C}) \to \mathcal{O}_{\tilde{C}}(K_{\tilde{C}}) \to 0$, gives the cohomology exact sequence:

$$\to H^0(X, \mathcal{O}(K_X + \tilde{C})) \to H^0(\tilde{C}, \mathcal{O}_{\tilde{C}}(K_{\tilde{C}}) \to H^1(X, \mathcal{O}(K_X)) = 0. \quad \square$$

Step 6. We can write $\mathcal{O}(K_X + \tilde{C}) \cong \pi^*\mathcal{O}(d-3) \otimes \mathcal{O}(-\sum(m_i - 1)E_i)$, where the E_i are total transforms of the exceptional curves and the m_i are the multiplicities of the curve C at the center of the blowing ups.

Step 7. In view of the results in Steps 5, 6, there exists a plane curve Γ of degree $d-3$ such that $\pi^*\Gamma|_{\tilde{C}} = A + \sum p_{ij} + \Delta$. Thus $\Gamma \supset \bigcup p_{ij}$. By Bezout's theorem, we infer that $\Gamma \supset H_1, \ldots, H_l$. So there exists a plane curve Γ_2 of degree $d-3-l$ such that $\pi^*\Gamma_2|_{\tilde{C}} = A + \Delta$.

Step 8. Since $(\varphi)_0 \sim \Delta = (\varphi)_\infty$, We can do the same process from Step 4 by replacing Δ with $(\varphi)_0$. Then we can find a plane curve Γ_1 of degree $d-3-l$ such that $\Gamma_1|_{\tilde{C}} = A + (\varphi)_0$. Letting Φ_1, Φ_2 be the defining polynomials of Γ_1, Γ_2, respectively, we define $\Phi = \Phi_1/\Phi_2$. We conclude that $\Phi|_{\tilde{C}} = \varphi$. Furthermore, we have $k = \deg \Phi \leq d-3-l$.

Step 9. By hypothesis, we have $l+1 \leq k$. So $l+1 \leq k \leq d-3-l$. Hence, we have $l+1 \leq (d-2)/2$. By Lemma 1, we have $r + \delta \geq Q(k)$. Combining this with the assumption, we obtain the inequality: $Q(k) < Q(l+1)$. By the property of the quadratic function $Q(x)$, we must have $k > d-(l+1)$. This contradicts the inequality $k \leq d-l-3$.

References

1. E.Arbarello, M.Cornalba, P.A.Griffiths, J.Harris, Geometry of Algebraic Curves I, *Grund. Math. Wiss.* **267**, Springer-Verlag, 1984.
2. M.Coppens, The gonality of general smooth curves with a prescribed plane nodal model, *Math. Ann.* **289**, 89–93 (1991).
3. M.Coppens, Free linear systems on integral Gorenstein curves, *J. Algebra* **145**, 209–218 (1992).
4. M.Coppens and T.Kato, The gonality of smooth curves with plane models, *Manuscripta Math.* **70**, 5–25 (1990).
5. M.Coppens and T.Kato, Correction to the gonality of smooth curves with plane models, *Manuscripta Math.* **71**, 337–338 (1991).
6. R.Hartshorn, Clifford index of ACM curves in $\mathbf{P}^3$, *Milan J. Math.* **70**, 209–221 (2002).
7. T.Meis, Die minimale Blätterzahl der Konkretisierung einer kompakten Riemannschen Fläche, *Schr. Math. Inst. Univ. Münster,* **16** (1960).
8. M.Namba, Families of meromorphic functions on compact Riemann surfaces, *Lecture Notes in Math.* **767**, Springer-Verlag, 1979.
9. M.Ohkouchi and F.Sakai, The gonality of singular plane curves, *Tokyo J. Math.* **27**, 137–147 (2004).
10. R.Paoletti, Free pencils on divisors, *Math. Ann.* **303**, 109–123 (1995).
11. A.H.W.Schmitt, Base point free pencils of small degree on certain smooth curves, *Arch. Math.* **74**, 104–110 (2000).
12. F.Serrano, Extension of morphisms defined on a divisor, *Math. Ann.* **277**, 395–413 (1987).

Examples of non-uniruled surfaces with pre-Tango structures involving non-closed global differential 1-forms

Yoshifumi Takeda

Department of Mathematics and Statistics,
Wakayama Medical University, Wakayama City 6418509, Japan
E-mail: ytakeda@wakayama-med.ac.jp

The pre-Tango structure is an ample invertible sheaf of locally exact differentials on a variety in positive characteristic, which often brings various sorts of pathological phenomena. We, however, know few examples of pre-Tango structures on non-uniruled varieties. In the present article, we explicitly construct non-uniruled surfaces with pre-Tango structures involving non-closed global differential 1-forms.

Keywords: Pathological phenomena, pre-Tango structures, non-uniruled surfaces, abelian surfaces.

1. Introduction

Let X be a projective algebraic variety over an algebraically closed field k of characteristic $p > 0$ and let $F_X : \widetilde{X} \to X$ be the relative Frobenius morphism over k. We then have a short exact sequence

$$0 \to \mathcal{O}_X \to F_{X*}\mathcal{O}_{\widetilde{X}} \to F_{X*}\mathcal{B}^1_{\widetilde{X}} \to 0,$$

where $\mathcal{B}^1_{\widetilde{X}}$ is the first sheaf of coboundaries of the de Rham complex of $\widetilde{X}$. Suppose that there exists an ample invertible subsheaf $\mathcal{L}$ of $F_{X*}\mathcal{B}^1_{\widetilde{X}}$ provided that $F_{X*}\mathcal{B}^1_{\widetilde{X}}$ is regarded as an $\mathcal{O}_X$-module. We call $\mathcal{L}$ a *pre-Tango structure* (see Takeda [10], see also Mukai [4]). Let us consider the exact sequence

$$0 \to \mathcal{L}^{-1} \to F_{X*}\mathcal{O}_{\widetilde{X}} \otimes_{\mathcal{O}_X} \mathcal{L}^{-1} \to F_{X*}\mathcal{B}^1_{\widetilde{X}} \otimes_{\mathcal{O}_X} \mathcal{L}^{-1} \to 0.$$

By taking cohomology, we have

$$0 \to H^0(X, \mathcal{L}^{-1}) \to H^0(X, F_{X*}\mathcal{O}_{\widetilde{X}} \otimes_{\mathcal{O}_X} \mathcal{L}^{-1}) \to H^0(X, F_{X*}\mathcal{B}^1_{\widetilde{X}} \otimes_{\mathcal{O}_X} \mathcal{L}^{-1}) \to H^1(X, \mathcal{L}^{-1}) \to \cdots.$$

Since $H^0(X, F_{X*}\mathcal{O}_{\widetilde{X}} \otimes_{\mathcal{O}_X} \mathcal{L}^{-1}) = 0$ and $H^0(X, F_{X*}\mathcal{B}^1_{\widetilde{X}} \otimes_{\mathcal{O}_X} \mathcal{L}^{-1}) \neq 0$, we know that $H^1(X, \mathcal{L}^{-1}) \neq 0$. Hence, if X is a smooth variety of dimension greater than one, then the pair $(X, \mathcal{L})$ is a counter-example to the Kodaira vanishing theorem in positive characteristic. It is, however, hard to find such a pair in dimension greater than one. Meanwhile, regarding in dimension one, we know that almost all smooth projective curves have pre-Tango structures (see Takeda and Yokogawa [11]). In fact, Raynaud's famous counter-example ([7]) is a uniruled surface constructed by using a certain pre-Tango structure on a smooth projective curve.

The uniruled surfaces which are constructed similarly to Raynaud's method are the only known examples of smooth surfaces which have pre-Tango structures, as far as the author knows. Hence the following problem seems interesting:

> Suppose that a smooth projective surface X has a pre-Tango structure. Then is X a uniruled surface?

Regrettably, the author does not know what the answer is. Meanwhile, it is known that, if a smooth non-uniruled projective variety X has an ample invertible sheaf $\mathcal{L}$ such that $\mathcal{L}^{p-1} \otimes_{\mathcal{O}_X} \omega_X^{-1}$ is ample, then we have $H^1(X, \mathcal{L}^{-1}) = 0$ (Corollary II.6.3 in Kollár [3]). On the other hand, in case of normal projective varieties, the answer is negative. Indeed, Mumford gave an example of a pre-Tango structure on a normal projective surface, which is not uniruled ([6]). It seems, however, hard to know whether its desingularization has a pre-Tango structure or not.

For any smooth proper variety over k which lifts over the ring of Witt-vectors of length 2, the Kodaira vanishing theorem holds on it. Furthermore, if it is of dimension $\leq p$, then its spectral sequence of Hodge to de Rham degenerates at E_1 (Deligne and Illusie [1]). So, it has no non-closed global differential 1-forms. In other words, the existence of non-closed global differential 1-forms is another typical pathological phenomenon in positive characteristic. Meanwhile, we know that, if a normal projective variety has non-closed global differential 1-forms, then so does its desingularization. Therefore, it seems appropriate to investigate normal projective surfaces with pre-Tango structures involving non-closed global differential 1-forms for the first step. In fact, we often see the normal uniruled surfaces, which are constructed similarly to Raynaud's method by using pre-Tango structures on curves, having non-closed global differential 1-forms (cf. [11]).

On the other hand, it is well-known that we can easily construct surfaces with non-closed global differential 1-forms by using Mumford's method, that is, by taking the composite of many Artin-Schreier coverings of base surfaces ([5]). We, however, hardly know their properties because of its elusive construction. Under the circumstances, the purpose of the present article is to give explicit and concrete examples of non-uniruled normal surfaces with pre-Tango structures involving non-closed global differential 1-forms in characteristic 2, 3. Precisely, we first consider a certain quotient of a superspecial abelian surface (the product of two supersingular elliptic curves) and take the composite finite covering of *two* suitable Artin-Schreier coverings of the quotient. On that finite covering, then we find out a pre-Tango structure with required attribute.

2. Case of characteristic $p = 2$

2.1. *A rational vector field on an abelian surface and the quotient*

Let E_1 be the elliptic curve defined by

$$y^2 + y = x^3,$$

which is the unique supersingular elliptic curve in characteristic 2. We then have

$$z + z^2 = w^3$$

near the point at infinity, where $z = y^{-1}$ and $w = xy^{-1}$. Moreover, we have

$$dy = x^2 dx \quad \text{and} \quad dz = w^2 dw.$$

Notc that

$$x + w = x + \frac{x}{y} = \frac{x(y^2 + y)}{y^2} = \frac{x \cdot x^3}{y^2} = x^2\,\frac{x^2}{y^2} = x^2 w^2.$$

We then know

$$dx = dw \quad \text{and} \quad \frac{\partial}{\partial x} = \frac{\partial}{\partial w}.$$

Take a copy $\widetilde{E}$ of E_1 and take the local parameters ξ_0 and ξ_∞ corresponding to x and w, respectively. We then have the same equations

$$\xi_0 + \xi_\infty = \xi_0^2 \xi_\infty^2, \quad d\xi_0 = d\xi_\infty, \quad \frac{\partial}{\partial \xi_0} = \frac{\partial}{\partial \xi_\infty}$$

as above. Let A be the product $E_1 \times \widetilde{E}$ and consider the p-closed rational vector field

$$D = \frac{\partial}{\partial x} + y^2 \frac{\partial}{\partial \xi_i} \quad (i = 0, \infty)$$

on A. We know that

$$D = \frac{1}{z^2}\left(z^2\frac{\partial}{\partial x} + \frac{\partial}{\partial \xi_i}\right) \quad (i = 0, \infty)$$

and that the divisor of D is

$$(D) = -6S,$$

where S is the fibre of the point at infinity of E_1, in other words, the curve defined by $w = 0$ on A. Besides S is an integral curve of D.

Take the quotient X of A by D, i.e., the underlying topological space is the same as A and the structure sheaf is the sheaf of the germs killed by D (see Rudakov and Shafarevich [8]). Since D has only divisorial singularities, we have that X is a nonsingular surface of Kodaira dimension 1 (see Katsura and Takeda [2]). Let Γ and Σ denote the images by the quotient morphism of S (the same as above) and $T = \{x = 0\}$, respectively. Since S is an integral curve of D and T is not, we have that $[k(S) : k(\Gamma)] = 2$ and $[k(T) : k(\Sigma)] = 1$. Consider the relative Frobenius morphisms $F_E : \widetilde{E} \to E$ and $F_1 : E_1 \to E_1^{(p)}$ over k. We then have two fibrations: one is an elliptic fibration $\psi : X \to E_1^{(p)}$ induced from the first projection $A \to E_1$; and the other is a fibration $\varphi : X \to E$ induced from the second projection $A \to \widetilde{E}$, each fibre of which is an elliptic curve with one cusp. By regarding the fibration ψ, we know that Γ is a fibre of multiplicity 2 and that Σ is a fibre of multiplicity 1, and by regarding the fibration φ, we know that Γ is a section and that Σ is a 2-section.

Let us consider local defining equations of X. Set $\eta_i = \xi_i^2$ for $i = 0, \infty$ and take the affine open subsets $U_0 = E - \{\eta_\infty = 0\}$, $U_\infty = E - \{\eta_0 = 0\}$. Take, furthermore, the affine open subsets

$$V_i = \varphi^{-1}(U_i) - \Gamma, \quad W_i = \varphi^{-1}(U_i) - \Sigma \quad (i = 0, \infty)$$

of X. Since $y^2 + y = x^3$, we know $\dfrac{\partial y}{\partial x} = x^2$. Hence we have

$$D(xy + \xi_i) = \left(\frac{\partial}{\partial x} + y^2\frac{\partial}{\partial \xi_i}\right)(xy + \xi_i) = y + x \cdot x^2 + y^2 = 0 \quad (i = 0, \infty).$$

Set $u = x^2$, $v = y^2$ and $t_i = xy + \xi_i$ for $i = 0, \infty$. We then know that $u, v, t_i \in k(X)$, $v^2 + v = u^3$, $t_0 + t_\infty = \eta_0\eta_\infty$ and

$$t_i^2 = uv + \eta_i,$$

which are local defining equations of V_i for $i = 0, \infty$. Next set $r = w^2$, $s = z^2$. We then know that $r, s \in k(X)$, $s + s^2 = r^3$, $s = v^{-1}$, $u + r = u^2r^2$

and $t_i^2 s^2 = r + \eta_i s^2$ by simple calculation. Therefore, by setting $q_i = t_i s$, we have local defining equations

$$q_i^2 = r + \eta_i s^2$$

of W_i for $i = 0, \infty$. By exterior differentiation on X, we obtain the relations:

$$du = dr, \quad dr = s^2 d\eta_i \quad (i = 0, \infty).$$

Let us consider the exact differential 1-form $\omega = du$. We then know that ω is regular on V_i for $i = 0, \infty$. Moreover, since $\omega = s^2 d\eta_i$ for $i = 0, \infty$, we know that ω is regular on W_i for $i = 0, \infty$. Therefore, we have that

$$\omega \in H^0(X, \mathcal{B}_X^1).$$

Since $s(1+s) = r^3$ and Γ is a fibre of multiplicity 2, we know that $(s^2) = 12\Gamma$. Hence the divisor of ω is

$$(\omega) = 12\Gamma$$

and that implies an inclusion

$$\mathcal{O}_X(12\Gamma)\omega \hookrightarrow \mathcal{B}_X^1.$$

By taking its adjoint, we obtain an injection

$$\mathcal{O}_X(6\Gamma) \hookrightarrow F_{X*}\mathcal{B}_{\tilde{X}}^1.$$

It is, however, *not* a pre-Tango structure because Γ is not ample.

2.2. *A pre-Tango structure on a finite covering of the quotient*

Let P_0 be the point defined by $\eta_0 = 0$ on E, and set $H = \varphi^{-1}(P_0)$. Consider the finite extension field $k(X)(\theta, \zeta)$ subjected to

$$\theta^2 + \eta_0^2\theta = u \quad \text{and} \quad \zeta^2 + \eta_0^2\zeta = \eta_0,$$

and take the normalization $\sigma : Y \to X$ in $k(X)(\theta, \zeta)$. We then know that *Y is not a uniruled surface.* Furthermore, we have

$$\eta_0^2 d\theta = du, \quad \eta_0^2 d\zeta = d\eta_0$$

on Y. Since $\omega = du = s^2 d\eta_0$, we obtain

$$\eta_0^2 d\theta = s^2\eta_0^2 d\zeta.$$

By regarding on Y, we have

$$(\omega) = \sigma^*(12\Gamma + 2H).$$

That induces an inclusion

$$\mathcal{O}_Y(\sigma^*(12\Gamma + 2H))\omega \hookrightarrow \mathcal{B}^1_Y.$$

By taking its adjoint, we obtain an injection

$$\mathcal{O}_Y(\sigma^*(6\Gamma + H)) \hookrightarrow F_{Y*}\mathcal{B}^1_{\tilde{Y}}.$$

Moreover, it is a *pre-Tango structure* because $6\Gamma + H$ is ample on X and so is $\sigma^*(6\Gamma + H)$ on Y.

Consider the differential 1-forms $d\theta$ (which is exact) and $t_0 d\theta$ (which is not closed) on Y. We have

$$d\theta = s^2 d\zeta \quad \text{and} \quad t_0 d\theta = q_0 s d\zeta.$$

Since $d\theta$ and $t_0 d\theta$ are regular on $\sigma^{-1}(V_0)$, and $s^2 d\zeta$ and $q_0 s d\zeta$ are so on $\sigma^{-1}(W_0)$, we have

$$d\theta,\ t_0 d\theta \in H^0(\sigma^{-1}(\varphi^{-1}(U_0)), \Omega^1_Y) = H^0(U_0, \varphi_*\sigma_*\Omega^1_Y).$$

On the other hand, since $\omega = du$ is regular on X, we have that du is regular on $\sigma^{-1}(\varphi^{-1}(U_\infty))$. Hence we obtain that

$$du \in H^0(\sigma^{-1}(\varphi^{-1}(U_\infty)), \mathcal{B}^1_Y) = H^0(U_\infty, \varphi_*\sigma_*\mathcal{B}^1_Y).$$

Next consider $t_\infty\omega$. We then know that $t_\infty\omega = t_\infty du$ is regular on V_∞. Besides, since $t_\infty du = q_\infty s d\eta_\infty$, we know that $t_\infty\omega$ is regular on W_∞. By regarding on Y, we have

$$t_\infty du \in H^0(\sigma^{-1}(\varphi^{-1}(U_\infty)), \Omega^1_Y) = H^0(U_\infty, \varphi_*\sigma_*\Omega^1_Y).$$

Now let us consider $\mathcal{O}_E$-submodules $\mathcal{R}$ and $\mathcal{S}$ of $\varphi_*\sigma_*\Omega^1_Y$ such that

$$\mathcal{R}|_{U_0} = \mathcal{O}_E|_{U_0} d\theta, \qquad \mathcal{R}|_{U_\infty} = \mathcal{O}_E|_{U_\infty} du,$$

$$\mathcal{S}|_{U_0} = \mathcal{O}_E|_{U_0} d\theta + \mathcal{O}_E|_{U_0} t_0 d\theta, \quad \mathcal{S}|_{U_\infty} = \mathcal{O}_E|_{U_\infty} du + \mathcal{O}_E|_{U_\infty} t_\infty du.$$

Note that the sections of $\mathcal{S}$ which are not contained in $\mathcal{R}$, are non-closed differential 1-forms. Meanwhile, since $\eta_0^2 d\theta = du$, we know that $\mathcal{R} \cong \mathcal{O}_E(2P_0)$. Moreover, since $t_0 + t_\infty = \eta_0\eta_\infty$, we have

$$\eta_0^2 t_0 d\theta = t_\infty du + \eta_0\eta_\infty du$$

and so,

$$(d\theta, t_0 d\theta) = (du, t_\infty du) \begin{pmatrix} \dfrac{1}{\eta_0^2} & \dfrac{\eta_\infty}{\eta_0} \\ 0 & \dfrac{1}{\eta_0^2} \end{pmatrix}.$$

Therefore, we obtain a short exact sequence

$$0 \to \mathcal{R} \to \mathcal{S} \to \mathcal{O}_E(2P_0) \to 0.$$

Since $H^1(E, \mathcal{R}) \cong H^1(E, \mathcal{O}_E(2P_0)) = 0$ and $H^0(E, \mathcal{O}_E(2P_0)) \neq 0$, we know $H^0(E, \mathcal{R}) \subsetneq H^0(E, \mathcal{S})$. Hence we conclude that Y *has non-closed global differential* 1*-forms.*

3. Case of characteristic $p = 3$

3.1. *A rational vector field on an abelian surface and the quotient*

Let E_1 be the elliptic curve defined by

$$y^2 = x^3 - x,$$

which is the unique supersingular elliptic curve in characteristic 3. We then have

$$z = w^3 - wz^2$$

near the point at infinity, where $z = y^{-1}$ and $w = xy^{-1}$. Moreover, we know $2ydy = -dx$ and so

$$dy = \frac{dx}{y},$$

which is an exact global differential 1-form. Set $\Delta = y\frac{\partial}{\partial x}$. We then have that Δ is a regular vector field on E_1 such that $\Delta^3 = 0$. Note that

$$\begin{aligned} z &= w^3 - wz^2 \\ z(1 + wz) &= w^3 \\ z(\frac{1}{w^3} + \frac{1}{w^3}wz) &= 1 \\ \frac{1}{w^3} + \frac{1}{w^3}wz &= y \\ \frac{1}{w^3} + \frac{1}{w^3}w\frac{w^3}{1+wz} &= y \\ \frac{1}{w^3} + \frac{w}{1+wz} &= y. \end{aligned}$$

We then obtain that

$$dy = d\frac{w}{1+wz}.$$

Meanwhile, we know that y (resp. $w/(1+wz)$) is a local parameter near the point over $x=0$ (resp. $x=\infty$).

Take a copy $\widetilde{E}$ of E_1 and take the local parameters ξ_0 and ξ_∞ corresponding to y and $w/(1+wz)$, respectively. We then have

$$d\xi_0 = d\xi_\infty \quad \text{and} \quad \frac{\partial}{\partial \xi_0} = \frac{\partial}{\partial \xi_\infty}.$$

Let A be the product $E_1 \times \widetilde{E}$ and consider the p-closed rational vector field

$$D = \Delta - x^3 \frac{\partial}{\partial \xi_i} \quad (i = 0, \infty)$$

on A. We know that

$$D = \frac{1}{z^2}\left(z^2\Delta - (1+wz)\frac{\partial}{\partial \xi_i}\right) \quad (i = 0, \infty)$$

and that the divisor of D is

$$(D) = -6S,$$

where S is the fibre of the point at infinity of E_1, in other words, the curve defined by $w=0$ on A. Besides S is an integral curve of D.

Take the quotient X of A by D, i.e., the underlying topological space is the same as A and the structure sheaf is the sheaf of the germs killed by D (see [8]). Since D has only divisorial singularities, we have that X is a nonsingular surface of Kodaira dimension 1 (see [2]). Let Γ and Σ denote the images by the quotient morphism of S (the same as above) and $T = \{x = 0\}$, respectively. Since S is an integral curve of D and T is not, we have that $[k(S) : k(\Gamma)] = 3$ and $[k(T) : k(\Sigma)] = 1$. Consider the relative Frobenius morphisms $F_E : \widetilde{E} \to E$ and $F_1 : E_1 \to E_1^{(p)}$ over k. We then have two fibrations: one is an elliptic fibration $\psi : X \to E_1^{(p)}$ induced from the first projection $A \to E_1$; and the other is a fibration $\varphi : X \to E$ induced from the second projection $A \to \widetilde{E}$, each fibre of which is an elliptic curve with one cusp. By regarding the fibration ψ, we know that Γ is a fibre of multiplicity 3 and that Σ is a fibre of multiplicity 1, and by regarding the fibration φ, we know that Γ is a section and that Σ is a 3-section.

Let us consider local defining equations of X. Set $\eta_i = \xi_i^3$ for $i = 0, \infty$ and take the affine open subsets $U_0 = E - \{\eta_\infty = 0\}$, $U_\infty = E - \{\eta_0 = 0\}$. Take, furthermore, the affine open subsets

$$V_i = \varphi^{-1}(U_i) - \Gamma, \quad W_i = \varphi^{-1}(U_i) - \Sigma \quad (i = 0, \infty)$$

of X. Since $\Delta(y) = 1$ and $\Delta(x) = y$, we obtain

$$D(xy + \xi_i) = \left(\Delta - x^3 \frac{\partial}{\partial \xi_i}\right)(xy + \xi_i) = y^2 + x - x^3 = 0 \quad (i = 0, \infty).$$

Set $u = x^3$, $v = y^3$ and $t_i = xy + \xi_i$ for $i = 0, \infty$. We then know that $u, v, t_i \in k(X)$, $v^2 = u^3 - u$ and

$$t_i^3 = uv + \eta_i,$$

which are local defining equations of V_i for $i = 0, \infty$. By exterior differentiation on X, we obtain the relations:

$$vdv = du, \quad u^3 dv = -d\eta_i \quad (i = 0, \infty).$$

Next set $r = w^3$, $s = z^3$. We then know that $r, s \in k(X)$, $s = r^3 - rs^2$, $s = v^{-1}$, $r = uv^{-1}$ and $t_i^3 s^3 = rs + \eta_i s^3$. Therefore, by setting $q_i = t_i s$, we have local defining equations

$$q_i^3 = rs + \eta_i s^3$$

of W_i for $i = 0, \infty$.

Let us consider the exact differential 1-form $\omega = dv$. We then know that ω is regular on V_i for $i = 0, \infty$. Moreover, by simple computation, we have $\omega = -\dfrac{s^2}{1 + rs} d\eta_i$ for $i = 0, \infty$. Hence we know that ω is regular on W_i for $i = 0, \infty$. Therefore, we have that

$$\omega \in H^0(X, \mathcal{B}_X^1).$$

Since $s(1 + rs) = r^3$ and Γ is a fibre of multiplicity 3, we know that $(s^2) = 18\Gamma$. Hence the divisor of ω is

$$(\omega) = 18\Gamma$$

and that implies an inclusion

$$\mathcal{O}_X(18\Gamma)\omega \hookrightarrow \mathcal{B}_X^1.$$

By taking its adjoint, we obtain an injection

$$\mathcal{O}_X(6\Gamma) \hookrightarrow F_{X*}\mathcal{B}_{\tilde{X}}^1.$$

It is, however, *not* a pre-Tango structure because Γ is not ample.

3.2. *A pre-Tango structure on a finite covering of the quotient*

Let P_0 be the point defined by $\eta_0 = 0$ on E, and set $H = \varphi^{-1}(P_0)$. Consider the finite extension field $k(X)(\theta, \zeta)$ subjected to

$$\theta^3 - \eta_0^3 \theta = v \quad \text{and} \quad \zeta^3 - \eta_0^3 \zeta = \eta_0,$$

and take the normalization $\sigma : Y \to X$ in $k(X)(\theta, \zeta)$. We then know that Y *is not a uniruled surface*. Furthermore, we have

$$-\eta_0^3 d\theta = dv, \quad -\eta_0^3 d\zeta = d\eta_0$$

on Y. Since $\omega = dv = -\dfrac{s^2}{1+rs} d\eta_0$, we obtain

$$\eta_0^3 d\theta = -\frac{s^2 \eta_0^3}{1+rs} d\zeta.$$

By regarding on Y, we have

$$(\omega) = \sigma^*(18\Gamma + 3H).$$

That induces an inclusion

$$\mathcal{O}_Y(\sigma^*(18\Gamma + 3H))\omega \hookrightarrow \mathcal{B}_Y^1.$$

By taking its adjoint, we obtain an injection

$$\mathcal{O}_Y(\sigma^*(6\Gamma + H)) \hookrightarrow F_{Y*}\mathcal{B}_{\tilde{Y}}^1.$$

Moreover, it is a *pre-Tango structure* because $6\Gamma + H$ is ample on X and so is $\sigma^*(6\Gamma + H)$ on Y.

Consider the differential 1-forms $d\theta$ (which is exact) and $t_0 d\theta$ (which is not closed) on Y. We have

$$d\theta = -\frac{s^2}{1+rs} d\zeta \quad \text{and} \quad t_0 d\theta = -\frac{q_0 s}{1+rs} d\zeta.$$

Since $d\theta$, $t_0 d\theta$ are regular on $\sigma^{-1}(V_0)$ and since $\dfrac{s^2}{1+rs} d\zeta$, $\dfrac{q_0 s}{1+rs} d\zeta$ are so on $\sigma^{-1}(W_0)$, we have

$$d\theta,\ t_0 d\theta \in H^0(\sigma^{-1}(\varphi^{-1}(U_0)), \Omega_Y^1) = H^0(U_0, \varphi_*\sigma_*\Omega_Y^1).$$

On the other hand, since $\omega = dv$ is regular on X, we have that dv is regular on $\sigma^{-1}(\varphi^{-1}(U_\infty))$. Hence we obtain that

$$dv \in H^0(\sigma^{-1}(\varphi^{-1}(U_\infty)), \mathcal{B}_Y^1) = H^0(U_\infty, \varphi_*\sigma_*\mathcal{B}_Y^1).$$

Next consider $t_\infty \omega$. We then know that $t_\infty \omega = t_\infty dv$ is regular on V_∞. Besides, since $t_\infty dv = -\dfrac{q_\infty s}{1+rs} d\eta_\infty$, we know that $t_\infty \omega$ is regular on W_∞. By regarding on Y, we have

$$t_\infty dv \in H^0(\sigma^{-1}(\varphi^{-1}(U_\infty)), \Omega_Y^1) = H^0(U_\infty, \varphi_*\sigma_*\Omega_Y^1).$$

Now let us consider $\mathcal{O}_E$-submodules $\mathcal{R}$ and $\mathcal{S}$ of $\varphi_*\sigma_*\Omega_Y^1$ such that

$$\mathcal{R}|_{U_0} = \mathcal{O}_E|_{U_0} d\theta, \qquad \mathcal{R}|_{U_\infty} = \mathcal{O}_E|_{U_\infty} dv,$$

$$\mathcal{S}|_{U_0} = \mathcal{O}_E|_{U_0} d\theta + \mathcal{O}_E|_{U_0} t_0 d\theta, \quad \mathcal{S}|_{U_\infty} = \mathcal{O}_E|_{U_\infty} dv + \mathcal{O}_E|_{U_\infty} t_\infty dv.$$

Note that the sections of $\mathcal{S}$ which are not contained in $\mathcal{R}$, are non-closed differential 1-forms. Meanwhile, since $-\eta_0^3 d\theta = dv$, we know that $\mathcal{R} \cong \mathcal{O}_E(3P_0)$. Recall that ξ_0, ξ_∞ are corresponding to y, $w/(1+wz)$, respectively. Therefore, the difference $\xi_0 - \xi_\infty$ is corresponding to $1/w^3$. Denote it by $b_{0\infty}$. We then have that $b_{0\infty}$ is a section in $\mathcal{O}_E(U_0 \cap U_\infty)$. Moreover, since $t_i = xy + \xi_i$ for $i = 0, \infty$, we know that $t_0 - t_\infty = b_{0\infty}$. Hence we have

$$-\eta_0^3 t_0 d\theta = t_\infty dv + b_{0\infty} dv$$

and so,

$$(-d\theta, -t_0 d\theta) = (dv, t_\infty dv) \begin{pmatrix} \dfrac{1}{\eta_0^3} & \dfrac{b_{0\infty}}{\eta_0^3} \\ 0 & \dfrac{1}{\eta_0^3} \end{pmatrix}.$$

Therefore, we obtain a short exact sequence

$$0 \to \mathcal{R} \to \mathcal{S} \to \mathcal{O}_E(3P_0) \to 0.$$

Since $H^1(E, \mathcal{R}) \cong H^1(E, \mathcal{O}_E(3P_0)) = 0$ and $H^0(E, \mathcal{O}_E(3P_0)) \neq 0$, we know $H^0(E, \mathcal{R}) \subsetneq H^0(E, \mathcal{S})$. Hence we conclude that Y *has non-closed global differential* 1*-forms.*

Acknowledgements

The author expresses his sincere gratitude to Professor Toshiyuki Katsura for his insightful comments.

References

1. P. Deligne and L. Illusie, Relèvements modulo p^2 et décomposition du complexe de de Rham, Invent. Math. 89 (1987), 247–270.
2. T. Katsura and Y. Takeda, Quotients of abelian and hyperelliptic surfaces by rational vector fields, J. Algebra 124 (1989), 472–492.
3. J. Kollár, Rational Curves on Algebraic Varieties, Ergebnisse der Mathematik und ihrer Grenzgebiete 3. Folge Band 32, Springer-Verlag, Berlin, Heidelberg, New York, 1996.

4. S. Mukai, Counterexamples of Kodaira's vanishing and Yau's inequality in positive characteristics, preprint RIMS-1736, Kyoto Univ., December 2011 (*Japanese Original* in: Proceedings of the Symposium on Algebraic Geometry, Kinosaki, 1979, pp. 9–31).
5. D. Mumford, Pathologies of modular algebraic surfaces, Amer. J. Math. 83 (1961), 339–342.
6. D. Mumford, Pathologies III, Amer. J. Math. 89 (1967), 94–104.
7. M. Raynaud, Contre-exemple au "Vanishing Theorem" en caractéristique $p > 0$, in: C. P. Ramanujan-A Tribute, Tata Institute of Fundamental Research, Studies in Mathematics No. 8, Springer-Verlag, Berlin, Heidelberg, New York, 1978.
8. A. Rudakov and I. Shafarevich, Inseparable morphisms of algebraic surfaces, Math. U.S.S.R. Izvestija, 10 (1976), 1205–1237.
9. Y. Takeda, Vector fields and differential forms on generalized Raynaud surfaces, Tôhoku Math. J. 44 (1992), 359–364.
10. Y. Takeda, Pre-Tango structures and uniruled varieties, Colloq. Math. 108 (2007), 193–216.
11. Y. Takeda and K. Yokogawa, Pre-Tango structures on curves, Tôhoku Math. J. 54 (2002), 227–237; Errata and Addenda 55 (2003), 611–614.
12. H. Tango, On the behavior of extensions of vector bundles under the Frobenius map, Nagoya Math. J. 48 (1972) 73–89.

Representations of $\mathbb{G}_a$ of codimension two

Ryuji Tanimoto

Faculty of Education, Shizuoka University,
836 Ohya, Suruga-ku, Shizuoka 422-8529, Japan
E-mail: ertanim@ipc.shizuoka.ac.jp

Dedicated to Professor Masayoshi Miyanishi on the occasion of his 70th birthday

Let k be an algebraically closed field of positive characteristic and let $\mathbb{G}_a$ be the additive group of k. We give a classification of representations of $\mathbb{G}_a$ of codimension two.

Keywords: Representations of additive groups, Weitzenböck problem.

1. Introduction

Let k be an algebraically closed field and let $\mathbb{G}_a$ be the additive group of k. Let $\rho : \mathbb{G}_a \to GL(n,k)$ be a representation of $\mathbb{G}_a$. We assume that the representation ρ is regular, i.e., the matrix $\rho(t)$ has the form $\rho(t) = (a_{ij}(T)|_{T=t})_{1\leq i,j\leq n}$, where each $a_{ij}(T)$ is a polynomial in T over k and $a_{ij}(T)|_{T=t}$ means the value of $a_{ij}(T)$ at $T = t$. For simplicity, we shall write each entry $a_{ij}(T)|_{T=t}$ of the matrix $\rho(t)$ by a_{ij}. Let V be the k-vector space k^n consisiting of column vectors of n elements of k and let $\mathbb{G}_a$ act via ρ on V from the left. We say that the representation ρ of $\mathbb{G}_a$ is *of codimension* r if the subspace

$$V^{\mathbb{G}_a} := \{v \in V \mid \rho(t)(v) = v \text{ for all } t \in \mathbb{G}_a\}$$

of all $\mathbb{G}_a$-fixed vectors of V is of codimension r in V as a k-vector space.

Let $A := k[x_1, \ldots, x_n]$ be the polynomial ring in n variables over k and let $\mathbb{G}_a$ act linearly on A. Then, if the characteristic of k is zero, the invariant ring $A^{\mathbb{G}_a}$ is finitely generated as a k-algebra (see Ref. 2). In its proof, the following property of a representation of $\mathbb{G}_a$ is used. Any representation of $\mathbb{G}_a$ in characteristic zero factors through $SL(2,k)$. On the other hand, if the characteristic of k is positive, we do not know whether the invariant ring

$A^{\mathbb{G}_a}$ is finitely generated as a k-algebra. Fauntleroy [1] constructed a representation of $\mathbb{G}_a$ in positive characteristic which does not factor through $SL(2,k)$ and is of codimension one. Fauntleroy [1] shows that, for any representation of $\mathbb{G}_a$ of codimension one in positive characteristic, the invariant ring $A^{\mathbb{G}_a}$ is finitely generated as a k-algebra.

In this article, we give a classification of representations of $\mathbb{G}_a$ of codimesion two in positive characteristic. Even for such a class of representations of $\mathbb{G}_a$, we do not know whether the invariant ring $A^{\mathbb{G}_a}$ is finitely generated as a k-algebra.

2. Representations of $\mathbb{G}_a$ of codimension two

From now on, we assume that the characteristic p of k is positive. A polynomial $f(T) \in k[T]$ is said to be a *p-polynomial* if $f(T)$ has the following form:

$$f(T) = \sum_{i=0}^{\ell} c_i T^{p^i}$$

for some $\ell \geq 0$ and $c_0, \ldots, c_\ell \in k$.

We define a representation $A : \mathbb{G}_a \to GL(n,k)$ $(n \geq 3)$ as follows:

$$A(t) := \left(\begin{array}{ccc|cc} & & & a_1 & \alpha_1 \\ & I_{n-2} & & \vdots & \vdots \\ & & & a_{n-2} & \alpha_{n-2} \\ \hline 0 & \cdots & 0 & 1 & 0 \\ 0 & \cdots & 0 & 0 & 1 \end{array} \right),$$

where $a_1(T), \ldots, a_{n-2}(T), \alpha_1(T), \ldots, \alpha_{n-2}(T)$ are p-polynomials, and the matrix I_{n-2} means the $(n-2) \times (n-2)$ identity matrix.

We give a criterion for the representation A to be of codimension two.

Lemma 2.1. *Let $A : \mathbb{G}_a \to GL(n,k)$ be as above. Then the following conditions are equivalent.*

(1) A is of codimension two.
(2) The column vectors ${}^t(a_1(T), \ldots, a_{n-2}(T))$ and ${}^t(\alpha_1(T), \ldots, \alpha_{n-2}(T))$ in $k[T]^{n-2}$ are linearly independent over k.

Proof. We denote by V^A the fixed point subspace of V by A. Let $\boldsymbol{a} := {}^t(a_1(T), \ldots, a_{n-2}(T))$ and $\boldsymbol{\alpha} := {}^t(\alpha_1(T), \ldots, \alpha_{n-2}(T))$ be column vectors in $k[T]^{n-2}$. Since k is an algebraically closed field, we have

$$V^A = \{v = {}^t(v_1, \ldots, v_n) \in V \mid v_{n-1}\boldsymbol{a} + v_n\boldsymbol{\alpha} = \boldsymbol{0}\}.$$

Thus, the column vectors $\boldsymbol{a}$ and $\boldsymbol{\alpha}$ in $k[T]^{n-2}$ are linearly independent over k if and only if the subspace V^A is of codimension two in V. □

If $p \neq 2$, we can define a representation $B : \mathbb{G}_a \to GL(n,k)$ $(n \geq 3)$ as follows:

$$B(t) := \left(\begin{array}{ccc|cc} & & & 0 & \beta_1 \\ & I_{n-2} & & \vdots & \vdots \\ & & & 0 & \beta_{n-3} \\ & & & b & \frac{1}{2}b^2 + \beta_{n-2} \\ \hline 0 & \cdots & 0 & 1 & b \\ 0 & \cdots & 0 & 0 & 1 \end{array}\right),$$

where $b(T), \beta_1(T), \ldots, \beta_{n-2}(T)$ are p-polynomials.

We give a criterion for B to be of codimension two.

Lemma 2.2. *Let $B : \mathbb{G}_a \to GL(n,k)$ be as above. Then the following conditions are equivalent.*

(1) B is of codimension two.
(2) $b(T) \neq 0$.

Proof. We denote by V^B the fixed point subspace of V by B. Let $\boldsymbol{b} := {}^t(0, \ldots, 0, b(T))$ and $\boldsymbol{\beta} := {}^t(\beta_1(T), \ldots, \beta_{n-3}(T), \frac{1}{2}b(T)^2 + \beta_{n-2}(T))$ be column vectors in $k[T]^{n-2}$. Since k is an algebraically closed field, we have

$$V^B = \{v = {}^t(v_1, \ldots, v_n) \in V \mid v_{n-1}\boldsymbol{b} + v_n\boldsymbol{\beta} = \boldsymbol{0},\ v_n b(T) = 0\}.$$

Thus, if $b(T) \neq 0$ then the subspace V^B is of codimension two in V, and if $b(T) = 0$ then the subspace V^B is of codimension zero or one in V. □

Now, we can state a classification of representations of $\mathbb{G}_a$ of codimension two.

Theorem 2.1. *Let $\rho : \mathbb{G}_a \to GL(n,k)$ be a representation of codimension two, where $n \geq 3$. Then, if $p = 2$ then ρ is equivalent to a representation of the form A, and if $p \neq 2$ then ρ is equivalent to a representation of the form A or B.*

Proof. Since $\rho : \mathbb{G}_a \to GL(n,k)$ is of codimension two, we may assume without loss of generality that $\rho(t)$ can be written as follows:

$$\rho(t) := \left(\begin{array}{ccc|cc} & & & u_1 & v_1 \\ & I_{n-2} & & \vdots & \vdots \\ & & & u_{n-2} & v_{n-2} \\ \hline 0 & \cdots & 0 & 1 & v_{n-1} \\ 0 & \cdots & 0 & 0 & 1 \end{array}\right),$$

where $u_1(T), \ldots, u_{n-2}(T), v_1(T), \ldots, v_{n-1}(T) \in k[T]$ and $u_{n-2}(T) \neq 0$. Since ρ is a homomorphism of groups, we have the following:

(1) $u_1(T), \ldots, u_{n-2}(T)$ are p-polynomials.
(2) For all $1 \le i \le n-2$, we have $v_i(s+t) = v_i(s) + v_{n-1}(s)u_i(t) + v_i(t)$ for all $s, t \in \mathbb{G}_a$.

One of the two cases where $v_{n-1}(T) = 0$ and $v_{n-1}(T) \neq 0$ can occur.

If $v_{n-1}(T) = 0$, then $v_i(T)$ $(1 \le i \le n-2)$ are p-polynomials. So, the representation ρ is equivalent to a representation of the form A, where the column vectors ${}^t(u_1(T), \ldots, u_{n-2}(T))$ and ${}^t(v_1(T), \ldots, v_{n-2}(T))$ in $k[T]^{n-2}$ are linearly independent over k.

If $v_{n-1}(T) \neq 0$, each $u_i(T)$ can be written as

$$u_i(T) = \lambda_i \cdot v_{n-1}(T) \qquad \text{for some } \lambda_i \in k$$

because $v_i(S+T) - v_i(S) - v_i(T) (= v_{n-1}(S)u_i(T))$ is symmetric in the two variables S and T over k. Let $\ell := \deg_T u_{n-2}(T) = \deg_T v_{n-1}(T)$, where $\deg_T f(T)$ denotes the degree of $f(T)$ in T. Note that $p \ge 3$. In fact, if $p = 2$ then taking the monomials of degree 2ℓ in S, T of the equality $v_{n-2}(S + T) - v_{n-2}(S) - v_{n-2}(T) = v_{n-1}(S)u_{n-2}(T)$, we have a contradiction. Let

$$\lambda_i^* := \begin{cases} \lambda_i \text{ if } & \lambda_i \neq 0 \\ 1 \text{ if } & \lambda_i = 0 \end{cases} \qquad (1 \le i \le n-2)$$

and let

$$P := \begin{pmatrix} \lambda_1^* & & & & \\ & \ddots & & & \\ & & \lambda_{n-2}^* & & \\ & & & 1 & \\ & & & & 1 \end{pmatrix}$$

be the diagonal matrix. Let $r_i := \lambda_i/\lambda_i^*$ $(1 \le i \le n-2)$ and let

$$Q := \left(\begin{array}{ccc|cc} & & & r_1 & 0 \\ & I_{n-2} & & \vdots & \vdots \\ & & & r_{n-2} & 0 \\ \hline 0 & \cdots & 0 & 1 & 0 \\ 0 & \cdots & 0 & 0 & 1 \end{array}\right).$$

Let

$$s_i := \begin{cases} 1 \text{ if } & \lambda_i \neq 0 \\ 0 \text{ if } & \lambda_i = 0 \end{cases} \qquad (1 \le i \le n-3)$$

and let

$$R := \left(\begin{array}{ccc|ccc} & & & s_1 & 0 & 0 \\ & I_{n-3} & & \vdots & \vdots & \vdots \\ & & & s_{n-3} & 0 & 0 \\ \hline 0 & \cdots & 0 & 1 & 0 & 0 \\ 0 & \cdots & 0 & 0 & 1 & 0 \\ 0 & \cdots & 0 & 0 & 0 & 1 \end{array}\right).$$

Then

$$R^{-1}Q^{-1}P^{-1}\rho(t)PQR = \left(\begin{array}{ccc|ccc} & & & 0 & 0 & w_1 \\ & I_{n-3} & & \vdots & \vdots & \vdots \\ & & & 0 & 0 & w_{n-3} \\ \hline 0 & \cdots & 0 & 1 & v_{n-1} & w_{n-2} \\ 0 & \cdots & 0 & 0 & 1 & v_{n-1} \\ 0 & \cdots & 0 & 0 & 0 & 1 \end{array}\right),$$

where

$$w_i(T) := \begin{cases} \dfrac{v_i(T)}{\lambda_i^*} - r_i \cdot v_{n-1}(T) - s_i\Big(\dfrac{v_{n-2}(T)}{\lambda_{n-2}^*} - r_{n-2} \cdot v_{n-1}(T)\Big) & \\ & \text{if } \ 1 \le i \le n-3 \\ \dfrac{v_{n-2}(T)}{\lambda_{n-2}^*} - r_{n-2} \cdot v_{n-1}(T) & \text{if } \ i = n-2. \end{cases}$$

Each $w_i(T)$ $(1 \le i \le n-3)$ is a p-polynomial because $R^{-1}Q^{-1}P^{-1}\rho(t)PQR$ is a homomorphism of groups. Now, from this representation, we have a subrepresentation

$$\begin{pmatrix} 1 & v_{n-1} & w_{n-2} \\ 0 & 1 & v_{n-1} \\ 0 & 0 & 1 \end{pmatrix}.$$

It follows that $v_{n-1}(T)$ is a p-polynomial and $w_{n-2}(S)+v_{n-1}(S)v_{n-1}(T)+w_{n-2}(T)=w_{n-2}(S+T)$, where S, T are variables over k. Thus, we have

$$w_{n-2}(S)-\frac{v_{n-1}(S)^2}{2}+w_{n-2}(T)-\frac{v_{n-1}(T)^2}{2}=w_{n-2}(S+T)-\frac{v_{n-1}(S+T)^2}{2}.$$

This equality implies that $w_{n-2}(T) - v_{n-1}(T)^2/2$ is a p-polynomial. Hence, the representation ρ is equivalent to the representation $R^{-1}Q^{-1}P^{-1}\rho(t)PQR$ of the form B. $\square$

Acknowledgments

The author would like to express his gratitude to the referees for careful reading. The author is supported in part by Grant-in-Aid for Scientific Research (Grant-in-Aid for Young Scientists (B) 22740009) from JSPS.

References

1. A. Fauntleroy, On Weitzenböck's theorem in positive characteristic, Proc. Amer. Math. Soc. **64** (1977), No. 2, 209–213.
2. C. S. Seshadri, On a theorem of Weitzenböck in invariant theory, J. Math. Kyoto Univ. **1** (1961/1962), 403–409.

The projective characterization of genus two plane curves which have one place at infinity

Keita Tono

Department of Mathematics, Graduate School of Science and Engineering, Saitama University,
Saitama-City, Saitama 338-8570, Japan
E-mail: ktono@rimath.saitama-u.ac.jp

In this paper we consider smooth affine plane curves of genus two having one place at infinity. We identify them with projective plane curves of genus two having only one cusp as their singular points and meeting with the line at infinity only at the cusp. We characterize such curves by the self-intersection number of the strict transform of them via the minimal embedded resolution of their cusp.

Keywords: Plane curve, one place at infinity, cusp.

1. Introduction

Let C be a plane algebraic curve on $\mathbf{P}^2 = \mathbf{P}^2(\mathbf{C})$. A singular point of C is said to be a *cusp* if it is a locally irreducible singular point. We say that C is *cuspidal* (resp. *unicuspidal*) if C has only cusps (resp. one cusp) as its singular points. Suppose that C is unicuspidal. Let $\sigma : V \to \mathbf{P}^2$ denote the minimal embedded resolution of the cusp of C. We denote by C' the strict transform of C via σ. For example, $(C')^2 = d$ if C is the rational unicuspidal plane curve defined by the equation: $y^d = x^{d-1}$, $d \geq 3$. This example shows that $(C')^2$ is not bounded from above if C is a rational unicuspidal plane curve. But it is bounded if C' is of genus $g \geq 1$.

Theorem 1 ([H, Theorem 3.5]). *If Γ is a smooth curve of genus $g \geq 1$ on a smooth rational projective surface S ($S \neq \mathbf{P}^2$ if $g = 1$), then $\Gamma^2 \leq 4g + 4$.*

Let n denote the largest integer such that the image of C' under the first n blow-downs of σ is smooth. Then we have $n \geq 2$, which shows the next corollary to the above theorem.

Corollary 2. *If C is a unicuspidal plane curve of genus $g \geq 1$, then $(C')^2 \leq 4g + 2$.*

In view of [AM, Su], we say that a unicuspidal plane curve C is of *Abhyankar-Moh-Suzuki type* (*AMS type*, for short) if there exists a line L such that $C \cap L = \{\text{the cusp}\}$. It was shown in [T] that an elliptic unicuspidal plane curve C is of AMS type if and only if $(C')^2 = 6$. One purpose of this paper is to give a similar characterization of unicuspidal plane curves C of AMS type for the case in which C is of genus two.

Theorem 3. *Let C be a unicuspidal plane curve of genus two. Suppose that $\deg C \geq 5$. Then $(C')^2 = 10$ if and only if C is of AMS type.*

Remark 4. The smallest degree of unicuspidal plane curves C of genus two is four. Such quartic curves C are not of AMS type and satisfy $(C')^2 = 10$. See Lemma 15.

It was shown in [T] that if an elliptic unicuspidal plane curve C is not of AMS type, then $(C')^2 \leq 3$. We prove a similar result for the case in which C is of genus two.

Theorem 5. *Let C be a unicuspidal plane curve of genus two. Suppose that $\deg C \geq 5$. If C is not of AMS type, then $(C')^2 \leq 7$. There exist such curves C with $(C')^2 = 7$.*

2. Preliminaries

Let C be a unicuspidal plane curve of genus two and P the cusp of C. We denote the *multiplicity sequence* of the cusp by $\overline{m}_P(C)$, or simply $\overline{m}_P$. We use the abbreviation m_k for a subsequence of $\overline{m}_P$ consisting of k consecutive m's. For example, (2_k) means an A_{2k} singularity. Let $\sigma : V \to \mathbf{P}^2$ denote the minimal embedded resolution of P. That is, σ is the composite of the shortest sequence of blow-ups such that the strict transform C' of C intersects $\sigma^{-1}(P)$ transversally. The dual graph of $D := \sigma^{-1}(C)$ has the following shape, where $h \geq 1$ and all A_i, B_i are not empty.

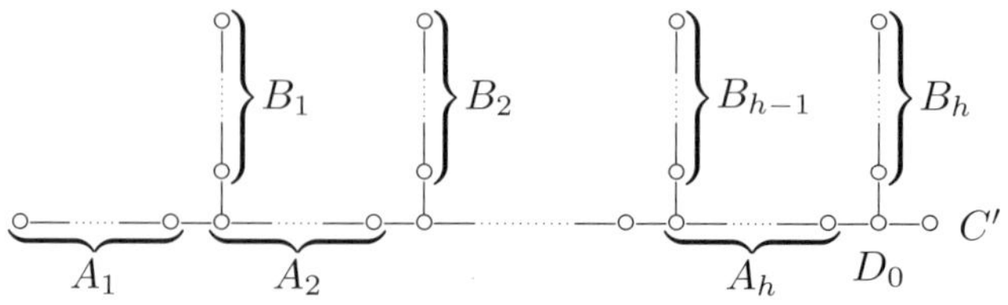

Here D_0 is the exceptional curve of the last blow-up and A_1 contains that of the first one. The morphism σ contracts $A_h + D_0 + B_h$ to a (-1)-curve E, $A_{h-1} + E + B_{h-1}$ to a (-1)-curve and so on. Every irreducible component E of A_i and B_i is a smooth rational curve with $E^2 < -1$. Each A_i contains an irreducible component E such that $E^2 < -2$. See [BK, MS]. We give weights to $A_1, \ldots, A_h, B_1, \ldots, B_h$ in the usual way. Put $D' = D - C'$. There are two irreducible components of $D' - D_0$ meeting with D_0. One of them must be a (-2)-curve and the other is not. Let D_1 denote the (-2)-curve and D_2 the remaining one.

Suppose that $n := (C')^2 \geq 7$. Perform $(n-2)$-times of blow-ups $\tau_0 : W_0 \to V$ over $C' \cap D_0$ in the following way, where $*$ (resp. $\bullet$) denotes a (-1)-curve (resp. (-2)-curve) and E_i is the exceptional curve of the i-th blow-up. The integer near $\circ$ denotes the self-intersection number of it. We use the same symbols to denote the strict transforms via blow-ups of τ_0 of C', E_i, etc.

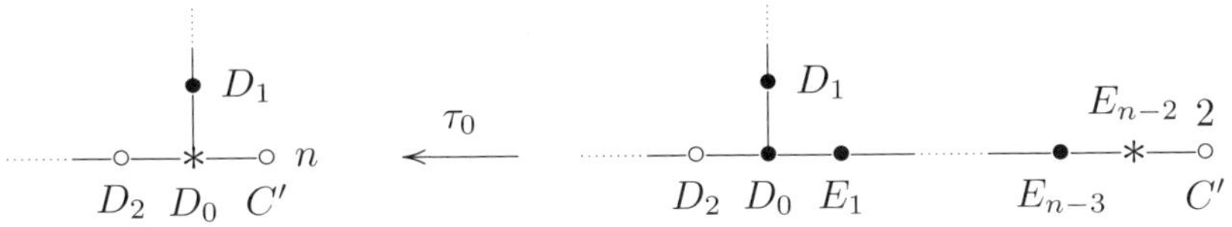

On W_0, there exists an exact sequence:

$$0 \longrightarrow H^0(\mathcal{O}_{W_0}) \longrightarrow H^0(\mathcal{O}_{W_0}(C')) \longrightarrow H^0(\mathcal{O}_{C'}(C')) \longrightarrow H^1(\mathcal{O}_{W_0}) = 0.$$

We have $h^0(\mathcal{O}_{C'}(C')) \geq 1$. Let $\Lambda \subset |C'|$ be a pencil such that $C' \in \Lambda$. Let $\tau_1 : W \to W_0$ be the minimal resolution of the base points of Λ. Set $\tau = \tau_0 \circ \tau_1$. The morphism τ_1 consists of 2-times of blow-ups. Let Q_i denote the center of the i-th blow-up of τ_1. By arranging the order of blow-ups, we may assume that the position of Q_1, Q_2 is one of those in the following figure.

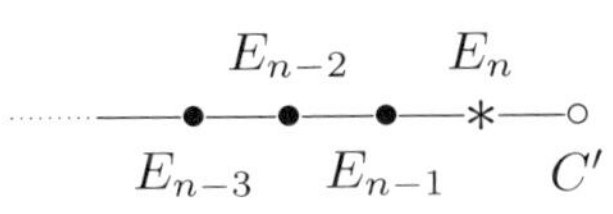

(I) $Q_1 \in E_{n-2}$, $Q_2 \in E_{n-1}$

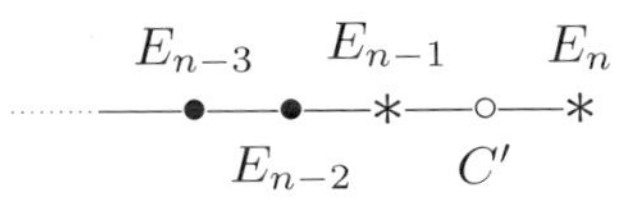

(II) $Q_1 \in E_{n-2}$, $Q_2 \notin E_{n-1}$

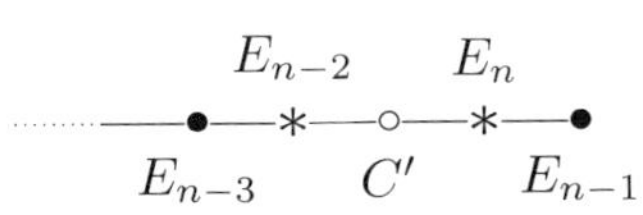

(III) $Q_1 \notin E_{n-2}$, $Q_2 \in E_{n-1}$

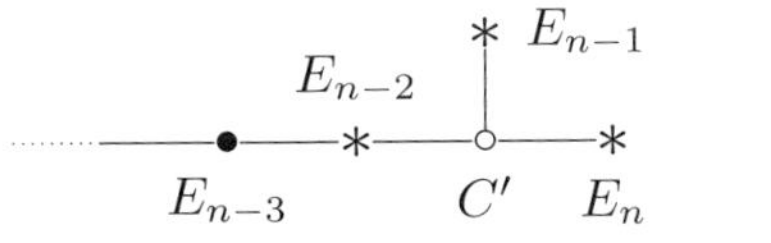

(IV) $Q_1 \notin E_{n-2}$, $Q_2 \notin E_{n-2} + E_{n-1}$

The morphism $f := \Phi_\Lambda \circ \tau_1 : W \to \mathbf{P}^1$ is a fibration, whose nonsingular fibers are smooth curves of genus two. Let S denote the sum of all (-1)-curves in $E_1, \ldots, E_n$. Then each $E_i \subset S$ is a section of f. That is, if $E_i \subset S$, then $E_i F = 1$ for a fiber F of f. The divisor $D' + E_1 + \cdots + E_n$ contains no other sections. All (-2)-curves in $E_1, \ldots, E_n$ are contained in a fiber $\overline{F}_0$ of f except for the case (III). We define two sub-cases of (III): (III$_\mathrm{a}$) $E_{n-1} \subset \overline{F}_0$, (III$_\mathrm{b}$) $E_{n-1} \not\subset \overline{F}_0$. We say that C is of type (I) (resp. (II), (III$_\mathrm{a}$), (III$_\mathrm{b}$), (IV)) if we are in the case (I) (resp. (II), (III$_\mathrm{a}$), (III$_\mathrm{b}$), (IV)).

Put $F_0 = D' + E_1 + \cdots + E_n - S$ (resp. $F_0 = D' + E_1 + \cdots + E_n - S - E_{n-1}$) if C is not of type (III) (resp. C is of type (III)). We have $F_0 \subset \overline{F}_0$. Let $\varphi : W \to X$ be successive contractions of (-1)-curves in the singular fibers of f such that the fibration $p := f \circ \varphi^{-1} : X \to \mathbf{P}^1$ is relatively minimal. Let $\rho = \rho(X)$ denote the Picard number of X. Put $r_0 = r(F_0)$, $\overline{r}_0 = r(\overline{F}_0)$ and $R_0 = r(\varphi(\overline{F}_0))$, where $r(F)$ denotes the number of irreducible components of a divisor F. We sometimes use the same symbols to denote the images under φ of C', E_i, etc. We show several lemmas to determine the structure of the fibration p.

Lemma 6. *The following assertions hold.*

- *(i) We have* $2 + \sum_F (r(F) - 1) \leq \rho(X) \leq 14$, *where F runs over all fibers of p.*
- *(ii)* $\varphi(E_i)$ *is a (-1)-curve for every* $E_i \subset S$.
- *(iii) The morphism φ does not contract D_0, D_1, D_2 and every E_i. Furthermore $\varphi(D_0)$, $\varphi(D_1)$ and every $\varphi(E_i)$ are (-2)-curves, where* $E_i \not\subset S$.

Proof. (i) follows from Theorem 3 in [Sh] and Theorem 2.8 in [SS]. (ii) We have $\varphi(E_i)^2 \geq -1$. Suppose $\varphi(E_i)^2 \geq 0$. There exists a pencil $\Lambda' \subset |\varphi(E_i)|$. Let $\psi : X' \to X$ be the resolution of the base points of Λ'. Then $\Phi_{\Lambda'} \circ \psi : X' \to \mathbf{P}^1$ is a $\mathbf{P}^1$-fibration. For a general nonsingular fiber F of $p \circ \psi$, $\Phi_{\Lambda'} \circ \psi|_F : F \to \mathbf{P}^1$ is of degree one, which is absurd. (iii) follows from (ii). See Lemma 6 in [T] and its proof. □

Lemma 7. *The following assertions hold.*

- *(i) Let $F \neq \overline{F}_0$ be a fiber of f. If C is of type* (III$_\mathrm{b}$) *and* $E_{n-1} \subset F$, *then all irreducible components of F meet with $S + E_{n-1}$. Otherwise all of them meet with S and* $r(F) \leq r(S)$.
- *(ii) The morphism φ only contracts irreducible components of* $\overline{F}_0$.

(iii) *We have* $\overline{r}_0 - r_0 + \rho - R_0 = 2$ *(resp.* 3, 4*) if* C *is of type (I) (resp. (II), (III) or (IV)). Moreover,* $\overline{r}_0 - r_0 \leq 1$ *(resp.* $\overline{r}_0 - r_0 \leq 2$, $\overline{r}_0 - r_0 \leq 3$*) if* C *is of type (I) (resp. (II), (III) or (IV)).*

Proof. (i) If C is of type (III), then $\mathbf{P}^2 \setminus C = W \setminus (F_0 \cup E_{n-1} \cup S \cup C')$. Otherwise $\mathbf{P}^2 \setminus C = W \setminus (F_0 \cup S \cup C')$. The assertions follow from the fact that $\mathbf{P}^2 \setminus C$ does not contain complete curves.

(ii) By the assertion (i), each irreducible component E of a fiber $F \neq \overline{F}_0$ meets with $S + E_{n-1}$. By Lemma 6, φ does not contract E.

(iii) We have $\rho(W) = r(D') + n + 1$ and

$$r_0 = \begin{cases} r(D') + n - 1 & \text{if } C \text{ is of type (I)},\\ r(D') + n - 2 & \text{if } C \text{ is of type (II)},\\ r(D') + n - 3 & \text{if } C \text{ is of type (III) or (IV)}.\end{cases}$$

The number of blow-downs of φ is equal to $\rho(W) - \rho(X) = \overline{r}_0 - R_0$. The first assertion follows from these facts. The second one follows from the first one and Lemma 6. □

Lemma 8. *The following assertions hold.*

(i) *Suppose that* $\varphi(D_1)$ *meets with only* $\varphi(D_0)$ *among the irreducible components of* $\varphi(D - D_1)$. *Then* D_1 *meets with only* D_0 *among the irreducible components of* $D - D_1$ *on* W. *Furthermore,* $D_2^2 = -3$ *on* W *and* $\varphi(D_2)^2 \geq -3$.

(ii) *Suppose that an irreducible curve* $E \not\subset \varphi(E_1 + \cdots + E_n + D_0)$ *meets with* $\varphi(E_i)$ *for some* i. *Then the strict transform* E' *of* E *on* W *meets with* E_i *and is not a component of* F_0.

(iii) *Let* E *be an irreducible component of* $\varphi(\overline{F}_0 - F_0)$ *and* $C_1, \ldots, C_r$ *the connected components of* $\varphi(\overline{F}_0) - E$. *Suppose that* $r > 1$ *and that* C_1 *contains an irreducible component of* $\varphi(F_0)$. *Then* $C_2, \ldots, C_r \subset \varphi(\overline{F}_0 - F_0)$.

Proof. (i) follows from the fact that φ does not contract the irreducible components of $\overline{F}_0$ meeting with D_1 by Lemma 6.

(ii) The curve E' must meet with E_i since φ does not perform blow-ups over $\varphi(E_i)$ by Lemma 6. By the definition of F_0, we have $E' \not\subset F_0$.

(iii) We have $\varphi(F_0) \subset C_1$ since F_0 is connected. The assertion follows from this fact. □

We will determine the weighted dual graph of $\varphi(\overline{F}_0)$. By Lemma 6, it must contain the following graph, where E_m intersects the section E_{m+1},

• denotes a (-2)-curve. The multiplicity of E_m in the fiber is equal to one. We have $m = n - 1$ (resp. $m = n - 2$, $m = n - 3$) if C is of type (I) (resp. (II), (III) or (IV)). In particular, $4 \leq m \leq 9$.

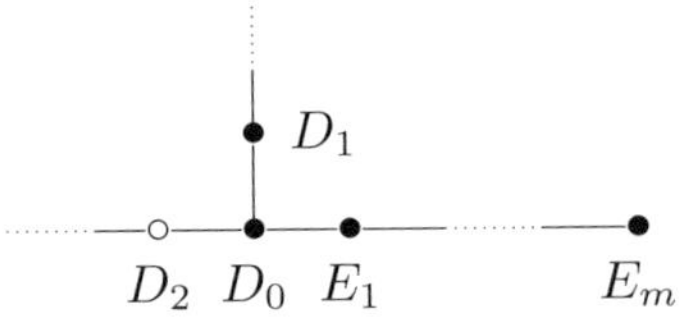

In [NU] (cf. [O]), the singular fibers in pencils of curves of genus two were classified. Among their results, we use the list of 44 numerical types of singular fibers (some of them contain sub-types). The following table gives the list of all types of the singular fibers which contain the above graph.

n	Types
10	1, 2, 5, 9, 14, 20, 22, 39, 41, 43
9	1, 2, 5, 9, 14, 22, 39, 41, 43
8	1, 2, 5, 9, 14, 19, 22, 25, 39, 41, 43
7	1, 2, 5, 9, 14, 19, 21, 25, 39, 41, 43, 44

Lemma 9. *The fiber $\varphi(\overline{F}_0)$ is not of type 1, 2, 5, 9, 14, 19, 21, 39, 44.*

Proof. We only show that $\varphi(\overline{F}_0)$ is not of type 1. We can similarly deal with the remaining cases. Suppose that $\varphi(\overline{F}_0)$ is of type 1. The weighted dual graph of $\varphi(\overline{F}_0)$ has the following shape, where we use the same symbols to denote the images under φ of E_i, D_j, etc.

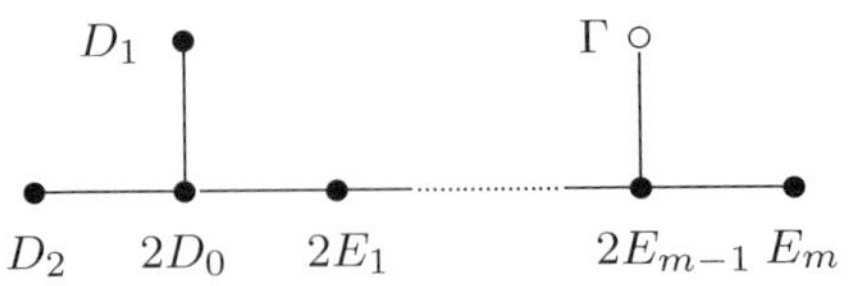

We have $R_0 = m + 4$. The curve Γ satisfies the conditions: $\Gamma^2 = -2$, $p(\Gamma) = 1 + \frac{1}{2}(K_X + \Gamma)\Gamma = 1$, where K_X is a canonical divisor of X. The position of D_1, D_2 is determined by Lemma 8 (i). By Lemma 8 (ii), we have $\Gamma \not\subset \varphi(F_0)$. Since $D_2^2 < -2$ on W, φ performs at least one blow-up over D_2. The exceptional curve of the last blow-up over D_2 is not a component of F_0. This shows that $\overline{r}_0 - r_0 \geq 2$. By Lemma 7 (iii), C is not of type (I). We have $\rho = R_0 + 1$ (resp. $\rho \leq R_0 + 2$) if C is of type (II) (resp. (III) or (IV)).

Suppose C is of type (II). We have $R_0 = n+2$. The section E_n intersects Γ. Let $\phi : X \to \mathbf{P}^2$ denote the contraction of $E_n + \cdots + E_1 + D_0 + D_1$. Then $\phi(D_2)^2 = 0$, which is a contradiction. If C is of type ($\mathrm{III_a}$), then $\varphi(\overline{F}_0)$ cannot contain $\varphi(E_{n-1})$ as its component. Suppose that C is of type ($\mathrm{III_b}$) or (IV). We have $R_0 = n+1$ and $\rho \leq n+3$. We infer that each section E_i $(i > m+1)$ intersects Γ. Let $\phi : X \to X'$ denote the contraction of $E_n + \cdots + E_1 + D_0 + D_1$. We have $\rho = n+3$, $X' = \mathbf{P}^2$ and $\phi(D_2)^2 = 0$, which is absurd. □

3. Proof of Theorem 3 and Theorem 5

In this section, we prove the "only if" part of Theorem 3. We also prove the first assertion of Theorem 5 by assuming the "if" part of Theorem 3. The remaining assertions of Theorem 3 and Theorem 5 will be proved in the next section. Let C be a unicuspidal plane curve of genus two. Suppose that $n := (C')^2 \geq 7$. We show that $n = 10$ or $n = 7$ and that C is of AMS type or $\deg C = 4$ if $n = 10$. We use the same notation as in the previous section. For simplicity, we sometimes use the same symbols to denote the images under φ of E_i, D_j, etc. By Lemma 9, $\varphi(\overline{F}_0)$ is of type 20, 22, 25, 41 or 43.

Lemma 10. *If $\varphi(\overline{F}_0)$ is of type 20 or 22, then $n = 10$. The curve C is of AMS type.*

Proof. We first show the assertion for the case in which $\varphi(\overline{F}_0)$ is of type 20. The weighted dual graph of $\varphi(\overline{F}_0)$ has the following shape. In the figure, • denotes a (-2)-curve. The integer near ∘ denotes the self-intersection number of it.

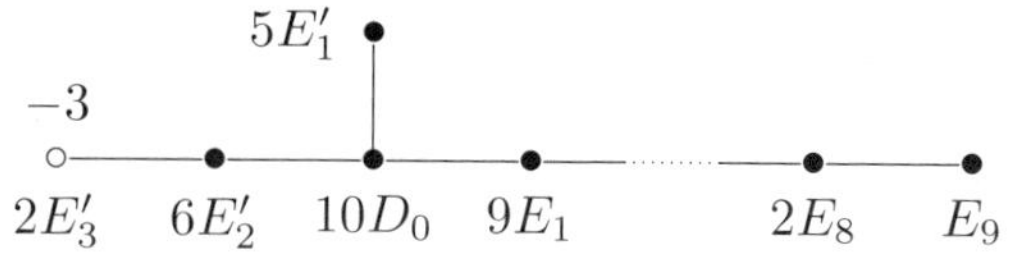

We have $R_0 = 13$. The divisor $D_1 + D_2$ coincides with $E'_1 + E'_2$. Since $m = 9$, only the case (I) with $n = 10$ occurs. Since $D_2^2 < -2$ on W, φ performs at least one blow-up over D_2. The exceptional curve E of the last blow-up over D_2 is not a component of F_0. This shows that $\overline{r}_0 - r_0 \geq 1$. We have $\rho = R_0 + 1 = 14$ by Lemma 7 (iii). It follows that $\overline{F}_0 = F_0 + E = D' + E_1 + \cdots + E_{n-1} + E$. The plane curve $\sigma(\tau(E))$ meets with C only at the cusp.

Let $\phi' : X' \to X$ denote the blow-up at a point on $E_1' \setminus D_0$ and E' its exceptional curve. We use the same symbols to denote the strict transform via ϕ' of E_i, E_j', etc. Let $\phi : X' \to \mathbf{P}^2$ denote the contraction of $E_{10} + \cdots + E_1 + D_0 + E_2' + E_1' + E_3'$. The curve $\phi(C')$ is a unicuspidal curve of degree 5. The multiplicity sequence of the cusp coincides with $(3, 2)$. The curve $\phi(E')$ is a line meeting with $\phi(C')$ only at the cusp. We have the isomorphism $\sigma \circ \tau \circ \varphi^{-1} \circ \phi' \circ \phi^{-1}|_{\mathbf{P}^2 \setminus \phi(E')} : \mathbf{P}^2 \setminus \phi(E') \to \mathbf{P}^2 \setminus \sigma(\tau(E))$. By [D, Proposition 4.1.3], we obtain $\mathbf{Z}/(\deg \sigma(\tau(E)))\mathbf{Z} \cong H_1(\mathbf{P}^2 \setminus \sigma(\tau(E)), \mathbf{Z}) \cong H_1(\mathbf{P}^2 \setminus \phi(E'), \mathbf{Z}) \cong \{0\}$. This shows that $\sigma(\tau(E))$ is a line. Thus C is of AMS type.

We next pass to the case in which $\varphi(\overline{F}_0)$ is of type 22. The weighted dual graph of $\varphi(\overline{F}_0)$ has the following shape.

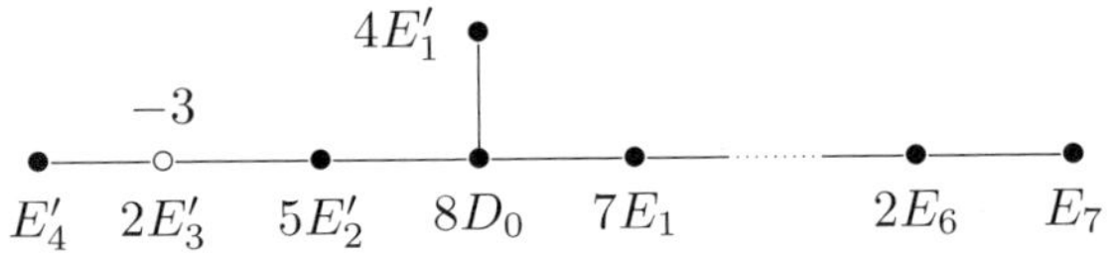

We see $R_0 = 12$. The divisor $D_1 + D_2$ coincides with $E_1' + E_2'$. We have $n = 10$ (resp. $n = 9$, $n = 8$) if C is of type (III) or (IV) (resp. (II), (I)). Since $D_2^2 < -2$ on W, φ performs at least one blow-up over D_2. The exceptional curve E of the last blow-up over D_2 is not a component of F_0. This proves $\overline{r}_0 - r_0 \geq 1$. The plane curve $\sigma(\tau(E))$ meets with C only at the cusp. Suppose that C is of type (I). We have $\rho = R_0 + 1 = 13$. Let $\phi' : X' \to X$ denote the blow-up at a point on $E_1' \setminus D_0$ and E' its exceptional curve. Let $\phi : X' \to \mathbf{P}^2$ denote the contraction of $E_8 + \cdots + E_1 + D_0 + E_2' + E_1' + E_3' + E_4'$. Then $\phi(E')^2 = 2$, which is a contradiction.

Suppose that C is not of type (I). Each section E_i ($i > 8$) meets with E_4'. If C is of type ($\mathrm{III_a}$), then $E_4' = E_{n-1} \not\subset \varphi(F_0)$. Since φ does not perform blow-ups over E_4', it intersects E_3' on W. It follows that $E_3' \not\subset \varphi(F_0)$. Thus $\overline{r}_0 - r_0 \geq 3$. Otherwise we have $E_4' \not\subset \varphi(F_0)$ and $\overline{r}_0 - r_0 \geq 2$. By Lemma 7 (iii), we see $\rho = R_0 + 1 = 13$ (resp. $\rho \leq R_0 + 2 = 14$) if C is of type (II) or ($\mathrm{III_a}$) (resp. ($\mathrm{III_b}$) or (IV)). The divisor $E_n + E_4' + E_8 + \cdots + E_1 + D_0 + E_2' + E_3'$ can be contracted to a point by blow-downs. Hence $\rho = 14$, $\overline{r}_0 - r_0 = 2$ and C is of type ($\mathrm{III_b}$) or (IV).

Let $\phi' : X' \to X$ denote the blow-up at a point on $E_1' \setminus D_0$ and E' its exceptional curve. Let $\phi : X' \to \mathbf{P}^2$ denote the contraction of $E_{10} + \cdots + E_1 + D_0 + E_2' + E_1' + E_3'$. The curve $\phi(C')$ is a unicuspidal curve of degree 5. The multiplicity sequence of the cusp coincides with $(3, 2)$. The curve $\phi(E')$ is a line meeting with $\phi(C')$ only at the cusp. On the other hand, since $E'_4 \not\subset \varphi(F_0)$ and $\overline{r}_0 - r_0 = 2$, we infer that $\overline{F}_0 = F_0 + E'_4 + E$, where E is the exceptional curve of the last blow-up of φ over D_2. We have the isomorphism $\sigma \circ \tau \circ \varphi^{-1} \circ \phi' \circ \phi^{-1}|_{\mathbf{P}^2 \setminus \phi(E')} : \mathbf{P}^2 \setminus \phi(E') \to \mathbf{P}^2 \setminus \sigma(\tau(E))$. By the same reason as in the proof of the previous case, we conclude that $\sigma(\tau(E))$ is a line meeting with C only at the cusp. □

Lemma 11. *If $\varphi(\overline{F}_0)$ is of type 25, then $n = 7$.*

Proof. The weighted dual graph of $\varphi(\overline{F}_0)$ has the following shape, where $b \geq 1$.

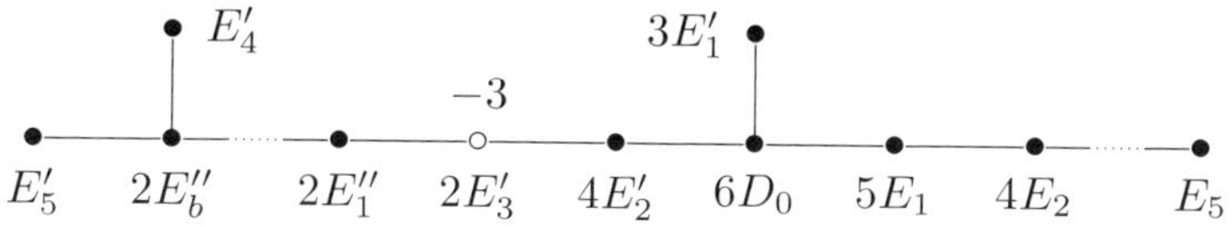

We have $m = 5$ and $R_0 = b + 11$. We see that C is not of type (I) and that $n = 7$ (resp. $n = 8$) if C is of type (II) (resp. (III) or (IV)). By Lemma 6 (i), $b = 1$ or 2. The divisor $D_1 + D_2$ coincides with $E'_1 + E'_2$. We have $\overline{r}_0 - r_0 \geq 1$ since φ performs a blow-up over D_2. Let S' denote the sum of all sections E_i $(i > m+1)$. The divisor S' intersects $E'_4 + E'_5$. Let $\phi' : X' \to X$ denote the blow-up at a point on $E'_1 \setminus D_0$ and E its exceptional curve. We use the same symbols to denote the strict transforms via ϕ' of E_i, E'_j, etc.

Suppose that S' intersects both E'_4 and E'_5. We see that $n = 8$ and that C is of type (IV). We have $\overline{r}_0 - r_0 \geq 3$ since $E'_4 \not\subset \varphi(F_0)$ and $E'_5 \not\subset \varphi(F_0)$. By Lemma 7 (iii), we get $\overline{r}_0 - r_0 = 3$ and $\rho = R_0 + 1 = b + 12$. Let $\phi : X' \to \mathbf{P}^2$ denote the contraction of $E_8 + \cdots + E_1 + D_0 + E'_2 + E'_1 + E'_3 + E''_1 + \cdots + E''_b$. Then $\phi(E'_4)^2 = 0$, which is impossible.

Suppose that S' intersects either E'_4 or E'_5. We may assume that S' intersects E'_5. Suppose C is of type (III$_\mathrm{a}$). We have $E'_5 = E_{n-1} \not\subset \varphi(F_0)$. The curve E'_5 intersects E''_b on W because φ does not perform blow-ups over E'_5 by Lemma 6 (iii). Thus $E''_b \not\subset \varphi(F_0)$. By Lemma 8 (iii), we infer $E'_4 \not\subset \varphi(F_0)$. This means that $\overline{r}_0 - r_0 \geq 4$, which contradicts Lemma 7 (iii).

Suppose C is of type (III$_\mathrm{b}$) or (IV). We have $n = 8$. We infer that $E'_5 \not\subset \varphi(F_0)$ since we assumed that S' intersects E'_5. Thus $\overline{r}_0 - r_0 \geq 2$. By Lemma 7 (iii), we obtain $\rho(X) \leq b + 13$. Let $\phi : X' \to X''$ denote the contraction of $E_8 + \cdots + E_1 + D_0 + E'_2 + E'_1 + E'_3 + E''_1 + \cdots + E''_b + E'_4$. The existence of $(b+13)$-times of blow-downs shows that $\rho(X') = b + 14$ and $X'' = \mathbf{P}^2$. We have $\phi(E'_5)^2 = 2$, which is impossible. Thus C must be of type (II) and $n = 7$. □

Lemma 12. *Suppose that $\varphi(\overline{F}_0)$ is of type 41. Then $n = 10$ or 7. If $n = 10$, then* $\deg C = 4$.

Proof. The weighted dual graph of $\varphi(\overline{F}_0)$ has the following shape, where $b \geq 0$.

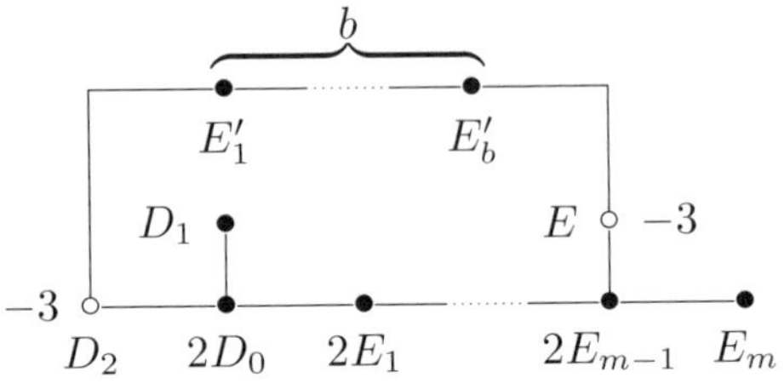

We have $R_0 = b + m + 4$. By Lemma 6 (i), $R_0 \leq 13$. By Lemma 8 (ii), $E \not\subset \varphi(F_0)$. Thus $\overline{r}_0 - r_0 \geq 1$. We consider each type of C separately.

(I) By Lemma 7 (iii), $\overline{r}_0 - r_0 = 1$. We have $\rho = R_0 + 1 = b + n + 4$. If φ is not the identity, then the exceptional curve of the last blow-up of φ is not a component of F_0, which contradicts $\overline{r}_0 - r_0 = 1$. Thus φ is the identity. The morphism $\sigma \circ \tau$ contracts $E_n + \cdots + E_1 + D_0 + D_1 + D_2 + E'_1 + \cdots + E'_b$. We have $\overline{m}_P(C) = (2_{b+1})$ and $(\deg C)^2 = n + 2 + 4(b + 1)$. If $n = 10$, then $b = 0$ since $13 \geq R_0 = b + 13$. Hence $\deg C = 4$. If $n = 9$, then $b = 0, 1$ since $13 \geq R_0 = b + 12$. We have $(\deg C)^2 = 15$ (resp. 19) if $b = 0$ (resp. $b = 1$), which is impossible. Similarly, we have $n \neq 8$.

(II) We have $R_0 = b + n + 2$. The section E_n intersects $E + E'_1 + \cdots + E'_b$. Suppose E_n intersects E. By Lemma 7 (iii), $\rho \leq R_0 + 2 = b + n + 4$. Let $\phi : X \to X'$ denote the contraction of $E_n + \cdots + E_1 + D_0 + D_1 + D_2 + E'_1 + \cdots + E'_b$. The existence of $(b + n + 3)$-times of blow-downs shows that $\rho = b + n + 4$ and $X' = \mathbf{P}^2$. We infer that φ is the identity and that $\phi = \sigma \circ \tau$. We have $\overline{m}_P(C) = (2_{b+1})$ and $(\deg C)^2 = n + 2 + 4(b + 1)$. Similar arguments as in the previous case show that C satisfies the assertion.

Suppose that E_n intersects E'_i for some $i = 1, \ldots, b$. By Lemma 8 (iii), $E'_i, \ldots, E'_b$ are not components of $\varphi(F_0)$. By Lemma 7 (iii), we have $i = b$, $\overline{r}_0 - r_0 = 2$ and $\rho = R_0 + 1 = b + n + 3$. Let $\phi : X \to \mathbf{P}^2$ denote the contraction of $E_n + \cdots + E_1 + D_0 + D_1 + D_2 + E'_1 + \cdots + E'_{b-1}$. Then $\phi(E'_b)^2 = 0$, which is absurd.

(III_a) We have $R_0 = b + n + 1$. The curve E_{n-1} coincides with E'_i for some i. By Lemma 6 (iii), φ does not perform blow-ups over E_{n-1}. Thus the irreducible components of $\varphi(\overline{F}_0)$ meeting with E'_i is not contained in $\varphi(F_0)$. By Lemma 7 (iii), we have $i = b$, $\overline{r}_0 - r_0 = 3$ and $\rho = R_0 + 1 = b + n + 2$. Since $D_2 \subset F_0$, we get $b \geq 2$. Let $\phi : X \to \mathbf{P}^2$ denote the contraction of

$E_n+\cdots+E_1+D_0+D_1+D_2+E'_1+\cdots+E'_{b-2}$. Then $\phi(E'_{b-1})^2=0$, which is a contradiction.

($\mathrm{III_b}$) We have $R_0=b+n+1$. The section E_n intersects $E+E'_1+\cdots+E'_b$. Suppose E_n intersects E. By Lemma 7 (iii), $\rho\le R_0+3=b+n+4$. Let $\phi: X\to X'$ denote the contraction of $E_n+\cdots+E_1+D_0+D_1+D_2+E'_1+\cdots+E'_b$. The existence of $(b+n+3)$-times of blow-downs shows that $\rho=b+n+4$ and $X'=\mathbf{P}^2$. We infer that φ is the identity and that $\phi=\sigma\circ\tau$. We have $\overline{m}_P(C)=(2_{b+1})$ and $(\deg C)^2=n+2+4(b+1)$. Similar arguments as in the case (I) show that C satisfies the assertion.

Suppose E_n intersects E'_i for some $i=1,\ldots,b$. We have $E'_i,\ldots,E'_b\not\subset\varphi(F_0)$. By Lemma 7 (iii) and Lemma 8 (iii), we have $i\ge b-1$ and $\rho\le R_0+2+i-b=i+n+3$. Let $\phi: X\to X'$ denote the contraction of $E_n+\cdots+E_1+D_0+D_1+D_2+E'_1+\cdots+E'_{i-1}$. The existence of $(i+n+2)$-times of blow-downs shows that $\rho=i+n+3$ and $X'=\mathbf{P}^2$. If $i=b-1$, then $\phi(E'_b)^2=-2$, which is impossible. Thus $i=b$. We infer that φ is the identity and that $\phi=\sigma\circ\tau$. We have $\overline{m}_P(C)=(2_b)$ and $(\deg C)^2=n+2+4b$. Similar arguments as in the case (I) show that C satisfies the assertion.

(IV) We have $R_0=b+n+1$. The sections E_{n-1}, E_n intersect $E+E'_1+\cdots+E'_b$. If both E_{n-1} and E_n intersects E, then the same arguments as in the previous case show that C satisfies the assertion. Suppose that $E_{n-1}+E_n$ does not intersect $E'_1,\ldots,E'_{i-1}$ but intersects E'_i for some i. We have $E'_i,\ldots,E'_b\not\subset\varphi(F_0)$. By Lemma 7 (iii) and Lemma 8 (iii), we have $i\ge b-1$ and $\rho\le R_0+2+i-b=i+n+3$. Thus similar arguments as in the previous case prove that C satisfies the assertion. □

Lemma 13. *If $\varphi(\overline{F}_0)$ is of type 43, then $n=10$ and $\deg C=4$.*

Proof. The weighted dual graph of $\varphi(\overline{F}_0)$ has the following shape.

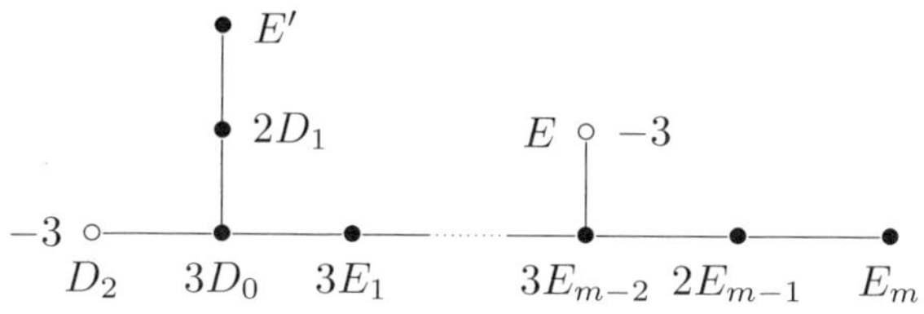

We have $R_0=m+5$. By Lemma 8 (ii), we see $E\not\subset\varphi(F_0)$. This shows that $\overline{r}_0-r_0\ge 1$. Suppose C is of type (I). By Lemma 7 (iii), $\overline{r}_0-r_0=1$. We infer that $E'\subset\varphi(F_0)$ and that φ is the identity. But σ cannot contract D' to a point in this case. Hence C is not of type (I).

Let S' denote the sum of all sections E_i $(i > m+1)$. The divisor S' intersects $E+E'$. Suppose that S' does not intersect E'. The curve C is not of type (III$_\mathrm{a}$) since $\varphi(\overline{F}_0)$ cannot contain E_{n-1} as its component. If φ is the identity, then σ cannot contract E'. Thus $E' \not\subset \varphi(F_0)$. This shows $\overline{r}_0 - r_0 \geq 2$. Otherwise the exceptional curve of the last blow-up of φ is not a component of F_0 and $\overline{r}_0 - r_0 \geq 2$. By Lemma 7 (iii), $\rho = R_0 + 1 = n+4$ (resp. $\rho \leq R_0 + 2 = n+4$) if C is of type (II) (resp. (III$_\mathrm{b}$) or (IV)) in both cases. Let $\phi : X \to \mathbf{P}^2$ denote the contraction of $E_n + \cdots + E_1 + D_0 + D_1 + D_2$. Then $\phi(E')^2 = 0$, which is a contradiction. Thus S' intersects E'.

Suppose C is of type (III$_\mathrm{a}$). We have $E' = E_{n-1}$. By Lemma 6 (iii), φ does not perform blow-ups over E_{n-1}. Thus E_{n-1} meets with D_1 on W, which is impossible. Hence C is of type (II), (III$_\mathrm{b}$) or (IV). We have $\overline{r}_0 - r_0 \geq 2$ since $E' \not\subset \varphi(F_0)$. By Lemma 7 (iii), $\rho = R_0 + 1$ (resp. $\rho \leq R_0 + 2$) if C is of type (II) (resp. (III$_\mathrm{b}$) or (IV)). We have $R_0 = n+3$ (resp. $R_0 = n+2$) if C is of type (II) (resp. (III$_\mathrm{b}$) or (IV)).

Let $\phi : X \to X'$ denote the contraction of $E_n + \cdots + E_1 + D_0 + D_1 + D_2$. The existence of the blow-downs ϕ of $n+3$ curves shows that $\rho = R_0 + 2$ if C is of type (III$_\mathrm{b}$) or (IV). Regardless of the type of C, we have $\overline{r}_0 - r_0 = 2$ and $X' = \mathbf{P}^2$. If φ is not the identity, then the exceptional curve of the last blow-up of φ is not contained in F_0. This contradicts $\overline{r}_0 - r_0 = 2$. Thus φ is the identity. We infer that $\phi = \sigma \circ \tau$. The multiplicity sequence of the cusp of $C = \phi(C')$ is equal to (2). We have $\phi(C')^2 = n+6$. Hence $n = 10$ and $\deg C = 4$. □

4. Proof of Theorem 3 and Theorem 5 — continued

In this section, we first show the "if" part of Theorem 3. Let C be a unicuspidal plane curve of genus two. Suppose that C is of AMS type. The following theorem was proved in [AO, M]. See also Lemma 16.

Theorem 14. *Let C be a unicuspidal plane curve of genus two. Suppose that C is of AMS type. Let L be the tangent line of C at the cusp. Then there exists a birational map $f : \mathbf{P}^2 \to \mathbf{P}^2$ such that $f|_{\mathbf{P}^2 \setminus L} \in \mathrm{Aut}(\mathbf{P}^2 \setminus L)$ and the strict transform of C via f is a unicuspidal quintic curve Γ_5 such that the multiplicity sequence of the cusp is equal to* $(3,2)$.

Let $f : \mathbf{P}^2 \to \mathbf{P}^2$ be a birational map given by Theorem 14. The strict transform of C via f is a unicuspidal quintic curve Γ_5. Let $\sigma : V \to \mathbf{P}^2$ denote the minimal embedded resolution of the cusp of Γ_5. Let Γ_5' denote the strict transform of Γ_5 via σ. We have $(\Gamma_5')^2 = 10$. If $f^{-1} \circ \sigma$ has base points, then let $\phi : V' \to V$ denote their minimal resolution. Otherwise we define ϕ as the identity. Then one can show that every blow-up of ϕ is done at a point $\notin \Gamma_5'$. See the proof of the "if" part of Theorem 1 in [T]. Every blow-up of $f^{-1} \circ \sigma \circ \phi : V' \to \mathbf{P}^2$ is done at a point $\in C$. It follows that $f^{-1} \circ \sigma \circ \phi : V' \to \mathbf{P}^2$ coincides with the minimal embedded resolution of the cusp of C. Thus $(C')^2 = (\Gamma_5')^2 = 10$. We completed the proof of Theorem 3.

In order to complete the proof of Theorem 5, we next construct examples of unicuspidal plane curves C of genus two such that $(C')^2 = 7$. Let Γ_4 be a unicuspidal plane curve of genus two such that $\deg \Gamma_4 = 4$. See Lemma 15. The multiplicity sequence of the cusp is equal to (2). There exists a line L passing through the cusp such that it is tangent to a smooth point Q of Γ_4. Perform 3-times of blow-ups $X \to \mathbf{P}^2$ over Q in the following way. Let E_i denote the exceptional curve of the i-th blow-up. We use the same symbols to denote the strict transforms of them and Γ_4, L. For $i = 2, 3$, we define the center of the i-th blow-up as the point of intersection of E_{i-1} and Γ_4. Then L is a (-1)-curve and E_1, E_2 are (-2)-curves on X.

Let C denote the image of Γ_4 under the contraction of $L + E_2 + E_1$. Then C is a unicuspidal quintic curve such that the multiplicity sequence of the cusp is equal to (2_4). We have $(C')^2 = 7$. If we replace Γ_4 in the above example with Γ_5 given by Lemma 16, then C is a unicuspidal plane curve of degree seven such that the multiplicity sequence of the cusp is equal to $(3_4, 2)$. We have $(C')^2 = 7$. We completed the proof of Theorem 5.

The following lemmas give defining equations of Γ_4 and Γ_5. They follow from [SST, Proposition 13].

Lemma 15. *Let Γ_4 be the plane curve defined by the following equation, where (x, y, z) are homogeneous coordinates of $\mathbf{P}^2$.*

$$y^2z^2 + 2G(x,y)z + \frac{G(x,y)^2 - \Delta(x,y)}{y^2} = 0.$$

Here $G(x,y) = a_0x^3 + a_1x^2y + a_2xy^2 + a_3y^3$, $\Delta(x,y) = \prod_{i=1}^{6}(x - \lambda_i y)$ and $\lambda_i \in \mathbf{C}$ are distinct. The coefficients $a_j \in \mathbf{C}$ of G satisfy $y^2 \mid (G(x,y)^2 - \Delta(x,y))$. For given distinct λ_i's, there exist a_j's satisfying the above condition. The curve Γ_4 has the following properties.

- *(i) Γ_4 is a unicuspidal plane curve of genus two.*
- *(ii) The multiplicity sequence of the cusp coincides with (2).*
- *(iii) Γ_4 is not of AMS type. We have $(\Gamma_4')^2 = 10$.*

Conversely, for a given unicuspidal quartic curve of genus two, there exist a_j's and distinct λ_i's satisfying the above condition such that it is projec-

tively equivalent to the curve defined by the above equation.

Lemma 16. *Let* Γ_5 *be the plane curve defined by the following equation, where* $\lambda_i, a_j \in \mathbf{C}$, λ_i*'s are distinct,* (x, y, z) *are homogeneous coordinates of* $\mathbf{P}^2$.

$$(yz + a_0x^2 + a_1xy + a_2y^2)^2y = \prod_{i=1}^{5}(x - \lambda_i y).$$

The curve Γ_5 *has the following properties.*

(i) Γ_5 *is a unicuspidal plane curve of genus two.*
(ii) *The multiplicity sequence of the cusp coincides with* $(3, 2)$.
(iii) Γ_5 *is of AMS type. We have* $(\Gamma_5')^2 = 10$.

Conversely, for a given quintic curve having the properties (i) and (ii), there exist a_j*'s and distinct* λ_i*'s such that it is projectively equivalent to the curve defined by the above equation.*

Acknowledgment

The author would like to express his thanks to the referees for giving him valuable comments on this paper. He also would like to express his thanks to Professor Fumio Sakai for his helpful advice. This work was supported by JSPS Grant-in-Aid for Scientific Research (22540040).

References

AM. Abhyankar, S. S. and Moh, T. T.: Embeddings of the line in the plane, J. Reine Angew. Math. **276** (1975), 148–166.

AO. A'Campo, N. and Oka, M.: Geometry of plane curves via Tschirnhausen resolution tower, Osaka J. Math. **33** (1996), 1003–1033.

BK. Brieskorn, E. and Knörrer, H.: Plane algebraic curves. Basel, Boston, Stuttgart: Birkhäuser 1986.

D. Dimca, A.: Singularities and topology of hypersurfaces, Universitext, Springer, 1992.

H. Hartshorne, R.: Curves with high self-intersection on algebraic surfaces, Publ. Math. I.H.E.S., **36** (1969), 111–125.

MS. Matsuoka, T. and Sakai, F.: The degree of rational cuspidal curves, Math. Ann. **285** (1989), 233–247.

M. Miyanishi, M.: Minimization of the embeddings of the curves into the affine plane, J. Math. Kyoto Univ. **36** (1996), 311–329.

NU. Namikawa, Y. and Ueno, K.: The complete classification of fibres in pencils of curves of genus two, Manuscripta Math. **9** (1973), 143–186.

O. Ogg, A. P.: On pencils of curves of genus two, Topology **5** (1966), 355–362.

SS. Saito, M.-H. and Sakakibara, K.: On Mordell-Weil lattices of higher genus fibrations on rational surfaces, J. Math. Kyoto Univ. **34** (1994), 859–871.

SST. Sakai, F. and Saleem, M. and Tono, K.: Hyperelliptic plane curves of type $(d, d-2)$, Beiträge zur Algebra und Geometrie **51** (2010), 31–44.

Sh. Shioda, T.: Mordell-Weil lattices for higher genus fibration over a curve. New trends in algebraic geometry (Warwick, 1996), 359–373, London Math. Soc. Lecture Note Ser., 264, Cambridge Univ. Press, Cambridge, 1999

Su. Suzuki, M.: Propriétés topologiques des polynômes de deux variables complexes, et automorphismes algébriques de l'espace $\mathbf{C}^2$, J. Math. Soc. Japan **26** (1974), 241–257.

T. Tono, K.: The projective characterization of elliptic plane curves which have one place at infinity, Saitama Math. J. **25** (2008), 35–46.

Sextic variety as Galois closure variety of smooth cubic

Hisao Yoshihara

Department of Mathematics, Faculty of Science, Niigata University,
Niigata 950-2181, Japan
E-mail: yosihara@math.sc.niigata-u.ac.jp

Let V be a nonsingular projective algebraic variety of dimension n. Suppose there exists a very ample divisor D such that $D^n = 6$ and $\dim \mathrm{H}^0(V, \mathcal{O}(D)) = n+3$. Then, (V, D) defines a D_6-Galois embedding if and only if it is a Galois closure variety of a smooth cubic in $\mathbb{P}^{n+1}$ with respect to a suitable projection center such that the pull back of hyperplane of $\mathbb{P}^n$ is linearly equivalent to D.

Keywords: Galois embedding, Galois closure variety, sextic variety.

1. Introduction

The purpose of this article is to generalize the following assertion (cf. [13, Theorem 4.5]) to n-dimensional varieties.

Proposition 1.1. *Let C be a smooth sextic curve in $\mathbb{P}^3$ and assume the genus is four. If C has a Galois line, then the group G is isomorphic to the cyclic or dihedral group of order six. Moreover, G is isomorphic to the latter one if and only if C is obtained as the Galois closure curve of a smooth plane cubic E with respect to a point $P \in \mathbb{P}^2 \setminus E$, where P does not lie on the tangent line to E at any flex.*

Before going into the details, we recall the definition of Galois embeddings of algebraic varieties and the relevant results. In this article a variety, a surface and a curve will mean a nonsingular projective algebraic variety, surface and curve, respectively.

Let k be the ground field of our discussion, we assume it to be an algebraically closed field of characteristic zero. Let V be a variety of dimension n with a very ample divisor D; we denote this by a pair (V, D). Let $f = f_D : V \hookrightarrow \mathbb{P}^N$ be the embedding of V associated with the complete linear system $|D|$, where $N + 1 = \dim \mathrm{H}^0(V, \mathcal{O}(D))$. Suppose W is a linear subvariety of $\mathbb{P}^N$ satisfying $\dim W = N - n - 1$ and $W \cap f(V) = \emptyset$. Consider the projection π_W from W to $\mathbb{P}^n$, i.e., $\pi_W : \mathbb{P}^N \dashrightarrow \mathbb{P}^n$. Restricting π_W onto $f(V)$, we get a surjective morphism $\pi = \pi_W \cdot f : V \longrightarrow \mathbb{P}^n$.

Let $K = k(V)$ and $K_0 = k(\mathbb{P}^n)$ be the function fields of V and $\mathbb{P}^n$ respectively. The covering map π induces a finite extension of fields $\pi^* : K_0 \hookrightarrow K$ of degree $\deg f(V) = D^n$, which is the self-intersection number of D. We denote by K_W the Galois closure of this extension and by $G_W = Gal(K_W/K_0)$ the Galois group of K_W/K_0. By [1] G_W is isomorphic to the monodromy group of the covering $\pi : V \longrightarrow \mathbb{P}^n$. Let V_W be the K_W-normalization of V (cf. [3, Ch. 2]). Note that V_W is determined uniquely by V and W.

Definition 1.2. In the above situation we call G_W and V_W the Galois group and the Galois closure variety at W respectively (cf. [14]). If the extension K/K_0 is Galois, then we call f and W a Galois embedding and a Galois subspace for the embedding respectively.

Definition 1.3. A variety V is said to have a Galois embedding if there exist a very ample divisor D satisfying that the embedding associated with $|D|$ has a Galois subspace. In this case the pair (V, D) is said to define a Galois embedding.

If W is a Galois subspace and T is a projective transformation of $\mathbb{P}^N$, then $T(W)$ is a Galois subspace of the embedding $T \cdot f$. Therefore the existence of Galois subspace does not depend on the choice of the basis giving the embedding.

Remark 1.4. If a variety V exists in a projective space, then by taking a linear subvariety, we can define a Galois subspace and Galois group similarly as above. Suppose V is not normally embedded and there exists a linear subvariety W such that the projection π_W induces a Galois extension of fields. Then, taking D as a hyperplane section of V in the embedding, we infer readily that (V, D) defines a Galois embedding with the same Galois group in the above sense.

By this remark, for the study of Galois subspaces, it is sufficient to consider the case where V is normally embedded.

We have studied Galois subspaces and Galois groups for hypersurfaces in [9], [10] and [11] and space curves in [13] and [15]. The method introduced in [14] is a generalization of the ones used in these studies.

Hereafter we use the following notation and convention:

- $\mathrm{Aut}(V)$: the automorphism group of a variety V
- $\langle a_1, \cdots, a_m \rangle$: the subgroup generated by $a_1, \cdots, a_m$
- D_{2m} : the dihedral group of order $2m$
- $|G|$: the order of a group G
- $\sim$: the linear equivalence of divisors
- $\mathbf{1}_m$: the unit matrix of size m
- $X * Y$: the intersection cycle of cycles X and Y in a variety.
- $(X_0 : \cdots : X_m)$: a set of homogeneous coordinates on $\mathbb{P}^m$
- $g(C)$: the genus of a smooth curve C
- For a mapping $\varphi : X \longrightarrow Y$ and a subset $X' \subset X$, we often use the same φ to denote the restriction $\varphi|_{X'}$.

2. Results on Galois embeddings

We state several properties concerning Galois embedding without the proofs, for the details, see [14]. By definition, if W is a Galois subspace, then each element σ of G_W is an automorphism of $K = K_W$ over K_0. Therefore it induces a birational transformation of V over $\mathbb{P}^n$. This implies that G_W can be viewed as a subgroup of $\mathrm{Bir}(V/\mathbb{P}^n)$, the group of birational transformations of V over $\mathbb{P}^n$. Further we can say the following:

Representation 1. Each birational transformation belonging to G_W turns out to be regular on V, hence we have a faithful representation

$$\alpha : G_W \hookrightarrow \mathrm{Aut}(V). \tag{1}$$

Remark 2.1. Representation 1 is proved by using transcendental method in [14], however we can prove it algebraically by making use of the results [7, Ch. I, 5.3. Theorem 7] and [2, Ch. V, Theorem 5.2].

Therefore, if the order of $\mathrm{Aut}(V)$ is smaller than the degree d, then (V, D) cannot define a Galois embedding. In particular, if $\mathrm{Aut}(V)$ is trivial, then V has no Galois embedding. On the other hand, in case V has an infinitely many automorphisms, we have examples such that there exist infinitely many distinct Galois embeddings, see Example 4.1 in [14].

When (V, D) defines a Galois embedding, we identify $f(V)$ with V. Let H be a hyperplane of $\mathbb{P}^N$ containing W and put $D' = V * H$. Since $D' \sim D$

and $\sigma^*(D') = D'$, for any $\sigma \in G_W$, we see σ induces an automorphism of $\mathrm{H}^0(V, \mathcal{O}(D))$. This implies the following.

Representation 2. We have a second faithful representation

$$\beta : G_W \hookrightarrow PGL(N+1, k). \tag{2}$$

In the case where W is a Galois subspace we identify $\sigma \in G_W$ with $\beta(\sigma) \in PGL(N+1, k)$ hereafter. Since G_W is a finite subgroup of $\mathrm{Aut}(V)$, we can consider the quotient V/G_W and let π_G be the quotient morphism, $\pi_G : V \longrightarrow V/G_W$.

Proposition 2.2. *If (V, D) defines a Galois embedding with the Galois subspace W such that the projection is $\pi_W : \mathbb{P}^N \dashrightarrow \mathbb{P}^n$, then there exists an isomorphism $g : V/G_W \longrightarrow \mathbb{P}^n$ satisfying $g \cdot \pi_G = \pi$. Hence the projection π is a finite morphism and the fixed loci of G_W consist of only divisors.*

Therefore, π turns out to be a Galois covering in the sense of Namba [6].

Lemma 2.3. *Let (V, D) be the pair in Proposition 2.2. Suppose $\tau \in G$ has the representation*

$$\beta(\tau) = [1, \ldots, 1, e_m], \ (m \geq 2)$$

where e_m is an m-th root of unity. Let p be the projection from $(0 : \cdots : 0 : 1) \in W$ to $\mathbb{P}^{N-1}$. Then, $V/\langle\tau\rangle$ is isomorphic to $p(V)$ if $p(V)$ is a normal variety.

We have a criterion that (V, D) defines a Galois embedding.

Theorem 2.4. *The pair (V, D) defines a Galois embedding if and only if the following conditions hold:*

1. *There exists a subgroup G of $\mathrm{Aut}(V)$ satisfying that $|G| = D^n$.*
2. *There exists a G-invariant linear subspace $\mathcal{L}$ of $\mathrm{H}^0(V, \mathcal{O}(D))$ of dimension $n+1$ such that, for any $\sigma \in G$, the restriction $\sigma^*|_{\mathcal{L}}$ is a multiple of the identity.*
3. *The linear system $\mathcal{L}$ has no base points.*

It is easy to see that $\sigma \in G_W$ induces an automorphism of W, hence we obtain another representation of G_W as follows. Take a basis $\{f_0, f_1, \ldots, f_N\}$ of $\mathrm{H}^0(V, \mathcal{O}(D))$ satisfying that $\{f_0, f_1, \ldots, f_n\}$ is a basis of $\mathcal{L}$ in Theorem 2.4. Then we have the representation

$$\beta_1(\sigma) = \begin{pmatrix} \lambda_\sigma & & & \vdots & \\ & \ddots & & \vdots & * \\ & & \lambda_\sigma & \vdots & \\ \cdots & \cdots & \cdots & \vdots & \cdots \\ & \mathbf{0} & & \vdots & M_\sigma \end{pmatrix}. \tag{3}$$

Since the projective representation is completely reducible, we get another representation using a direct sum decomposition:

$$\beta_2(\sigma) = \lambda_\sigma \cdot \mathbf{1}_{n+1} \oplus M'_\sigma.$$

Thus we can define

$$\gamma(\sigma) = M'_\sigma \in PGL(N-n, k).$$

Therefore σ induces an automorphism on W given by M'_σ.

Representation 3. We get a third representation

$$\gamma : G_W \longrightarrow PGL(N-n, k). \tag{4}$$

Let G_1 and G_2 be the kernel and image of γ respectively.

Theorem 2.5. *We have an exact sequence of groups*

$$1 \longrightarrow G_1 \longrightarrow G \xrightarrow{\gamma} G_2 \longrightarrow 1,$$

where G_1 is a cyclic group.

Corollary 2.6. *If $N = n+1$, i.e., $f(V)$ is a hypersurface, then G is a cyclic group.*

This assertion has been obtained in [11]. Moreover we have another representation.

Suppose that (V, D) defines a Galois embedding and let G be a Galois group at some Galois subspace W. Then, take a general hyperplane W_1 of $\mathbb{P}^n$ and put $V_1 = \pi^*(W_1)$. The divisor V_1 has the following properties:

(i) If $n \geq 2$, then V_1 is a smooth irreducible variety.
(ii) $V_1 \sim D$.
(iii) $\sigma^*(V_1) = V_1$ for any $\sigma \in G$.
(iv) V_1/G is isomorphic to W_1.

Put $D_1 = V_1 \cap H_1$, where H_1 is a general hyperplane of $\mathbb{P}^N$. Then (V_1, D_1) defines a Galois embedding with the Galois group G (cf. Remark 1.4). Iterating the above procedures, we get a sequence of pairs (V_i, D_i) such that

$$(V, D) \supset (V_1, D_1) \supset \cdots \supset (V_{n-1}, D_{n-1}). \tag{5}$$

These pairs satisfy the following properties:

(a) V_i is a smooth subvariety of V_{i-1}, which is a hyperplane section of V_{i-1}, where $D_i = V_{i+1}$, $V = V_0$ and $D = V_1$ $(1 \leq i \leq n-1)$.
(b) (V_i, D_i) defines a Galois embedding with the same Galois group G.

Definition 2.7. The above procedure to get the sequence (5) is called the Descending Procedure.

Letting C be the curve V_{n-1}, we get the next fourth representation.

Representation 4. We have a fourth faithful representation

$$\delta : G_W \hookrightarrow \mathrm{Aut}(C), \tag{6}$$

where C is a curve in V given by $V \cap L$ such that L is a general linear subvariety of $\mathbb{P}^N$ with dimension $N - n + 1$ containing W.

Since the Inverse Problem of Galois Theory over $k(x)$ is affirmative ([4]), we can prove the following.

Remark 2.8. Giving any finite group G, there exists a smooth curve and very ample divisor D such that (C, D) defines a Galois embedding with the Galois group G.

3. Statement of results

Let V be a variety of dimension n. We say that V has the property $(\P_n)$ if

(1) there exists a very ample divisor D with $D^n = 6$, and
(2) $\dim \mathrm{H}^0(V, \ \mathcal{O}(D)) = n + 3$.

An example of such a variety is a smooth $(2,3)$-complete intersection, where D is a hyperplane section. In particular, in case $n = 1$, V is a non-hyperelliptic curve of genus four and D is a canonical divisor. In case $n = 2$, V is a $K3$ surface such that there exists a very ample divisor D with $D^2 = 6$. However, the variety with the property $(\P_n)$ is not necessarily the complete intersection, see Remark 3.10 below.

We will study the Galois embedding of V for the variety with the property $(\P_n)$. Clearly the Galois group is isomorphic to the cyclic group of order six or D_6. In the latter case we say that (V, D) defines a D_6-embedding or, more simply V has a D_6-embedding.

Theorem 3.1. *Assume V has the property $(\P_n)$. If V has a D_6-embedding, then V is obtained as the Galois closure variety of a smooth cubic Δ in $\mathbb{P}^{n+1}$ with respect to a suitable projection center.*

Next we consider the converse assertion. Let Δ be a smooth cubic of dimension n in $\mathbb{P}^{n+1}$. Take a non-Galois point $P \in \mathbb{P}^{n+1} \setminus \Delta$. Note that, for a smooth hypersurface $X \subset \mathbb{P}^{n+1}$, the number of Galois points is at most $n+2$. The maximal number is attained if and only if X is projectively equivalent to the Fermat variety (cf. [11]).

Define the set Σ_P of lines as

$\Sigma_P = \{\ell \mid \ell$ is a line passing through P such that $\ell * \Delta$ can be expressed as $2P_1 + P_2$, where $P_i \in \Delta$ $(i = 1, 2)$ and $P_1 \neq P_2\}$

The closure of the set $\bigcup_{\ell \in \Sigma_P} \ell$ is a cone, we denote it by $C_P(\Delta)$. Then we have the following.

Lemma 3.2. *The cone $C_P(\Delta)$ is a hypersurface of degree six.*

We can express as $\Delta * C_P(\Delta) = 2R_1 + R_2$, where R_1 and R_2 are different divisors on Δ.

Definition 3.3. We call P a good point if

(1) R_2 is smooth and irreducible in case $n \geq 2$, or
(2) R_2 consists of six points in case $n = 1$.

Proposition 3.4. *If P is a general point for Δ, then P is a good point.*

To some extent the converse assertion of Theorem 3.1 holds as follows.

Theorem 3.5. *If Δ_P is a Galois closure variety of a smooth cubic $\Delta \subset \mathbb{P}^{n+1}$, where the projection center P is a good point, then Δ_P is a smooth $(2, 3)$-complete intersection in $\mathbb{P}^{n+2}$ with D_6-embedding.*

Remark 3.6. In the assertion of Theorem 3.5, the construction of the Galois closure is closely related to the one in [8, Tokunaga]. In case $n = 2$, the Galois closure surface is a $K3$ surface.

Applying the Descending Procedure to the variety of Theorem 3.5, we get the following.

Proposition 3.7. *If a variety V is a smooth $(2,3)$-complete intersection and has a D_6-embedding, then there exists the following sequence of varieties V_i, where V_i has the same properties as V does, i.e.,*

(i) *V_i is a subvariety of V_{i-1} $(i \geq 1)$, where $V_0 = V$,*
(ii) *V_i is also a smooth $(2,3)$-complete intersection of hypersurfaces in $\mathbb{P}^{n+2-i}$, $0 \leq i \leq n-1$,*
(iii) *V_i has the property $(\P_{n-i})$,*
(iv) *V_i has a D_6-embedding.*

The situation above is illustrated as follows:

$$
\begin{array}{ccccccccc}
\mathbb{P}^{n+2} & \dashrightarrow & \mathbb{P}^{n+1} & \dashrightarrow & & \dashrightarrow & \mathbb{P}^4 & \dashrightarrow & \mathbb{P}^3 \\
\cup & & \cup & & & & \cup & & \cup \\
V & \supset & V_1 & \supset & \cdots & \supset & V_{n-2} & \supset & V_{n-1} \\
\downarrow & & \downarrow & & & & \downarrow & & \downarrow \\
\mathbb{P}^n & \dashrightarrow & \mathbb{P}^{n-1} & \dashrightarrow & & \dashrightarrow & \mathbb{P}^2 & \dashrightarrow & \mathbb{P}^1,
\end{array}
$$

where $\dashrightarrow$ is a point projection, $\downarrow$ is a triple covering, V_{n-2} and V_{n-1} are a $K3$ surface and a sextic curve, respectively.

Here we present examples.

Example 3.8. Let Δ be the smooth cubic in $\mathbb{P}^3$ defined by

$$F(X_0, X_1, X_2, X_3) = X_0^3 + X_1^3 + X_2^3 + X_0^2 X_3 + X_1 X_3^2 + X_3^3. \tag{7}$$

Let π_P be the projection from $P = (0:0:0:1)$ to the hyperplane $\mathbb{P}^2$. Taking the affine coordinates $x = X_0/X_3$, $y = X_1/X_3$ and $z = X_2/X_3$, we get the defining equation of the affine part

$$f(x, y, z) = x^3 + y^3 + z^3 + x^2 + y + 1.$$

Put $x = at$, $y = bt$ and $z = ct$. Computing the discriminant $D(f)$ of $f(at, bt, ct) = (a^3 + b^3 + c^3)t^3 + a^2t^2 + bt + 1$ with respect to t, we obtain

$$\begin{aligned} D(f) = -(&31a^6 - 18a^5b - a^4b^2 + 58a^3b^3 - 18a^2b^4 \\ &+ 31b^6 + 54a^3c^3 - 18a^2bc^3 + 58b^3c^3 + 27c^6). \end{aligned} \tag{8}$$

This yields the branch divisor of $\pi_P : \Delta \longrightarrow \mathbb{P}^2$. From (7) and (8) we infer that the defining equation of $2R_1$ is

$$F(X_0, X_1, X_2, X_3) = 0 \text{ and } (X_0^2 + 2X_1X_3 + 3X_3^2)^2 = 0,$$

and that of R_2 is

$$F(X_0, X_1, X_2, X_3) = 0 \text{ and } 4X_0^2 - X_1^2 + 2X_1X_3 + 3X_3^2 = 0.$$

It is not difficult to check that R_2 is smooth and irreducible, hence P is a good point for Δ. By taking a double covering along this curve [8], we get the $K3$ surface Δ_P in $\mathbb{P}^4$ defined by $F = 0$ and $X_4^2 = 4X_0^2 - X_1^2 + 2X_1X_3 + 3X_3^2$, which is a $(2,3)$-complete intersection. The Galois line is given by $X_0 = X_1 = X_2 = 0$.

How is the Galois closure variety when the projection center is not a good point? Let us examine the following example.

Example 3.9. For a projection with some center $P \in \mathbb{P}^3 \setminus \Delta$, the Galois closure surface Δ_P is not necessarily a $K3$ surface. Indeed, let Δ be the smooth cubic defined by

$$F(X_0, X_1, X_2, X_3) = X_0^3 + X_1^3 + X_2^3 + X_0X_3^2 - X_3^3. \tag{9}$$

Clearly the point $P = (0:0:0:1)$ is not a Galois one. Taking the same affine coordinates as in Example 3.8, we get the defining equation of the affine part

$$f(x, y, z) = x^3 + y^3 + z^3 + x - 1.$$

Put $x = at$, $y = bt$ and $z = ct$. Computing the discriminant $D(f)$ of $f(at, bt, ct) = (a^3 + b^3 + c^3)t^3 + at - 1$ with respect to t, we obtain

$$D(f) = -(31a^3 + 27b^3 + 27c^3)(a^3 + b^3 + c^3). \tag{10}$$

This yields the branch divisor of $\pi_P : \Delta \longrightarrow \mathbb{P}^2$. From (9) and (10) we infer that the defining equation of $2R_1$ is $C_1 + C_2$, where C_1 (resp. C_2) is given by $X_0^3 + X_1^3 + X_2^3 = 0$ (resp. $31X_0^3 + 27X_1^3 + 27X_2^3 = 0$). Hence the defining equation of the sextic $C_P(V)$ is

$$(X_0^3 + X_1^3 + X_2^3)(31X_0^3 + 27X_1^3 + 27X_2^3) = 0.$$

Let Δ_P be the double covering of Δ branched along the divisor R_2, where $R_2 = R_{21} + R_{22}$ such that R_{21} (resp. R_{22}) is given by the intersection of $F = 0$ and $X_0 - X_3 = 0$ (resp. $F = 0$ and $X_0 - 3X_3 = 0$). The R_{2i} $(i = 1, 2)$ is a smooth curve on Δ satisfying that $R_{2i}^2 = 3$, $R_2^2 = 12$ and $R_{21}.R_{22} = 3$. We infer that Δ_P is a normal surface, therefore it is a Galois closure surface at P (Definition 1.2). However, it has three singular points of type A_1, so that it is not a $K3$ surface. The minimal resolution of Δ_P turns out to be a $K3$ surface.

Remark 3.10. The variety with the property ($\P_n$) is not necessarily a $(2,3)$-complete intersection. For example, in case $n = 1$, Take $V = C$ as the Galois closure curve of a smooth cubic $\Delta \subset \mathbb{P}^2$ obtained as follows: let T be a tangent line to Δ at a flex. Choose a point $P \in T$ satisfying the following condition: if ℓ_P is a line passing through P and $\ell_P \neq T$, then ℓ_P does not tangent to Δ at any flex. Let C be the Galois closure curve for the point projection $\pi_P : \Delta \longrightarrow P^1$, i.e., $\widetilde{\pi} : C \longrightarrow \Delta$ is a double covering, which has four branch points (see, for example [5, pp. 287–288]), hence $g(C) = 3$. Let D be the divisor $\widetilde{\pi}^*(\ell * \Delta)$, where ℓ is a line passing through P. Clearly we have $\deg D = 6$, the complete linear system $|D|$ has no base point and $\dim \mathrm{H}^0(C, \mathcal{O}(D)) = 4$. Let $f : C \longrightarrow C'$ be the morphism associated with $|D|$. The double covering $\widetilde{\pi}$ factors as $\widetilde{\pi} = \widetilde{\pi}' \cdot f$, where $\widetilde{\pi}' : C' \longrightarrow \Delta$ is a restriction of the projection $\mathbb{P}^3 \dashrightarrow \mathbb{P}^2$. Since $g(C') \geq 1$, we see $\deg C' \neq 2$ and 3. Hence $\deg C' = 6$ and f is a birational morphism. Further, we have the projection $\widetilde{\pi}' : C' \longrightarrow \Delta$ and Δ is nonsingular, hence C' is smooth. Therefore f is an isomorphism. Since $g(C) = 3$, C is not a $(2,3)$-complete intersection.

4. Proof

First we prove Theorem 3.1. The case $n = 1$ have been proved ([13]). So that we will restrict ourselves to the case $n \geq 2$.

Since V is embedded into $\mathbb{P}^{n+2}$ associated with $|D|$, where D is a very ample divisor with $D^n = 6$, we can apply the results in Section 2. By assumption V has a Galois line ℓ such that the Galois group $G = G_\ell$ is isomorphic to D_6. We can assume $G = \langle \sigma, \tau \rangle$ where $\sigma^3 = \tau^2 = 1$ and $\tau\sigma\tau = \sigma^{-1}$. Let $\rho_1 : V \longrightarrow V^\tau = V/\langle \tau \rangle$. We see $\rho_2 : V^\tau \longrightarrow V^\tau/G \cong \mathbb{P}^n$ turns out a morphism. Then, we have $\pi = \rho_2\rho_1 : V \longrightarrow V/G \cong \mathbb{P}^n$. Note that ρ_2 is a non-Galois triple covering. By taking suitable coordinates, we can assume ℓ is given by $X_0 = X_1 = \cdots = X_n = 0$. As we see in Section 2, we have the representation $\beta : G \hookrightarrow PGL(n+3, k)$. Since the characteristic of k is zero, the projective representation is completely reducible, hence $\beta(\sigma)$ and $\beta(\tau)$ can be represented as

$$\mathbf{1}_{n+1} \oplus M_2(\sigma) \quad \text{and} \quad \mathbf{1}_{n+1} \oplus M_2(\tau),$$

respectively, where $M_2(\sigma)$ and $M_2(\tau)$ are in $GL(2,k)$. Since $G \cong D_6$, we have the representation

$$\beta(\sigma) = \begin{pmatrix} 1 & & & & \\ & \ddots & & & 0 \\ & & 1 & & \\ & & & -\frac{1}{2} & \omega + \frac{1}{2} \\ 0 & & & \omega + \frac{1}{2} & -\frac{1}{2} \end{pmatrix} \quad \text{and} \quad \beta(\tau) = \begin{pmatrix} 1 & & & 0 \\ & \ddots & & \\ & & \ddots & \\ & & & 1 & \\ 0 & & & & -1 \end{pmatrix},$$

where ω is a primitive cubic root of 1. Therefore, the fixed locus of τ is given by $f(V) \cap H$, where H is the hyperplane defined by $X_{n+2} = 0$. Put $Z = f(V) \cap H$, i.e., $Z \sim D$. Since Z is ample, it is connected. Looking at the representation $\beta(\tau)$, we see Z is smooth, hence it is a smooth irreducible variety. Take the point $P = (0 : \cdots : 0 : 1) \in \ell$ and an arbitrary point Q in V. Let ℓ_{PQ} be the line passing through P and Q. Then we have $\tau(\ell_{PQ}) = \ell_{PQ}$ and $\tau(V) = V$. Let π_P be the projection from the point P to the hyperplane H. Since Z is smooth, $\pi_P(V)$ is smooth. By Lemma 2.3, $\pi_P(V)$ is isomorphic to $V/\langle\tau\rangle$ and we may assume $\pi_P = \rho_1$. Therefore we see V is contained in the cone consisting of the lines passing through P and the points in V. Since $\deg V = 6$ and $\deg p = 2$, we conclude the variety V^{τ} is a smooth cubic in $\mathbb{P}^{n+1}$. This proves Theorem 3.1.

Next we prove Lemma 3.2. Let H_2 be a linear variety of dimension two and passing through P. If H_2 is general, then $\Delta \cap H_2$ is a smooth cubic in the plane $H_2 \cong \mathbb{P}^2$. Thus $C_P(\Delta) \cap H_2$ consists of six lines, hence we have $\deg C_P(\Delta) = 6$.

The proof of Proposition 3.4 is as follows. Suppose P is a general point for Δ and let π_P be the projection from P to the hyperplane $\mathbb{P}^n$. Put $B = \pi_P(R_2)$.

Claim 4.1. *The divisor B is irreducible.*

Proof. It is sufficient to check in a general affine part. Put $x_i = X_i/X_0$ ($i = 1, \dots, n+1$) and let $f(x_1, \dots, x_{n+1})$ be the defining equation of an affine part $X_0 \neq 0$ of Δ and $P = (u_1, \dots, u_{n+1}) \in \mathbb{A}^{n+1}$. Put

$$g(u_1, \dots, u_{n+1}, t_0, \dots, t_n, x) = f(u_1 + xt_0, \dots, u_{n+1} + xt_n),$$

where $(t_0, \dots, t_n) \in \mathbb{P}^n$. Let $D(g) = D(u_1, \dots, u_{n+1}, t_0, \dots, t_n)$ be the discriminant of g with respect to x. Owing to [9, Lemma 3] and [11, Claim 1], we see $D(g)$ is reduced and irreducible. Therefore for a general value $u_1 = a_1, \dots, u_{n+1} = a_{n+1}$, $D(a_1, \dots, a_{n+1}, t_0, \dots, t_n)$ is irreducible. This implies B is irreducible. □

Claim 4.2. *The divisor R_2 is irreducible and smooth.*

Proof. Suppose R_2 is decomposed into irreducible components $R_{21}+\cdots+R_{2r}$. Since B is irreducible, we have $\pi_P(R_{2i}) = B$ for each $1 \leq i \leq r$. However, since $\Delta * \ell$ has an expression $2P_1+P_2$, the r must be 1. Thus R_2 is irreducible. Since $\Delta * \ell$ has an expression $2P_1 + P_2$, where $P_i \in \Delta$ ($i = 1, 2$), Δ and ℓ has a normal crossing at P_2 if $P_1 \neq P_2$. In case $P_1 = P_2$, the intersection number of Δ and ℓ at P_1 is three. Since $R_1 \ni P_1$ and $\Delta * C_P(\Delta) = 2R_1 + R_2$, we see that R_2 is smooth at P_1. □

This completes the proof of Proposition 3.4. The proof of Theorem 3.5 is as follows.

First note that P is not a Galois point. So we consider the Galois closure variety. The ramification divisor of $\pi_P : \Delta \longrightarrow \mathbb{P}^n$ is $2R_1 + R_2$. The divisor R_2 is smooth and irreducible by assumption. Let Φ be the equation of the branch divisor of π_P. As we see in Example 3.8 ($a = X_0/tX_3$, $b = X_1/tX_3$, $c = X_2/tX_3$) the discriminant is given by the homogeneous equation of $X_0, \ldots, X_n$, hence we infer that $\pi_P^*(\Phi)$ has the expression as $\Phi_1^2 \cdot \Phi_2$, where $\Phi_2 = 0$ defines R_2. Since $\deg \Phi_2 = 2$, we can define the variety in $\mathbb{P}^{n+2}$ by $F = 0$ and $X_{n+2}^2 = \Phi_2$, which is smooth and turns out to be the Galois closure variety. This proves Theorem 3.5.

We go to the proof of Proposition 3.7. Let H be a general hyperplane containing the Galois line ℓ for V in Theorem 3.1. Put $V_1 = V \cap H$ and $D_1 = D \cap H$. Since we are assuming $n \geq 2$, the V_1 is irreducible and nonsingular by Bertini's theorem. Thus, we have $\dim V_1 = n-1$, $D_1^{n-1} = 6$ and V_1 is also a smooth $(2,3)$-complete intersection. Note that $V_1 \sim D$ on V. Thus we have the exact sequence of sheaves

$$0 \longrightarrow \mathcal{O}_V \longrightarrow \mathcal{O}_V(V_1) \longrightarrow \mathcal{O}_{V_1}(D_1) \longrightarrow 0.$$

Taking cohomology, we get a long exact sequence

$$\begin{aligned} 0 \longrightarrow \mathrm{H}^0(V,\, \mathcal{O}_V) \longrightarrow \mathrm{H}^0(V,\, \mathcal{O}_V(V_1)) \longrightarrow \mathrm{H}^0(V_1,\, \mathcal{O}_{V_1}(D_1)) \\ \longrightarrow \mathrm{H}^1(V,\, \mathcal{O}_V) \longrightarrow \cdots. \end{aligned}$$

Since V is the complete intersection, we have $\mathrm{H}^1(V,\, \mathcal{O}_V) = 0$ (cf. [2, III, Ex. 5.5]). Then V_1 has the same properties as V does, i.e., $\dim V_1 = n-1$, $D_1^{n-1} = 6$, $\dim \mathrm{H}^0(V_1,\, \mathcal{O}(D_1)) = n+2$ and ℓ is a Galois line for V_1 and the Galois group is isomorphic to D_6. Continuing the Descending Procedure, we get the sequence of Proposition 3.7.

There are a lot of problems concerning our theme, we pick up some of them.

Problems.

(1) For each finite subgroup G of $GL(2, k)$, does there exist a pair (V, D) which defines the Galois embedding with the Galois group G such that $D^n = |G|$, $\dim V = n$ and $\dim \mathrm{H}^0(V, \mathcal{O}(D)) = n + 3$?

(2) How many Galois subspaces do there exist for one Galois embedding? In case a smooth hypersurface V in $\mathbb{P}^{n+1}$, there exist at most $n + 2$. Further, it is $n + 2$ if and only if V is Fermat variety [11].

(3) Does there exist a variety V on which there exist two divisors D_i $(i = 1, 2)$ such that they give Galois embeddings and $D_1^n \neq D_2^n$?

For the detail, please visit our website
http://mathweb.sc.niigata-u.ac.jp/ yosihara/openquestion.html

Acknowledgment

The author expresses his thanks to the referee(s) for carefully reading the manuscript and giving the suitable suggestions for improvements.

References

1. J. Harris, Galois groups of enumerative problems, *Duke Math. J.*, **46** (1979), 685–724.
2. R. Hartshorne, Algebraic Geometry, *Graduate Texts in Mathematics*, **52** Springer-Verlag.
3. S. Iitaka, Algebraic Geometry, An introduction to birational geometry of algebraic varieties, *Graduate Texts in Mathematics*, **76** Springer-Verlag.
4. G. Malle and B. H. Matzat, Inverse Galois Theory, *Springer Monogr., Math.*, Springer-Verlag, New York, Heidelberg, Berlin, 1999.
5. K. Miura and H. Yoshihara, Field theory for function fields of plane quartic curves, *J. Algebra*, **226** (2000), 283–294.
6. M. Namba, Branched coverings and algebraic functions, *Pitman Research Notes in Mathematics*, Series 161.
7. R. Shafarevich, Basic Algebraic Geometry I, Second, Revised and Expanded Edition, Springer-Verlag.
8. H. Tokunaga, Triple coverings of algebraic surfaces according to the Cardano formula, *J. Math. Kyoto University* , **31** (1991), 359–375.
9. H. Yoshihara, Function field theory of plane curves by dual curves, *J. Algebra*, **239** (2001), 340–355.
10. H. Yoshihara, Galois points on quartic surfaces, *J. Math. Soc. Japan*, **53** (2001), 731–743.
11. H. Yoshihara, Galois points for smooth hypersurfaces, *J. Algebra*, **264** (2003), 520–534.

12. H. Yoshihara, Families of Galois closure curves for plane quartic curves, *J. Math. Kyoto Univ.*, **43** (2003), 651–659.
13. H. Yoshihara, Galois lines for space curves, *Algebra Colloquium*, **13** (2006), 455–469.
14. H. Yoshihara, Galois embedding of algebraic variety and its application to abelian surface, *Rend. Sem. Mat. Univ. Padova.* **117** (2007), 69–86.
15. H. Yoshihara, Galois lines for normal elliptic space curves, II, *Algebra Colloquium* **19**, 867–876 (2012).

Invariant hypersurfaces of endomorphisms of the projective 3-space

De-Qi Zhang*

Department of Mathematics, National University of Singapore,
10 Lower Kent Ridge Road, Singapore 119076
E-mail: matzdq@nus.edu.sg

Dedicated to Prof. Miyanishi on the occasion of his 70th birthday

We consider surjective endomorphisms f of degree > 1 on the projective n-space $\mathbb{P}^n$ with $n = 3$, and f^{-1}-stable hypersurfaces V. We show that V is a hyperplane (i.e., $\deg(V) = 1$) but with four possible exceptions; it is conjectured that $\deg(V) = 1$ for any $n \geq 2$; cf. [7], [3].

Keywords: Endomorphism, iteration, projective 3-space.

1. Introduction

We work over the field $\mathbb{C}$ of complex numbers. In this paper, we study properties of f^{-1}-stable prime divisors of X for endomorphisms $f : \mathbb{P}^3 \to \mathbb{P}^3$. Below is our main result.

Theorem 1.1. *Let* $f : \mathbb{P}^3 \to \mathbb{P}^3$ *be an endomorphism of degree* > 1 *and* V *an irreducible hypersurface such that* $f^{-1}(V) = V$*. Then either* $\deg(V) = 1$*, i.e.,* V *is a hyperplane, or* V *equals one of the four cubic hypersurfaces* $V_i = \{S_i = 0\}$*, where* S_i*'s are as follows, with suitable projective coordinates:*

(1) $S_1 = X_3^3 + X_0X_1X_2$;
(2) $S_2 = X_0^2X_3 + X_0X_1^2 + X_2^3$;
(3) $S_3 = X_0^2X_2 + X_1^2X_3$;
(4) $S_4 = X_0X_1X_2 + X_0^2X_3 + X_1^3$.

2000 *Mathematics Subject Classification*: 37F10, 32H50, 14E20, 14J45.
*The author is supported by an ARF of NUS.

We are unable to rule out the four cases in Theorem 1.1 and do not know whether there is any endomorphism $f_{V_i} : V_i \to V_i$ of $\deg(f_{V_i}) > 1$ for $i = 2, 3$ or 4, but see Examples 1.5 (for V_1) and 1.2 below.

Example 1.2. There are many endomorphisms $f_{V'} : V' \to V'$ of $\deg(f_{V'}) > 1$ for the normalization V' of $V = V_i$ $(i = 3, 4)$, where $V' \simeq \mathbb{F}_1$ in either case (cf. Remark 1.3 below). Conjecture 1.4 below asserts that $f_{V'}$ is *not* lifted from any endomorphism $f : \mathbb{P}^3 \to \mathbb{P}^3$ restricted to the non-normal cubic surface V. Indeed, consider the endomorphism $f_{\mathbb{P}^2} : \mathbb{P}^2 \to \mathbb{P}^2$ $([X_0, X_1, X_2] \to [X_0^q, X_1^q, X_2^q])$ with $q \geq 2$. It lifts to an endomorphism $f_{\mathbb{F}_1} : \mathbb{F}_1 \to \mathbb{F}_1$ of $\deg(f_{\mathbb{F}_1}) = q^2$, where $\mathbb{F}_1 \to \mathbb{P}^2$ is the blowup of the point $[0, 0, 1]$ fixed by $f_{\mathbb{P}^2}^{-1}$.

Remark 1.3. Below are some remarks about Theorem 1.1.

(1) The non-normal locus of V_i $(i = 3, 4)$ is a single line C and stabilized by f^{-1}. Let $\sigma : V_i' \to V_i$ $(i = 3, 4)$ be the normalization. Then V_i' is the (smooth) Hirzebruch surface $\mathbb{F}_1$ (i.e., the one-point blowup of $\mathbb{P}^2$; see [1, Theorem 1.5], [15]) with the conductor $\sigma^{-1}(C) \subset V_i'$ a smooth section at infinity (for V_3), and the union of the negative section and a fibre (for V_4), respectively. $f|V_i$ lifts to a (polarized) endomorphism $f_{V_i'} : V_i' \to V_i'$.

(2) V_1 (resp. V_2) is unique as a normal cubic (or degree three del Pezzo) surface of Picard number one and with the singular locus $\operatorname{Sing} V_1 = 3A_2$ (resp. $\operatorname{Sing} V_2 = E_6$); see [16, Theorem 1.2], and [8, Theorem 4.4] for the anti-canonical embedding of V_i in $\mathbb{P}^3$. The V_1 contains exactly three lines (triangle-shaped) whose three vertices form the singular locus of V_1. And V_2 contains a single line on which lies its unique singular point. f^{-3} fixes the singular point(s) of V_i $(i = 1, 2)$.

(3) f^{-1} (or its positive power) does not stabilize the only line L on V_2 by using [13, Theorem 4.3.1] since the pair (V_2, L) is not log canonical at the singular point of V_2. For V_1, we do not know whether f^{-1} (or its power) stabilizes the three lines.

1.4. **A motivating conjecture.** Here are some motivations for our paper. It is conjectured that *every hypersurface* $V \subset \mathbb{P}^n$ *stabilized by the inverse* f^{-1} *of an endomorphism* $f : \mathbb{P}^n \to \mathbb{P}^n$ *of* $\deg(f) > 1$, *is linear.* This conjecture is still open when $n \geq 3$ and V is singular, since the proof of [3] is incomplete as we were informed by an author (and see his homepage). The smooth hypersurface case was settled in the affirmative in any dimension by Cerveau-Lins Neto [4] and independently by Beauville [2]. See also [14, Theorem 1.5 in arXiv version 1], [17] and [18] for related results.

By Theorem 1.1, this conjecture is true when $n = 3$ but with four exceptional cubic surfaces V_i which we could not rule out.

From the dynamics point of view, as seen in Dinh-Sibony [5, Theorem 1.3, Corollary 1.4], $f : \mathbb{P}^n \to \mathbb{P}^n$ behaves nicely *exactly* outside those f^{-1}-stabilized subvarieties. We refer to Fornaess-Sibony [7], and [5] for further references.

A smooth hypersurface X in $\mathbb{P}^{n+1}$ with $\deg(X) \geq 3$ and $n \geq 2$, has no endomorphism $f_X : X \to X$ of degree > 1 (cf. [4], [2, Theorem]). However, singular X may have plenty of endomorphisms f_X of arbitrary degrees as shown in Example 1.5 below. Conjecture 1.4 asserts that such f_X cannot be extended to an endomorphism of $\mathbb{P}^{n+1}$.

Example 1.5. We now construct many polarized endomorphisms for some degree $n+1$ hypersurface $X \subset \mathbb{P}^{n+1}$, with X isomorphic to the V_1 in Theorem 1.1 when $n = 2$. Let

$$f = (F_0, \ldots, F_n) : \mathbb{P}^n \to \mathbb{P}^n$$

$(n \geq 2)$, with $F_i = F_i(X_0, \ldots, X_n)$ homogeneous, be any endomorphism of degree $q^n > 1$, such that $f^{-1}(S) = S$ for a reduced degree $n+1$ hypersurface $S = \{S(X_0, \ldots, X_n) = 0\}$. So S must be normal crossing and linear: $S = \sum_{i=0}^n S_i$ (cf. [14, Theorem 1.5 in arXiv version 1]). Thus we may assume that $f = (X_0^q, \ldots, X_n^q)$ and $S_i = \{X_i = 0\}$. The relation $S \sim (n+1)H$ with $H \subset \mathbb{P}^n$ a hyperplane, defines

$$\pi : X = Spec \oplus_{i=0}^n \mathcal{O}(-iH) \to \mathbb{P}^n$$

which is a Galois $\mathbb{Z}/(n+1)$-cover branched over S so that $\pi^* S_i = (n+1)T_i$ with the restriction $\pi|T_i : T_i \to S_i$ an isomorphism.

This X is identifiable with the degree $n+1$ hypersurface

$$\{Z^{n+1} = S(X_0, \ldots, X_n)\} \subset \mathbb{P}^{n+1}$$

and has singularity of type $z^{n+1} = xy$ over the intersection points of S locally defined as $xy = 0$. Thus, when $n = 2$, we have $\operatorname{Sing} X = 3A_2$ and X is isomorphic to the V_1 in Theorem 1.1 (cf. Remark 1.3). We may assume that

$$f^* S(X_0, \ldots, X_n) = S(X_0, \ldots, X_n)^q$$

after replacing $S(X_0, \ldots, X_n)$ by a scalar multiple, so f lifts to an endomorphism

$$g = (Z^q, F_0, \ldots, F_n)$$

of $\mathbb{P}^{n+1}$ (with homogeneous coordinates $[Z, X_0, \ldots, X_n]$), stabilizing X, so that

$$g_X := g|X : X \to X$$

is a polarized endomorphism of $\deg(g_X) = q^n$ (cf. [14, Lemma 2.1]). Note that $g^{-1}(X)$ is the union of q distinct hypersurfaces

$$\{Z^{n+1} = \zeta^i S(X_0, \ldots, X_n)\} \subset \mathbb{P}^{n+1}$$

(all isomorphic to X), where $\zeta := \exp(2\pi\sqrt{-1}/q)$.

This X has only Kawamata log terminal singularities and $\operatorname{Pic} X = (\operatorname{Pic} \mathbb{P}^{n+1}) \,|\, X$ ($n \geq 2$) is of rank one, using Lefschetz type theorem [11, Example 3.1.25] when $n \geq 3$. We have $f^{-1}(S_i) = S_i$ and $g_X^{-1}(T_i) = T_i$, where $0 \leq i \leq n$.

When $n = 2$, the relation $(n+1)(T_1 - T_0) \sim 0$ gives rise to an étale-in-codimenion-one $\mathbb{Z}/(n+1)$-cover

$$\tau : \mathbb{P}^n \simeq \widetilde{X} \to X$$

so that $\sum_{i=0}^{n} \tau^* T_i$ is a union of $n+1$ normal crossing hyperplanes; indeed, τ restricted over $X \setminus \operatorname{Sing} X$, is its universal cover (cf. [12, Lemma 6]), so that g_X lifts up to $\widetilde{X}$. A similar result *seems* to be true for $n \geq 3$, by considering the 'composite' of the $\mathbb{Z}/(n+1)$-covers given by $(n+1)(T_i - T_0) \sim 0$ $(1 \leq i < n)$.

2. Proofs of Theorem 1.1 and Remark 1.3

We use the standard notation in Hartshorne's book and [10].

2.1. We now prove Theorem 1.1 and Remark 1.3. By [14, Theorem 1.5 in arXiv version 1], we may assume that $V \subset \mathbb{P}^3$ is an irreducible rational *singular* cubic hypersurface.

We first consider the case where V is non-normal. Such V is classified in [6, Theorem 9.2.1] to the effect that either $V = V_i$ $(i = 3, 4)$ or V is a cone over a nodal or cuspidal rational planar cubic curve B. The description in Remark 1.3 on V_3, V_4 and their normalizations, is given in [15, Theorem 1.1], [1, Theorem 1.5, Case (C), (E1)]; the f^{-1}-invariance of the non-normal locus C is proved in [14, Proposition 5.4 in arXiv version 1].

We consider and will rule out the case where V is a cone over B. Since V is normal crossing in codimension 1 (cf. [14, Theorem 1.5 or Proposition 5.4 in arXiv version 1]), the base B of the cone V is nodal. Let P be the vertex of the cone V, and $L \subset V$ the generating line lying over the node of B. Then $f_V := f|V$ satisfies the assertion that $f_V^{-1}(P) = P$. Indeed, the normalization V' of V is a cone over a smooth rational (twisted) cubic curve (in $\mathbb{P}^3$), i.e., the contraction of the (-3)-curve on the Hirzebruch surface $\mathbb{F}_3$ of degree 3; f_V lifts to an endomorphism $f_{V'}$ of V' so that the conductor $C' \subset V'$ is preserved by $f_{V'}^{-1}$ (cf. [14, Proposition 5.4 in arXiv version 1]) and consists of two distinct generating lines L_i (lying over L). Thus $f_{V'}^{-1}$ fixes the vertex $L_1 \cap L_2$ (lying over P). Hence $f_V^{-1}(P) = P$ as asserted.

By [14, Lemma 5.9 in arXiv version 1], $f : \mathbb{P}^3 \to \mathbb{P}^3$ (with $\deg(f) = q^3 > 1$ say) descends, via the projection $\mathbb{P}^3 \dashrightarrow \mathbb{P}^2$ from the point P, to an endomorphism $h : \mathbb{P}^2 \to \mathbb{P}^2$ with $\deg(h) = q^2 > 1$ so that $h^{-1}(B) = B$. This and $\deg(B) = 3 > 1$ contradict the linearity property of h^{-1}-stable curves in $\mathbb{P}^2$ (see e.g. Theorem 1.5 and the references in [14, arXiv version 1]).

2.2. Next we consider the case where $V \subset \mathbb{P}^3$ is a normal rational *singular* cubic hypersurface. By the adjunction formula,

$$-K_V = -(K_{\mathbb{P}^3} + V)|V \sim H|V$$

which is ample, where $H \subset \mathbb{P}^3$ is a hyperplane. Since V is rational by [14, Theorem 1.5 in arXiv version 1] and since K_V is a Cartier divisor, V has only Du Val (or rational double, or ADE) singularities (cf. [8]). Let

$$\sigma : V' \to V$$

be the minimal resolution. Then

$$K_{V'} = \sigma^* K_V \sim \sigma^*(-H|V).$$

For $f : \mathbb{P}^3 \to \mathbb{P}^3$, we can apply the result below to $f_V := f|V$.

Lemma 2.3. *Let $V \subset \mathbb{P}^3$ be a normal cubic surface, and $f_V : V \to V$ an endomorphism such that $f_V^*(H|V) \sim qH|V$ for some $q > 1$ and the hyperplane $H \subset \mathbb{P}^3$. Let*

$$S(V) = \{G \,|\, G : \textit{irreducible curve on } V,\ G^2 < 0\}$$

be the set of negative curves on V, and set

$$E_V := \sum_{E \in S(V)} E.$$

Replacing f_V by its positive power, we have:

(1) *If $f_V^* G \equiv aG$ for some Weil divisor $G \not\equiv 0$, then $a = q$. We have*

$$f_V^*(L|V) \sim q(L|V)$$

for every divisor L on $\mathbb{P}^3$. Especially, $\deg(f_V) = q^2$; $K_V \sim -H|V$ satisfies $f_V^* K_V \sim qK_V$.

(2) *$S(V)$ is a finite set. $f_V^* E = qE$ for every $E \in S(V)$. So $f_V^* E_V = qE_V$.*

(3) *A curve $E \subset V$ is a line in $\mathbb{P}^3$ if and only if E is equal to $\sigma(E')$ for some (-1)-curve $E' \subset V'$.*

(4) *Every curve $E \in S(V)$ is a line in $\mathbb{P}^3$.*

(5) *We have*

$$K_V + E_V = f_V^*(K_V + E_V) + \Delta$$

for some effective divisor Δ containing no line in $S(V)$, so that the ramification divisor

$$R_{f_V} = (q-1)E_V + \Delta.$$

In particular, the cardinality $\#S(V) \leq 3$, and the equality holds exactly when $K_V + E_V \sim_{\mathbb{Q}} 0$; in this case, f_V is étale outside the three lines of $S(V)$ and $f_V^{-1}(\operatorname{Sing} V)$.

Proof. For (1) and (2), we refer to [14, Lemma 2.1] and [13, Proposition 3.6.8] and note that $L \sim bH$ for some integer b.

(3) We may assume that $E' \simeq \mathbb{P}^1$, where $E' := \sigma'(E)$ is the proper transform of E. (3) is true because E is a line if and only if

$$1 = E.H|V \; (= E'.\sigma^*(H|V) = E'.(-K_{V'})),$$

and by the genus formula $-2 = 2g(E') - 2 = (E')^2 + E'.K_{V'}$.

(4) $E' := \sigma'(E)$ satisfies $E'.K_{V'} = E.K_V < 0$ and $(E')^2 \leq E'.\sigma^* E = E^2 < 0$. Hence E' is a (-1)-curve by the genus formula. Thus (4) follows from (3).

(5) The first part is true because, by (2), the ramification divisor $R_{f_V} = (q-1)E_V +$ (other effective divisors). Also, by (1) and (2), $\Delta \sim (1-q)(K_V + E_V)$. Since $K_V.E = -1$ for every $E \in S(V)$ (by (4)), we have

$$0 \leq -K_V.\Delta = -K_V.(1-q)(K_V + E_V) = (q-1)(3 - \#S(V)).$$

Now the second part of (5) follows from this and the fact that $\Delta = 0$ if and only if $-K_V.\Delta = 0$ since $-K_V$ is ample. The last part of (5) follows from the purity of branch loci and the description of R_{f_V} in (5). □

2.4. We now prove Theorem 1.1 and Remark 1.3 for the normal cubic surface V. We use the notation in Lemma 2.3. Suppose that the Picard number

$$\rho := \rho(V) \geq 3.$$

Since K_V is not nef, the minimal model program for klt surfaces in [10, Theorem 3.47] and the fact that a Du Val singularity is the contraction of (-2)-curves of Dynkin type A_n, D_n ($n \geq 4$) or E_n ($n = 6, 7, 8$), imply the existence of a composition

$$V = V_\rho \stackrel{\tau_\rho}{\to} V_{\rho-1} \cdots \stackrel{\tau_3}{\to} V_2$$

of birational extremal contractions such that

$$\rho(V_i) = i.$$

Let

$$E_i \subset V_i$$

be the exceptional (irreducible) divisor of $\tau_i : V_i \to V_{i-1}$. Since V is Du Val, either E_i is contained in the smooth locus $V_i \setminus \mathrm{Sing}(V_i)$ and is a (-1)-curve, or E_i contains exactly one singular point

$$P_i \in \mathrm{Sing}\, V_i$$

of type A_{n_i} so that $\tau_i(E_i) \in V_{i-1}$ is a smooth point. In particular, every V_i is still Du Val. Let

$$V_i' \to V_i$$

be the minimal resolution. Since $-K_{V_i}$ is the pushforward of the ample divisor $-K_V$, it is ample. So V_i is still a Gorenstein del Pezzo surface. Noting that $K_{V_i'}$ is the pullback of K_{V_i}, we have

$$(K^2_{V'_{i-1}} =)\, K^2_{V_{i-1}} = K^2_{V_i} + (n_i + 1) \geq 3 + (0 + 1) = 4$$

for all $3 \leq i \leq \rho$.

Note that the proper transform

$$E_i(V) \subset S(V)$$

of $E_i \subset V_i$ is a negative curve. Since f_V^{-1} stabilizes every negative curve in $S(V)$ and especially $E_i(V)$ (when f is replaced by its positive power, as seen in Lemma 2.3), f_V descends to

$$f_i : V_i \to V_i.$$

The V_2' and $S(V_2')$, the set of negative curves on V_2', are classified in [16, Figure 6]. Since $K_{V_2}^2 \geq 4$, $(V_2', S(V_2'))$ is as described in one of the last 10 cases in [ibid.]. For example, we write

$$V_2 = V_2(2A_2 + A_1)$$

if Sing V_2 consist of two points of type A_2 and one point of type A_1.

Except the four cases

$$V_2(D_4),\ V_2(4A_1),\ V_2(A_3),\ V_2(2A_1)$$

in [ibid.], exactly two (-1)-curves in $S(V_2')$ map to intersecting negative curves

$$M_i \in S(V_2)$$

so that

$$S(V) = \{E_\rho,\ M_1(V),\ M_2(V)\}$$

(using the fact that $\#S(V) \leq 3$ in Lemma 2.3) with

$$M_i(V) \subset V$$

the proper transform of M_i, so

$$K_V + E_\rho + M_1(V) + M_2(V) \sim_{\mathbb{Q}} 0$$

(cf. Lemma 2.3) and hence $(*)$ $K_{V_2} + M_1 + M_2 \sim_{\mathbb{Q}} 0$, which is impossible by a simple calculation after blowing down V_2' to its relative minimal model, a Hirzebruch surface $\mathbb{F}_d$ for some $d \leq 2$. Indeed, the relation $(*)$ induces a similar one called $(**)$ on $\mathbb{F}_d$ and we use the description of the canonical divisor and divisor classes of irreducible curves on $\mathbb{F}_d$ as can be found from Hartshorne's GTM book, to deduce a contradiction to $(**)$.

For each of the above four exceptional cases, we may assume that f_2^{-1} stabilizes both extremal rays $\mathbb{R}_+[M_i]$ of the closed cone $\overline{\mathrm{NE}}(V_2)$ of effective 1-cycles, with

$$M_i \subset V_2$$

the image of some (-1)-curve on V_2', where both extremal rays are of fibre type in the cases $V_2(D_4)$ and $V_2(4A_1)$, where the first (resp. second) is of fibre type (resp. birational type) in the cases $V_2(A_3)$ and $V_2(2A_1)$. Let

$$F_i \ (\sim 2M_i)$$

with $i = 1, 2$, or with $i = 1$ only, be the fibre of the extremal fibration

$$\varphi_i = \Phi_{|2M_i|} : V \to B_i \simeq \mathbb{P}^1$$

passing through the point $(\tau_3 \circ \cdots \circ \tau_\rho)(E_\rho)$. Then the proper transform

$$F_i(V) \subset V$$

of F_i is a negative curve so that (using the fact that $\#S(V) \leq 3$ in Lemma 2.3)

$$E_V = F_1(V) + F_2(V) + E_\rho$$

in the cases $V_2(D_4)$ and $V_2(4A_1)$, and

$$E_V = F_1(V) + M_2(V) + E_\rho$$

in the cases $V_2(A_3)$ and $V_2(2A_1)$ (cf. Lemma 2.3). Then $K_V + E_V \sim_{\mathbb{Q}} 0$, and hence $K_{V_2} + F_1 + F_2 \sim_{\mathbb{Q}} 0$ or $K_{V_2} + F_1 + M_2 \sim_{\mathbb{Q}} 0$ where the latter is impossible by a simple calculation as in the early paragraph. Thus

$$K_{V_2} + F_1 + F_2 \sim_{\mathbb{Q}} 0.$$

By making use of Lemma 2.3 (1) or (2), $f_2^* F_i = qF_i$, and f_2 descends to an endomorphism

$$f_{B_1} : B_1 \to B_1$$

of degree q. Thus the ramification divisor of f_{B_1} is of degree $2(q-1)$ by the Hurwitz formula, and is hence equal to

$$(q-1)P + \sum (b_i - 1)P_i$$

with

$$\sum (b_i - 1) = q - 1$$

where $P \in B_1$ so that F_1 lies over P. But then

$$R_{f_2} \geq (q-1)(F_1 + F_2) + \sum (b_i - 1)F_i'$$

where F_i' are fibres of φ_1 lying over P_i, so that

$$K_{V_2} + F_1 + F_2 \geq f_2^*(K_{V_2} + F_1 + F_2) + \sum (b_i - 1)F_i'$$

which is impossible since $K_{V_2} + F_1 + F_2 \sim_{\mathbb{Q}} 0$.

2.5. Consider the case $\rho(V) = 2$. Then the minimal resolution

$$V' \to V$$

and its negative curves are described in one of the first five cases in [16, Figure 6].

For the case $V = V(A_5)$, two (-1)-curves on V' map to two negative curves

$$M_1, \ M_2$$

on V. Note that $f_V^*(M_i) = q^*M_i$ (see Lemma 2.3). There is a contraction

$$V \to \mathbb{P}^2$$

of M_1 so that the image of M_2 is a plane conic preserved by f_P^{-1} where

$$f_P : \mathbb{P}^2 \to \mathbb{P}^2$$

is the descent of f_V (of degree $q^2 > 1$), contradicting [14, Theorem 1.5(4) in arXiv version 1].

For the case $V(2A_2 + A_1)$, there are exactly five (-1)-curves

$$M_i' \subset V' \ (1 \leq i \leq 5)$$

with $M_i \subset V$ their images. Moreover, $M_1'.M_2' = 1$ and both M_i $(i = 1, 2)$ are negative curves on V; each M_j' $(j = 3, 4)$ meets the isolated (-2)-curve; M_1' and M_3' (reap. M_2' and M_4') meet the same component of one (resp. another) (-2)-chain of type A_2. We have

$$M_1 + M_2 \sim 2L$$

for some integral Weil divisor L (the image M_5 of the fifth (-1)-curve M_5') by considering a relative minimal model of V'. In fact, $M_1 + 3M_2 \sim 4M_3$ (which will not be used). Since $f_V^*(M_1 + M_2) = q(M_1 + M_2)$ (see Lemma 2.3), f_V lifts to some

$$g : U \to U.$$

Here the double cover (given by the relation $M_1 + M_2 \sim 2L$)

$$\pi : U = Spec \oplus_{i=0}^{1} \mathcal{O}(-iL) \to V$$

is branched along $M_1 + M_2$. Indeed, when $2 \not| q$, the normalization $\hat{U}$ of the fibre product of $\pi : U \to V$ and $f_V : V \to V$ is isomorphic to U and we take g to be the first projection $\hat{U} \to U$; when $2 \,|\, q$, we have $\hat{U} = V \coprod V$ and let g be the composite of $\pi : U \to V$, the inclusion $V \cup \emptyset \to V \coprod V$ and the first projection $\hat{U} \to U$. Now $\operatorname{Sing} U$ consists of a type A_1 singularity lying over $M_1 \cap M_2$ and four points in $\pi^{-1}(\mathrm{Sing}\, V)$ of type $A_1, A_1, \frac{1}{3}(1,1)$ and $\frac{1}{3}(1,1)$ lying over $M_3 \cap M_4$, $M_1 \cap M_3$ and $M_2 \cap M_4$; every M_j $(j = 3, 4)$ splits into two negative curves on U which are hence preserved by g^{-1} (as in Lemma 2.3 after f_V is replaced by its positive power). Thus $f_V^{-1}(M_i) = M_i$ $(1 \le i \le 4)$. Hence f_V is totally ramified along four curves. This leads to a contradiction as in the proof of Lemma 2.3 for the fact that $\#S(V) \le 3$ (where the observation $E'.K_{V'} = E.K_V = -1$ is used).

For $V = V(A_4 + A_1)$, there are exactly four (-1)-curves

$$M_i' \subset V' \ (1 \le i \le 4)$$

with $M_i \subset V$ their images. We may label them so that the five (-2)-curves and M_j' $(j = 1, 2, 3)$ form a simple loop:

$$M_1' - (-2) - (-2) - (-2) - (-2) - M_2' - M_3' - (-2) - M_1';$$

both M_j $(j = 2, 3)$ are negative curves on V. We can verify that $M_3 + 2M_2 \sim 3(M_5 - M_2)$, and $M_1 + M_2 \sim M_5$, where M_5 is the image of a curve $M_5' \simeq \mathbb{P}^1$ and M_5' is the smooth fibre (passing through the intersection point of M_3' and the isolated (-2)-curve) of the $\mathbb{P}^1$-fibration on V' with a singular fibre consisting of M_1', M_2' and the (-2)-chain of type A_4 sitting in between them. As in the case $V(2A_2 + A_1)$, f_V lifts to

$$g : U \to U$$

on the triple cover U defined by the relation $M_3 + 2M_2 \sim 3(M_5 - M_2)$, so that each M_i $(i = 4, 5)$ splits into three negative curves on U preserved by g^{-1}. Hence $f_V^{-1}(M_i) = M_i$ $(i = 2, \ldots, 5)$. As in the case $V(2A_2 + A_1)$, this leads to a contradiction as in the proof of Lemma 2.3 for the fact that $\#S(V) \le 3$.

For $V = V(D_5)$, the lonely (-1)-curve M_1' and the intersecting (-1)-curves $M_2' \cup M_3'$ on V' satisfy $2M_1 \sim M_2 + M_3$ where

$$M_i \subset V \ (1 \le i \le 5)$$

denotes the image of M_i' (indeed, the three M_i' together with the five (-2)-curves form the support of two singular fibres and a section in a $\mathbb{P}^1$-fibration). As in the case $V(2A_2 + A_1)$, f_V lifts to

$$g : U \to U$$

on the double cover U defined by the relation $2M_1 \sim M_2 + M_3$, so that M_1 splits into two negative curves on U preserved by g^{-1}. Thus $f_V^{-1}(M_1) = M_1$. Hence (V, M_1) is log canonical (cf. [13, Theorem 4.3.1]). But (V, M_1) is not log canonical because M_1' meets the (-2)-tree of type-D_5 in a manner different from the classification of [9, Theorem 9.6]. We reach a contradiction.

For $V = V(A_3 + 2A_1)$, let $M_1' \subset V'$ be the (-1)-curve meeting the middle component of the (-2)-chain of type A_3, let M_3' be the (-1)-curve meeting two isolated (-2)-curves, and let M_2' be the (-1)-curve meeting both M_1' and M_3'. Then the images

$$M_i \subset V$$

of M_i' satisfy $2M_1 \sim 2M_3$ (indeed, M_1', M_3' and the five (-2)-curves form the support of two singular fibres in some $\mathbb{P}^1$-fibration). The relation $2(M_1 - M_3) \sim 0$ defines a double cover

$$\pi : U = Spec \oplus_{i=0}^{1} \mathcal{O}(-i(M_1 - M_3)) \to V$$

étale over $V \setminus \operatorname{Sing} V$. In fact, π restricted over $V \setminus \operatorname{Sing} V$, is the universal cover over it, so U is again a Gorenstein del Pezzo surface and hence the irregularity $q(U) = 0$. Thus f_V lifts to

$$g : U \to U.$$

Now $\pi^{-1}(\operatorname{Sing} V)$ consists of two smooth points and the unique singular point of U (of type A_1), and each $\pi^* M_i$ $(i = 1, 2)$ splits into two negative curves $M_i(1)$, $M_i(2)$ on U preserved by g^{-1}; thus $f_V^* M_i = q M_i$ and $g^* M_i(j) = q M_i(j)$ (as in Lemma 2.3 (1)).

We assert that f_V^{-1} permutes members of the pencil

$$\Lambda := |M_1 + M_2|.$$

It suffices to show that g^{-1} permutes members of the irreducible pencil Λ_U (parametrized by $\mathbb{P}^1$ for $q(U) = 0$) which is the pullback of Λ. Now $\pi^*(M_1 + M_2)$ splits into two members

$$M_1(j) + M_2(j) = \operatorname{div}(\xi_j)$$

(in local equation; $j = 1, 2$) which are preserved by g^{-1} and span Λ_U. We may assume that $g^* \xi_j = \xi_j^q$ after replacing the equation by a scalar multiple. Then the g^*-pullback of every member $\operatorname{div}(a\xi_1 + b\xi_2)$ in Λ_U is equal to $\operatorname{div}(a\xi_1^q + b\xi_2^q)$ and hence is the union of members in Λ_U because we can factorize $a\xi_1^q + b\xi_2^q$ as a product of linear forms in ξ_1, ξ_2. This proves the assertion.

By the assertion and since $f_V^*(M_1 + M_2) = q(M_1 + M_2)$, f_V descends to an endomorphism

$$f_B : B \to B$$

of degree q on the curve $B \simeq \mathbb{P}^1$ parametrizing the pencil Λ. We have $f_B^* P_0 = qP_0$ for the point parametrizing the member $M_1 + M_2$ of Λ. Write $K_B = f_B^* K_B + R_{f_B}$, where the ramification divisor

$$R_{f_B} = (q-1)P_0 + \Delta_B$$

with $\Delta_B = \sum (b_i - 1) Q_j$ of degree $q-1$ for some $b_i \geq 2$. Thus the ramification divisor

$$R_{f_V} = (q-1)(M_1 + M_2) + \Delta_V$$

with $\Delta_V = \sum (b_i - 1) F_i +$ (other effective divisor), where

$$F_i \in \Lambda$$

is parametrized by Q_i. On the other hand, one can verify that $-K_V \sim 2M_1 + M_2$, by blowing down to a relative minimal model of V'; indeed, M_2 is a double section of the $\mathbb{P}^1$-fibration

$$\varphi := \Phi_{|2M_1|} : V \to \mathbb{P}^1.$$

So $-M_1 \sim K_V + M_1 + M_2 = f_V^*(K_V + M_1 + M_2) + \Delta_V$ and hence, by Lemma 2.3 (1),

$$(b_1 - 1)F_1 \leq \Delta_V \sim (1-q)(K_V + M_1 + M_2) \sim (q-1)M_1.$$

This is impossible because F_1 is horizontal to the half fibre M_1 of φ. Indeed, $F_1.M_1 = (M_1 + M_2).M_1 = M_2.M_1 = 1$.

2.6. Consider the last case $\rho(V) = 1$. Since $K_V^2 = 3$, we have

$$V = V(3A_2),\ V(E_6),\ \text{or}\ V(A_1 + A_5),$$

and the minimal resolution

$$V' \to V$$

and the negative curves on V' are described in [16, Figure 5]. For the first two cases, V is isomorphic to V_i ($i = 1$, or 2) in Theorem 1.1 by the uniqueness result in [16, Theorem 1.2] and by [8, Theorem 4.4].

For $V = V(A_1 + A_5)$, the images

$$M_i \subset V$$

of the two (-1)-curves $M_i' \subset V'$ satisfy $2M_1 \sim 2M_2$; indeed, M_i' together with the six (-2)-curves form the support of two singular fibres and a section in some $\mathbb{P}^1$-fibration. Let

$$\pi : U \to V$$

be the double cover given by the relation $2(M_1 - M_2) \sim 0$. In fact, π restricted over $V \setminus \operatorname{Sing} V$, is the universal cover over it. So f_V lifts to

$$g : U \to U.$$

As in the case $V(A_3 + 2A_1)$, if we let M_1' be the one meeting the second component of the (-2)-chain of type A_5, then $\pi^* M_1$ splits into two negative curves on U preserved by g^{-1}. Thus $f_V^{-1}(M_1) = M_1$, and, as in the case $V(D_5)$ above, contradicts the results in [13, Theorem 4.3.1] and [9, Theorem 9.6].

This completes the proof of Theorem 1.1 for normal cubic surfaces and hence the whole of Theorem 1.1. To determine the equations of V_i $(i = 1, 2)$, we can check that the equations in Theorem 1.1 possess the right combination of singularities and then use the very ampleness of $-K_{V_i}$ to embed V_i in $\mathbb{P}^3$ as in [8] and the uniqueness of V_i up to isomorphism, and hence up to projective transformation by [8] (cf. [16, Theorem 1.2]).

2.7. Now we prove Remark 1.3. From Lemma 2.3 till now, we did not assume the hypothesis $(*)$ that f_V is the restriction of some $f : \mathbb{P}^3 \to \mathbb{P}^3$ whose inverse stabilizes V. From now on till the end of the paper, we assume this hypothesis $(*)$.

For $V = V(E_6)$ or $V(3A_2)$, the relation $V \sim 3H$ defines a triple cover

$$\pi : X = Spec \oplus_{i=0}^{2} \mathcal{O}(-iH) \to \mathbb{P}^3$$

branched along V. Then

$$X = \{Z^3 = V(X_0, \ldots, X_3)\} \subset \mathbb{P}^4$$

is a cubic hypersurface, where we let $V(X_0, \ldots, V_3)$ be the cubic form defining $V \subset \mathbb{P}^3$. Our π^{-1} restricts to a bijection $\pi^{-1} : \operatorname{Sing} V \to \operatorname{Sing} X$. As in Example 1.5, f lifts to $(Z^q, f) : \mathbb{P}^4 \to \mathbb{P}^4$ stabilizing X, so that the restriction

$$g = (Z^q, f)|X : X \to X$$

is also a lifting of f. By the Lefschetz type theorem [11, Example 3.1.25], $\operatorname{Pic}(X) = (\operatorname{Pic}(\mathbb{P}^4))|X$.

For $V = V(E_6)$, V' contains only one (-1)-curve M' (cf. [16, Figure 5]) and hence V contains only one line M (the image of M') by Lemma 2.3. Note that

$$\{Q\} := \operatorname{Sing} V \subset M.$$

Let

$$\Pi_{MM} \subset \mathbb{P}^3$$

(say given by $X_3 = 0$) be the unique plane such that

$$\Pi_{MM}|V = 3M.$$

Indeed, $3M$ belongs to the complete linear system $|H|V|$, and the exact sequence

$$0 \to \mathcal{O}(-2H) \to \mathcal{O}(H) \to \mathcal{O}_V(H) \to 0$$

and the vanishing of $H^1(\mathbb{P}^3, -2H)$ (e.g. by the Kodaira vanishing) imply

$$H^0(\mathbb{P}^3, \mathcal{O}(H)) \simeq H^0(V, \mathcal{O}_V(H)).$$

Our $\pi^*\Pi_{MM}$ is a union of three 2-planes

$$L_i \subset \mathbb{P}^4$$

because the restriction of π over Π_{MM} is given by the equation

$$Z^3 = V(X_0, \ldots, X_3) \,|\, \Pi_{MM} = M(X_0, \ldots, X_2)^3$$

where $M(X_0, \ldots, X_2)$ is a linear equation of $M \subset \Pi_{MM}$. This and the fact that $\pi^*\Pi_{MM}$ is a generator of $\mathrm{Pic}(X) = (\mathrm{Pic}(\mathbb{P}^3)) \,|\, X$, imply that the Weil divisor L_1 is not a Cartier divisor on X. Since $\mathrm{Sing}\, X$ consists of a single point P lying over $\{Q\} = \mathrm{Sing}\, V$, L_1 is not Cartier at P and hence X is not factorial at P. Thus X is not $\mathbb{Q}$-factorial at P because the local π_1 of P is trivial by a result of Milnor (cf. the proof of [9, Lemma 5.1]). Hence $g^{-1}(P)$ contains no smooth point (cf. [10, Lemma 5.16]) and must be equal to $\mathrm{Sing}\, X = \{P\}$. Thus $f^{-1}(Q) = Q$ because $\pi^{-1}(Q) = P$.

2.8. Before we treat the case $V(3A_2)$, we make some remarks. Up to isomorphism, there is only one $V(3A_2)$ (cf. [16, Theorem 1.2]). Set $V := V(3A_2)$. There is a Gorenstein del Pezzo surface W such that $\rho(W) = 1$, $\mathrm{Sing}\, W$ consists of four points β_i of Du Val type A_2,

$$\pi_1(W \setminus \mathrm{Sing}\, W) = (\mathbb{Z}/(3))^{\oplus 2}$$

and there is a Galois triple cover $V \to W$ étale over $W \setminus \{\beta_1, \beta_2, \beta_3\}$ so that a generator

$$h \in \mathrm{Gal}(V/W)$$

permutes the three singular points of V lying over β_4 (cf. [12, Figure 1, Lemma 6]). Since the embedding $V \subset \mathbb{P}^3$ is given by the complete linear system $|-K_V|$ (cf. [8]) and $h^*(-K_V) \sim -K_V$, our h extends to a projective transformation of $\mathbb{P}^3$, also denoted as h. Since $h(V) = V$,

$$h^*V(X_0, \ldots, X_3) = cV(X_0, \ldots, X_3)$$

for some nonzero constant c. This h lifts to a projective transformation of $\mathbb{P}^4$, also denoted as h, stabilizing the above triple cover $X \subset \mathbb{P}^4$ of $\mathbb{P}^3$ by defining $h^*Z = \sqrt[3]{c}Z$. Then this h permutes the three singular points of X lying over $\operatorname{Sing} V$.

2.9. For $V = V(3A_2)$, V' has exactly three (-1)-curves M_i' (cf. [16, Figure 5]) and their images M_i are therefore the only lines on V (cf. Lemma 2.3). The graph $\sum M_i$ is triangle-shaped whose vertices (the intersection $M_i \cap M_j$) are the three points in $\operatorname{Sing} V$. The sum of the three (-1)-curves M_i' and three (-2)-chains of type A_2 is linearly equivalent to $-K_{V'}$ and hence $\sum M_i \sim -K_V$; also $2M_a \sim M_b + M_c$ so long $\{a, b, c\} = \{1, 2, 3\}$; indeed, the three M_i' and the six (-2)-curves form the support of two singular fibres and two cross-sections of some $\mathbb{P}^1$-fibration. Thus $3M_i \sim -K_V \sim H|V$. As argued in the case $V(E_6)$, there is a unique hyperplane Π_i such that

$$\Pi_i|V = 3M_i;$$

our $\pi^*\Pi_i$ is a union of three 2-planes L_{ij} in $\mathbb{P}^4$ (sharing a line lying over $M_i \subset \mathbb{P}^3$), L_{i1} is not a Cartier divisor on X, the X is not $\mathbb{Q}$-factorial at least at one of the two points (and hence at both points, since the above h permutes $\operatorname{Sing} X$) in $L_{i1} \cap \operatorname{Sing} X$ (lying over $M_i \cap \operatorname{Sing} V$), and $g^{-1}(\operatorname{Sing} X) = \operatorname{Sing} X$. Thus $f^{-1}(\operatorname{Sing} V) = \operatorname{Sing} V$. Hence f^{-3} fixes each point in $\operatorname{Sing} V$.

This completes the proof of Remark 1.3 for normal cubic surfaces and hence the whole of Remark 1.3.

Remark 2.10. The proof of Theorem 1.1 actually shows: if $f_V : V \to V$ is an endomorphism (not necessarily the restriction of some $f : \mathbb{P}^3 \to \mathbb{P}^3$) of $\deg(f_V) > 1$ of a Gorenstein normal del Pezzo surface with $K_V^2 = 3$ (i.e., a normal cubic surface), then V is equal to V_1 or V_2 in Theorem 1.1 in suitable projective coordinates.

References

1. M. Abe and M. Furushima, On non-normal del Pezzo surfaces, Math. Nachr. **260** (2003), 3–13.

2. A. Beauville, Endomorphisms of hypersurfaces and other manifolds, Internat. Math. Res. Notices 2001, no. 1, 53–58.
3. J.-V. Briend, S. Cantat and M. Shishikura, Linearity of the exceptional set for maps of $\mathbb{P}_k(\mathbb{C})$, Math. Ann. **330** (2004), 39–43.
4. D. Cerveau and A. Lins Neto, Hypersurfaces exceptionnelles des endomorphismes de $CP(n)$, Bol. Soc. Brasil. Mat. (N.S.) **31** (2000), no. 2, 155–161.
5. T.-C. Dinh and N. Sibony, Equidistribution speed for endomorphisms of projective spaces, Math. Ann. **347** (2010), no. 3, 613–626.
6. I. V. Dolgachev, Topics in Classical Algebraic Geometry. Part I, January 19, 2009, at: http://www.math.lsa.umich.edu/ idolga/CAG.pdf
7. J. E. Fornaess and N. Sibony, Complex dynamics in higher dimension. I, *Complex analytic methods in dynamical systems* (Rio de Janeiro, 1992), pp. 201–231, Astérisque **222**, Soc. Math. France, 1994.
8. F. Hidaka and K. Watanabe, Normal Gorenstein surfaces with ample anti-canonical divisor, Tokyo J. Math. **4** (1981), no. 2, 319–330.
9. Y. Kawamata, Crepant blowing-ups of 3-dimensional canonical singularities and its application to degeneration of surfaces, Ann. of Math. **127** (1988), 93–163.
10. J. Kollár and S. Mori, Birational geometry of algebraic varieties, Cambridge Tracts in Math. **134**, Cambridge Univ. Press, 1998.
11. R. Lazarsfeld, Positivity in algebraic geometry. I. Classical setting: line bundles and linear series. Ergebnisse der Mathematik und ihrer Grenzgebiete. 3. Folge. A Series of Modern Surveys in Mathematics, Vol. **48**. Springer-Verlag, Berlin, 2004.
12. M. Miyanishi and D.-Q. Zhang, Gorenstein log del Pezzo surfaces of rank one, J. Algebra, **118** (1988), 63-84.
13. N. Nakayama, On complex normal projective surfaces admitting non-isomorphic surjective endomorphisms, Preprint 2 September 2008.
14. N. Nakayama and D.-Q. Zhang, Polarized endomorphisms of complex normal varieties, Math. Ann. **346** (2010), no. 4, 991–1018; also: arXiv:0908.1688v1.
15. M. Reid, Nonnormal del Pezzo surfaces, Publ. Res. Inst. Math. Sci. Kyoto Univ. **30** (1994), 695–727.
16. Q. Ye, On Gorenstein log del Pezzo surfaces, Japan. J. Math. **28** (2002), no. 1, 87–136.
17. D.-Q. Zhang, Polarized endomorphisms of uniruled varieties (with an Appendix by Y. Fujimoto and N. Nakayama), Compos. Math. **146** (2010) 145-168.
18. D.-Q. Zhang, Invariant hypersurfaces of endomorphisms of projective varieties, Preprint 2009.